CAMBRIDGE PHYSICAL SERIES

A TREATISE

ON

THE THEORY OF
ALTERNATING CURRENTS

A TREATISE

ON

THE THEORY OF
ALTERNATING CURRENTS

by

ALEXANDER RUSSELL, M.A., D.Sc., M.I.E.E.

Principal of Faraday House, London; Vice-President of the
Physical Society of London; Member of the Council
of the Institution of Electrical Engineers

Volume I

SECOND EDITION

Cambridge:

at the University Press

1914

CAMBRIDGE
UNIVERSITY PRESS

University Printing House, Cambridge CB2 8BS, United Kingdom

Published in the United States of America by Cambridge University Press, New York

Cambridge University Press is part of the University of Cambridge.

It furthers the University's mission by disseminating knowledge in the pursuit of education, learning and research at the highest international levels of excellence.

www.cambridge.org
Information on this title: www.cambridge.org/9781107671331

© Cambridge University Press 1914

First edition 1904
Second edition 1914
First published 1914
First paperback edition 2014

A catalogue record for this publication is available from the British Library

ISBN 978-1-107-67133-1 Paperback

PREFACE TO THE FIRST EDITION

THE great progress that has been made in the application of alternating currents to industrial purposes makes it desirable to collect together and examine the mathematical theorems which electricians use in their everyday work. In this volume the more general theorems are collected, and proofs are given of the more important of them, due stress being laid on the assumptions that it is sometimes necessary to make, and on the consequent limitations in the use of the formulae. In the second volume the theory of alternators, motors, transformers, converters, and of the transmission of power by polyphase currents will be given.

The reader is supposed to be familiar with the elementary theory of electricity and magnetism, and to have a working knowledge of trigonometry and of the elements of the calculus. A knowledge of De Moivre's theorem, De Moivre's property of the circle and of hyperbolic sines and cosines will be found essential. Lord Kelvin's bei and ber functions given in Chapter XVI will be understood at once from their definitions.

The references, given at the end of nearly every chapter, are to the books or papers I have consulted when writing the chapter. In many cases they contain more detailed discussions of the same or similar problems, and will be helpful to the student. The proofs of several of the problems could have been considerably shortened by deducing them from the general equations of electro-magnetism, but, in my opinion, the proofs given bring out the physical meaning more clearly to the electrical student.

The early chapters deal mainly with inductance and capacity, and it is hoped that the practical formulae in them will be helpful to the working electrician. In Chapter V, illustrations are given of methods by means of which the capacities of polyphase cables and overhead wires can be calculated. It is also shown how the inductances of these combinations of conductors for the case of surface currents can be found. The definition and the theory of the power factor given in Chapter VI are now almost universally adopted by electricians. It was thought necessary to define terms like 'watt current' and 'wattless current'—'le courant watté' and 'le courant dewatté' of French writers—as they are so widely used. In Chapter IX, the test room methods of measuring power are given, and in Chapter X the method, when discussing practical problems, of replacing an air-core transformer by its equivalent net-work is explained. In Volume II this method will be extended to iron-core transformers.

In Chapter XII some problems in two phase theory are discussed graphically and illustrate how the theorems of solid geometry can be applied usefully in this case. In Chapter XIII the main problems in the theory of phase indicators and induction type watt-hour meters are stated, and approximate solutions are obtained. The theory of rotating magnetic fields, given in Chapter XIV, is founded on a paper which I wrote for the *Electrical Review* in 1893. In Chapter XV the interesting problem of the nature of the magnetic field round parallel wires, carrying polyphase currents, is discussed. Although complete solutions of the problem of the eddy currents in magnetic metals have not yet been obtained, the approximate solutions obtained by J. J. Thomson and Oliver Heaviside, which are given in Chapter XVI, will be found most helpful in practical work. Heaviside's solution is given in terms of bei and ber functions, and so, in given cases, numerical values can be found readily by means of the tables given in this volume. In Chapter XVII a slight sketch is given of the useful method of duality.

I have to thank Mr G. F. C. Searle, M.A., of St Peter's College, Cambridge, University Lecturer in Experimental Physics, for the many suggestions and emendations he has made, during the printing of the work. Of these suggestions and emendations I have freely availed myself. Particularly I have to acknowledge the valuable help he has given me in the problems connected with capacity, eddy currents and inductance, more especially the inductance when the currents are confined to the surface of the conductors. I have to thank him also for his unwearying kindness in reading the proofs and for checking nearly all the formulae given.

I take this opportunity of thanking Dr Charles Chree, F.R.S., the Superintendent of the Observatory Department of the National Physical Laboratory, for his help and encouragement. I am also indebted to Mr W. C. Dampier Whetham, F.R.S., who has edited this work, for many valuable suggestions and criticisms.

Finally, I have to thank the Council of the Institution of Electrical Engineers and *The Electrician* Printing and Publishing Company for their kind permission to make what use I pleased, in the preparation of this volume, of any of my papers which they have published.

A. R.

2 BELLEVUE PLACE,
   RICHMOND, SURREY.
     October, 1904.

PREFACE TO THE SECOND EDITION

IN this volume Chapters III, VII, VIII and IX are entirely new and many additions have been made in the earlier chapters. The recent improvements in methods of measuring inductance and capacity and the importance of a knowledge of their values in practical work have induced the author to devote a large amount of space to perfecting formulae for computing them. In Chapter III the self and mutual inductance formulae for circular and helical coils are proved by the straightforward method of linkages given in the first edition. Many other examples might have been added, but it is hoped that the cases discussed will enable the reader to see how similar problems can be attacked by this method. It was found convenient to use a few elementary theorems in elliptic integrals, but anyone with a working knowledge of the Calculus will be able to follow the proofs.

Several of the readers of the first edition who are specially interested in telephony and radio-telegraphy have suggested to the author that he should discuss high frequency currents at length in this volume. This has been done in Chapter VII. The many auxiliary mathematical theorems and tables given not only enable the variations in the resistance and inductance of the wires to be easily and accurately computed in certain cases, but also considerably simplify the solution of the problem of the eddy current losses in a metal cylinder discussed in Chapter XX. They are thus of considerable importance in everyday work. The theoretical results obtained have been experimentally verified by several physicists, notably by Dr J. A. Fleming.

Owing to the great value of the theory of two coupled oscillation circuits in radio-telegraphy a brief *résumé* of the theory is given in Chapter IX.

In connection with the effects produced by the high voltages at which electrical power is now transmitted and the possible utilization of these effects for measuring purposes, the interesting and historical problem of two spherical electrodes is fully considered. The limited use that has hitherto been made by engineers and physicists of the important results obtained by Poisson more than a hundred years ago and by Kelvin forty years later is probably due to the excessive numerical labour involved in using their solutions. In Chapter VIII convenient working formulae are obtained and several tables given by means of which the capacity coefficients, the attractions or repulsions of the spheres, and the maximum electric stress between them can be readily found in all cases. The importance of knowing the numerical values of these quantities is now recognised. It is confidently hoped that they will be found still more useful in the future.

Amongst other new matter may be mentioned the methods of finding the effective capacities of horizontal antennae described in Chapter VI and the proofs given of the circle diagrams of the air core transformer in Chapter XIV.

For some of the theorems and tables included the author has pleasure in acknowledging his indebtedness to recent papers by the Physicists of the Bureau of Standards, Washington. He has to thank the editor, Mr T. G. Bedford, M.A., for reading all the manuscript and for the heavy labour of checking the mathematical calculations. He is indebted also to the officials of the University Press for their valuable help in reading the proofs.

A. R.

Faraday House,
London, W.C.
March, 1914.

CONTENTS

SYMBOLS

A, effective value of an alternating current; a constant.

$A_1, A_2, ...,$ effective values of the currents in polyphase mains.

B, magnetic induction.

$B_{max.}$, maximum value of the magnetic induction.

C, direct current; a constant.

E, maximum value of the alternating voltage; direct voltage.

$$E = \int_0^{\pi/2} (1 - k^2 \sin^2 \phi)^{1/2} \, \partial \phi.$$

$E_1, E_2, ...,$ effective values of star voltages.

F, force.

G, average torque.

H, magnetic force; constant in theory of equations (p. 267).

I, intensity of magnetisation; maximum value of an alternating current when it follows the harmonic law; constant (p. 267).

$I_{max.}$, maximum value of an alternating current.

$I_1, I_2, ...,$ effective values of the mesh currents.

$I(x)$, Bessel's function of the first kind.

J, constant (p. 267).

K, capacity between two conductors; capacity of a conductor.

$K_{1.1}, K_{1.2}, ...,$ Maxwell's coefficients of self and mutual induction for electrostatic charges.

K_q, capacity when the charges are constant.

K_v, capacity when the potentials are constant.

$$K = \int_0^{\pi/2} \frac{\partial \phi}{(1 - k^2 \sin^2 \phi)^{1/2}}.$$

$K(x)$, Bessel's function of the second kind.

L, self inductance.

$L_{1.1}, L_{1.2}, ...,$ coefficients of induction in electro-magnetism.

M, mutual inductance; magnetic moment.

P, power.

Q, quantity of electricity.

R, resultant electrostatic force; resistance; resistance of primary circuit; radial force; potential gradient.

$R_{max.}$, the maximum value of the potential gradient in a given space.

R_1, resistance of primary circuit.

R_2, resistance of secondary circuit.

$$S_m = \sum_1^\infty (1/n)^m.$$

S, area of cross section.

T, periodic time; tangential force; longitudinal tension.

V, electrostatic potential; magnetic potential; volume; effective value of potential difference.

$V_{1.2}, V_{1.3}, \ldots,$ effective values of the mesh voltages.

W, average value of the power; power; energy.

Z, impedance.

$S(x)$; $T(x)$; $V(x)$; $V_1(x)$; $W(x)$; $X(x)$; $X_1(x)$; $Z(x)$; mathematical functions, *see* p. 210.

c, thermal capacity per unit volume.

e, instantaneous value of the electromotive force.

$[e]$, $E \cos \omega t + \sqrt{-1} E \sin \omega t.$

f, force; frequency; fault resistance.

$f_1, f_2, \ldots,$ fault resistances of the mains.

g, strength of magnetic shell; instantaneous value of the torque; acceleration produced by gravity.

i, instantaneous value of the current.

$[i]$, $I \cos \omega t + \sqrt{-1} I \sin \omega t.$

k, capacity between two conductors per unit length; thermal conductivity; modulus of elliptic functions.

$k_{1.1}, k_{1.2}, \ldots,$ Maxwell's capacity coefficients.

l, length.

$l_{1.1}, l_{1.2}, \ldots,$ coefficients of induction in electro-magnetism.

m, mass; strength of magnetic pole; $2\pi \sqrt{2\mu f/\rho}$.

n, number of turns.

p, $2\pi \sqrt{\mu f/\rho}$.

$p_{1.1}, p_{1.2}, \ldots,$ Maxwell's potential coefficients.

q, electrostatic charge; instantaneous value of quantity of electricity.

r, resistance.

t, time.

v, instantaneous value of P.D.; temperature.

w, instantaneous value of power.

a, $\log 2 - \gamma = 0 \cdot 1159315 \ldots$

γ, Euler's constant $= 0 \cdot 5772157 \ldots$

$a, \beta, \gamma, \delta,$ angles.

$a_{1.2}, \theta, \phi, \psi,$ phase differences.

ϵ, base of Neperian logarithms.

η, Steinmetz's coefficient; efficiency.

ι, $\sqrt{-1}$.

xiv

SYMBOLS

$\cos\phi$, $\cos\psi$,	power factor.
λ,	dielectric coefficient; constant.
$\lambda_{1.1}$, $\lambda_{1.2}$, ...,	current coefficients.
μ,	magnetic permeability.
π,	$3\cdot14159...$
ρ,	volume density; resistivity.
$[\rho]$,	$R + L\omega\sqrt{-1}$.
σ,	surface density.
τ,	time.
ϕ,	instantaneous value of the flux; phase difference.

$$\psi(x) = \frac{\partial}{\partial x}\{\log \Gamma(x)\}.$$

ω,	angular velocity; $2\pi \times$ frequency; solid angle.
$\Gamma(x)$,	the gamma function.
Δ,	constant in Theory of Equations (p. 267).
Φ,	maximum value of flux, when it follows the sine law.
$\Phi_{max.}$,	maximum value of flux.
Ω,	solid angle.
$\mathfrak{R}$,	reluctance.

ABBREVIATIONS

c.g.s. system,	centimetre-gramme-second system of units.
e.m.f.	electromotive force.
p.d.	potential difference.
r.m.s.	root-mean-square.

CHAPTER I

THE theory of alternating currents of electricity is founded on results arrived at in electrostatics, electrodynamics and magnetism. We shall first, therefore, give a *résumé* of the more important theorems and formulae in these sciences, before using them in subsequent chapters. It will be seen that there are many analogies in these sciences, and it will be

Introduction.

found useful to bear such analogies in mind. In Chapter XXI an elementary sketch is given of a method of duality founded on these analogies, and Maxwell has shown how important they are in advanced theory.

Experiments show that the electric charges generated when certain bodies are rubbed together are of two kinds,
Electrostatics. which on account of their properties are conveniently distinguished as positive and negative. It is found that a body charged with positive electricity will repel a body charged with the same kind of electricity and will attract a body negatively charged. By means of a torsion balance Coulomb verified that when the charges are suitably measured the force of repulsion f between two small bodies in air possessing electric charges q and q' of like sign is given by the formula

$$f = k \frac{qq'}{r^2},$$

where r is the distance between the bodies and k is a constant. If we define unit charge to be that charge which if concentrated at a point would repel an equal like charge concentrated at a point one centimetre away with a force of one dyne, then k is unity and the formula becomes

$$f = \frac{qq'}{r^2}.$$

If the bodies be not immersed in air we must write

$$f = \frac{1}{\lambda} \frac{qq'}{r^2},$$

where λ is a quantity depending on the medium in which the bodies are placed. This quantity is sometimes called the "specific inductive capacity" and sometimes the "dielectric constant" or "dielectric coefficient" of the medium. As the quantity generally varies with the temperature we shall call it the "dielectric coefficient."

Faraday mapped out the electric field which surrounds a charged body by means of lines drawn so that the direction of the

resultant electric force at any point on these lines is the direction
of the tangent at that point. By the electric force at a point,
sometimes called the "electric intensity" or the "intensity of the
electric force," is meant the force with which a unit positive
charge placed at the point would be urged if it could be placed
there without disturbing the electrical distribution elsewhere.
These lines he called lines of force. We shall see later on that we
can map out by means of tubes of force not only the direction of
the field, but also its strength.

The electric potential at a point P due to any electrified
bodies in the neighbourhood is the amount of work
Potential. in ergs that has to be done on a unit of positive
electricity to bring it from the boundary of the field to P, the
electric distribution being supposed to be undisturbed during
the process. If the electric potentials at two neighbouring points
P and P' be V and $V + \partial V$ respectively, and if PP' equal ∂s,
then the work done by the electric forces while unit of positive
electricity is moved from P to P' will be $V - (V + \partial V)$, and if F
be the average electric force in dynes per unit charge from P to P',
which is ultimately equal to the electric force at P when PP' is
diminished indefinitely, the work done will also be represented
by $F\partial s$.

Therefore $$F\partial s = -\partial V,$$

and $$F = -\frac{\partial V}{\partial s}.$$

Hence, if we know the mathematical expression for the potential at
a point, this equation completely specifies the electric force in any
direction at the point. We can therefore describe the electric field
at each point by means of a single symbol V instead of having to
give the components of the force at the point in three directions mutually at right angles. This is one of the advantages
of the potential method of treating problems in attractions and
repulsions. We should notice that the potential function itself is
an undirected quantity; it possesses merely magnitude. In practice,
the potential of a body is generally reckoned from that of the earth
as zero.

By the potential gradient at a given point in a dielectric is meant the rate at which the potential is diminishing in the direction of the line of force through the given point. It is therefore equal to the electric force at the point.

The electromotive force between two points P and Q is defined as the work done on unit of positive electricity by the electric forces while it moves from P to Q. Thus if V_1 and V_2 be the potentials at P and Q, then

the electromotive force between P and $Q = V_1 - V_2$

$$= \int_1^2 F \partial s.$$

It will be seen that the dimensions of electromotive force, or as it is generally written E.M.F., are the same as those of work divided by electrical quantity.

Suppose that we have q units of electricity concentrated at a point O and that we wish to find the potential V at a point P distant r from O.

By definition $\qquad V = \int_r^\infty \dfrac{q}{\lambda r^2}\, \partial r = \dfrac{q}{\lambda r}.$

If we have n charges $q_1, q_2 \dots$ at distances $r_1, r_2 \dots$ from O, then

$$V = \frac{q_1}{\lambda r_1} + \frac{q_2}{\lambda r_2} + \dots$$

$$= \Sigma \frac{q}{\lambda r}.$$

A surface at every point of which V is constant is called an equipotential surface. Since $\dfrac{\partial V}{\partial s}$ is obviously zero along such a surface, there will be no force tangential to it, and hence the lines of force must always cut it at right angles. Two equipotential surfaces cannot cut one another unless they are both at the same potential. In this case the electric force at all points on the curve of intersection must be zero.

Gauss's Theorem.

Let R (Fig. 1) be the intensity of the electric force, *i.e.* the force on unit quantity of electricity, at a small element ∂S of a surface completely enclosing a charged body q. Let α be the angle which R makes with PN the outward normal to ∂S. Then $R \cos \alpha \partial S$ is the normal force across ∂S. If $\partial \omega$ be the solid angle which ∂S subtends at q, then

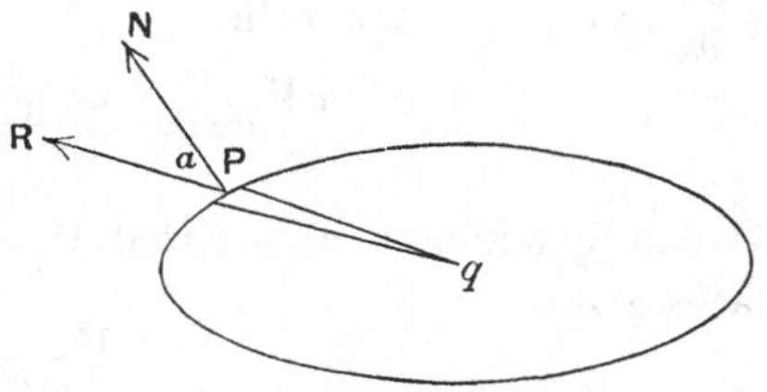

Fig. 1. The integral of the normal force over the surface $= \Sigma R \cos \alpha \partial S = 4\pi q$.

$$\partial \omega = \frac{\partial S \cos \alpha}{r^2}.$$

Now, if we take the integral of $R \cos \alpha \partial S$ over the whole surface, we get

$$\begin{aligned} \Sigma R \cos \alpha \partial S &= \Sigma \frac{q}{r^2} \cos \alpha \partial S \\ &= \Sigma q \partial \omega \\ &= 4\pi q, \end{aligned}$$

for q is constant, and the sum of all the solid angles at a point is the ratio of the surface of a sphere to the square of its radius, *i.e.* 4π. If we had n particles with charges q_1, q_2 ... inside the closed surface, then we have by addition

$$\begin{aligned} \Sigma R \cos \alpha \partial S &= \Sigma R_1 \cos \alpha_1 \partial S + \Sigma R_2 \cos \alpha_2 \partial S + \dots \\ &= 4\pi (q_1 + q_2 + \dots + q_n). \end{aligned}$$

This is true whatever the shape of the surface may be, provided that it embraces all the charges. It follows that, if we know $\Sigma R \cos \alpha \partial S$ over a closed surface, then we can find the sum of the charges enclosed by dividing this sum by 4π. It also follows that if there are no charges within the surface, then

$$\Sigma R \cos \alpha \partial S = 0.$$

Poisson's Equation.

If we choose three axes OX, OY and OZ at right angles to one another, and take the surface integral of the normal force over a small rectangular parallelepiped $\partial x \partial y \partial z$, we get an important equation due to Poisson. Consider the part of the surface integral contributed by the two faces of the parallelepiped parallel to the plane YOZ. One side contributes $+\dfrac{\partial V}{\partial x}\partial y\,\partial z$, and the other

$$-\frac{\partial V}{\partial x}\partial y\,\partial z-\frac{\partial}{\partial x}\left\{\frac{\partial V}{\partial x}\partial y\,\partial z\right\}\partial x.$$

Hence by addition we see that the surface integral over these two faces gives

$$-\frac{\partial^2 V}{\partial x^2}\partial x\partial y\partial z.$$

Proceeding similarly for the other four faces, and equating the total sum to $4\pi\rho\partial x\partial y\partial z$, where ρ is the volume density of the distribution, we get Poisson's equation

$$\frac{\partial^2 V}{\partial x^2}+\frac{\partial^2 V}{\partial y^2}+\frac{\partial^2 V}{\partial z^2}+4\pi\rho=0.$$

This equation is generally written

$$\nabla^2 V+4\pi\rho=0,$$

or

$$\rho=-\frac{1}{4\pi}\nabla^2 V.$$

In the particular case when V is the potential at a point in free space, we get Laplace's equation

$$\nabla^2 V=0.$$

If the electric field is uniform and the lines of force are all parallel to the axis of x, then this becomes

$$\frac{\partial^2 V}{\partial x^2}=0,$$

and so,

$$V=A+Bx,$$

where A and B are constants.

If the field be symmetrical about an axis and if r be the distance from the axis of a point P in air at which the potential is V, then we can easily show that

$$\frac{\partial}{\partial r}\left(2\pi r\,\frac{\partial V}{\partial r}\right)=0,$$

and therefore,

$$V=A+B\log r.$$

If we imagine a tube which starts from a positively electrified
body and the sides of which are formed by lines of
force, we get a tube of force. As we move along a
line of force, the potential continually diminishes, and hence a
line of force can never be a closed curve. We imagine then a tube
of force as starting from a positively electrified surface, and ending
at a negatively electrified surface. Consider the portion of a tube
of force intercepted between two equipotential surfaces. Let ∂S
and $\partial S'$ be the intercepts on them, and let F and F' be the
intensities of the electric force at the two surfaces respectively.
Applying Gauss's equation to this element, and noting that the
sides of the tube add nothing to the integral $\Sigma R \cos \alpha \partial S$, since $\cos \alpha$
is zero over the sides, we see that

$$F\partial S - F'\partial S' = 0.$$

Hence along a tube of force the product of the intensity of the
force at a point and the area of the section of the equipotential surface
through the point intercepted by the sides of the tube is constant.
If the element of the positive surface from which the tube starts
has a unit quantity of electricity upon it, we get a unit tube.
This unit tube is called a Faraday tube.

Tubes of force. (margin)

Coulomb found experimentally that the electric force near
an electrified conductor was at right angles to its
surface and that the magnitude of the force was
proportional to the surface density. The exact
relation between these two quantities can be found
from the following considerations.

Coulomb's Law for the intensity of the force nea conductor. (margin)

In order that the electricity on a conducting body may be in
equilibrium, the E.M.F. between any two points must be zero;
hence the potential must be constant throughout and equal to its
surface value. Thus the bounding surface of the conductor is an
equipotential surface, the electric force at points outside the
surface and infinitely near it is normal to the surface, and the
electric force at all points in the conductor is zero. Consider now
the surface integral of the normal force over a small closed surface
formed by an element ∂S of an equipotential surface very close to
the conductor, the tube of force through the boundary of ∂S, and a

surface inside the conductor continuous with the tube and closing it. The tube of force is supposed to enclose a quantity $\sigma \partial S'$ of electricity spread over an area $\partial S'$ on the surface of the conductor. We see that $F \partial S$ is the value of the surface integral, for the sides of the tube and the surface in the conductor contribute nothing to it. Hence by Gauss's theorem

$$F\partial S = 4\pi\sigma\partial S'.$$

When the equipotential surface is infinitely close to the charged body $\partial S = \partial S'$; and hence

$$F = 4\pi\sigma.$$

This numerical relation was proved by Poisson.

If V be the potential at a point near the surface, and ∂n be an element of the normal to ∂S drawn outwards, then

$$\sigma = \frac{F}{4\pi}$$

$$= -\frac{1}{4\pi}\frac{\partial V}{\partial n}.$$

If the body be immersed in a medium of which the dielectric coefficient is λ, then basing our theory on experiment, we have

$$\sigma = -\frac{\lambda}{4\pi}\frac{\partial V}{\partial n}.$$

Since $F\partial S$ is constant along a tube of force, we see that the quantities of electricity at each end of the tube are equal in magnitude but opposite in sign. If a conductor be placed in an electric field, it clearly follows from the principles we have developed that those parts of it where tubes of induction stop must be negatively electrified and those parts where they begin must be positively electrified.

The potential V due to a body cannot have a maximum or a minimum value at a point in free space, for if it had then $\Sigma F \cos \alpha \partial S$ taken round a small sphere enclosing the point would not be zero. Hence, if the potential be constant round a closed surface it will be constant at all points in that space, as otherwise it would have a maximum or a minimum value at some point in it.

If we have various charged bodies enclosed in a metallic envelope, each tube of force starting from one of the charged bodies will terminate either on the inside surface of the envelope or on another of the charged bodies. Hence the charge induced on the inside of the metallic envelope will be exactly equal and of opposite sign to the sum of the charges on the bodies.

The number of Faraday tubes which start from an element $\partial S'$ of an electrified conductor is $\sigma \partial S'$, where σ denotes

Polarization. the surface density of the electricity. Since, by Gauss's theorem, $F \partial S = 4\pi\sigma \partial S'$, we see that F is 4π times the number of Faraday tubes passing through unit area of the equipotential surface at P. If N denote this number and the medium be air, we have $F = 4\pi N$.

In a medium the dielectric coefficient of which is λ,

$$F = (4\pi/\lambda)\, N.$$

Maxwell calls N the electric displacement and J. J. Thomson calls it the polarization of the dielectric.

We have previously seen that the electric force F at a point distant r from a point charge q is $q/\lambda r^2$. Hence the polarization N at this point

$$= (\lambda/4\pi)\,(q/\lambda r^2) = q/4\pi r^2.$$

When the charge remains constant, therefore, the polarization is independent of the nature of the medium. It is therefore simply measured by the aggregate strength of the Faraday tubes per unit area at the point.

Green proved mathematically that if we suppose an equipotential surface replaced by a conductor having the

Replacing a system of charged bodies by a conductor of which the boundary is an equipotential surface. same boundary as the surface and if it be electrified so that the surface density is given by

$$\sigma = -\frac{1}{4\pi}\frac{\partial V}{\partial n},$$

then the electricity on this conductor will be in equilibrium, and will produce the same potential at external points as the charges enclosed by the equipotential surface originally did.

A physical proof of this theorem can be given as follows. Suppose the equipotential surface replaced by a metal sheet coincident with it. If the metal be connected to earth, it will be at zero potential, and all points external to it will also be at zero potential. Let V be the potential at an external point due to the enclosed charges and V_1 be that due to the bound charge on the sheet, then $V + V_1 = 0$, and therefore, $V_1 = - V$. Also at a point on the conductor itself $v + v_1 = 0$, and $v_1 = - v =$ constant by hypothesis; hence the electricity on the conductor would be in equilibrium. Therefore, if we suppose this conductor charged with electricity of the opposite sign to what it has in this case, it will produce a potential V at all external points and have itself a constant potential v. Hence the theorem follows.

The method of images is due to Lord Kelvin. It is of great
Electric images. practical value in solving problems connected with the forms of the lines of force round conductors suspended parallel to the earth, etc. Suppose for example we have a wire parallel to the earth at a distance h above it. Let the wire be charged with q units of electricity. We imagine an equal parallel wire at a depth h in the earth with a charge of $-q$ units. The surface of the earth will then be an equipotential surface of this system, and the distribution of the lines of force in the air can be found by solving the simple problem of two parallel wires at a distance $2h$ apart, and having equal and opposite charges.

As an example of the method of images we shall show how to
The image of an electrified point in a sphere. find the electric field round a spherical conductor maintained at zero potential when there is a charge of electricity at a point R (Fig. 2) near the sphere. Let the centre of the sphere be at O and join OR cutting the sphere at A. Find a point R' in OR such that

$$OR \cdot OR' = OA^2.$$

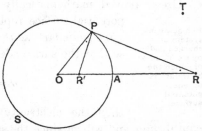

Fig. 2.   When $OR' \cdot OR = OA^2$, R' is the image of R.

Let P be any point on the surface of the sphere. Then since $OP = OA$, we have

$$OR'/OP = OP/OR.$$

The angle POR also is common to the two triangles POR and POR'. They are therefore similar, and so

$$OP/PR' = OR/PR.$$

Hence
$$\frac{-q(OP/OR)}{PR'} + \frac{q}{PR} = 0.$$

We see, therefore, that if the sphere were removed, the spherical surface PAS would be the zero equipotential surface of the two point charges $-q(OP/OR)$ at R' and q at R. Hence the equipotential surfaces outside the sphere PAS due to the two point charges are the same as for the solid sphere and the point charge at R. It follows that the potential at any point T is

$$\frac{q}{RT} - \frac{q(OP/OR)}{R'T},$$

and that the charge on the surface of the given sphere is

$$-q(OP/OR).$$

The point R' is called the electric image of the point R. Problems of this nature can often be greatly simplified by considering merely electrified points and their images.

Let us suppose that the field is due to two electrified conductors **The capacity between two electrified conductors.** which have equal and opposite charges Q and $-Q$ respectively. We suppose that the dielectric by which they are surrounded is uniform and that there are no other bodies in the neighbourhood. We picture tubes of force starting at right angles from the positive conductor and ending on the negative conductor which also they meet at right angles. If F is the electric force at any point in a tube and s the area of the cross section of the tube at that point, the product Fs is constant. Now the work done in taking a unit charge from the negative to the positive conductor is $\int_0^l F\partial x$, that is, $Fs\int_0^l \frac{\partial x}{s}$, where ∂x is an element of a line of force. Hence if the potentials of the conductors be V_1 and V_2, we have

$$V_1 - V_2 = Fs \int_0^l \frac{\partial x}{s} = \frac{4\pi\sigma}{\lambda} \partial S \int_0^l \frac{\partial x}{s} = \frac{4\pi\partial Q}{\lambda} \int_0^l \frac{\partial x}{s},$$

where σ is the surface density of the electricity, and ∂S the elementary area at the positive end of the tube, so that

$$\sigma\partial S = \partial Q = \text{the elementary charge.}$$

We see that
$$\frac{\partial Q}{V_1 - V_2} = \frac{\lambda}{4\pi} \Big/ \int_0^l \Big(\frac{\partial x}{s}\Big).$$

Integrating over the whole conductor we get

$$\frac{Q}{V_1 - V_2} = \frac{\lambda}{4\pi} \Sigma 1 \Big/ \Big\{\int_0^l \frac{\partial x}{s}\Big\} \quad\ldots\ldots\ldots\ldots(a).$$

If we were now to give further charges, Q and $-Q$, to the two conductors the charges would distribute themselves so as to double the surface densities at every point of the two conductors. The mutual actions between the two coincident charges on each conductor will be perpendicular to the surface and so, provided that the dielectric can withstand the increased stress to which it is subjected, doubling the density of the layer will not disturb the equilibrium. Although the Faraday tubes are twice as many in the second case, yet the lines of force are identical, and thus the value of the integral on the right hand side of (a) is the same. The ratio of Q to $V_1 - V_2$ is therefore unaltered when the charge is doubled. Similarly it is unaltered if the charge is increased n times. It is therefore a constant. This constant is called the capacity between the two conductors.

It is instructive to consider the analogous problem in the theory of the conduction of heat. Let us consider the case of two bodies which are maintained at temperatures θ_1, θ_2 respectively and let us suppose that the thermal conductivity of the medium in which they are placed is k. By comparing Laplace's equation (p. 6) with the corresponding equation for the steady flow of heat we see that the lines of flow of the heat will be the same as the lines of force in the analogous electrostatic problem

Thermal conductance.

and the isothermal surfaces will coincide with the equipotential surfaces. If ∂Q_1 denote the flow of heat per second along a tube of flow, we have

$$\partial Q_1 = k\frac{s}{\partial x}\cdot\partial\theta$$

and so, $\qquad\qquad \partial\theta = \partial Q_1\,(1/k)\,(\partial x/s),$

and therefore, $\qquad \theta_1 - \theta_2 = \partial Q_1\,(1/k)\displaystyle\int_0^l\frac{\partial x}{s},$

since, in the case considered, ∂Q_1 is constant along a tube of flow. Hence if Q_1 be the total flow of heat per second leaving one body and arriving at the other, we have

$$\frac{Q_1}{\theta_1 - \theta_2} = k\,\Sigma\,1\Big/\left\{\int_0^l\frac{\partial x}{s}\right\}.$$

$Q_1/(\theta_1 - \theta_2)$ is the thermal conductance K_1 between the two bodies, and so

$$K_1 = \frac{4\pi}{\lambda}kK,$$

where K is the electrostatic capacity between the two bodies when the dielectric in which they are immersed has a coefficient λ.

This formula is useful in practical work.

When an electrified body, surrounded by air, is at a very great distance from all other conductors the ratio of its charge to its potential is called its capacity. If we assume that the total charge on the body added to the total charge on the earth is zero, and that its dimensions are very minute compared with those of the earth, then the capacity of the body is practically equal to the capacity between the body and the earth. We shall discuss this question further in Chapter v.

The capacity of a conductor.

We have already seen that the potential v at a distance a from a point charge q immersed in a medium the dielectric coefficient of which is λ is given by

$$v = q/(\lambda a).$$

Hence the sphere which has the given point for its centre and its radius equal to a, is an equipotential surface of the given charge. If this sphere, therefore, were a solid conductor having a charge q its potential would be v. Its capacity q/v, therefore, is equal to λa.

The capacity of a parallel plate condenser. When we have two metal plates very close together the lines of force between them, except near the edges of the plates, are straight lines and the capacity per unit area of the surfaces is given by

$$K = \frac{\lambda}{4\pi} \; \Sigma \; \frac{1}{\displaystyle\int_0^h \partial x} = \frac{\lambda}{4\pi h},$$

approximately, where λ is the dielectric coefficient of the medium between them. If the area of each plate be S and we neglect the curvature of the lines of force near the edges, we see that the capacity is $\lambda S/4\pi h$ approximately. When h is small the capacity will be large. An arrangement of this kind is called a condenser.

The capacity of two orthogonal spheres. The capacity of a conductor formed by portions of two spheres intersecting at right angles can be readily found. Let A and B (Fig. 3) be the centres of the two spheres and let a and b be their radii. Since the spheres intersect at right angles ADB is a right angle, where D is a point of intersection, and it is easy to see that

$$AC \cdot AB = a^2, \quad BC \cdot BA = b^2$$

and

$$CD^2 = AC \cdot CB = (ab/\sqrt{a^2 + b^2})^2.$$

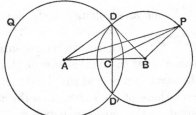

Fig. 3.   Surface formed by two orthogonal spheres.

Let us now suppose that point charges va, $-v \cdot CD$, and vb, are placed at the points A, C, and B respectively. If P be any point on the portion DPD' of the spherical surface whose centre is at B, since $BC \cdot BA = BP^2$, the triangles ABP and CBP are similar, and so

$$AP/CP = AB/BP = a/CD.$$

Hence $$a/AP = CD/CP.$$

The potential at P

$$= v\left\{\frac{a}{AP} + \frac{b}{b} - \frac{CD}{CP}\right\} = v.$$

Similarly the potential at any point of the spherical surface DQD' is v.

The capacity of a conductor, therefore, having the figure of 8 shape shown in Fig. 3 is the sum of the three charges at A, C and B divided by v. Its capacity is therefore equal to

$$a + b - ab/\sqrt{a^2 + b^2}.$$

By considerations in connection with the energy stored in the field round an electrified conductor it can be shown that the capacity of a conductor of given surface is a maximum when the shape of the conductor is spherical.

When $a = b$, the capacity of the conductor shown in Fig. 3 is $1\cdot293a$. The capacity of a sphere having the same surface is $1\cdot307a$. It will be seen therefore that the numerical value of the capacity is not much altered by changing the shape of the surface. It will also be shown in Chapter VIII that the capacity of the conductor formed by two equal spheres touching one another and having the same total surface as either of the two conductors we have just considered is $1\cdot281a$. This is a further illustration of the small change in the capacity due to deforming the surface of conductors.

We shall now find an expression for the potential energy of two electrified conductors, one of which has a charge q and the other a charge $-q$. If K be the capacity between them and their potentials be v_1 and v_2 respectively, we have $q = K(v_1 - v_2)$. Let now a charge ∂q be given to the positively charged conductor and a charge $-\partial q$ to the negatively charged conductor.

The potential energy of two electrified bodies.

The work ∂W done during this operation will be $v_1\,\partial q - v_2\,\partial q$, and hence,

$$\partial W = (v_1 - v_2)\,\partial q$$
$$= K(v_1 - v_2)\,\partial(v_1 - v_2),$$

and therefore,
$$W = K\,(v_1 - v_2)^2/2$$
$$= q^2/(2K)$$
$$= q\,(v_1 - v_2)/2.$$

The energy stored in a tube of force. If ∂q be the electricity at the positive end of the tube and $\partial S'$ be the area of the cross section at this end, we have, as before,
$$\partial q = \sigma\,\partial S' = (\lambda/4\pi)\,F\partial S.$$

By the preceding paragraph,
$$W = (v_1 - v_2)\,\partial q/2$$
$$= (\Sigma F\partial x)\,(\lambda/8\pi)\,F\partial S,$$

but along a tube of force $F\partial S$ is constant and $\partial S\partial x = \partial v = $ an element of volume of the tube. Hence
$$W = \frac{\lambda}{8\pi}\Sigma F^2\,\partial v.$$

When the field in the tube is uniform we may suppose that the energy is distributed in such a way that $\lambda F^2/(8\pi)$ is the energy per unit volume.

Electric current. When we have an E.M.F. between two points in a conductor, a current is produced. A current is measured by the rate at which quantity of electricity flows through any cross section of the conductor; it is therefore $\dfrac{\partial q}{\partial t}$ and will be denoted by i. The work done in taking q from a point where the potential is v_1 to a point where it is v_2 is
$$q\,(v_1 - v_2) = wt \text{ ergs,}$$

where w is the work done per second in ergs, and t seconds is the time of working, the rate of working being uniform.

Hence differentiating
$$i\,(v_1 - v_2) = w.$$

In this equation i is the current in electrostatic units, $v_1 - v_2$ is the potential difference (P.D.) in electrostatic units and w is the power in ergs per second.

If any magnet be supported in such a way that it is free to

Magnetism. turn about its centre of gravity in the earth's magnetic field, it is found that a particular line through the centre of gravity of the magnet always tends to point in the same direction. This line is called the magnetic axis of the magnet. If we have a long thin cylindrical magnet with its magnetic axis coincident with the axis of the cylinder, then the centres of the circular faces are the poles of the magnet. To a first approximation we can suppose that these poles are centres of force, and the action of the magnet can be calculated by supposing positive magnetism m concentrated at the north pole of the magnet, *i.e.* the pole which points to the north when the magnet is suspended, and negative magnetism $- m$ concentrated at the south pole, the rest of the magnet consisting of inert matter.

It is found that two north poles or two south poles repel one another, while a north pole attracts a south pole. The strengths of poles or "the quantities of magnetism" on them are compared by the forces which they exert on another pole.

We define the unit pole to be that pole which repels an equal like pole one centimetre away with a force of one dyne. A pole of strength m is one which repels unit pole (or is repelled by unit pole) with a force of m dynes when the poles are one centimetre apart.

It is found by experiments with the torsion balance and otherwise that like magnetic poles repel one another and unlike attract with a force which is directly proportional to the strengths of the poles and inversely proportional to the square of the distance between them. Hence, in air or other non-magnetic medium, the law of repulsion of like magnetic poles is

$$f = \frac{mm'}{r^2}.$$

If $2l$ be the distance between the poles, $2lm$ is called the magnetic moment (M) of the magnet.

Following the electrostatic analogy, we define the magnetic potential at a point to be the work done in ergs in taking unit north pole from the boundary of the field to the point in question.

Hence if there was only one pole of strength m, the potential at a distance r from it would be given by

$$V = \frac{m}{r}.$$

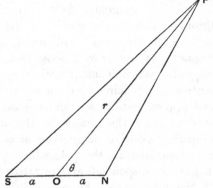

Potential near a bar magnet. Let N and S (Fig. 4) be the poles of a bar magnet. Let their strengths be m and $-m$ respectively and let $2a$ be the distance between them. Let O be the middle point of SN and let OP equal r. Then if V be the potential at P due to this magnet we have

Fig. 4.   Magnetic Potential at P
$$= \frac{M\cos\theta}{r^2} + \frac{M\,(5\cos^3\theta - 3\cos\theta)\,a^2}{2r^4} + \dots$$

$$V = \frac{m}{\sqrt{a^2 + r^2 - 2ar\cos\theta}} - \frac{m}{\sqrt{a^2 + r^2 + 2ar\cos\theta}} \quad \dots\dots(1).$$

If r is greater than $(\sqrt{2}+1)\,a$ we have, by the binomial theorem,

$$\frac{1}{\sqrt{a^2 + r^2 - 2ar\cos\theta}}$$
$$= \frac{1}{r}\left\{1 - \left(2\frac{a}{r}\cos\theta - \frac{a^2}{r^2}\right)\right\}^{-\frac{1}{2}}$$
$$= \frac{1}{r}\left\{1 + \frac{1}{2}\left(2\frac{a}{r}\cos\theta - \frac{a^2}{r^2}\right) + \frac{3}{8}\left(2\frac{a}{r}\cos\theta - \frac{a^2}{r^2}\right)^2 + \dots\right\}$$
$$= \frac{1}{r}\left\{1 + \cos\theta\,\frac{a}{r} + \left(\frac{3}{2}\cos^2\theta - \frac{1}{2}\right)\frac{a^2}{r^2} + \left(\frac{5}{2}\cos^3\theta - \frac{3}{2}\cos\theta\right)\frac{a^3}{r^3} + \dots\right\}.$$

It may be shown that if r is greater than a this expansion can be used. The coefficients of the various powers of a/r are called Legendre's coefficients.

Substituting in (1) and simplifying we get

$$V = \frac{2ma\cos\theta}{r^2} + \frac{5\cos^3\theta - 3\cos\theta}{r^4}\,ma^3 + \dots$$
$$= \frac{M\cos\theta}{r^2} + \frac{M\,(5\cos^3\theta - 3\cos\theta)\,a^2}{2r^4} + \dots,$$

where M is the magnetic moment of the magnet. Hence if r be large compared with a,

$$V = \frac{M \cos \theta}{r^2}.$$

It follows that the force in the direction OP will be

$$-\frac{\partial V}{\partial r} = \frac{2M \cos \theta}{r^3},$$

and in a direction at right angles to OP, in the direction in which θ increases, it will be

$$-\frac{\partial V}{r \partial \theta} = \frac{M \sin \theta}{r^3}.$$

Hence we can easily find the direction and the magnitude of the magnetic force at a point P when r is large compared with the length of the magnet.

The equipotential surfaces round an infinitely small magnet will be found from the equation

$$Vr^2 = M \cos \theta,$$

or

$$V (x^2 + y^2)^{\frac{3}{2}} = Mx,$$

by giving various values to V.

We can prove in exactly the same way as in electrostatics (Gauss's Theorem) that

Tubes of force.
Tubes of induction.

$$\Sigma R \cos \alpha \partial S = 4\pi m,$$

where the integral is taken over a closed surface enclosing poles the sum of whose strengths is m. R is the intensity of the magnetic force at ∂S and α is the angle between the direction of the force and the normal to the surface. Applying this theorem to the part of a tube of force intercepted between two equipotential surfaces, we find that $F\partial S = F'\partial S'$, where F is the intensity of the force at a point on ∂S. Hence we can suppose the magnetic field divided up into tubes of force in exactly the same way as we divided up the electrostatic field, and the tubes of force map out the intensity and direction of the magnetic force at any point.

The product $R \cos \alpha \partial S$ is called the flux of force through the area ∂S and we see from Gauss's theorem that the total flux of force from a unit magnetic pole is 4π.

2—2

If we break into two portions a long thin bar magnet which
has been uniformly magnetised and has poles of
strength m and $-m$ respectively, it will be found that
each portion is a magnet with poles of practically the
same strength as the original magnet. The axes of the magnets
are the two portions of the original axis. This is true also when
we divide the magnet into many parts, and hence we can regard
such a magnet as made up of a large number of little magnets of
which the axes coincide with the axis of the original magnet and
with ends perpendicular to that axis. Such a magnet is called
a solenoidal magnet, and we see that no great error is made in
assuming that its ends are covered with a layer of strength m of
attracting and repelling matter respectively, and that the rest of
the magnet is inert.

Intensity of Magnetisation. Permeability.

The intensity of the magnetisation of a solenoidal magnet is
defined as the magnetic moment per unit volume, and is generally
denoted by I.

Hence
$$I = \frac{M}{V} = \frac{ml}{Sl} = \frac{m}{S},$$

where S is the cross sectional area of the magnet. We see that I
may also be defined as the pole strength per square centimetre of
the area of the cross section.

If the magnet were not uniformly magnetised, I would be
different at different points, and so we should have to define it as

$$\frac{\partial M}{\partial V}.$$

Consider now a circular iron ring uniformly magnetised, the
cross section of which is circular. If I be the intensity of the
magnetisation, then, if this ring be sawn through we should have
what may be regarded as a layer of attracting matter on one side
of the air gap and a layer of repelling matter on the other, the
surface density being in each case I. If a unit magnetic pole be
placed in the air gap on the axis of the ring and at a distance a
from either circular face, the force of attraction F to one face will
be obviously along the axis, and therefore

$$F = \int_0^R \frac{2\pi r I \partial r}{a^2 + r^2} \times \cos\theta,$$

where θ is the angle made by the axis with a line drawn from the pole to an element of the ring with a radius r. Now $r = a \tan \theta$,

and therefore $\qquad \partial r = a \sec^2 \theta \partial\theta = \frac{1}{a} (a^2 + r^2) \partial\theta.$

Therefore $\qquad\qquad F = 2\pi I \int_0^\phi \sin \theta \partial\theta$

$$= 2\pi I (1 - \cos \phi),$$

where ϕ is the value of θ when R is the radius of the cross section. Hence, when a is small, F is $2\pi I$, and since the repelling face will repel the pole with an equal force, we see that the intensity of the field in the air gap is $4\pi I$.

If we saw through the ring at some other point, and then imagined it stretched out straight, the ends of the first gap still remaining the same distance apart, we see that our unit magnetic pole will now be subjected to the attractions and repulsions of the poles at the other ends of the bar; hence the field in which it is situated will be weakened. If the bar were very long however we could neglect the demagnetising effects of the ends. If the bar were placed in a magnetising field of intensity H, the unit pole would be subjected to forces H and $4\pi I$, where I is the new intensity of the magnetisation. If B is the resultant of H and $4\pi I$, then B will be the strength of the field in the air gap. If H and $4\pi I$ are in the same direction, then

$$B = H + 4\pi I.$$

This quantity B is a measure, on the electromagnetic system, of what is called the magnetic induction in the iron, a quantity upon whose variations the electromotive force of induction in a circuit enclosing a magnetic material depends.

The ratio of I to H is called the susceptibility of the iron, and the ratio of B to H is called the permeability. It is this latter ratio that is generally wanted in practice; it is always denoted by μ, so that

$$B = \mu H.$$

When iron is magnetised to a certain induction density B, it is found that when the force is withdrawn a certain quantity B_0 is left remanent in the iron, and a certain coercive force has to be applied to get rid of B_0. We will return to this point when we discuss magnetic tests.

If we have a thin sheet of iron made up of an infinite number of little magnets with their axes perpendicular to the sheet and their like poles all pointing in the same direction, we get what is called a magnetic shell.

Magnetic shell. Mutual potential energy of two shells.

A study of the properties of these shells is most helpful in understanding the theory of electrodynamics. The strength of a magnetic shell is its magnetic moment per unit area, so that, if g be its strength and h its thickness, $\frac{g}{h}$ is the polar strength of the face per unit area.

We will now find the mutual potential energy of two such shells, A and B, *i.e.* the work done in bringing A and B, initially at an infinite distance apart, into their actual positions. Consider an elementary small magnet of the shell A; its potential at a point P is $\frac{g_1 \, \partial S_1 \cos \theta}{r^2}$, *i.e.* $g_1 \, \partial \omega$, where $\partial \omega$ is the solid angle subtended at P by the element. Integrating over the whole shell A, we see that $g_1 \Omega_1$ is the potential at P due to the shell, where Ω_1 is the solid angle subtended by its boundary at P. Ω_1 is considered to be positive when the lines from P to the boundary of A first meet the north side of A. If the point P be on the shell B which has a strength g_2, then the polar strength of an element ∂S_2 at P would be $\frac{g_2}{\partial n} \, \partial S_2$, where ∂n is the thickness of the shell, and we have shown that the potential at P is $g_1 \Omega_1$. Hence the work done in taking the polar element ∂S_2 from infinity to its position at P against the repulsion of the shell A would be

$$g_1 g_2 \frac{\partial S_2}{\partial n} \Omega_1,$$

and this expression gives the mutual potential energy of ∂S_2 and the shell A. Similarly, assuming that the positive direction of n is from the south to the north side of the shell, the energy of the other pole of the elementary magnet the end of which is ∂S_2 will be

$$-g_1 g_2 \frac{\partial S_2}{\partial n} \left(\Omega_1 - \frac{\partial \Omega_1}{\partial n} \partial n \right).$$

Hence the total potential energy of the elementary magnet is

$$g_1 g_2 \frac{\partial \Omega_1}{\partial n} \partial S_2.$$

Therefore, integrating over the whole shell B, we find that the mutual potential energy is given by

$$g_2 \int \frac{\partial g_1 \Omega_1}{\partial n} \partial S_2.$$

Now $-\dfrac{\partial g_1 \Omega_1}{\partial n}$ is the resultant force measured normally at the shell B due to the shell A. Hence the integral gives the surface integral of magnetic force ψ_1 over a side of the shell B due to the shell A. Similarly if ψ_2 were the surface integral of magnetic force over a side of the shell A caused by B, $-g_1\psi_2$ would be the mutual potential energy.

Therefore $g_2\psi_1 = g_1\psi_2,$

which is a remarkable reciprocal relation of great practical importance. When the strengths of the shells are equal, we see that the surface integral of the magnetic force over a side of the shell A caused by B equals the corresponding surface integral over B caused by A.

Oersted showed that when a wire carrying an electric current produced by a battery was brought near a magnetic needle, the needle was deflected. Hence an electric current produces a magnetic field.

Electrodynamics.

We have seen that an infinitely small magnet, the magnetic moment of which is M, produces a potential $\dfrac{M \cos \theta}{r^2}$ at points at a distance r from it, where θ is the angle which r makes with the length of the bar. Weber proved experimentally that a small closed plane circuit carrying a current not only produced the same field but also was acted on by the same forces as a small solenoidal magnet with an axis perpendicular to the plane of the coil, provided that a certain relation held between the magnetic moment of the magnet, the area of the circuit, and the current flowing in it. If S be the area of the circuit, and i the strength of the current, then the potential at any point is $k \dfrac{Si \cos \theta}{r^2}$, where

k is a constant. If the fields produced by the small magnet and the small circuit are the same, we must have $M = kSi$. In the system of units known as the electromagnetic system, in order to simplify our formulae as much as possible, we choose our unit of current so that k is unity, and therefore

$$M = Si.$$

Hence also
$$V = \frac{Si \cos \theta}{r^2}.$$

The potential V of the small circuit at a point P, distant r from its centre, may be written

$$V = i\omega,$$

where ω is the small solid angle subtended by S at P. If we now suppose that an infinite number of these small circuits are all crowded together, forming a network, and that they are all carrying equal currents i flowing in the same sense round the meshes, then it is easy to see that

$$V = i\Sigma\omega = i\Omega,$$

where Ω is the solid angle subtended by the boundary of all the small circuits. Where the elementary circuits touch one another we have equal currents flowing in opposite directions, and hence they are neutralised. We see then that this arrangement is equivalent to a circuit coinciding with the boundary of the small circuits and carrying a current i. It follows that the potential V at any point due to an electric circuit is always equal to $i\Omega$, where Ω is the solid angle which the circuit subtends at the point.

We have shown above that the potential due to a magnetic shell of strength g at a point P is given by

$$V = g\Omega,$$

where Ω is the solid angle subtended by the boundary of the shell at the point. Hence Ampère's theorem follows, namely, that a closed circuit and a magnetic shell the boundary of which coincides with the circuit are equivalent if $g = i$.

When the circuit carrying the current and the magnetic shell "equivalent" to it are placed in another medium, the magnetic force at any point P due to the shell has $(1/\mu)$ times its former value, while the magnetic force due to the current retains its former value. Hence the potential at P due to the current g_2 is μ times that due to the shell of strength g_2. Thus the mutual potential energy of two shells is $-\mu g_2 \psi_1 = -g_2 \phi_1$, where ϕ_1 is the flux of magnetic induction through the first shell. It also follows that the mutual potential energy of two circuits carrying currents i_1 and i_2 respectively may be expressed either by $i_1 \phi_2$ or by $-i_2 \phi_1$.

Consider the case of a circle of wire of radius r carrying a

Circular current. current i. The potential V at a point P on its axis perpendicular to its plane will be $i\omega$. With centre P (Fig. 5) describe a sphere passing through the circle. The area of the spherical cap intercepted by this circle is $2\pi Rh$, where R is the radius of the sphere and h the height of the cap, hence

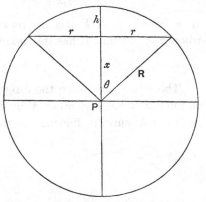

$$V = i \cdot \frac{2\pi Rh}{R^2}$$

$$= 2\pi i (1 - \cos\theta)$$

$$= 2\pi i \left(1 - \frac{x}{\sqrt{x^2 + r^2}}\right),$$

where x is the distance of the point P from the plane of the wire. Hence if H be the force

Fig. 5.   Potential at $P = 2\pi i (1 - \cos\theta)$.

on unit pole at P along the axis, we have

$$H = -\frac{\partial V}{\partial x}$$

$$= \frac{2\pi i r^2}{(r^2 + x^2)^{\frac{3}{2}}}.$$

We see that if x be constant, H has its maximum value

$$4\pi i/(3\sqrt{3}x)$$

when $r = x\sqrt{2}.$

When x is zero $H = \dfrac{2\pi i}{r}.$

In this formula H is in dynes per unit pole, r in centimetres and i in C.G.S. units. We can thus define the unit current as the current which, flowing in a circle of radius r centimetres, produces a magnetic force of $2\pi/r$ at its centre perpendicular to its plane.

If we place a magnet of length $2l$ and pole strength m with its centre at P and its axis at right angles to the plane of the wire, the force F acting on it is given by

$$F = \frac{2\pi m i r^2}{\{r^2 + (x-l)^2\}^{\frac{3}{2}}} - \frac{2\pi m i r^2}{\{r^2 + (x+l)^2\}^{\frac{3}{2}}}.$$

When l is very small, we have

$$F = \frac{6\pi M r^2 x i}{(r^2 + x^2)^{\frac{5}{2}}}, \text{ approximately.}$$

If r be constant F has its maximum value when $x = r/2$, and when x is constant it has its maximum value when

$$r = x\sqrt{2/3}.$$

The relation between the direction of the current and the lines of force can be remembered by the diagrams shown in Figs. 6 and 7. A current flowing in the direction of the arrowheads

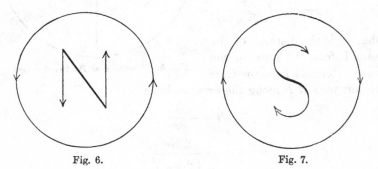

Fig. 6.   Fig. 7.

Relation between polarity and direction of current.

shown in Fig. 6 produces magnetic lines upwards through the paper.

In practice the unit current adopted is the ampere, which is one-tenth of the absolute C.G.S. unit defined above. A current of electricity in a conductor is the rate at which a quantity q of electricity is flowing through the conductor. In symbols

$$i = \frac{\partial q}{\partial t}.$$

We may, therefore, define the unit quantity of electricity as the quantity which flows past any cross section of the conductor every second when unit current is flowing. The practical unit of quantity, in the electromagnetic system, is the coulomb, which is sometimes called the ampere-second.

When the number N of the windings on the coil is large, we
The magnetic force at the axis of a cylindrical coil. may consider the field at the axis of the coil to be practically the same as the field produced by N circular currents evenly spaced over the surface of the coil and each carrying a current i.

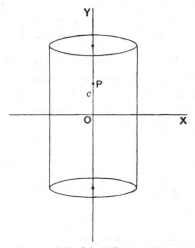

Fig. 8. Cylindrical Current Sheet.

Hence, by the preceding section, the magnetic force H at a point P on the axis of the coil (Fig. 8) the centre of which is O, is given by

$$H = \int_{-h}^{+h} \frac{2\pi r^2 i}{\{r^2 + (c - y)^2\}^{\frac{3}{2}}} \cdot \frac{N \partial y}{2h},$$

where $2h$ is the axis of the coil, r is its radius, and OP is c.

Put $c - y = r \tan \theta$; then $\partial y = - r \sec^2 \theta \partial \theta$, and

$$H = \frac{N\pi i}{h} \int \cos \theta \partial \theta,$$

where the limits are $\tan^{-1} \dfrac{c - h}{r}$ to $\tan^{-1} \dfrac{c + h}{r}$.

Hence

$$H = \frac{N\pi i}{h} \left[\frac{c + h}{\{r^2 + (c + h)^2\}^{\frac{1}{2}}} - \frac{c - h}{\{r^2 + (c - h)^2\}^{\frac{1}{2}}} \right].$$

If n be the number of turns per unit length $N = 2hn$.

At O, we have

$$H = \frac{4\pi n h i}{(r^2 + h^2)^{\frac{1}{2}}},$$

since $c = 0$.

At the end of the coil $c = h$, and

$$H = 4\pi n h i / (r^2 + 4h^2)^{\frac{1}{2}}.$$

We see that when $r^2/(2h^2)$ can be neglected compared with unity, the force at the centre of the coil is nearly $4\pi n i$ and at the centres of the ends of the coil it is nearly $2\pi n i$.

Let us suppose that there are N circular coils arranged on the **A spherical current sheet.** surface of a sphere of radius a with their planes perpendicular to a given diameter. Let the current in each be i and let the coils be so arranged that the number of

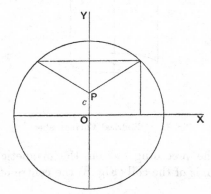

Fig. 9. Spherical Current Sheet.

them between any two planes perpendicular to the given diameter and at unit distance apart is n, so that $N = 2an$. When the number of coils is indefinitely great such an arrangement is called a spherical current sheet. We shall now calculate the magnetic force F at a point P on the axis at a distance c from the centre O of the sphere (Fig. 9).

Since the total current between two planes parallel to OX at a distance ∂y apart is $ni\partial y$ we easily see that

$$H = \int_{-a}^{+a} \frac{2\pi nix^2 \partial y}{\{x^2 + (y-c)^2\}^{\frac{3}{2}}},$$

where x is the radius of the circle of wire at a distance y from the centre. Noticing that $x^2 + y^2 = a^2$, and changing the variable by writing $a^2 + c^2 - 2cy = z^2$, we get

$$H = \frac{8\pi ni}{3} = \frac{4\pi Ni}{3a}.$$

It is interesting to notice that the value of H is independent of c, and therefore has the same value at every point of the axis of the spherical current sheet. This is a particular case of the more general theorem that the magnetic force at every point inside a spherical current sheet is constant in magnitude and parallel to the axis of the sheet. This theorem may be easily proved by replacing the circular currents by their equivalent magnetic circular plates. We thus get a spherical magnet, from the well known properties of which the above theorem follows. In practice this theorem is sometimes utilised in order to obtain a uniform magnetic field.

The electromagnetic system of units is based on the unit magnetic pole. This pole is defined to be the pole which when placed one centimetre away from an equal magnetic pole repels it with a force of one dyne. The unit current is determined by finding the magnitude of a current which when flowing in a circle of one centimetre radius acts with a force of 2π dynes on a unit pole placed at its centre. The unit of quantity can then be determined by finding the quantity of electricity which passes in one second across any section of a circuit carrying unit current.

The electromagnetic and the electrostatic systems of units.

In the electrostatic system of units the fundamental unit is the unit of electrostatic charge or quantity defined on p. 2. By measuring the attraction between two spheres having equal and opposite charges of electricity, and using formulae which we shall give later, the magnitude of each charge in electrostatic units can be readily calculated. Connecting the two spheres through a ballistic galvanometer the electromagnetic measure of the charge is obtained. It is found that the ratio of this latter number to the former is $1/v$, where v is very nearly equal to 3×10^{10}. One electromagnetic unit of charge is therefore equal to v electrostatic units.

The value of v has been experimentally determined by many physicists. Rosa and Dorsey from the results of a series of elaborate and careful tests found that $v = 2\cdot997 \times 10^{10}$ very approximately.

It can be shown that if light is an electromagnetic disturbance propagated through the ether, its velocity should be numerically equal to the ratio v. The results of the electrical determinations of v are therefore in excellent agreement with the value $2\cdot9986 \times 10^{10}$ which is the mean of the best astronomical determinations of the velocity of light in centimetres per second.

In addition to the electromagnetic and electrostatic systems of units there is the practical system. In this system the unit of charge is the coulomb, the unit of pressure the volt, the unit of current the ampere, the unit of resistance the ohm, and the unit of capacity the microfarad.

The ampere and the coulomb are both one-tenth of the corresponding electromagnetic units. The volt is 10^8, the ohm is 10^9, and the microfarad is 10^{-15} the corresponding units in the electromagnetic system.

The quantity of electricity conveyed by a steady current of
Joule's Law. i amperes flowing for t seconds in a conductor
Ohm's Law.   between two points with a potential difference e, will be it. Hence the work done is

$$eit.$$

Now the practical electrical unit of work is the Joule or 10^7 ergs. The work done can be measured by the number H of units of heat (calories) generated in the conductor, and if the dynamical equivalent of heat be denoted by J (in joules per calorie), then

$$e = \frac{JH}{it} \qquad \dots\dots\dots\dots\dots\dots\dots(1).$$

The unit in which e is measured is called the volt. If i were in centimetre-gramme-second units and JH in ergs the number obtained for e would be 10^8 times larger. The volt equals 10^8 c.g.s. electromagnetic units. Equation (1) gives Joule's law in symbols.

A difference of potential between two points can be maintained by means of a battery; a current will flow between them, work being done in the process. Now Ohm proved experimentally that the ratio of the difference of potential to the current was constant so long as the conductor between the two points remained in the same physical state. This constant ratio is called the resistance of the conductor between the points, and Ohm's law may be written

$$i = \frac{e}{r} \qquad \dots\dots\dots\dots\dots\dots\dots\dots(2).$$

Hence if W be the number of joules done in time t,

$$W = JH = eit = \frac{e^2}{r}t = i^2 rt.$$

If P be the rate at which work is being done between the two points, in joules per second, *i.e.* if P be the power in watts,

$$P = ei = \frac{e^2}{r} = i^2 r.$$

Resistivity. The resistivity ρ of a conducting material at a given temperature is the resistance of a centimetre cube of the substance to a flow of electricity parallel to an edge of the cube. If R be the resistance of a prism of length l and cross section S to a flow of electricity parallel to its length we have

$$R = \rho \frac{l}{S}.$$

Current flow in a solid conductor. If two surfaces, A and B, immersed in a conducting medium be maintained at potentials V_1 and V_2 respectively and if V_1 be greater than V_2, there will be a flow of current from A to B. The stream lines of flow will be the same as the lines of force in the corresponding electrostatic problem.   Hence if i be the current in a tube of flow, we have, by Ohm's law,

$$i = \frac{\partial V}{\rho\,(\partial x/S)},$$

where ∂x is an element of the length of the tube of flow and S the area of its cross section.

Hence, since i is constant along a tube of flow,

$$V_1 - V_2 = \rho i \int_0^l (\partial x/S),$$

where l is the length of the tube.

If, therefore, $I = \Sigma i =$ the total current, we have

$$\frac{I}{V_1 - V_2} = \Sigma\,\frac{1}{\rho \int_0^l (\partial x/S)} = K \times \frac{4\pi}{\rho\lambda},$$

where K is the electrostatic capacity between the two bodies when the dielectric coefficient of the medium in which they are placed is λ.   Hence, if R be the electric resistance between the two bodies,

$$RK = \rho\lambda/(4\pi).$$

For example, the insulation resistance R per unit length of a concentric main is given by

$$R = \frac{\rho}{2\pi} \log_e \frac{b}{a},$$

a theorem often used in practical work.   It follows that the capacity per unit length of a concentric main is $\lambda/\{2 \log_e (b/a)\}$, a result which we shall prove later.

In the atomic theory of electricity a current is supposed to consist of a flow of infinitesimal particles.   Each of these particles is called an electron and each carries a definite charge of electricity.   In a solid body some of the electrons are supposed bound to its molecules and the others move about in the spaces between the molecules.   The latter are called "free" electrons and their velocities are continually being altered by collisions.   When no electric forces act on the body the free electrons move at random.   But when an electric force is applied, a velocity in the direction of the force is superposed on

The electron theory of conduction.

the natural velocity of each, the continual collisions that occur between the free electrons preventing their velocities from growing indefinitely.

Let us suppose that there are n electrons each of mass m, per unit length of a tube of flow, and let us suppose that the charge on each is q. If v be the average velocity of an electron in the direction of the electric force, nv of them will pass through a cross section of the tube of flow per second, and so the current $i = nqv$. The momentum per unit length in the direction of the force is nmv. The rate at which this is being checked by the collisions will be proportional to the number n of the particles and to their average velocity v. Hence the retarding force equals αnv where α is a constant. The force acting per unit length in the direction of the flow is the potential gradient $-\partial V/\partial x$ multiplied by the charge nq. Hence the equation of motion is

$$\frac{\partial}{\partial t}(nmv) + \alpha nv = -nq\frac{\partial V}{\partial x},$$

that is,

$$\frac{\partial}{\partial t}\left(\frac{mi}{q}\right) = -nq\frac{\partial V}{\partial x} - \frac{\alpha i}{q}.$$

Therefore, when

$$-\frac{\partial V}{\partial x} = \frac{\alpha i}{nq^2},$$

the current will have attained its steady state. Integrating from a section P to a section Q of the tube of flow we get

$$\frac{V_P - V_Q}{i} = \int_P^Q \frac{\alpha}{nq^2}\partial x$$

$$= \text{a constant.}$$

Hence

$$R = \int_P^Q \frac{\alpha}{nq^2}\partial x,$$

where R is the resistance of the tube.

Therefore, if n be large, as in conductors, the resistance is small, but if n be small, as in insulating materials, the resistance is great.

We shall now consider the value of the electromotive force set up in a circuit when the number of tubes of magnetic induction linked with it alters. To simplify the problem as much as possible let us take the case of a

Electromagnetic induction.

single pole of strength m on the north side of a circuit carrying a current i and let this circuit subtend an angle ω at the pole. The total flux from the pole is $4\pi m$ and hence the portion ϕ of this flux linked with the circuit is $4\pi m\,(\omega/4\pi) = m\omega$. The mutual potential energy of the circuit and pole is $mi\omega$. Let the pole m now move so that ω diminishes and as it moves let the current i be maintained constant by varying the electromotive force in the circuit by means of a suitable generator. In the time ∂t let ω become $\omega - \partial\omega$. The rate at which work is being done by the circuit on the pole is $-\partial\,(mi\omega)/\partial t$. Now suppose that E is the initial electromotive force and that it has to be increased to $E + e$ in the time ∂t so as to maintain i constant. At this instant the additional power being supplied by the generator is ei, and this, by the Conservation of Energy, must equal the power expended on the pole. Hence $ei = -\,mi\,(\partial\omega/\partial t)$, and therefore,

$$e = -\frac{\partial\,(m\omega)}{\partial t} = \frac{\partial\phi}{\partial t},$$

if ϕ be positive when it is in the same direction as the flux due to i. The electromotive force of induction therefore, which is equal and opposite to e, is equal to $-\,(\partial\phi/\partial t)$. We see that the electromotive force induced in the circuit by the motion of the pole equals the rate at which the number of tubes of induction linked with the circuit is diminishing. This is known as Faraday's Law. It is to be noticed that the induced electromotive force is independent of i. It has therefore the same value when i is zero, *i.e.* when there is no initial current.

It follows by Ohm's law that if e be the electromotive force of a battery in a circuit, R the resistance of the whole circuit, ϕ the flux linked with it and i the current,

$$Ri = e - \frac{\partial\phi}{\partial t}, \text{ and so } e = Ri + \frac{\partial\phi}{\partial t}.$$

Lenz's law, as modified by Maxwell, can be stated as follows.

Lenz's Law. The E.M.F. generated in a circuit always tends to produce a current which opposes any change in the value of the flux. If we bring a positive magnetic pole near a circuit, we diminish the number of lines of force coming through it towards the spectator, and hence the induced current must flow

in such a direction that the positive face of the equivalent magnetic shell may face the positive pole. Therefore the current
must flow in the opposite direction to the hands of a watch (Fig. 6).

If the flux be increased, the induced current must tend to
maintain the initial state of affairs, hence the direction of the
induced E.M.F. is with the hands of a watch. To get the positive
direction of rotation the flux must diminish, and hence the induced
E.M.F. must be written $-\partial\phi/\partial t$. Hence if e be the counter E.M.F.
in volts,

$$e = -\frac{\partial\phi}{\partial t} 10^{-8}.$$

If ϕ traverse n turns of wire, then

$$e = -n\frac{\partial\phi}{\partial t} 10^{-8}.$$

If we have two circuits A and B carrying currents i_1 and i_2
respectively, then their mutual potential energy may
be represented by $-i_1\phi_2$ or by $-i_2\phi_1$, where ϕ_2 is the
flux of induction through A due to B, and ϕ_1 is the flux through
B due to A. Now ϕ_2 is in direct proportion to i_2; we may therefore write

$$\phi_2 = Mi_2,$$

where M is a constant depending only on the positions of the
circuits. Similarly we can write $\phi_1 = M'i_1$, and, noting that the
two expressions for the potential energy must be equal, we see
that $M = M'$. Hence the flux of induction through A due to unit
current in B equals the flux of induction through B due to unit
current in A. If the current in B alters, the flux ϕ_2 through A
alters also, and the induced E.M.F. set up is $-\frac{\partial\phi_2}{\partial t}$ or $-M\frac{\partial i_2}{\partial t}$. It is
to be noted that M may be positive or negative, depending on
the directions taken as positive for the two currents. The
constant M is called the coefficient of mutual induction, or simply,
the mutual inductance between the two circuits.

In practice the wires carrying the current have finite thickness
and the preceding definition only applies strictly when their cross
section is negligibly small. To make the definition rigorous therefore we must take into account the thickness of the wires. Let us

first find the mutual potential energy between an elementary tube of flow ∂i_2 of the current in B, and the current i, in A. Let m_1 be the mutual inductance between ∂i_2 and an elementary tube ∂i_1 of the current in A. The mutual potential energy between ∂i_2 and i_1 can then be written $\partial i_2 \Sigma m_1 \partial i_1$, the summation being taken over the cross section of the wire forming the circuit A. But ∂i_1 is directly proportional to i_1 and thus the mutual energy may be written $M_1 i_1 \partial i_2$ where M_1 is a constant depending on the geometrical dimensions of the circuit A. Hence the total mutual energy of the two circuits is $i_1 \Sigma M_1 \partial i_2$, the summation being taken over the cross section of the circuit B, and since ∂i_2 is proportional to i_2, the mutual energy can be written $M i_1 i_2$ where M is a constant. We may therefore define the mutual inductance between two circuits as the sum of the linkages of the flux due to the current in one circuit with the current in the other when each of the currents is unity. This definition still applies when the circuits consist of many turns of wire.

As the flux linked with the various elementary tubes ∂i_2 into which we may suppose the current i_2 divided is in general different, we see that as i_1 changes, the electromotive forces set up in the elementary circuits which form B are in general different, and this makes the rigorous solution of the problem of finding the current flow difficult. Solutions, sufficiently accurate for practical work, can in many cases be readily obtained by neglecting the thickness of the wires and this we shall do when discussing subsequent problems. In certain problems, however, this assumption is not permissible.

In what precedes we have also assumed that the permeability of the medium in which the coils are immersed and of the metal of which the wires are made is constant.

Let us first consider a circuit made of very thin wire. We Self inductance. have already seen that if an electromotive force e be applied to this circuit the current in it is determined by the equation

$$e = Ri + \partial \phi / \partial t,$$

where R is the resistance of the circuit and ϕ is the flux of induction linked with it. In obtaining this equation we have supposed that none of the flux cuts the wire. We may write it in the form

$$ei = Ri^2 + i\partial\phi/\partial t.$$

Now ei is the rate at which electric energy is being supplied to the circuit, Ri^2 is the rate at which energy is expended in heating the circuit and $i\partial\phi/\partial t$ is the rate at which it is being stored up in the field. Let us suppose that the current attains the value i_1 in the time T and let ϕ_{11} be the value of the flux linked with this current. The value of the energy stored up in the field in the time T is $\int_0^T i\,(\partial\phi/\partial t)\,\partial t = \int_0^{\phi_{11}} i\partial\phi$. Since ϕ is proportional to i we may write $\phi = ki$ where k is a constant. Hence the energy stored in the field is $(1/2)\,ki_1^2 = (1/2)\,i_1\phi_{11}$. The thinner the wire the more accurate will be the value of the energy given by this expression.

Let us now consider a circuit the cross section of which is appreciable. Let the total current in the circuit be i and let us suppose that it is divided up into a large number of circuits carrying currents i_1, i_2, etc., so that $i = i_1 + i_2 + \dots$. If ϕ_{nm} be the flux of induction due to the current i_n which is linked with the current i_m we see by the preceding paragraph that the self energy of the current i equals

$$(1/2)\,i_1\,(\phi_{11} + \phi_{12} + \dots) + (1/2)\,i_2\,(\phi_{21} + \phi_{22} + \dots) + \dots.$$

Now ϕ_{nm} is proportional to i_n when the permeability of the medium and the substance of the conductor is constant. The expression for the self energy may therefore be written

$$(1/2)\,k_1 i_1^2 + (1/2)\,k_2 i_2^2 + \dots$$

where $k_1, k_2, \dots$ are constants. But $i_1, i_2, \dots$ are each proportional to i and hence the self energy may be written $(1/2)\,Li^2$ where L is a constant which is variously called the coefficient of self induction, the self inductance, or simply the inductance of the circuit. The value of L depends on how the current density varies over the cross section of the circuit. We shall see later that with alternating currents of high frequency the current density varies very appreciably over the cross section of the circuit. Strictly speaking therefore we should call L the self inductance of the current and not the self inductance of the circuit.

We may define the self inductance of unit current flowing in a circuit as the sum of the linkages of the flux of induction due to this current with the current itself. This definition applies whatever may be the shape of the circuit. In symbols we may write $L = (1/i^2)\, \Sigma \phi \partial i$, where ϕ is the flux linked with the tube of flow carrying the current ∂i.

When the current varies the flux linked with the various tubes of flow varies also. The back electromotive force set up in these tubes by this variation will in general be different and hence the solution of the problem of determining the value of the current density over the cross section is difficult. In many practical cases, however, the current density is very approximately uniform over the cross section, and so, we can write

$$ei = Ri^2 + \frac{\partial}{\partial t}(\tfrac{1}{2} Li^2),$$

where e is the applied E.M.F. and R the resistance of the circuit. Hence $\qquad\qquad e - L\,(\partial i/\partial t) = Ri.$

Thus the effect of self induction is to induce a back E.M.F. $L\,(\partial i/\partial t)$ in the circuit.

Suppose that we have two circuits A and B respectively, and

Electromagnetic energy. that L_1, L_2 are their self inductances and M their mutual inductance. Then if i_1 and i_2 be the currents through the circuits, the energy stored up in the field is

$$\tfrac{1}{2}\,L_1 i_1^2 + M i_1 i_2 + \tfrac{1}{2}\,L_2 i_2^2.$$

This expression may be written in the form

$$\tfrac{1}{2}\,L_1 \left(i_1 + \frac{M}{L_1}\, i_2 \right)^2 + \tfrac{1}{2}\,i_2^2 \frac{L_1 L_2 - M^2}{L_1}\ ;$$

and since it must be positive for all values of i_1 and i_2, and therefore when $i_1 = -\dfrac{M}{L_1}\, i_2$, we see that M^2 cannot be greater than $L_1 L_2$.

When the cross section of the circuit A is very small $Li_1 + Mi_2$ is very approximately the value of the flux linked with i_1, and so we have

$$e_1 = R_1 i_1 + \frac{\partial}{\partial t}(L_1 i_1 + M i_2) = R_1 i_1 + L_1 \frac{\partial i_1}{\partial t} + M \frac{\partial i_2}{\partial t},$$

where R_1 is the resistance of the circuit A and e_1 is the E.M.F. applied to it.

The magnetic potential at a point due to a closed tube of

Self energy in terms of magnetic force. Kelvin's Formula. current i is $i\omega$, where ω is the solid angle subtended at the point by the tube. If we take a unit positive pole once round a line of force embracing the tube, the change of potential energy is $4\pi i$, since 4π is the difference between the initial and final values of the solid angle. Hence if H be the magnetic force at any point

$$4\pi i = \int H \partial s,$$

where ∂s is an element of the line of force round which the integral has been taken.

If $\partial\phi$ be the flux of induction over an element of an equipotential surface through the point

$$\partial\phi = B\partial S.$$

Hence $$i\partial\phi = \frac{1}{4\pi}\int H\partial s \times B\partial S.$$

But $H\partial S$ is constant along a tube of force, and $\partial s \times \partial S = \partial v =$ an element of volume of the tube of force.

Thus $$i\partial\phi = \frac{1}{4\pi}\int HB\partial v = \frac{1}{4\pi}\int \mu H^2\partial v,$$

the integration being taken along a tube of force.

Hence, integrating over all the equipotential surface, we get

$$i\phi = \frac{1}{4\pi}\Sigma\mu H^2\partial v,$$

where the summation is taken throughout all space.

Hence the expression $\frac{1}{2}i\phi$ for the self energy of the system may be written

$$\Sigma\frac{\mu H^2}{8\pi}\partial v.$$

To simplify the problem let us consider the case of two

The electromagnetic attraction between coils carrying currents. coaxial cylindrical coils which are screened from all external magnetic forces. If they are not concentric and the currents i_1 and i_2 flowing in them are in the same sense, we see by replacing them by their equivalent magnetic shells that the force between them is attractive. If they are free to move therefore their centres will approach one another. More tubes of induction will be linked with both circuits and so the energy stored in the field will be increased. As the self inductances of the coils remain constant the increase of the electromagnetic energy when the distance between their centres diminishes by ∂x is $i_1 i_2 (\partial M/\partial x)\, \partial x$, and if F be the attractive force the mechanical work done by the electromagnetic forces will be $F\partial x$. During the motion a back electromotive force is induced in each circuit. If the currents be maintained constant, therefore, the electromotive force of the generators has to be increased so as to neutralise these induced electromotive forces. It follows from the principle of the conservation of energy that the sum of the increased outputs of the generators must equal the sum of the increase of the field energy and the mechanical work done by the attractive forces. We shall now prove that the attractive force F is exactly equal to

$$(\partial M/\partial x)\, i_1 i_2.$$

Let ϕ_1, ϕ_2 be the values of the fluxes of induction linked with the two coils, so that

$$\phi_1 = L_1 i_1 + M i_2, \text{ and } \phi_2 = L_2 i_2 + M i_1.$$

Instead of supposing that the currents are maintained constant during the motion of the coils, let us suppose that the flux linkages ϕ_1 and ϕ_2 are maintained constant by varying the currents. No back electromotive forces will be induced in either coil in this case and so the generators will give no work to the field. By the conservation of energy the work $F\partial x$ done by the attractive forces must equal the diminution of the stored electromagnetic energy, and F has the same value as in the last paragraph. We have

$$F\partial x = \tfrac{1}{2}\, \phi_1 i_1 + \tfrac{1}{2}\, \phi_2 i_2 - \tfrac{1}{2}\, \phi_1 (i_1 + \partial i_1) - \tfrac{1}{2}\, \phi_2 (i_2 + \partial i_2)$$

$$= -\tfrac{1}{2}\, \phi_1 \partial i_1 - \tfrac{1}{2}\, \phi_2 \partial i_2.$$

But since ϕ_1 and ϕ_2 are constants,

$$L_1 \partial i_1 + M \partial i_2 = -i_2 \partial M, \quad L_2 \partial i_2 + M \partial i_1 = -i_1 \partial M$$

and thus

$$F\partial x = -\tfrac{1}{2}\, (L_1 \partial i_1 + M \partial i_2)\, i_1 - \tfrac{1}{2}\, (L_2 \partial i_2 + M \partial i_1)\, i_2$$

$$= i_1 i_2 \partial M,$$

or

$$F = i_1 i_2\, (\partial M/\partial x).$$

Thus $F\partial x$ the mechanical work done by the system when the currents remain constant is equal to the concurrent increase of the electromagnetic energy.

The above is a simple case of a very general theorem due to Kelvin. He proved that if we maintain the currents in all the circuits in a given field constant and they move under their mutual attractions and repulsions, the total work done by the generators in the circuits equals twice the mechanical work done on the circuits and also equals twice the increase in the energy stored up in the field.

Suppose that we have a wire sliding with uniform velocity v on

The E.M.F. generated in a conductor cutting tubes of induction. two parallel wires which are joined by another wire at one end. Let us suppose also that the magnetic induction is constant in the neighbourhood of the moving wire and that the tubes of induction cut by the moving wire make an angle θ with the plane of the wires. If B be the magnetic induction and the slider be perpendicular to the parallel wires, then the number of tubes of induction cut by it in t seconds is $B\sin\theta . lvt$, where l is the length of the slider the ends of which are on the parallel wires. If the slider be supposed to move so as to increase the area of the circuit the increase in the flux through the closed circuit is $B\sin\theta . lvt$. Therefore e, the E.M.F. generated, is given by

$$e = -\frac{\partial}{\partial t}(B\sin\theta . lvt) = -B\sin\theta . lv,$$

the positive direction round the circuit being that direction in which a current must flow in order to cause tubes of induction to thread the circuit in the same direction as those of the field. In this formula, e and B are in C.G.S. units, l is in centimetres and v is in centimetres per second. If e is in volts, we have

$$e = -B\sin\theta . lv \times 10^{-8}.$$

In Fig. 10 is shown the relation between the motion of a wire, the magnetic induction, and the direction of the induced E.M.F. and current, when the three vectors representing these quantities are mutually at right angles. If we place the fingers of the right hand so that the thumb is in the direction of the motion, the first

finger in the direction of the magnetic induction, and the second finger in the direction of the wire,
the second finger will point out the direction of the induced E.M.F. When the motion of the wire is parallel to the magnetic induction there is no induced E.M.F.

Force on a moving wire. If f be the force in dynes in the direction of v exerted on the slider by an external agent the power expended on it will be fv ergs per second. Let e be the value of the E.M.F. induced in the slider and let i be the value of the induced current.

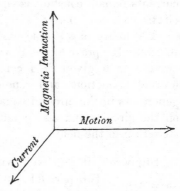

Fig. 10.   Relation between the motion of a wire and the induced current.

Then if the resistance of the parallel wires is negligible

$$ei = fv.$$

As e and i are always of the same sign f and v are always of the same sign. The electromagnetic force on the wire, therefore, always resists the motion. Hence Lenz's law is applicable to the motion of parts of a circuit. Thus if f' be the electromagnetic force on the wire

$$f' = B \sin \theta \,.\, li.$$

If B be in C.G.S. units, l in centimetres and i in amperes

$$f' = B \sin \theta \,.\, li/10 \text{ dynes.}$$

This force acts in such a direction as to make the wire move so as to increase the number of lines of induction through the circuit in the direction of those due to the current.

If γ be the angle between the plane of i and B and the plane of the circuit and ϕ be the angle which B makes with the wire, we have $\sin \theta = \sin \phi \sin \gamma$. Therefore for fixed directions of the wire and of B, the electromagnetic force on the wire is greatest when $\gamma = \pi/2$, *i.e.* when the plane of B and i is at right angles to the plane of the circuit. Assuming that the force on the wire depends only on B and l and not on the position of the rest of the circuit, it follows that the electromagnetic force on the element is at right angles to B and l.

In practice in order to find out the direction of the electro-magnetic force on a conductor Fleming's left-hand rule will be found convenient.

If the fore-finger of the left hand point in the direction of the field, the second finger in the direction of the current in the wire, then the thumb will point in the direction of the force on the wire, if the two fingers be held at right angles to one another and also to the thumb.

It will be sufficient for our next purpose to suppose that the

The magnetic force due to an element of current. Laplace's Formula. current is in a plane. We will calculate the magnetic force at a point O (Fig. 11) due to a current i in a closed circuit. Suppose that we have a positive pole of strength m at O and that OA is a fixed line. The

strength of the magnetic field at P due to O will be $\dfrac{m}{r^2}$. Hence if

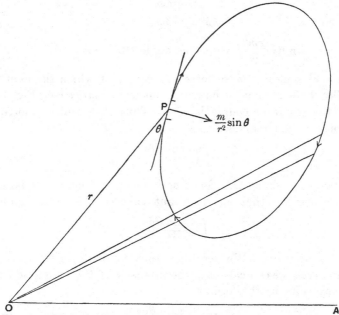

Fig. 11. Magnetic force at O due to the current i in a plane closed curve is upwards from the plane of the paper and

$$= \int \frac{i \sin \theta \partial s}{r^2} = \int \frac{i \partial \phi}{r},$$

where $\phi =$ the angle POA.

the arrow indicate the direction of the current we see by the preceding paragraph that the force ∂F on ∂s will be given by

$$\partial F = \frac{m}{r^2} \sin \theta i \partial s \text{ dynes,}$$

and the force, by the left-hand rule, tends to move ∂s downwards at right angles to the paper. Using a well-known artifice in Statics, we may replace this force by a force ∂F at O acting downwards, and a couple with a moment $r\partial F$ round an axis lying in the plane of the paper and perpendicular to OP. It is easy to see that

$$r\partial F = \frac{m}{r} \sin \theta i \partial s$$

$$= \frac{m}{r} \frac{r\partial \phi}{\partial s} i \partial s$$

$$= mi\partial \phi,$$

and hence $\Sigma r \partial F = \int mi \partial \phi,$

for $\sin \theta = \dfrac{r \partial \phi}{\partial s}$, where the angle $POA = \phi$.

But round a closed curve for every couple of which the moment is $mi\partial\phi$ there is another having a moment $-mi\partial\phi$ (see Fig. 11); hence the resultant couple is zero. Thus the resultant force of the pole on the circuit is a force,

$$\int \frac{m \sin \theta i \partial s}{r^2},$$

acting downwards at O. Since action and reaction are equal and opposite, we see that the resultant force of the circuit on the pole is $\displaystyle\int \frac{mi \sin \theta \partial s}{r^2},$

and acts upwards. We see then that, so long as we consider closed curves, we can calculate the intensity of the magnetic force H at any point by the formula

$$H = \int \frac{i \sin \theta \partial s}{r^2}$$

$$= \int \frac{i \partial \phi}{r}.$$

If i be in amperes, then the magnetic force is given by

$$H = \frac{1}{10} \int \frac{i \partial \phi}{r} .$$

In proving the formula above we have considered the circuit as a whole. By assuming, however, that the current consists of a flow of electrons we shall prove later on that the force at P due to the current in an element ∂s of the wire is $(i \partial s \sin \theta)/r^2$. Hence we can use this result to compute the force at a point due to part of a circuit.

For example, the force H at a point P due to a current i in the wire BC is given by

$$H = \int_{OB}^{OC} \frac{i \sin \theta \partial s}{r^2} .$$

If PO be drawn perpendicular to BC and if the angle PRO be θ, we have $s = OR = a \cot \theta$, if $OP = a$. Hence $\partial s = -a \operatorname{cosec}^2 \theta \partial \theta$, and

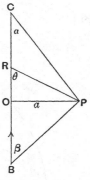

$$H = -\int_{\pi - \beta}^{a} (i/a) \sin \theta \partial \theta$$

$$= (i/a)(\cos \alpha + \cos \beta),$$

where α and β are the angles PCO and PBO respectively. For an infinitely long wire α and β are both zero, and so,

$$H = 2i/a.$$

Fig. 12. The magnetic force at a point due to the current in BC.

As a further example, let us find the magnetic force at the focus of an elliptic wire carrying a current i. The polar equation to the ellipse referred to the focus as the origin and the major axis as the initial line is

$$l/r = 1 - e \cos \theta,$$

where l is the semi-latus rectum and e is the eccentricity. Hence if H be the magnetic force at the focus

$$H = \int \frac{i \partial \theta}{r} = \frac{i}{l} \int_0^{2\pi} (i - e \cos \theta) \partial \theta = \frac{2\pi i}{l},$$

and $l = a(1 - e^2)$ where a is the semi-major axis.

The force at the centre of a circle of radius R can easily be

Formulae for the magnetic forces inside circles and rectangles.

found, for $r = R$ and is constant, and $\Sigma \partial \phi = 2\pi$, and hence

$$H = \frac{2\pi i}{R},$$

which agrees with our former result on page 25.

If the point O be not at the centre of the circle (Fig. 13) let $OC = a$, $CP = R$ and the angle $POC = \theta$, then

$$r = OP = a \cos \theta + \sqrt{R^2 - a^2 \sin^2 \theta}.$$

Therefore

$$\frac{1}{r} = \frac{-a \cos \theta + \sqrt{R^2 - a^2 \sin^2 \theta}}{R^2 - a^2}.$$

But

$$H = i \int_0^{2\pi} \frac{\partial \theta}{r}.$$

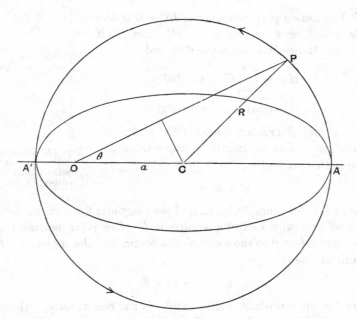

Fig. 13. The magnetic force at O due to a current i in the circle is perpendicular to the plane of the paper and equals

$$\frac{i \times \text{circumference of ellipse}}{R^2 - a^2}.$$

The ellipse has O for a focus. The force acts towards the reader.

Thus
$$H = \frac{i}{R^2 - a^2} \int_0^{2\pi} \sqrt{R^2 - a^2 \sin^2 \theta} \, \partial \theta$$

$$= \frac{i \times \text{circumference of ellipse}}{R^2 - a^2} \quad \dots\dots\dots\dots(3).$$

The ellipse referred to in the formula has C for its centre, O for a focus, and has its major axis equal to a diameter of the circle. When a is zero this reduces to $2\pi i/R$.

If we expand $R\sqrt{1 - a^2 \sin^2 \theta/R^2}$ by the Binomial theorem and integrate each term separately we find that

$$H = \frac{2\pi i R}{R^2 - a^2} \left[1 - \left(\frac{1}{2}\right)^2 \frac{a^2}{R^2} - \frac{1}{3}\left(\frac{1 \cdot 3}{2 \cdot 4}\right)^2 \frac{a^4}{R^4} - \cdots \right.$$
$$\left. - \frac{1}{2n - 1}\left(\frac{1 \cdot 3 \dots 2n - 1}{2 \cdot 4 \dots 2n}\right)^2 \frac{a^{2n}}{R^{2n}} - \cdots \right]$$

$$= \frac{2\pi i}{R} \left[1 + 3\left(\frac{1}{2}\right)^2 \frac{a^2}{R^2} + 5\left(\frac{1 \cdot 3}{2 \cdot 4}\right)^2 \frac{a^4}{R^4} + \cdots \right.$$
$$\left. + (2n + 1)\left(\frac{1 \cdot 3 \dots 2n - 1}{2 \cdot 4 \dots 2n}\right)^2 \frac{a^{2n}}{R^{2n}} + \cdots \right].$$

It will be seen that the first series is the more 'rapidly convergent.

As an example, let us take the case when $a = R/2$. Using the first formula, we get

$$H = \frac{8\pi i}{3R} \left[1 - 0 \cdot 0625 - 0 \cdot 0029 - 0 \cdot 0003 - \cdots \right]$$
$$= \frac{2\pi i}{R} \times 1 \cdot 2457 \text{ approximately.}$$

Let us now suppose that the point O (Fig. 14) is outside the circular current.

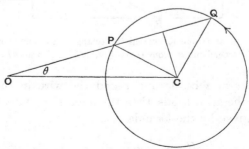

Fig. 14. Magnetic force at a point O due to the circular current.

Then
$$H = 2i \int_0^{\sin^{-1}(R/a)} \left(\frac{1}{OP} - \frac{1}{OQ} \right) \partial \theta.$$

But
$$OQ - OP = PQ = 2\sqrt{R^2 - a^2 \sin^2 \theta}.$$

Hence
$$H = \frac{4i}{a^2 - R^2} \int_0^{\sin^{-1}(R/a)} \sqrt{R^2 - a^2 \sin^2 \theta}\ \partial \theta.$$

Writing $R \sin \phi = a \sin \theta$, and then expanding and integrating, we get

$$H = \frac{4R^2 i}{a(a^2 - R^2)} \int_0^{\pi/2} \frac{\cos^2 \phi}{\sqrt{1 - (R/a)^2 \sin^2 \phi}}\ \partial \phi,$$

$$= \frac{\pi R^2 i}{a(a^2 - R^2)} \left[1 + \frac{1}{2} \left(\frac{1}{2} \right)^2 \frac{R^2}{a^2} + \frac{1}{3} \left(\frac{1.3}{2.4} \right)^2 \frac{R^4}{a^4} + \dots \right.$$
$$\left. + \frac{1}{n+1} \left(\frac{1.3 \dots 2n-1}{2.4 \dots 2n} \right)^2 \frac{R^{2n}}{a^{2n}} + \dots \right].$$

For example, we easily find that $H = 2\pi i/R$, nearly, when $a = 1\cdot21R$. When $a = 1\cdot345R$, $H = \pi i/R$, and when $a = 2R$, $H = 0\cdot542i/R$. For large values of a, the force at a point varies inversely as the cube of its distance from the centre of the circle.

The case of a rectangular wire (Fig. 15) also admits of an easy solution.

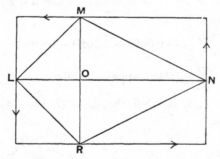

Fig. 15.   Force at O due to current in rectangle $= i \times$ sum of reciprocals of perpendiculars from O on LM, MN, NR and RL.

If the current i be flowing round the wire in the direction indicated by the arrowheads, then the force H on unit pole placed at O will be given by the formula

$$H = i \left\{ \frac{1}{p_1} + \frac{1}{p_2} + \frac{1}{p_3} + \frac{1}{p_4} \right\} \quad \dots\dots\dots\dots\dots(4),$$

where p_1, p_2, p_3 and p_4 are the perpendiculars from O on the lines LM, MN, NR and RL respectively, and ROM and NOL are drawn parallel to the sides of the rectangle. It is an instructive exercise to find graphically by means of this construction the density of the lines of force at various points inside a rectangle.

Let three of the sides of the rectangle move to infinity, then **Force near a** the formula becomes
long straight
wire carrying
a current.
$$H = \frac{2i}{r}$$

which agrees with the result obtained by direct calculation. The force H is perpendicular to r, and the directions of force and current are related in the same way as the directions of current and force in Figs. 6 and 7. If the current be upwards through the plane of the paper, then the lines of force are circles, and act in the direction $\not{N}$. If the current be downwards through the paper, then the lines of force act in the direction S.

By Ampère's theorem, we could have replaced the wire by a plane magnetic shell bounded by it and extending to infinity. Suppose that r makes an angle θ with the plane of this shell. Draw two planes through the point, one passing through the wire and the other parallel to the shell. The area of the lune intercepted on a sphere of radius R by these planes is $2R^2\theta$. Therefore the solid angle ω subtended by the infinite plane at the point is
$$\frac{2R^2\theta}{R^2} = 2\theta.$$

Hence the potential V at the point is given by
$$V = 2i\theta.$$

Thus $-\dfrac{\partial V}{\partial r}$, the force in the direction of r, must be zero, and $-\dfrac{\partial V}{r\partial\theta}$, the force perpendicular to r in the direction in which θ increases, must be $-\dfrac{2i}{r}$. We see again that the work done in taking unit pole round the wire is
$$2i\,(2\pi + \theta) - 2i\theta = 4\pi i.$$

We shall now find the formula for the magnetic force at a point
Helical O on the axis of a helical current (Fig. 16).
current.   Let OZ be the axis of the helix and NA a section
of the cylinder on which we may suppose that it is wound. Consider
the force at O due to an element of the current at P. Draw PN

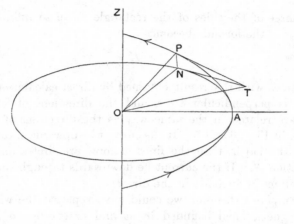

Fig. 16. Magnetic force at O due to the current in the helix AP.

perpendicular to the plane NA. The tangent at P to the helix
will meet the tangent at N to the circle NA at a point T. If the
angle NOA be ϕ, we have $NT = a\phi$, where a is the radius of the
cylinder. By Laplace's law, the force at O due to the element at
P is $i\partial s \sin OPT / OP^2$, and hence, if ∂H be the component of this
force parallel to OZ, we have

$$\partial H = \frac{i\partial s \sin OPT}{OP^2} \cos \gamma,$$

where γ is the angle between the perpendicular to the plane OPT
and OZ. Now the ratio of the area of the triangle ONT to the
triangle OPT is $\cos \gamma$, and thus

$$\frac{ON \cdot NT}{OP \cdot PT \sin OPT} = \cos \gamma,$$

and $\sin OPT \cos \gamma = \cos \psi \cos \delta,$

where ψ and δ are the angles PON and PTN respectively. We
also have $\partial s = a\partial\phi / \cos \delta.$

Hence
$$H = \int \frac{ia\,\partial\phi \cos \psi}{a^2 + z^2}$$

where $z = PN = a\phi \tan \delta$.

Therefore
$$H = \frac{ia}{\tan \delta} \int_{-h_2}^{h_1} \frac{\partial z}{(a^2 + z^2)^{\frac{3}{2}}}$$

$$= \frac{i}{a \tan \delta} \left[\frac{z}{(a^2 + z^2)^{\frac{1}{2}}} \right]_{-h_2}^{h_1}$$

where h_1 and h_2 are the distances of the ends of the axis of the helix from O.

Thus
$$H = \frac{i}{a \tan \delta} \left[\frac{h_1}{(a^2 + h_1^2)^{\frac{1}{2}}} + \frac{h_2}{(a^2 + h_2^2)^{\frac{1}{2}}} \right].$$

If there are n_1 complete turns of the helix above O and n_2 complete turns below O, then since $h = a\phi \tan \delta = 2n\pi a \tan \delta$, we get

$$H = \frac{2\pi i}{a} \left[\frac{n_1}{(1 + 4\pi^2 n_1^2 \tan^2 \delta)^{\frac{1}{2}}} + \frac{n_2}{(1 + 4\pi^2 n_2^2 \tan^2 \delta)^{\frac{1}{2}}} \right].$$

When n_1 and n_2 are large numbers,

$$H = 4\pi \frac{n_1 + n_2}{l} i = 4\pi n i,$$

where l is the length of the axis of the helical coil and n is the number of turns per unit length. The force at either end of the axis in this case is very nearly $2\pi n i$.

Let us first suppose that the thin wire carrying the current
intercepts the circle. If the current is flowing as in
the diagram (Fig. 17) the flux in the sector $LAKO$
will be coming towards the reader and the flux in the
sector $LOKB$ will be away from him. If the arc
$LA'K$ equals LAK in all respects the algebraical sum of all the
tubes of induction taken over the area of the whole circle will
equal the flux in $LBKA'$. Let $OP = r$, $PQ = \partial r$, and the angle
$POB = \theta$. Then the area of the element $PQP'Q'$ will be $r\partial\theta\partial r$,
and hence the flux through it equals $(2i/r \cos \theta)\, r\partial\theta\partial r$.

The flux due to a straight current linked with a circle in its plane.

Hence the flux $\partial\Phi$ in the element $TT'RR'$ is given by

$$\partial\Phi = 2i \int_{r_1}^{r_2} \frac{\partial\theta}{\cos \theta} \partial r = \frac{2i}{\cos \theta} (r_2 - r_1)\, \partial\theta,$$

where $OR = r_2$, $OT = r_1$. But if $OC = b$, where C is the centre of the circle and a is its radius, we have,

$$a^2 = r_2{}^2 + b^2 - 2r_2 b \cos \theta$$
$$a^2 = r_1{}^2 + b^2 + 2r_1 b \cos \theta,$$

and therefore, $\quad r_2{}^2 - r_1{}^2 - 2(r_2 + r_1) b \cos \theta = 0,$

and so, $\qquad\qquad r_2 - r_1 = 2b \cos \theta.$

Substituting in the formula for $\partial \Phi$ we get

$$\partial \Phi = 4bi\partial\theta.$$

Finally integrating from $\theta = -\pi/2$ to $+\pi/2$, we see that

$$\Phi = 4\pi bi.$$

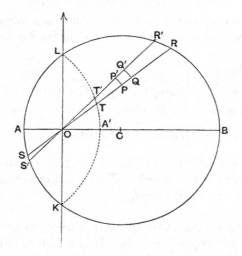

Fig. 17.   Linkages of magnetic flux due to the linear current
in KL with the circle ALB.

Now if M be the mutual inductance between the circle and the wire $\Phi = Mi$. The mutual inductance is therefore $4\pi b$. If there is a second wire parallel to the first, at a distance c from the centre of the circle, where c is less than a, the mutual inductance between the circuit formed by the two parallel wires and the circle is given by

$$M = 4\pi (b - c).$$

We thus deduce the curious result that the mutual inductance between a circuit formed by a pair of parallel wires and any circle which they both intersect is independent both of the magnitude of the radius and the position of the centre of the circle.

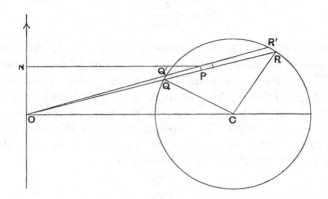

Fig. 18. Linkages of magnetic flux due to the linear current in ON with the circle QR.

In the case when the wire does not cut the circle (Fig. 18), we have

$$\Phi = i \int_{-a}^{+a} \int_{r_1}^{r_2} \frac{2}{r \cos \theta}\, r \partial \theta \partial r$$

$$= 2i \int_{-a}^{+a} \frac{r_2 - r_1}{\cos \theta}\, \partial \theta,$$

where $r_2 - r_1 = QR = 2\sqrt{a^2 - b^2 \sin^2 \theta}.$

Hence $\Phi = 4i \int_{-a}^{+a} \frac{\sqrt{a^2 - b^2 \sin^2 \theta}}{\cos \theta}\, \partial \theta.$

Writing $a \sin \phi = b \sin \theta,$

$$\Phi = 8a^2 b i \int_0^{\pi/2} \frac{\cos^2 \phi \, \partial \phi}{b^2 - a^2 \sin^2 \phi}.$$

Expanding by the Binomial theorem and integrating

$$\Phi = 4\pi a i \left[\frac{1}{2} \cdot \frac{a}{b} + \frac{1 \cdot 1}{2 \cdot 4} \frac{a^3}{b^3} + \frac{1 \cdot 1 \cdot 3}{2 \cdot 4 \cdot 6} \cdot \frac{a^5}{b^5} + \dots \right]$$

$$= 4\pi i \,(b - \sqrt{b^2 - a^2}).$$

We shall generalise these theorems in the next chapter.

When a is small compared with b, the labour of computation is lessened by writing the formula in the form

$$\Phi = \frac{4\pi a^2 i}{b + \sqrt{b^2 - a^2}}.$$

The magnetic force inside an infinite cylindrical tube carrying a current parallel to its axis.

Let C (Fig. 19) be the centre of the section of the tube the radius of which is a. Let us first find the magnetic force at a point O inside the tube. Draw a chord POP' making an angle ϕ with OC and let the angle POQ be $\partial\phi$. Let i be the total current flowing in the tube. The current flowing in the section PQ is $i \cdot PQ/(2\pi a)$. The force H at O due to the element PQ is perpendicular to OP and its magnitude is

$$\frac{2\left(i \dfrac{PQ}{2\pi a}\right)}{OP}.$$

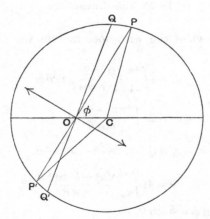

Fig. 19.   The magnetic force inside a hollow cylindrical conductor is zero.

Similarly the force due to the current through $P'Q'$ is

$$i \cdot P'Q'/(\pi a \cdot OP'),$$

and acts in the opposite direction to H. But from similar triangles $PQ/OP = P'Q'/OP'$ and hence the forces are equal and neutralise one another. As the whole circumference can be divided into pairs of elements in this manner, the resultant force at O and therefore at any point inside the cylinder must be zero.

Let O (Fig. 20) be the point at which the force is required.

The magnetic
force outside
an infinite
cylindrical
tube carrying
a current
parallel to its
axis.
Draw OPP' cutting the circle and let the angle POQ be $\partial\phi$. Let the angles PCO and $P'CM$ be θ_1 and θ_2 respectively and let $OC = d$.

If ∂H be the component force perpendicular to OC, due to currents through PQ and $P'Q'$ respectively, we have

$$\partial H = \frac{2\left(i\dfrac{\partial\theta_1}{2\pi}\right)}{OP}\cos\phi + \frac{2\left(i\dfrac{\partial\theta_2}{2\pi}\right)}{OP'}\cos\phi.$$

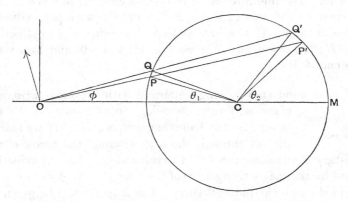

Fig. 20.   The magnetic force outside a hollow cylindrical conductor is the same as if the current were concentrated along its axis.

Noticing that by similar triangles, $a\partial\theta_1/OP = a\partial\theta_2/OP'$, and therefore also $= a\partial\,(\theta_1+\theta_2)/(OP+OP')$, we get

$$\partial H = \frac{i}{\pi}\cos\phi\,\frac{2\partial\,(\theta_1+\theta_2)}{OP+OP'}.$$

But $\qquad OP + OP' = 2d\cos\phi$, and so

$$\partial H = \frac{i}{\pi d}\,\partial\,(\theta_1+\theta_2),$$

and hence $\qquad\displaystyle H = \frac{2i}{\pi d}\int_0^\pi \partial\,(\theta_1+\theta_2)$

$$= \frac{2i}{d}.$$

Hence the cylindrical conductor acts as if its current were concentrated along its axis. As we can suppose that a solid cylindrical conductor is made up of an infinite number of coaxial tubes, we see that a solid cylinder also acts on external points as if its current were concentrated at the axis.

The theorems above can also be proved by remembering that the line integral of the magnetic force round a current which gives the work done in taking unit magnetic pole once round the current equals 4π multiplied by the current. Let us consider for instance the magnetic force H at a point at a distance d from the axis of the wire. The line integral of H round a circle of radius d having its centre on the axis of the cylinder and its plane perpendicular to it is $H \cdot 2\pi d$. If d be greater than the radius of the cylindrical shell $H \cdot 2\pi d = 4\pi i$, but if it be less $H \cdot 2\pi d = 0$, and hence the theorems follow.

Magnetic tests of iron. Hysteresis. Let us wind an iron ring uniformly with insulated wire, and place a further winding on it connected to the terminals of a ballistic galvanometer. If we start a current through the first winding, the throw of the ballistic galvanometer enables us to calculate the flux of induction caused by the magnetising force of the current. If ϕ be the total flux in the core the induced current i in amperes will be given by

$$i = \frac{-n\dfrac{\partial \phi}{\partial t}}{R} \times 10^{-8},$$

if there are n turns of wire in series with the galvanometer, and R is the total resistance of this circuit in ohms.

Thus we have $$\int i \partial t = -\frac{n\,10^{-8}}{R} \int \partial \phi,$$

and $$Q = \frac{n\,10^{-8}}{R}(\phi_1 - \phi_2),$$

where Q is the number of coulombs that traverse the circuit when the flux increases from ϕ_1 to ϕ_2. The throw of the galvanometer needle gives the value of Q.

If we divide the flux by the cross section in square centimetres, we get the mean flux density B, and we shall prove in the next chapter that we can calculate the mean value of H by the formula

$$H = \frac{4\pi n_1 i}{10l},$$

where n_1 is the number of turns in the primary winding which carries a current of i amperes, and l is the mean circumference of the iron ring.

If the magnetisation of the iron ring be reversed several times, until it gets into what is called its steady cyclic state, we get curves like those shown in Fig. 21.

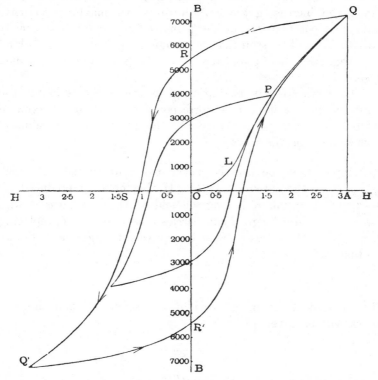

Fig. 21. Curves of Magnetisation of steel strips. $OLPQ$ = Permeability Curve. Hysteresis loops with $B_{\text{max.}} = 3900$ and $B_{\text{max.}} = 7300$.

Suppose that, when the current has its maximum positive value, OA is the value of H, and QA is the corresponding value of B. Then, as the current diminishes, the extremity of the ordinate representing B moves along QR. When the current is zero, the density of the flux left in the ring is represented by OR, and is called the remanence. When the flux is zero, H is OS, and this quantity is called the coercive force.

An inspection of the figure will show that, when the current diminishes from its highest value to zero, the change in the flux density is only $AQ - OR$, but in increasing numerically from zero to its maximum negative value it changes from OR to $-AQ$ or by an amount $AQ + OR$. On diminishing the current again, the extremity of the ordinate representing B moves along $Q'R'$. Finally on increasing the current up to its maximum positive value, the point arrives at the point Q from which it originally started. It will be seen that the magnetic induction lags behind the magnetic force producing it. This phenomenon is called hysteresis.

The curve OPQ in the figure gives the positions of the top points of the cyclic curves corresponding to various maximum values of H. The permeability μ of the iron for various values of H is found from this curve.

Work done in taking iron through a magnetic cycle. Suppose that the iron is placed in a uniform field of which the intensity is H. If this field be produced by an infinitely long solenoid which has n turns of wire in a length l and carries a current i, then, replacing the turns of wire by equivalent magnetic shells, we see that

$$H = 4\pi \frac{n}{l} i.$$

The induced E.M.F. in n turns of the solenoid when the flux varies will be given by

$$e = n \frac{\partial \phi}{\partial t}$$

$$= nS \frac{\partial B}{\partial t},$$

where S is the area of the cross section. Hence we see that

$$E - nS \frac{\partial B}{\partial t} = Ri$$

and the work W done by the generator in time T, during which the current and consequently the magnetisation of the iron go through a complete cycle of changes, is given by

$$W = \int_0^T E i \, \partial t = \int_0^T i^2 R \partial t + n S \int_0^T i \frac{\partial B}{\partial t} \, \partial t.$$

If, in the absence of the iron, the E.M.F. of the generator had been made to vary in such a way that i varied with the time in the same way as it did when the iron was present, the work done by the generator would have been $\int_0^T i^2 R \partial t$. The additional work, therefore, due to the presence of the iron is

$$n S \int_0^T i \frac{\partial B}{\partial t} \, \partial t = \frac{l S}{4\pi} \int H \partial B$$

$$= \frac{V}{4\pi} \int H \partial B,$$

where V is the volume of the iron in cubic centimetres, and the integral is taken over a whole cycle.

From Fig. 21 we see that $\int H \partial B$ gives the area of the hysteresis loop corresponding to a given maximum value of B. If B be measured in C.G.S. units, W will be in ergs.

The following empirical formula due to Steinmetz is found of great use in practice,

Steinmetz's Formula.

$$W = \eta \, V B_{max.}^{1\cdot6},$$

where W represents the ergs lost in taking V cubic centimetres of iron from a maximum induction density of $B_{max.}$ to a density $-B_{max.}$, and then back again to $B_{max.}$. The constant η depends on the kind of iron used.

If f be the frequency of the alternating current which magnetises the iron, the formula may be written

$$W = \eta f \, V B_{max.}^{1\cdot6} \times 10^{-7},$$

where W now represents the power lost in joules per second, or in watts owing to hysteresis. For values of $B_{max.}$ between 2000 and 14,000, this formula is often used in practical work. For good soft iron and steel η is usually less than $0\cdot0015$, and is sometimes less than $0\cdot001$.

The following figures give the result of a test on steel strips, the curves obtained for these strips being shown in Fig. 21.

Permeability Table.

H	B	μ
1	1580	1580
2	4930	2465
3	7000	2333
10	11,800	1180
20	13,840	692
50	15,800	316
100	16,600	166
160	17,980	112

Hysteresis.

Maximum B	Ergs per c.c. per cycle	Watts per c.c. at 100~per sec.	Remanence	Coercive force	Steinmetz's coefficient η
3900	864	0·00864	2920	0·82	0·00152
7300	2386	0·02386	5470	1·08	0·00157

Taking $\eta = 0·0015$ we see that the loss in watts per kilogramme of iron of which the specific gravity is 7·8, when $B_{max.} = 4000$ and $f = 100$, would be

$$= \eta f V B_{max.}^{1·6} \times 10^{-7}$$

$$= 0·0015 \times 100 \times \frac{1000}{7·8} \times (4000)^{1·6} \times 10^{-7}$$

$$= 1·1 \text{ nearly.}$$

Number N	$\dfrac{N^{1 \cdot 50}}{1000}$	$\dfrac{N^{1 \cdot 55}}{1000}$	$\dfrac{N^{1 \cdot 60}}{1000}$	$\dfrac{N^{1 \cdot 65}}{1000}$	$\dfrac{N^{1 \cdot 70}}{1000}$
1000	31·62	44·67	63·10	89·13	125·9
2000	89·44	130·8	191·3	279·7	409·0
3000	164·3	245·2	365·9	546·1	814·9
4000	253·0	383·0	579·8	877·8	1329
5000	353·6	541·3	828·6	1269	1942
6000	464·8	718·0	1109	1714	2648
7000	585·7	911·8	1420	2210	3441
8000	715·6	1122	1758	2755	4318
9000	853·8	1346	2122	3345	5275
10,000	1000	1585	2512	3981	6310
11,000	1154	1837	2926	4659	7419
12,000	1315	2103	3363	5378	8602
13,000	1482	2380	3822	6138	9856
14,000	1657	2670	4303	6936	11,180
15,000	1837	2971	4806	7772	12,570
16,000	2024	3284	5328	8646	14,030
17,000	2217	3607	5871	9555	15,550
18,000	2415	3942	6433	10,500	17,130
19,000	2619	4286	7015	11,480	18,790
20,000	2828	4641	7615	12,490	20,500
21,000	3043	5005	8233	13,540	22,270
22,000	3263	5380	8869	14,620	24,110
23,000	3488	5763	9523	15,730	26,000
24,000	3718	6156	10,190	16,880	27,950
25,000	3953	6559	10,880	18,060	29,960

A more accurate way of stating the formula for the hysteresis loss is

$$W = \eta f V B^{n}_{\text{max.}} \times 10^{-7},$$

where n is a number which varies slightly over different parts of the curve of W and B and also varies for different kinds of iron and steel. For example Ewing and Klaassen found experimentally for a particular specimen of iron that n was nearly 1·55 from B equal to 1000 to B equal to 2000, and that it was only 1·475 from B equal to 2000 to B equal to 8000. For values of B between 8000 and 14,000 they found that for this specimen n was 1·7.

F. Stroude using a slow cyclic method tested this law for transformer iron and for an alloy of iron called 'stalloy.' For transformer iron when the flux density was between 14,000 and 16,000 and also when it was between 16,000 and 18,000, he got

the same value of 1·71 for the Steinmetz index. The values of
the index in other cases are given in the following table.

Value of B	500—1000	1000—2000	2000—4000	4000—6000
Transformer iron	2·00	1·71	1·71	1·71
Stalloy	1·82	1·78	1·72	1·66

Value of B	6000—8000	8000—10,000	10,000—12,000	12,000—14,000
Transformer iron	1·71	1·71	1·71	1·71
Stalloy	1·66	1·66	1·66	—

Tables of the 1·5th, 1·55th, 1·6th, 1·65th and 1·7th powers of
the numbers representing the induction densities usually wanted
in practice are given above.

REFERENCES

G. CAREY FOSTER and A. W. PORTER, *Elementary Treatise on Electricity and Magnetism.*

J. J. THOMSON, *Elements of the Mathematical Theory of Electricity and Magnetism.*

J. H. JEANS, *The Mathematical Theory of Electricity and Magnetism.*

CLERK MAXWELL, *Electricity and Magnetism.*

C. P. STEINMETZ, *Journ. of the Am. Inst. of Elect. Engin.* Vol. 9, p. 3, Jan. 19, 1892.

F. STROUDE, 'The Steinmetz Index for Transformer Iron, Stalloy and Cast Iron,' *Proc. Phys. Soc.* p. 238, 1912.

CHAPTER II

The growth of the current in an inductive coil. Simple harmonic waves. Periodic waves. The mean magnetic force inside a circular ring of rectangular section. The mean magnetic force inside a ring of circular cross section. The magnetic analogy of Ohm's law. Reluctance. Inductance formulae for anchor rings. The mutual inductance between two parallel wires and a rectangle. The linkages of the magnetic flux inside the wire. The inductance of a straight wire. The inductance of a rectangle. The mutual inductance between two equal, parallel and coaxial rectangles. The mutual inductance between a straight wire and a circle. The self inductance of a concentric main. The self inductance of two parallel cylindrical wires. The self inductance of a circuit formed by three equal parallel cylinders, the axes of which lie along the edges of an equilateral prism. Triple concentric main. Minimum self inductance. The repulsive force between two parallel wires carrying equal currents in opposite directions. References.

IF a coil of inductance L is put in series with a battery of constant electromotive force E, the current does not attain instantaneously its final value I, which equals E/R, where R is the resistance of the whole circuit. If it did the energy $LI^2/2$ would have been taken from the battery and stored in the field in an infinitely short time—which is contrary to physical principles. When the current is growing in the circuit we have seen that a reverse electromotive force $L(\partial i/\partial t)$ is generated in the circuit. Hence, by Ohm's law, the value of the current i is determined by

The growth of the current in an inductive circuit.

$$i = \frac{E - L\dfrac{\partial i}{\partial t}}{R},$$

i.e. by

$$E = Ri + L\frac{\partial i}{\partial t}.$$

This is a linear differential equation with constant coefficients and its solution is the sum of two functions i_1 and i_2, where i_1 is the particular integral of this equation and i_2 is the complementary function which satisfies the equation

$$Ri_2 + L\frac{\partial i_2}{\partial t} = 0.$$

We see at once that the value of i_1 is E/R, as this obviously satisfies the equation, and it is easy to verify that $i_2 = Ae^{-Rt/L}$. Hence the complete solution is given by

$$i = E/R + Ae^{-Rt/L}.$$

Now when t is zero, i must be zero and so $A = -E/R$, and therefore

$$i = \frac{E}{R}\left(1 - e^{-\frac{R}{L}t}\right).$$

This equation gives the law of the growth of the current in the circuit.

L/R is called the time constant of the coil, and we see that when $t = L/R$, $i = (E/R)(1 - e^{-1}) = 0.6321\,(E/R)$ approximately. We also have

$$\int_0^\infty \left(\frac{E}{R} - i\right)\partial t = \frac{L}{R}\cdot\frac{E}{R}.$$

Hence the action of inductance prevents a quantity of electricity LE/R^2 from flowing in the circuit.

If the coil were short circuited, it is easy to show that the law giving the decay of the current in the coil is

$$i = \frac{E}{R}e^{-\frac{R}{L}t},$$

and the quantity of electricity that flows after the short circuit is LE/R^2.

If instead of an inductive coil we had a condenser K and a non-inductive resistance R in series, then the equation to give the growth of the current flowing into the condenser would be

$$E = Ri + \frac{\int i\partial t}{K}.$$

The solution of this differential equation for $\int i \partial t$ is

$$\int i \partial t = KE \left(1 - \epsilon^{-\frac{t}{KR}} \right),$$

assuming that there is initially zero charge in the condenser. If the condenser be charged so that the potential difference between its terminals is E, and its terminals be then connected through a resistance R, the law of discharge is

$$i = \frac{E}{R} \epsilon^{-\frac{t}{KR}}.$$

The product KR is called the time constant of the circuit consisting of a capacity K and a resistance R. In the time KR the condenser will have lost 0·6321... of its charge through the resistance R.

Let us suppose that the potential difference e between the supply mains follows the simple harmonic law. We can then write it in the form $E \sin \omega t$, where E is the maximum value of the potential difference, t the time and $2\pi/\omega$ the periodic time T of the potential difference. The frequency f is the number of complete periods per second and so

$$f = 1/T \text{ and } \omega = 2\pi f.$$

Simple harmonic waves.

Now let an inductive coil (R, L) be connected with the mains. By Ohm's law, the current i must satisfy the equation

$$i = \frac{e - L \dfrac{\partial i}{\partial t}}{R},$$

i.e.

$$E \sin \omega t = Ri + L \frac{\partial i}{\partial t}.$$

Writing D for $\partial/\partial t$, we get for the particular integral

$$i_1 = \frac{E}{R + LD} \sin \omega t$$

$$= \frac{E(R - LD)}{R^2 - L^2 D^2} \sin \omega t.$$

But $D^2 \sin \omega t = -\omega^2 \sin \omega t$, and hence writing $-\omega^2$ for D^2, we get

$$i_1 = \frac{ER \sin \omega t - EL \omega \cos \omega t}{R^2 + L^2 \omega^2}.$$

If we now put $\tan \alpha = L\omega/R$, we get after a little reduction

$$i_1 = \frac{E}{\sqrt{R^2 + L^2\omega^2}} \sin(\omega t - \alpha).$$

The complementary function is $A\epsilon^{-(R/L)t}$ and so the complete solution is

$$i = \frac{E}{\sqrt{R^2 + L^2\omega^2}} \sin(\omega t - \alpha) + A\epsilon^{-(R/L)t}.$$

Let us now suppose that the switch is closed when t is equal to t_1.

Then since the circuit cannot receive finite energy in an infinitely short time we see that the initial value of i must be zero and the general solution becomes in this case

$$i = \frac{E}{\sqrt{R^2 + L^2\omega^2}} \left\{ \sin(\omega t - \alpha) - \sin(\omega t_1 - \alpha)\epsilon^{-\frac{R}{L}(t-t_1)} \right\} \quad \ldots\ldots(a),$$

where

$$\tan \alpha = \frac{\omega L}{R}.$$

When t is very large the initial disturbance caused by closing the switch will be negligible as the damping factor $\epsilon^{-(R/L)(t-t_1)}$ will be practically zero, and hence we can write

$$i = \frac{E}{\sqrt{R^2 + L^2\omega^2}} \sin(\omega t - \alpha).$$

We see, in this case, that whenever t equals $\alpha/\omega + nT$, then i is zero, and

$$\frac{\partial i}{\partial t} = \frac{E\omega}{\sqrt{R^2 + L^2\omega^2}} = \frac{E\sin\alpha}{L}.$$

Now at the moment of closing the switch

$$E\sin\omega t_1 = Ri + L\frac{\partial i}{\partial t},$$

therefore

$$\frac{\partial i}{\partial t} = \frac{E\sin\omega t_1}{L},$$

since i is zero at this instant. Hence comparing the initial value of $\partial i/\partial t$ with its value for $t = \alpha/\omega + nT$ after the current has become purely alternating, we see that if $\omega t_1 = \alpha$, i.e. if $t_1 = (\alpha/2\pi)T$, then there will be no initial disturbance and the current will at once become purely alternating. This can also be seen from equation (a) above, the coefficient of the exponential term vanishing when t_1 is α/ω. It follows from this equation that the initial disturbance is a maximum when

$$t_1 = \frac{\alpha}{\omega} + \frac{T}{4} = \frac{\alpha}{2\pi} T + \frac{T}{4}.$$

It is easy to see also that the maximum value of the current after switching on can never be as great as

$$\frac{2E}{\sqrt{R^2 + L^2\omega^2}}.$$

If t_1 lies in value between $(\alpha/2\pi) T$ and $(\alpha/2\pi) T + T/2$, then the maximum values of i on the positive side are smaller initially than

$$\frac{E}{\sqrt{R^2 + L^2\omega^2}},$$

but the maximum values of i on the negative side are greater than

$$\frac{E}{\sqrt{R^2 + L^2\omega^2}}.$$

The number of coulombs that circulate during the first interval T after closing the circuit is

$$\int_{t_1}^{t_1 + T} i\,\partial t = -\frac{L}{R}\frac{E\sin(\omega t_1 - \alpha)}{\sqrt{R^2 + L^2\omega^2}}\left\{1 - \epsilon^{-\frac{R}{L}T}\right\}.$$

Similarly for the second period

$$\int_{t_1 + T}^{t_1 + 2T} i\,\partial t = -\frac{L}{R}\frac{E\sin(\omega t_1 - \alpha)}{\sqrt{R^2 + L^2\omega^2}}\left\{\epsilon^{-\frac{R}{L}T} - \epsilon^{-\frac{2R}{L}T}\right\}.$$

Hence we see that this number rapidly gets smaller and smaller and the current soon becomes purely alternating.

When the coil has an iron core, the back E.M.F. is no longer proportional to $\partial i/\partial t$, and the equation to determine the initial disturbance is much more complicated. It is easy to see that the remanence of the iron in the core initially is a principal factor in determining the magnitude of the disturbance on closing the switch. In practice this factor is generally unknown. What happens on closing the switch can be determined experimentally by getting a record of the current with an oscillograph. It is found that, when we have iron in the core, the initial fluctuations of the current are sometimes very large.

We have seen that when the steady state is assumed the current i follows the simple harmonic law but the current wave lags by a time α/ω behind the potential difference wave. The amplitude also of the current wave is less than its value with continuous current in the ratio of R to $(R^2 + L^2\omega^2)^{\frac{1}{2}}$. The quantity $(R^2 + L^2\omega^2)^{\frac{1}{2}}$ is called the impedance of the coil and α is called the angle of time lag.

In practical electrical work the assumption that the potential difference wave is a pure sine wave is seldom admissible. We shall therefore now consider what happens in the general case.

The rotation of the armature of an alternator causes a rapid
Periodic waves. periodic change in the flux of induction through its coils and hence the potential difference between the collector rings is a periodic function of the time. We may therefore express the difference of potential mathematically by the expression $f(t)$ where t is the time in seconds, and if T be the time required for the P.D. to go through all its values, i.e. if it is the period, then

$$f(t) = f(t + T) = f(t + 2T) = \dots.$$

Also, since the north and south poles of the field magnets in actual machines are always made alike, it follows that the wave for the first half period is the same as for the second half with the sign changed, so that

$$f(t) = -f\left(t + \frac{T}{2}\right).$$

By Fourier's formula, $f(t)$ may be expressed in this case by the series

$$f(t) = A_1 \sin\left(\frac{2\pi}{T} t + \alpha_1\right) + A_3 \sin\left(3\frac{2\pi}{T} t + \alpha_3\right) + \dots,$$

where A_1, A_3, ... etc. are the amplitudes of the first, third, ... harmonics which can be determined by the usual formulae when $f(t)$ is known. We express $2\pi/T$ by ω, and we see that ω is the angular velocity of a line rotating $1/T$ times a second. If f be the frequency, i.e. the number of times per second that the potential difference goes through all its values, then $\omega = 2\pi f$.

Now suppose that an alternating P.D., of which the instantaneous value is e, is applied to a circuit of resistance R and self inductance L. Then if i be the instantaneous value of the current in the circuit $-L\,(\partial i/\partial t)$ will be the E.M.F. due to induction, and we have by Ohm's law

$$i = \frac{e - L\dfrac{\partial i}{\partial t}}{R}.$$

Hence

$$L\frac{\partial i}{\partial t} + Ri = e,$$

and

$$(LD + R)\,i = A_1 \sin(\omega t + \alpha_1) + A_3 \sin(3\omega t + \alpha_3) + \ldots,$$

and

$$i = \frac{A_1}{R + LD}\sin(\omega t + \alpha_1) + \frac{A_3}{R + LD}\sin(3\omega t + \alpha_3) + \ldots.$$

Now

$$\frac{A_{2n+1}}{R + LD}\sin\{(2n+1)\,\omega t + \alpha_{2n+1}\}$$

$$= \frac{A_{2n+1}(R - LD)}{R^2 - L^2 D^2}\sin\{(2n+1)\,\omega t + \alpha_{2n+1}\}.$$

Also

$$D\sin\{(2n+1)\,\omega t + \alpha_{2n+1}\} = (2n+1)\,\omega\cos\{(2n+1)\,\omega t + \alpha_{2n+1}\},$$

and

$$D^2\sin\{(2n+1)\,\omega t + \alpha_{2n+1}\} = -(2n+1)^2\,\omega^2\sin\{(2n+1)\,\omega t + \alpha_{2n+1}\}.$$

Hence, using symbolic methods, we can replace

$$D^2 \text{ by } -(2n+1)^2\,\omega^2.$$

Substituting and simplifying we find that

$$\frac{A_{2n+1}}{R + LD}\sin\{(2n+1)\,\omega t + \alpha_{2n+1}\}$$

$$= \frac{A_{2n+1}\sin\{(2n+1)\,\omega t + \alpha_{2n+1} - \beta_{2n+1}\}}{\sqrt{R^2 + (2n+1)^2 L^2 \omega^2}},$$

where

$$\tan\beta_{2n+1} = \frac{(2n+1)\,L\omega}{R}.$$

Hence the complete solution of the equation is

$$i = \Sigma\,\frac{A_{2n+1}\sin\{(2n+1)\,\omega t + \alpha_{2n+1} - \beta_{2n+1}\}}{\sqrt{R^2 + (2n+1)^2 L^2 \omega^2}} + B\epsilon^{-\frac{R}{L}t}.$$

In this formula B is a constant which depends on the initial conditions. Now as R/L is in most cases very great, we see that in these cases the current after a fraction of a second assumes a steady periodic value which is given by

$$i = \frac{A_1\sin(\omega t + \alpha_1 - \beta_1)}{\sqrt{R^2 + L^2 \omega^2}} + \frac{A_3\sin(3\omega t + \alpha_3 - \beta_3)}{\sqrt{R^2 + 3^2 L^2 \omega^2}} + \ldots.$$

The mean
magnetic force
inside a cir-
cular ring of
rectangular
section. Let R be the mean radius of a circular ring of rectangular section, a its radial depth, b its breadth, and N the total number of turns of wire wrapped round it in such a way that we may, without appreciable error, consider the turns to be in planes passing through the axis of the ring. We will also suppose that everything is symmetrical, so that the whole flux is inside the ring, and that all parts of the core of the ring at the same distance x from its axis are subjected to the same magnetising force.

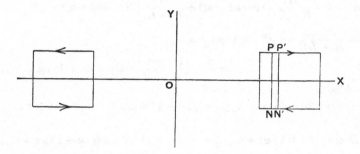

Fig. 22. The mean magnetic force inside a ring of rectangular
cross section.

If H be the magnetic force at the distance x, then $H . 2\pi x =$ the work done in taking unit magnetic pole round the circumference of the circle whose radius is x, and we know that this work equals $4\pi Ni$. Hence the mean value H_m of H over the cross section of the ring is given by

$$H_m = \frac{1}{ab} \int_{R-a/2}^{R+a/2} Hb\,\partial x$$

$$= \frac{2Ni}{a} \log_\epsilon \frac{1 + a/2R}{1 - a/2R}$$

$$= \frac{4Ni}{a} \left\{ \frac{a}{2R} + \frac{1}{3}\left(\frac{a}{2R}\right)^3 + \dots \right\}$$

$$= \frac{4\pi Ni}{2\pi R} \left\{ 1 + \frac{a^2}{12R^2} + \frac{a^4}{80R^4} + \dots \right\}$$

$$= 4\pi ni \left\{ 1 + \frac{a^2}{12R^2} + \frac{a^4}{80R^4} + \dots \right\},$$

where n is the number of turns of wire per centimetre length of the mean circumference. If R were infinite in comparison with a, H_m would equal $4\pi ni$, and since, in this case, H_m is independent of a, it is constant whatever may be the shape of the cross section. If i is in amperes,

$$H = \frac{4\pi ni}{10}.$$

When the core of the ring is magnetic, since

$$H = \frac{4\pi Ni}{2\pi x},$$

we see that the magnetising force varies at different points of the cross section and therefore, from the properties of iron, it follows that μ also varies. If however x only vary very little, that is, if a/R be small, then the ratio of B_m to $4\pi Ni/l$, where l is the length of the mean circumference of the ring, gives us an approximate value of μ corresponding to the value of H at the centre of the cross section of the ring.

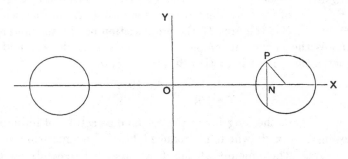

Fig. 23. The mean magnetic force inside a ring of circular cross section.

Let us now suppose that the radius of the cross section is a circle of radius a (Fig. 23) then, making the same assumptions as before,

The mean magnetic force inside a ring of circular cross section.

$$H \cdot 2\pi x = 4\pi Ni,$$

and noticing that $PN^2 = a^2 - (R - x)^2$, we get

$$H_m = \frac{4Ni}{\pi a^2} \int_{R-a}^{R+a} \frac{\sqrt{a^2 - (R - x)^2}}{x} \, dx.$$

Writing $x = R - a \sin \theta$, we get

$$H_m = \frac{4Ni}{\pi R} \int_{-\frac{\pi}{2}}^{+\frac{\pi}{2}} \cos^2 \theta \left\{1 - \frac{a}{R} \sin \theta\right\}^{-1} \delta\theta$$

$$= \frac{4Ni}{R} \left\{\frac{1}{2} + \frac{1}{2.4}\frac{a^2}{R^2} + \frac{1.3}{2.4.6}\frac{a^4}{R^4} + \cdots\right\}$$

$$= \frac{4Ni}{a^2}\{R - \sqrt{R^2 - a^2}\}$$

$$= \frac{4\pi Ni}{2\pi R}\left\{1 + \frac{1}{4}\frac{a^2}{R^2} + \frac{1}{8}\frac{a^4}{R^4} + \cdots\right\}.$$

In this case also, if a/R be small and i is in amperes, we can assume in practical work that

$$H_m = \frac{4\pi Ni}{10l}.$$

The number of lines of force embraced by an infinitely long solenoid, if N be the number of turns on a length l of it, is by what we have shown $4\pi NiS/(10l)$, where S is the area of the cross section of the solenoid and i denotes the current in amperes. If the core of the solenoid be magnetic, the flux of induction Φ in it is given by

The magnetic analogy of Ohm's law. Reluctance.

$$\Phi = \mu\frac{4\pi NiS}{10l} = \frac{(4\pi/10)\,Ni}{l/(\mu S)}.$$

We may regard the length l of the solenoid as a tube of induction enclosing a flux Φ which is produced by the magnetising force $(4\pi/10)\,Ni$. The amount of the flux produced depends on the value of $l/\mu S$, which is called the reluctance of the length l of the solenoid. We see that the reluctance R of a portion of a tube of induction varies as the length of the portion, and varies inversely as the area of the cross section and as the permeability. Hence we get the magnetic analogy to Ohm's law

$$\text{Flux} = \frac{\text{Magnetomotive Force}}{\text{Reluctance}}.$$

The magnetomotive force $(4\pi/10)\,Ni$ is the difference of magnetic potential between the two ends of the tube, and it is important to notice that it is independent of the shape of its section. It merely depends on the number of turns and the amperes. Electricians often take ampere turns as their unit.

Let us suppose that we have an iron ring of circular cross section uniformly wound with N turns of wire and that a steady current i is flowing in the wire. If the radius of the cross section of the ring be small compared with the radius of the aperture, H and therefore also B will be practically uniform in the iron. Let us now suppose that the ring is sawn across so that a narrow air-gap of breadth d is made. Then, since the gap is narrow, the flux density in it will be practically equal to the flux density in the iron. The magnetic force in the air will therefore equal B but in the iron it will equal B/μ, where μ is the permeability of the iron. Hence, if we take unit magnetic pole round the circuit the work done is $Bd + (B/\mu)\,l$, where l is the length of the path in the iron, and so we have

$$(B/\mu)\,l + Bd = 4\pi Ni/10.$$

If Φ be the magnetic flux in the circuit, we have $\Phi = BS$, where S is the area of the cross section of the ring, and hence

$$\Phi\,(l/\mu S + d/S) = 4\pi Ni/10$$

and
$$\Phi = \frac{4\pi Ni/10}{l/\mu S + d/S}.$$

This is the equation used in the calculation of the magnetic circuits of dynamos, and is of great importance to electricians. It has to be noticed that it is only accurate when the air-gap is very narrow and when the iron is uniformly magnetised. In some cases the term $l/\mu S$ can be neglected in comparison with d/S.

If we have a composite circuit made up of wires having different conductivities (κ), lengths (l) and cross sectional areas (S), and if there is an electromotive force E in the circuit, the current I, by Ohm's law, is determined by

$$I\,\Sigma\,(l/\kappa S) = E.$$

Hence by analogy we can write

$$\Phi\,\Sigma\,(l/\mu S) = 4\pi Ni/10.$$

In practical work we regard this as the fundamental magnetic equation.

Let us imagine an anchor ring wound uniformly with two windings of N_1 and N_2 turns respectively. Let its mean radius be R centimetres and the radius of its cross section a centimetres. Then if we can take a as the radius of the inner coil, since all the flux generated by unit current in the inner coil passes through the outer coil,

$$M = 4\pi N_1 N_2 \{R - \sqrt{R^2 - a^2}\},$$

when the core is of non-magnetic material. If the section is rectangular, of breadth b and radial depth a, we have

$$M = 2b N_1 N_2 \log_\epsilon \frac{R + a/2}{R - a/2}.$$

In the above formulae M is in centimetres; to obtain the value in henrys we multiply by 10^{-9}.

To get the self inductance of the rings when both are wound with N turns of wire, put $N_1 = N_2 = N$ in the above formulae. For a ring of circular cross section this gives

$$L = 4\pi N^2 \{R - \sqrt{R^2 - a^2}\}.$$

And for a ring of rectangular cross section

$$L = 2b N^2 \log_\epsilon \frac{R + a/2}{R - a/2}.$$

The most important practical cases, however, in which formulae are required for self inductance, are for concentric cylinders and for cylindrical wires parallel to one another. We will therefore give complete proofs of the formulae for these cases.

In order to simplify the problem we shall suppose that the plane containing the axes of the parallel wires, in which equal currents are flowing in opposite directions, is parallel to the plane of the rectangle. We shall also suppose that both planes are at right angles to the paper, the sides of the rectangle intersecting it at C and D (Fig. 24), and that the wires intersect it at A and B. Let $AB = 2c$, $CD = 2a$, and let the coordinates of C be b and d. Then since M is the flux linked with the rectangle when unit current flows in the wires, we have, if l be the length of the rectangle.

The mutual
inductance
between two
parallel wires
and a rect-
angle.

$$M = l \int_{b}^{b+2a} \left[\frac{2}{AP} \cos PAB - \frac{2}{PB} \cos PBX \right] \partial x$$

$$= l \int_{b}^{b+2a} \left[\frac{2(x+c)}{(x+c)^2 + d^2} - \frac{2(x-c)}{(x-c)^2 + d^2} \right] \partial x$$

$$= l \left[\log \frac{(x+c)^2 + d^2}{(x-c)^2 + d^2} \right]_{b}^{b+2a}$$

$$= l \log \frac{(b+2a+c)^2 + d^2}{(b+2a-c)^2 + d^2} \cdot \frac{(b-c)^2 + d^2}{(b+c)^2 + d^2}.$$

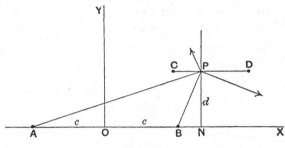

Fig. 24.

It is easy to see that when $c^2 + d^2 = b(2a+b)$, M vanishes. Hence, putting $a = 0$, we see that $c^2 + d^2 = b^2$ gives the locus of the points where the vertical component of the magnetic force due to the currents in the two parallel wires is zero. In the plane of the paper, therefore, these points lie on an equiaxial hyperbola.

If F be the component of the attractive force between the two circuits perpendicular to their planes when they carry currents i_1 and i_2 respectively, we have

$$F = -\frac{\partial M}{\partial d} i_1 i_2$$

$$= 8lcd \left[\frac{2a+b}{X^4 - 4c^2(2a+b)^2} - \frac{b}{Y^4 - 4b^2c^2} \right] i_1 i_2,$$

where $X^2 = (2a+b)^2 + c^2 + d^2$, and $Y^2 = b^2 + c^2 + d^2$. When M is zero,

$$F = -\frac{4acdl}{b(2a+b)(a^2+d^2)} i_1 i_2.$$

Let us now suppose that the wires and the rectangle are placed so that AB and CD (Fig. 25) bisect one another, where A and B are points on the wires and the angle BOD is γ. In this case the component perpendicular to CD of the force F at a point P on CD due to unit current in the parallel wires is given by

$$F = \frac{2}{AP} \cdot \frac{r + c \cos \gamma}{AP} - \frac{2}{BP} \cdot \frac{r - c \cos \gamma}{BP},$$

and hence

$$M = 2l \int_{-a}^{+a} \left\{ \frac{r + c \cos \gamma}{(r + c \cos \gamma)^2 + c^2 \sin^2 \gamma} - \frac{r - c \cos \gamma}{(r - c \cos \gamma)^2 + c^2 \sin^2 \gamma} \right\} \partial r$$

$$= 2l \log \frac{(a + c \cos \gamma)^2 + c^2 \sin^2 \gamma}{(a - c \cos \gamma)^2 + c^2 \sin^2 \gamma}$$

$$= 2l \log \frac{a^2 + c^2 + 2ac \cos \gamma}{a^2 + c^2 - 2ac \cos \gamma}.$$

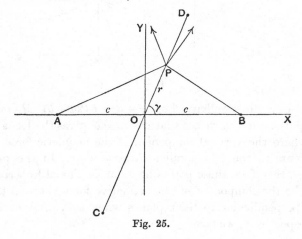

Fig. 25.

If i_1 and i_2 be the currents in the two circuits and if the currents in B and D are flowing in the same direction, the mutual potential energy of the two circuits is $- M i_1 i_2$. The couple tending to increase γ is

$$-\frac{\partial}{\partial \gamma}(- M i_1 i_2) = \frac{\partial M}{\partial \gamma} \cdot i_1 i_2$$

$$= -\frac{8lac(a^2 + c^2) \sin \gamma}{a^4 + c^4 - 2a^2 c^2 \cos 2\gamma} \cdot i_1 i_2.$$

If a be greater than $c(2^{1/2}-1)$ but less than $c(2^{1/2}+1)$, this couple has maximum numerical values

$$\{2l\,(a^2+c^2)/(a^2-c^2)\}\,i_1 i_2 \quad \text{when} \quad \sin^2\gamma = (a^2-c^2)^2/(4a^2c^2),$$

and minimum values when $\gamma = -\pi/2$ and $\pi/2$. For values of a outside these limits the couple is a maximum when γ is $-\pi/2$ and $\pi/2$.

In calculating accurate formulae for the self inductance of **The linkages of the mag-netic flux inside the wire.** circuits it is necessary to compute the linkages with the current of the magnetic flux inside the wire. Let us consider, for instance, a very long cylindrical wire of radius r, carrying a current i the density of which is uniform over the cross section of the wire. In calculating the magnetic flux density inside the wire at a distance x from its axis we have only to take into account the magnetic effect produced by the current flowing in the coaxial cylinder the radius of the cross section of which is x, as we have already seen that the magnetic force produced by the current flowing in the outer cylindrical shell is zero at points inside it. The current in the inner cylinder equals $i\,(x^2/r^2)$, and hence, if μ the permeability of the wire be constant, the linkages per unit length are given by

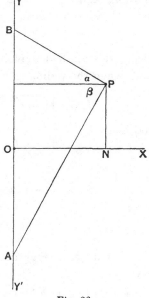

$$\mu \int_0^r \frac{2i\,(x^2/r^2)}{x} \, i\,\frac{x^2}{r^2}\, \partial x = \tfrac{1}{2}\,\mu i^2.$$

The part of the self inductance, therefore, due to linkages inside a length l of the wire will be $\mu l/2$.

We shall now consider the induct- **The induct-ance of a straight cylin-drical wire.** ance of a length l (AB in Fig. 26) of a straight cylindrical wire YY'. Let us take the middle point O of the wire as origin and the axis of the wire as axis of Y. If the radius of the wire be very small compared with its length,

Fig. 26.

then when the current in it is unity the flux linkages inside the wire by the preceding section will be approximately equal to $\mu l/2$. Hence taking μ equal to unity we have

$$\Phi = \frac{l}{2} + \int_r^\infty \int_{-l/2}^{+l/2} \frac{(\sin \alpha + \sin \beta)}{x} \partial x \partial y,$$

where Φ is the flux due to and linked with the current in the portion AB of the wire, the radius of which is r. Hence

$$\Phi = \frac{l}{2} + \int_r^\infty \frac{1}{x} \left[\{x^2 + (l/2 + y)^2\}^{1/2} - \{x^2 + (l/2 - y)^2\}^{1/2} \right]_{-l/2}^{+l/2} \partial x$$

$$= \frac{l}{2} + 2 \left[(x^2 + l^2)^{1/2} - x + l \log x - l \log \{(x^2 + l^2)^{1/2} + l\} \right]_r^\infty$$

$$= 2 \left[l \log \{l + (l^2 + r^2)^{1/2}\}/r - (l^2 + r^2)^{1/2} + r + l/4 \right].$$

We shall now find the inductance of a rectangular circuit of wire of circular cross section. Let the lengths of the sides of the rectangle be a and b. Then if L_a and L_b denote the inductances of sides of lengths a and b, and M_a and M_b denote the mutual inductances between the parallel sides, we have

The inductance of a rectangle.

$$L = 2 (L_a + L_b - M_a - M_b).$$

Using the method explained in the last section we see at once that, when the radius r of the wire is small,

$$M_a = 2 \left[a \log \frac{a + (a^2 + b^2)^{1/2}}{b} - (a^2 + b^2)^{1/2} + b \right],$$

and hence

$$L = 4 \left[a \log \frac{b}{r} + b \log \frac{a}{r} + 2 \{(a^2 + b^2)^{1/2} + r\} \right.$$

$$\left. - a \log \frac{a + (a^2 + b^2)^{1/2}}{a + (a^2 + r^2)^{1/2}} - b \log \frac{b + (a^2 + b^2)^{1/2}}{b + (b^2 + r^2)^{1/2}} \right.$$

$$\left. - (a^2 + r^2)^{1/2} - (b^2 + r^2)^{1/2} - (3/4) (a + b) \right].$$

When $(r/a)^2$ and $(r/b)^2$ can be neglected compared with unity, we have

$$L = 4 \left[(a + b) \log (2ab/r) + 2 \{(a^2 + b^2)^{1/2} + r\} \right.$$

$$\left. - a \log \{a + (a^2 + b^2)^{1/2}\} - b \log \{b + (a^2 + b^2)^{1/2}\} - (7/4) (a + b) \right].$$

Let us suppose that we have two equal rectangles with their
The mutual planes parallel and the centre of one vertically over
inductance that of the other and at a distance d from it. If the
between two
equal, coaxial lengths of the sides be a and b and we neglect the
and parallel
rectangles. thickness of the wire we get, as in the preceding
problem,

$$M = 4\left[a \log \frac{a + \sqrt{a^2 + d^2}}{d} - \sqrt{a^2 + d^2} + d \right.$$

$$\left. - a \log\frac{a + \sqrt{a^2 + b^2 + d^2}}{\sqrt{b^2 + d^2}} + \sqrt{a^2 + b^2 + d^2} - \sqrt{b^2 + d^2}\right] + 4\left[\quad \right],$$

where the expression inside the second bracket is obtained from
the first by interchanging a and b.

Hence

$$M = 4\left[a \log \left\{ \frac{a + \sqrt{a^2 + d^2}}{a + \sqrt{a^2 + b^2 + d^2}} \cdot \frac{\sqrt{b^2 + d^2}}{d} \right\} \right.$$

$$\left. + b \log \left\{ \frac{b + \sqrt{b^2 + d^2}}{b + \sqrt{a^2 + b^2 + d^2}} \cdot \frac{\sqrt{a^2 + d^2}}{d} \right\} \right]$$

$$+ 8\left[\sqrt{a^2 + b^2 + d^2} - \sqrt{a^2 + d^2} - \sqrt{b^2 + d^2} + d \right].$$

When $(d/a)^2$ and $(d/b)^2$ can be neglected compared with unity, we
may write

$$M = 4\left[(a + b) \log \frac{2ab}{d} - a \log (a + \sqrt{a^2 + b^2}) - b \log (b + \sqrt{a^2 + b^2}) \right.$$

$$\left. + 2d + 2\sqrt{a^2 + b^2} - 2(a + b) \right], \text{approximately,}$$

and when d is large compared with a or b,

$$M = \frac{2a^2b^2}{d^3} - \frac{a^2b^2(a^2 + b^2)}{d^5}, \text{approximately.}$$

If F be the force of attraction between the two rectangles when
one is carrying a current i_1 and the other a current i_2 flowing in the
same sense, we have

$$F = 4\left[\left(1 + \frac{a^2}{d^2}\right)^{1/2} + \left(1 + \frac{b^2}{d^2}\right)^{-1/2} \right.$$

$$\left. - \left(1 + \frac{b^2}{d^2}\right)^{-1}\left(1 + \frac{a^2 + b^2}{d^2}\right)^{1/2} - 1 \right] i_1 i_2 + 4\left[\quad \right] i_1 i_2,$$

where the expression inside the second bracket is obtained by interchanging a and b in the first. When d is very small

$$F = 4 \left[\frac{a+b}{d} - 2 \right] i_1 i_2, \text{ approximately,}$$

and when d is large

$$F = \left[\frac{6a^2b^2}{d^4} - \frac{5a^2b^2(a^2+b^2)}{d^6} \right] i_1 i_2,$$

approximately.

We shall now find the mutual inductance between a long straight wire and a circular wire, the former cutting a diameter of the latter at right angles and being inclined to its plane at an angle α.

The mutual inductance between a straight wire and a circle.

Let RO (Fig. 27) be the wire, and let SCK be the diameter of the circle to which it is perpendicular. Let OX and OY be at right angles to one another in the plane of the circle and let OZ be perpendicular to both. OR will be in the

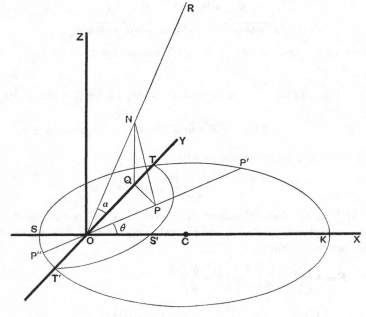

Fig. 27.

plane ZOY and ROY is the angle α. Let the segment $TS'T'$ in the plane of the circle be exactly similar to the segment TST'. It is obvious that the component flux passing through TST' and perpendicular to its plane due to unit current in OR will equal the corresponding component flux in $TS'T'$ but will be in the opposite direction. Hence in calculating the linkages we have only to consider the flux passing through $TS'T'K$. Let

$$OP = r, \; POC = \theta, \; OC = c, \; CP' = a, \; OP' = r_1 \text{ and } OP'' = r_2.$$

We have $\qquad a^2 = r_1^2 + c^2 - 2r_1 c \cos \theta,$

and $\qquad a^2 = r_2^2 + c^2 + 2r_2 c \cos \theta,$

and therefore $\qquad r_1 - r_2 = 2c \cos \theta.$

Consider an elementary area $r\partial\theta\partial r$ at P. The component of the magnetic flux through this element and perpendicular to it is given by

$$\frac{2}{r \sin \gamma} \cdot r\partial\theta\partial r \cos \delta,$$

where γ is the angle NOP and δ is the angle between the planes NOP and YOP.

Now draw PQ at right angles to OP and QN at right angles to OY. Join PN.

We have

$$ON^2 = OQ^2 + QN^2 = OP^2 + PQ^2 + QN^2 = OP^2 + PN^2.$$

Hence OPN is a right angle and the angle $NPQ = \delta$.

We have also $ON \cos \alpha \cos \theta = OQ \cos \theta = PQ,$

and $\qquad ON \sin \gamma \cos \delta = PN \cos \delta = PQ,$

and thus $\qquad \cos \alpha \cos \theta = \sin \gamma \cos \delta.$

Also $\qquad ON \cos \gamma = OP = OQ \sin \theta = ON \cos \alpha \sin \theta,$

and therefore $\qquad \cos \gamma = \cos \alpha \sin \theta.$

Thus the linkages with the flux through the element $r\partial\theta\partial r$ equal $2 \cos \alpha \cos \theta \, \partial\theta\partial r / \sin^2 \gamma$, and hence

$$M = 2 \int_{-\pi/2}^{\pi/2} \int_{r_2}^{r_1} \frac{\cos \alpha \cos \theta}{1 - \cos^2 \alpha \sin^2 \theta} \partial\theta\partial r$$

$$= 4c \int_{-\pi/2}^{\pi/2} \frac{\cos \alpha \cos^2 \theta \, \partial\theta}{1 - \cos^2 \alpha \sin^2 \theta}.$$

Expanding and integrating we get

$$M = 4\pi c \cos\alpha \left[\frac{1}{2} + \frac{1\cdot1}{2\cdot4}\cos^2\alpha + \frac{1\cdot1\cdot3}{2\cdot4\cdot6}\cos^4\alpha + \dots \right]$$

$$= 4\pi c \cos\alpha \left[\frac{1 - \sqrt{1 - \cos^2\alpha}}{\cos^2\alpha} \right]$$

$$= 4\pi c\,(\sec\alpha - \tan\alpha)$$

$$= 4\pi c \tan(\pi/4 - \alpha/2).$$

Thus M is independent of the radius of the circle.

When O is outside the circle, we get in a similar way

$$M = 2\cos\alpha \int_{r_1}^{r_2} \int_{-\sin^{-1}(a/c)}^{\sin^{-1}(a/c)} \frac{\cos\theta}{\sin^2\gamma}\,\partial\theta\partial r$$

and $\qquad r_2 - r_1 = 2\sqrt{a^2 - c^2\sin^2\theta}.$

Hence $\qquad M = 4\cos\alpha \int_{-\sin^{-1}(a/c)}^{\sin^{-1}(a/c)} \frac{\cos\theta\sqrt{a^2 - c^2\sin^2\theta}}{1 - \cos^2\alpha\sin^2\theta}\,\partial\theta.$

Changing the variable by writing $c\sin\theta = a\sin\phi$ we get

$$M = \frac{8a^2\cos\alpha}{c} \int_0^{\pi/2} \frac{\cos^2\phi\,\partial\phi}{1 - k^2\sin^2\phi},$$

where $k^2 = a^2\cos^2\alpha/c^2$.

Therefore, expanding and integrating,

$$M = \frac{4\pi a^2\cos\alpha}{c} \left[\frac{1}{2} + \frac{1\cdot1}{2\cdot4}k^2 + \frac{1\cdot1\cdot3}{2\cdot4\cdot6}k^4 + \dots \right]$$

$$= 4\pi\left[c\sec\alpha - \sqrt{c^2\sec^2\alpha - a^2}\right].$$

If the currents in the straight wire and the circle be i_1 and i_2 respectively, and if they be flowing in the directions OR and STK respectively, the torque G tending to bring the wire and the circle into the same plane is given by

$$G = -\frac{\partial M}{\partial\alpha}\cdot i_1 i_2$$

$$= \frac{4\pi c\sin\alpha}{\cos^2\alpha}\left\{ \frac{c}{(c^2 - a^2\cos^2\alpha)^{1/2}} - 1 \right\} i_1 i_2,$$

when c is greater or equal to a.

When c is less than a,

$$G = \frac{4\pi c}{1 + \sin\alpha}\,i_1 i_2.$$

It also readily follows that if we have parallel wires forming a circuit, the plane of the wires cutting the circle in a diameter which the wires intersect at right angles at points inside the circle, then we have

$$M = 4\pi b \tan(\pi/4 - \alpha/2),$$

and

$$G = \frac{4\pi b i_1 i_2}{1 + \sin \alpha},$$

where b is the distance between the wires.

Both M and G, therefore, are independent of the radius of the circle and the position of its centre.

Self induct-ance of a con-centric main. A concentric main consists of two hollow copper cylinders, one inside and coaxial with the other, and insulated from it. The outer copper conductor is generally protected by a lead sheath which is insulated from it. The alternating current flows in one cylinder and comes back by the other, and *vice versâ*. Hence in practice the cross sectional areas of the two copper cylinders are made equal to one another as each has to carry the same current.

We shall use Kelvin's formula $\Sigma \dfrac{H^2}{8\pi} \partial v$ to calculate the self energy of the main. Suppose that the current I flows along the inner copper cylinder and returns by the concentric outer cylinder. Now the magnetic force produced by a cylindrical current sheet at a point outside is the same as if the current were concentrated at its axis, and hence, since the currents in the inner and outer cylinders are equal and flowing in opposite directions, there will be no magnetic force outside the main. Again by page 54 the outside cylindrical current sheet produces no magnetic force inside, and hence, if H be the magnetic force between the two conductors at a distance x from the axis, then

$$H = \frac{2I}{x}.$$

Suppose that the outer and inner radii of the outer and inner cylinders are b_2, b_1, a_2 and a_1 respectively, and that the current density is uniform over their cross sections. Then if H_2 and H_1

6—2

are the magnetic forces at points in the copper of the outer and inner conductors, we have

$$H_2 = \frac{2I}{x} - \frac{2I}{x}\frac{\pi(x^2 - b_1^2)}{\pi(b_2^2 - b_1^2)} = \frac{2I}{b_2^2 - b_1^2}\left(\frac{b_2^2}{x} - x\right),$$

$$H_1 = \frac{2I}{x}\frac{\pi(x^2 - a_1^2)}{\pi(a_2^2 - a_1^2)} = \frac{2I}{a_2^2 - a_1^2}\left(x - \frac{a_1^2}{x}\right).$$

Hence if L be the self inductance of a length l of the concentric cable,

$$\frac{L}{l} = \frac{2}{I^2}\int_{a_1}^{a_2}\frac{H_1^2}{8\pi}\cdot 2\pi x\partial x + \frac{2}{I^2}\int_{a_2}^{b_1}\frac{H^2}{8\pi}\cdot 2\pi x\partial x + \frac{2}{I^2}\int_{b_1}^{b_2}\frac{H_2^2}{8\pi}\cdot 2\pi x\partial x.$$

Therefore

$$\frac{L}{l} = \frac{2a_1^4}{(a_2^2 - a_1^2)^2}\log_\epsilon\frac{a_2}{a_1} + \frac{1}{2}\frac{a_2^2 - 3a_1^2}{a_2^2 - a_1^2} + 2\log_\epsilon\frac{b_1}{a_2}$$
$$+ \frac{2b_2^4}{(b_2^2 - b_1^2)^2}\log_\epsilon\frac{b_2}{b_1} - \frac{1}{2}\frac{3b_2^2 - b_1^2}{b_2^2 - b_1^2}.$$

Since in practice the cross sectional areas of the cylinders are equal, we have

$$\pi(b_2^2 - b_1^2) = \pi(a_2^2 - a_1^2).$$

Therefore

$$\frac{L}{l} = 2\log\frac{b_1}{a_2} + \frac{2}{(a_2^2 - a_1^2)^2}\left\{a_1^4\log\frac{a_2}{a_1} + b_2^4\log\frac{b_2}{b_1}\right.$$
$$\left. - \frac{1}{2}(a_1^2 + b_2^2)(a_2^2 - a_1^2)\right\}.$$

If the inner cylinder were solid, a_1 would be zero and b_2^2 would equal $b_1^2 + a_2^2$. Using these values in the formula and noting that $a_1^4\log a_1$ is zero when a_1 is zero we get

$$\frac{L}{l} = 2\log\frac{b_1}{a_2} + \left(1 + \frac{b_1^2}{a_2^2}\right)^2\log\left(1 + \frac{a_2^2}{b_1^2}\right) - \frac{b_1^2}{a_2^2} - 1$$
$$= \left(2\log\frac{b_1}{a_2} + \frac{1}{2}\right) + \frac{1}{3}\frac{a_2^2}{b_1^2} - \frac{1}{12}\frac{a_2^4}{b_1^4} + \frac{1}{30}\frac{a_2^6}{b_1^6} - \cdots$$
$$+ \frac{2(-)^{n+1}}{n(n+1)(n+2)}\cdot\frac{a_2^{2n}}{b_1^{2n}} + \cdots.$$

We see that the least possible value of L is when b_1 equals a_2; in this case

$$L = (4\log_\epsilon 2 - 2)l = 0\cdot7726\,l.$$

If L be in henrys and l in miles, the general formula is

$$\frac{L}{l} = 0\cdot000741 \log_{10}\frac{b_1}{a_2} + 0\cdot000161 \left(\frac{1}{2} + \frac{1}{3}\frac{a_2{}^2}{b_1{}^2} - \frac{1}{12}\frac{a_2{}^4}{b_1{}^4} + \ldots\right).$$

For example, suppose that b_1 and a_2 are $0\cdot406$ and $0\cdot192$ inch respectively, then the inductance in henrys per mile is given by

$$L = 0\cdot000241 + 0\cdot000161\,(0\cdot5 + 0\cdot075 - 0\cdot004 + \ldots)$$
$$= 0\cdot000241 + 0\cdot000092$$
$$= 0\cdot000333.$$

The self inductance of two parallel cylindrical wires. Maxwell calculates the self inductance of two parallel wires by finding the value of $\Sigma\,(H^2/8\pi)\,\partial v$ and equating it to $(1/2)\,LI^2$. It is easier however in this case to calculate the self energy of the circuit from the formula $(1/2)\Sigma\phi i$ (page 37), where i denotes the number of tubes of current linked with the flux ϕ.

Consider the case of two parallel hollow cylindrical conductors (Fig. 28) and let a_1, a_2 and b_1, b_2 be the inner and outer radii of the two cylinders respectively. We shall consider the value of $\Sigma\phi i$ for the lines of force due to the current in each conductor separately and then add the results together.

In the cylinder of which the centre is O and radii a_1, a_2 (Fig. 28) we have first to calculate $\Sigma\phi i$ for the lines of force in the substance of the conductor itself, i being the total current embraced by ϕ. We know that the lines of force due to the current in this cylinder are circles and that

$$\phi = \frac{2I}{a_2{}^2 - a_1{}^2}\left(x - \frac{a_1{}^2}{x}\right)\partial x$$

for a tube of thickness ∂x and length unity, just as in the inner conductor of a concentric main. Also inside the cylindrical hollow ϕ is zero. The value of ϕi per unit length in the conductor itself is

$$\frac{2I^2}{(a_2{}^2 - a_1{}^2)^2}\int_{a_1}^{a_2}\frac{(x^2 - a_1{}^2)^2}{x}\partial x,$$

since

$$i = \frac{x^2 - a_1{}^2}{a_2{}^2 - a_1{}^2}I,$$

the current being supposed to be uniform over the cross section.

Hence the value of $\Sigma\phi i$ in the conductor itself is

$$\frac{2a_1{}^4 I^2}{(a_2{}^2 - a_1{}^2)^2} \log\frac{a_2}{a_1} + \frac{I^2}{2}\frac{a_2{}^2 - 3a_1{}^2}{a_2{}^2 - a_1{}^2} \quad\ldots\ldots\ldots\ldots(1).$$

If d be the distance between the axes of the cylinders, then the value of $\Sigma\phi i$ from $x = a_2$ up to $x = d - b_2$ is

$$\int_{a_2}^{d-b_2} \frac{2I}{x}.I\partial x = 2I^2 \log\frac{d - b_2}{a_2} \quad\ldots\ldots\ldots\ldots(2).$$

Now consider the circular lines of force of radius x due to the current in the cylinder a, where x lies between $d - b_2$ and $d + b_2$. They will embrace all the current in the first cylinder and part of the current in the second. As the direction of the current in the second cylinder is opposite to its direction in the first, the part of $\Sigma\phi i$ due to it will be negative. The part of $\Sigma\phi i$ due to values of x between $d - b_2$ and $d + b_2$ and the current in the first cylinder is

$$\int_{d-b_2}^{d+b_2} \frac{2I}{x}.I\partial x = 2I^2 \log\frac{d + b_2}{d - b_2} \quad\ldots\ldots\ldots\ldots(3).$$

To find the linkages with the tubes of current in the second cylinder, divide its section into a series of concentric rings and consider one of them with radius r and thickness ∂r (Fig. 28).

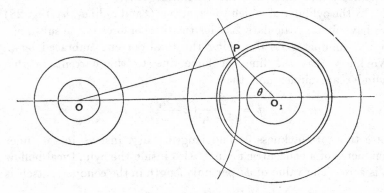

Fig. 28.   The self inductance of two parallel hollow cylinders.

Let $OP = x$, $O_1P = r$ and the angle $PO_1O = \theta$.   Then

$$x^2 = r^2 + d^2 - 2rd \cos\theta,$$

and therefore $x\partial x = rd \sin\theta . \partial\theta$

when r is kept constant. The current i_1 in the cylinder of which the radius is r and thickness ∂r is equal to $\dfrac{2\pi r \partial r I}{\pi (b_2{}^2 - b_1{}^2)}$. Hence the part of $\Sigma \phi i$ due to the current in this cylinder is

$$- \int_{d-r}^{d+r} \frac{\theta}{\pi} \, i_1 \frac{2I}{x} \, \partial x,$$

for $\dfrac{\theta}{\pi} i_1$ is the number of tubes of current in this elementary cylinder intercepted by a circle of radius x. Since

$$x \partial x = rd \sin \theta \,.\, \partial \theta,$$

this may be written

$$- \frac{Ii_1}{\pi} \int_0^\pi \frac{2rd \sin \theta \,.\, \theta \partial \theta}{r^2 + d^2 - 2rd \cos \theta}$$

$$= - \frac{Ii_1}{\pi} \left[\theta \log (r^2 + d^2 - 2rd \cos \theta) \right]_0^\pi$$

$$+ \frac{Ii_1}{\pi} \int_0^\pi \log (r^2 + d^2 - 2rd \cos \theta) \, \partial \theta$$

$$= - 2Ii_1 \log (r + d) + 2Ii_1 \log d$$

$$+ \frac{Ii_1}{\pi} \int_0^\pi \log \left(1 - \frac{2r}{d} \cos \theta + \frac{r^2}{d^2} \right) \partial \theta.$$

Now $\displaystyle \int_0^\pi \log \left(1 - \frac{2r}{d} \cos \theta + \frac{r^2}{d^2} \right) \partial \theta$

$$= \int_0^\pi \left\{ \log \left(1 - \frac{r}{d} \epsilon^{\theta \sqrt{-1}} \right) + \log \left(1 - \frac{r}{d} \epsilon^{- \theta \sqrt{-1}} \right) \right\} \partial \theta$$

$$= - 2 \int_0^\pi \left(\frac{r}{d} \cos \theta + \frac{1}{2} \frac{r^2}{d^2} \cos 2\theta + \frac{1}{3} \frac{r^3}{d^3} \cos 3\theta + \ldots \right) \partial \theta$$

$$= 0.$$

Therefore up to $d + r$ this elementary cylinder contributes

$$- 2Ii_1 \log (d + r) + 2Ii_1 \log d \ldots\ldots\ldots\ldots\ldots\ldots(\alpha)$$

to $\Sigma \phi i$. It also contributes from $d + r$ up to $d + b_2$

$$- \int_{d+r}^{d+b_2} i_1 \frac{2I}{z} \, \partial z = - 2Ii_1 \log (d + b_2) + 2Ii_1 \log (d + r) \ldots (\beta).$$

Hence the total contribution by this elementary cylinder is $(\alpha)+(\beta)$, *i.e.*

$$- 2Ii_1 \log (d + b_2) + 2Ii_1 \log d.$$

Substituting $\dfrac{2r\partial rI}{b_2^2-b_1^2}$ for i_1 and integrating from $r=b_1$ to $r=b_2$ we get

$$- 2I^2 \log (d + b_2) + 2I^2 \log d \dots\dots\dots\dots(4)$$

as the contribution of the second cylinder.

For values of x greater than $d + b_2$, every tube of force will embrace both the currents I and $- I$ respectively, and hence will contribute nothing to $\Sigma\phi i$.

By adding (1), (2), (3) and (4) together we find that $\Sigma\phi i$ per unit length, where ϕ is due to the current in the first cylinder, is

$$2I^2 \log \frac{d}{a_2} + \frac{2a_1^4 I^2}{(a_2^2 - a_1^2)^2} \log \frac{a_2}{a_1} + \frac{I^2}{2} \frac{a_2^2 - 3a_1^2}{a_2^2 - a_1^2}.$$

We can write down from symmetry the value of $\Sigma\phi i$ for the second cylinder; and hence adding the two quantities together and dividing by I^2, we find that

$$\frac{L}{l} = 2 \log \frac{d^2}{a_2 b_2} + \frac{2a_1^4}{(a_2^2 - a_1^2)^2} \log \frac{a_2}{a_1} + \frac{1}{2} \frac{a_2^2 - 3a_1^2}{a_2^2 - a_1^2}$$
$$+ \frac{2b_1^4}{(b_2^2 - b_1^2)^2} \log \frac{b_2}{b_1} + \frac{1}{2} \frac{b_2^2 - 3b_1^2}{b_2^2 - b_1^2}.$$

If we put $h^2 = a_2^2 - a_1^2$ and $k^2 = b_2^2 - b_1^2$, this may be written in the form

$$\frac{L}{l} = 2 \log_\epsilon \frac{d^2}{a_2 b_2} + \frac{1}{3}\left(\frac{h^2}{a_2^2} + \frac{k^2}{b_2^2}\right) + \frac{1}{12}\left(\frac{h^4}{a_2^4} + \frac{k^4}{b_2^4}\right) + \dots$$
$$+ \frac{2}{n(n+1)(n+2)}\left(\frac{h^{2n}}{a_2^{2n}} + \frac{k^{2n}}{b_2^{2n}}\right) + \dots.$$

It has to be remembered that this formula is calculated on the assumption that the current density is uniform over the cross sections of the two conductors.

When the cylinders are solid and have equal radii, then noting that $a_1^4 \log a_1$ is zero when a_1 is zero we have

$$L = l\left\{4 \log_\epsilon \frac{d}{a} + 1\right\},$$

where l is the length and a the radius of each cylinder, and d is the distance between their axes.

In these formulae, if l is in centimetres so also is L. To reduce to henrys we must multiply by 10^{-9}. If l be in statute miles, the following formula will be found useful in practice,

$$L = l \left\{ 0 \cdot 00148 \log_{10} \frac{d}{a} + 0 \cdot 000161 \right\} \text{ henrys.}$$

For example if $d/a = 40$ then the self inductance would be $0 \cdot 00253$ henrys per mile.

The minimum value of L is obtained when $d = 2a$. In this case
$$L = 2 \cdot 7726\, l + l$$
$$= 3 \cdot 7726\, l,$$

where L and l are in centimetres.

We may now consider three mains connected at their far ends, and suppose that a current i_1 flows into No. 1 main, and that currents i_2 and i_3 come back by No. 2 and No. 3 mains respectively, so that

The self inductance of a circuit formed by three equal parallel cylinders the axes of which lie along the edges of an equilateral prism.

$$i_1 = i_2 + i_3.$$

Let the radius of each cylinder be a, let d be the distance between their axes and suppose that the currents are evenly distributed over their cross sections. Forming $\Sigma \phi i$ as in the last problem, and equating it to $\dfrac{L}{l} i_1{}^2$, where l is the length of each cylinder, we get

$$\frac{L}{l} i_1{}^2 = 2 i_1{}^2 \log \frac{d}{a} + \tfrac{1}{2} i_1{}^2$$

$$+ 2 i_2{}^2 \log \frac{d}{a} + \tfrac{1}{2} i_2{}^2$$

$$+ 2 i_3{}^2 \log \frac{d}{a} + \tfrac{1}{2} i_3{}^2,$$

therefore
$$L = l \left(4 \log \frac{d}{a} + 1 \right) \left(1 - \frac{i_2 i_3}{i_1{}^2} \right).$$

We see that L has a maximum value $l \left(4 \log \dfrac{d}{a} + 1 \right)$ when either i_2 or i_3 is zero, and a minimum value $\tfrac{3}{4} l \left(4 \log \dfrac{d}{a} + 1 \right)$ when

$$i_2 = i_3 = \tfrac{1}{2} i_1.$$

Consider the case of three hollow coaxial cylinders of negligible thickness with radii a, b and c respectively, a being that of the smallest. Now if a current i_1 flow in the inner, and return by the two outer cylinders, the currents in which are i_2 and i_3 respectively, and if $L_{1\cdot23}$ denote the self inductance of the system in this case, then

Triple concentric main.

$$\tfrac{1}{2} L_{1\cdot23}\, i_1{}^2 = \frac{l}{8\pi} \left[\int_a^b \frac{4 i_1{}^2}{x^2} . 2\pi x \partial x + \int_b^c \frac{4 (i_1 - i_2)^2}{x^2} . 2\pi x \partial x \right]$$

$$= i_1{}^2 l \log \frac{b}{a} + l\,(i_1 - i_2)^2 \log \frac{c}{b}.$$

Also $i_1 = i_2 + i_3.$

Therefore $L_{1\cdot23} = 2l \log \dfrac{b}{a} + 2l \left(\dfrac{i_3}{i_1}\right)^2 \log \dfrac{c}{b}.$

Similarly $L_{2\cdot31} = 2l \left(\dfrac{i_1}{i_2}\right)^2 \log \dfrac{b}{a} + 2l \left(\dfrac{i_3}{i_2}\right)^2 \log \dfrac{c}{b},$

and $L_{3\cdot12} = 2l \left(\dfrac{i_1}{i_3}\right)^2 \log \dfrac{b}{a} + 2l \log \dfrac{c}{b}.$

The above formulae could also be calculated as follows:

$$L_{1\cdot23}\, i_1{}^2 = \Sigma \phi i$$

$$= l i_1 \int_a^b \frac{2 i_1}{x} \partial x + l\,(i_1 - i_2) \int_b^c \frac{2 i_1}{x} \partial x - l\,(i_1 - i_2) \int_b^c \frac{2 i_2}{x} \partial x.$$

Thus $L_{1\cdot23} = 2l \log \dfrac{b}{a} + 2l \left(\dfrac{i_3}{i_1}\right)^2 \log \dfrac{c}{b},$

which is the same result as before.

In general testing work it is often important to arrange a circuit so that its self inductance may be as small as possible. The formula $(1/I^2)\,\Sigma \phi i$ for the self inductance shows us that we have to make $\Sigma \phi i$ a minimum. If the flux ϕ contained in a tube of force embrace both the outgoing and return currents, then ϕi for this tube will be zero. Hence in order to make the self inductance as small as possible, it is necessary that the outgoing and return currents should be very close together. When we consider also that the tubular filaments of current in the conductors themselves add an appreciable amount to the total sum of ϕi, we see that the conductors should be flat

Minimum self inductance.

strips of metal separated from one another by the thinnest possible insulating material. In this case however the circuit possesses considerable electrostatic capacity.

From the equivalence of current circuits to magnetic shells, we see that two straight parallel wires in which currents flow in opposite directions repel one another. Let F be the force of repulsion per length l of one of the wires, when they are at a distance x apart and carry currents i and $-i$ respectively. If the wires are cylindrical, then, as shown in Chapter I, the magnetic force due to the current in one of them is the same as if the current were concentrated along the axis. It is easy to show in a similar way that the resultant mechanical force on the other wire is the same as if the current in this wire were concentrated along its axis. Now the magnetic force at a point on the axis of the second wire due to the current in the first is perpendicular to the plane of the two wires. Therefore (p. 42) the mechanical force on length l of the second wire is a repulsion F in the plane of the two wires and is perpendicular to them, and since θ is $= \pi/2$, it is given by

The repulsive force between two parallel wires carrying equal currents in opposite directions.

$$F = (2i/x)\, il = 2li^2/x.$$

This formula was given by Ampère.

The same result can be deduced from the formula for the self inductance of two parallel wires forming part of one circuit. For suppose that we keep the current i constant while x becomes $x + \partial x$, then

the rate of working of the E.M.F. per unit length $= i\dfrac{\partial}{\partial t}(Li) = i^2\dfrac{\partial L}{\partial t}$,

the rate of working against the external mechanical forces $= F\dfrac{\partial x}{\partial t}$,

and the rate of increase of the magnetic energy of the system is

$$\frac{\partial}{\partial t}(\tfrac{1}{2}Li^2),$$

and therefore $\qquad i^2\dfrac{\partial L}{\partial t} = F\dfrac{\partial x}{\partial t} + \dfrac{\partial}{\partial t}(\tfrac{1}{2}Li^2),$

and hence $\quad F = \tfrac{1}{2}i^2\dfrac{\partial L}{\partial x} = \tfrac{1}{2}i^2\dfrac{\partial}{\partial x}\left(4l\log\dfrac{x}{a} + l\right) = \dfrac{2li^2}{x}.$

If the two wires form part of a circuit, there will also be a tension along each of them due to the action of the magnetic field on the conductors which must close the circuit. If $\frac{1}{2}T$ denote the tension along each conductor, then, proceeding as above, we get

$$T = \tfrac{1}{2} i^2 \frac{\partial L}{\partial l} = i^2 \left(2 \log \frac{x}{a} + \tfrac{1}{2} \right),$$

where a is the radius of either wire and x the distance between them.

For example, suppose that i is 1000 amperes or 100 C.G.S. units, and that $\dfrac{x}{a}$ is 10, then

$$T = 100^2 \, (2 \log_e 10 + 0\cdot5)$$
$$= 100^2 \times 5\cdot105$$
$$= 51,050 \ \text{dynes}$$
$$= 52 \ \text{grammes weight nearly.}$$

Hence the pull along each conductor will be only equal to the weight of 26 grammes.

If the wires are surrounded by a magnetic medium, the forces will be much greater.

REFERENCES

CLERK MAXWELL, *Electricity and Magnetism*, Vol. 2.
LORD RAYLEIGH, *Scientific Papers*.
E. B. ROSA AND F. W. GROVER, *Bulletin of the Bureau of Standards*, Vol. 8, 1912.

CHAPTER III

The inductance of circular and helical currents. Mathematical formulae. Series for computing the elliptic functions. Force due to a circular current. Transforming elliptic integrals. The mutual inductance of two circles. The attraction between two circular currents. Coplanar circles. Inductance of a circular current. Maximum inductance. Rayleigh's formula. Maximum inductance. Weinstein's formula. Numerical examples. The electrostatic capacity of a circular ring. The mutual inductance of two coaxial coils. The self inductance of a single layer cylindrical coil. The coil of maximum inductance. Mutual inductance from self inductance. The kinetic energy of electrons. Table of the elliptic integrals K and E. Table of $\log_{10} K$ and $\log_{10} E$ for large values of k. Table of Neperian logarithms. References.

IT is convenient to collect together for reference the mathematical
Mathematical formulae. definitions and theorems which we shall require in proving the formulae in this chapter. The definitions of the complete elliptic integrals, which we have already met with in Chapter I (pp. 47 and 48), are

$$E = \int_0^{\pi/2} (1 - k^2 \sin^2 \phi)^{1/2} \, \partial\phi, \text{ and } K = \int_0^{\pi/2} \frac{\partial\phi}{(1 - k^2 \sin^2 \phi)^{1/2}}.$$

The latter function K is sometimes denoted by F. The quantity k is called the modulus of the functions. The values of E and K are tabulated at the end of this chapter.

By performing easy algebraical operations and by differentiating E and K, the following theorems can be readily proved:

$$\frac{\partial E}{\partial k} = -\frac{1}{k}(K - E) \quad\quad\quad\quad\quad (1),$$

$$\frac{\partial K}{\partial k} = \frac{1}{k k_0^2}(E - k_0^2 K) \quad\quad\quad\quad (2),$$

where $k_0^2 = 1 - k^2$, and k_0 is called the complementary modulus.

Denoting $(1 - k^2 \sin^2 \phi)^{1/2}$ by Δ, we also get

$$\int_0^{\pi/2} \frac{\sin^2 \phi}{\Delta} \partial\phi = \frac{1}{k^2}(K - E) \quad\ldots\ldots\ldots\ldots\ldots\ldots(3),$$

$$\int_0^{\pi/2} \frac{\cos 2\phi}{\Delta} \partial\phi = \frac{2}{k^2}(E - K) + K \quad\ldots\ldots\ldots\ldots(4),$$

$$\int_0^{\pi/2} \frac{\sin^2 \phi \cos^2 \phi}{\Delta} \partial\phi = \frac{2 - k^2}{3k^4} E - \frac{2 - 2k^2}{3k^4} K \quad\ldots\ldots(5),$$

$$\int_0^{\pi/2} \frac{\partial\phi}{\Delta^3} = \frac{E}{1 - k^2} \quad\ldots\ldots\ldots\ldots\ldots\ldots(6),$$

$$\int_0^{\pi/2} \frac{\sin^2 \phi}{\Delta^3} \partial\phi = \frac{E}{k^2(1 - k^2)} - \frac{K}{k^2} \quad\ldots\ldots\ldots\ldots(7),$$

$$\int_0^{\pi/2} \sin^2 \phi \Delta \partial\phi = \frac{2k^2 - 1}{3k^2} E + \frac{1 - k^2}{3k^2} K \quad\ldots\ldots(8).$$

For example, to prove (1), we notice that

$$\frac{\partial E}{\partial k} = -\int_0^{\pi/2} \frac{k \sin^2 \phi \partial\phi}{(1 - k^2 \sin^2 \phi)^{3/2}}$$

$$= \frac{1}{k}\int_0^{\pi/2} (1 - k^2 \sin^2 \phi)^{1/2} \partial\phi - \frac{1}{k}\int_0^{\pi/2} \frac{\partial\phi}{(1 - k^2 \sin^2 \phi)^{1/2}}$$

$$= -\frac{1}{k}(K - E).$$

We may prove (6) by expanding the integral on the left hand side in a series proceeding by powers of k^2. Then multiplying the series for E given in (9), below, by $1/(1 - k^2)$, that is, by $1 + k^2 + k^4 + \ldots$, it is easy to see that the two series are identical.

To prove (2), we have

$$\frac{\partial K}{\partial k} = \frac{1}{k}\int_0^{\pi/2} \frac{(k^2 \sin^2 \phi - 1) + 1}{(1 - k^2 \sin^2 \phi)^{3/2}} \partial\phi.$$

Hence, by means of (6), (2) follows.

Expanding by the binomial theorem and integrating we obtain at once

Series for computing the elliptic functions.

$$E = \frac{\pi}{2}\left\{1 - \left(\frac{1}{2}\right)^2 k^2 - \left(\frac{1.3}{2.4}\right)^2 \frac{k^4}{3}\right.$$
$$\left. - \left(\frac{1.3.5}{2.4.6}\right)^2 \frac{k^6}{5} - \ldots\right\}\ldots(9),$$

and $\quad K = \frac{\pi}{2}\left\{1 + \left(\frac{1}{2}\right)^2 k^2 + \left(\frac{1.3}{2.4}\right)^2 k^4 + \left(\frac{1.3.5}{2.4.6}\right)^2 k^6 + \ldots\right\}\ldots(10).$

Hence, when k is small, the numerical values of E and K can be readily found.

When however the value of k is nearly unity the computation of the values of E and K by the above formulae would be very laborious. When k is unity we see at once by integrating directly that the value of E is unity and the value of K is infinitely great. We shall now find series by means of which K and E can be very readily computed when k is nearly equal to unity.

We have always

$$K = \int_0^{\pi/2 - \epsilon} \frac{\partial \phi}{(\cos^2 \phi + k_0^2 \sin^2 \phi)^{1/2}} + \int_{\pi/2 - \epsilon}^{\pi/2} \frac{\partial \phi}{(\cos^2 \phi + k_0^2 \sin^2 \phi)^{1/2}} \quad (11),$$

where $k_0^2 = 1 - k^2$.

Now let us suppose that ϵ is a very small quantity but that it is very great compared with k_0.

Writing $\pi/2 - x$ for ϕ in the second integral it becomes

$$\int_0^\epsilon \frac{\partial x}{(\sin^2 x + k_0^2 \cos^2 x)^{1/2}}.$$

Now since x in this integral cannot be greater than ϵ and ϵ is very small, we can write $\sin x = x$ and $\cos x = 1$, without appreciable error. Hence this integral becomes

$$\int_0^\epsilon \frac{\partial x}{(k_0^2 + x^2)^{1/2}} = \left[\log \{ x + (k_0^2 + x^2)^{1/2} \} \right]_0^\epsilon$$
$$= \log (2\epsilon/k_0) \text{ approximately.}$$

In the first integral of (11) we can neglect $k_0^2 \sin^2 \phi$ in comparison with $\cos^2 \phi$, and thus

$$\int_0^{\pi/2 - \epsilon} \frac{\partial \phi}{\cos \phi} = \left[-\log \left\{ \tan \left(\frac{\pi}{4} - \frac{\phi}{2} \right) \right\} \right]_0^{\pi/2 - \epsilon}$$
$$= -\log \tan (\epsilon/2) = \log (2/\epsilon) \text{ approximately,}$$

since $\tan (\epsilon/2) = \epsilon/2$, when ϵ is small.

Hence when k is nearly equal to unity,

$$K = \log (2\epsilon/k_0) + \log (2/\epsilon) = \log (4/k_0) \text{ approximately.}$$

We shall next find an approximate formula for E when k is nearly unity. From (1) we get

$$k \frac{\partial E}{\partial k} = E - K \quad \dots\dots\dots\dots\dots\dots\dots\dots(12).$$

Hence $$k \frac{\partial E}{\partial k} = E - \log (4/k_0) \text{ approximately.}$$

Therefore, since $k^2 = 1 - k_0^2$, we get with the help of (2)

$$\frac{\partial E}{\partial k_0} = k_0 \log (4/k_0) - k_0,$$

and thus $E = A + (k_0^2/2) \{\log (4/k_0) - 1/2\},$

where A is a constant. But when $k_0 = 0$, $E = 1$, and thus $A = 1$.

Substituting this value of E in (2) and integrating we get K to a second approximation. Substituting these new values of E and K on the right hand side of (12) and integrating we get the value of E to a third approximation. Proceeding in this way we get finally the following series for E and K, which are very useful for computing purposes when k is nearly unity :

$$E = 1 + \tfrac{1}{2}k_0^2 \left(\log \frac{4}{k_0} - \frac{1}{1 \cdot 2}\right)$$

$$+ \frac{1^2 \cdot 3}{2^2 \cdot 4} k_0^4 \left(\log \frac{4}{k_0} - \frac{2}{1 \cdot 2} - \frac{1}{3 \cdot 4}\right)$$

$$+ \frac{1^2 \cdot 3^2 \cdot 5}{2^2 \cdot 4^2 \cdot 6} k_0^6 \left(\log \frac{4}{k_0} - \frac{2}{1 \cdot 2} - \frac{2}{3 \cdot 4} - \frac{1}{5 \cdot 6}\right) + \ldots(13),$$

and

$$K = \log \frac{4}{k_0} + \frac{1^2}{2^2} k_0^2 \left(\log \frac{4}{k_0} - \frac{2}{1 \cdot 2}\right)$$

$$+ \frac{1^2 \cdot 3^2}{2^2 \cdot 4^2} k_0^4 \left(\log \frac{4}{k_0} - \frac{2}{1 \cdot 2} - \frac{2}{3 \cdot 4}\right) + \ldots(14).$$

In all the formulae given above, the logarithms are Neperian.

In Fig. 29 BR is the circular conducting filament and OZ' is

Force due to a circular current.
its axis. It is obvious from symmetry that the lines of force will all be in planes passing through OZ'. We shall first find the component forces Z and X, parallel to OZ' and OX' respectively, at the point P in the plane $Z'OX'$. Let the radius of the circular filament be a, let PO be r, and let the angles POX' and $X'OR$ be θ and ϕ respectively. We shall consider the force at P due to the current in an element $a\partial\phi$ of the filament at R. Let RS be the tangent to the circle at R, and draw PK and KL perpendicular to OX' and RS respectively. Then, noticing that $KL = a - r \cos \theta \cos \phi$, and that

$$\cos POR = \cos \theta \cos \phi, \quad \text{so that} \quad PR^2 = a^2 + r^2 - 2ar \cos \theta \cos \phi,$$

we get by Laplace's formula,

$$\partial Z = \frac{ia\partial\phi}{PR^2}\sin PRL \cdot \sin KPL$$

$$= \frac{ia\partial\phi}{PR^2} \cdot \frac{PL}{PR} \cdot \frac{KL}{PL}$$

$$= \frac{ia\,(a - r\cos\theta\cos\phi)\,\partial\phi}{(a^2 + r^2 - 2ar\cos\theta\cos\phi)^{3/2}},$$

and hence
$$Z = 2ia \int_0^\pi \frac{(a - r\cos\theta\cos\phi)\,\partial\phi}{(a^2 + r^2 - 2ar\cos\theta\cos\phi)^{3/2}} \quad \dots\dots\dots(15).$$

Similarly we find that
$$X = 2iar \int_0^\pi \frac{\sin\theta\cos\phi\,\partial\phi}{(a^2 + r^2 - 2ar\cos\theta\cos\phi)^{3/2}} \quad \dots\dots(16).$$

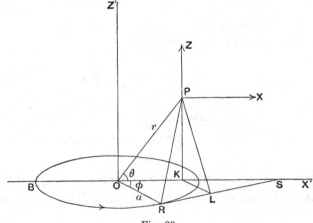

Fig. 29.

Now (15) may be written

$$Z = i \int_0^\pi \frac{\partial\phi}{(a^2 + r^2 - 2ar\cos\theta\cos\phi)^{1/2}}$$

$$+ i\,(a^2 - r^2) \int_0^\pi \frac{\partial\phi}{(a^2 + r^2 - 2ar\cos\theta\cos\phi)^{3/2}}.$$

Hence, writing $\phi = \pi - 2\phi'$, we get by (6) after a little reduction,

$$Z = \frac{2i}{r_1}\left\{\frac{a^2 - r^2}{r_2{}^2}E + K\right\} \dots\dots\dots\dots\dots\dots(17),$$

and
$$X = \frac{2i\tan\theta}{r_1}\left\{\frac{a^2 + r^2}{r_2{}^2}E - K\right\} \dots\dots\dots\dots(18),$$

R. I.   7

where $r_1^2 = a^2 + r^2 + 2ar \cos \theta$ and $r_2^2 = a^2 + r^2 - 2ar \cos \theta$, and the modulus k of the elliptic functions is given by $k^2 = 1 - r_2^2/r_1^2$.

If R and T be the component magnetic forces along and perpendicular to OP (Fig. 29) we have

$$R = Z \sin \theta + X \cos \theta = \frac{4ia^2 \sin \theta}{r_1 r_2^2} E \qquad \dots\dots\dots\dots\dots(19),$$

and

$$T = Z \cos \theta - X \sin \theta = \frac{2i}{r_1 r_2^2} \left\{ \frac{a^2 \cos 2\theta - r^2}{\cos \theta} E + \frac{r_2^2}{\cos \theta} K \right\} \dots(20).$$

In rectangular coordinates, we have

$$X = \frac{2i}{r_1} \left\{ \frac{2az}{r_2^2} E - \frac{z}{x}(K - E) \right\} \qquad \dots\dots\dots\dots(21),$$

$$Z = \frac{2i}{r_1} \left\{ \frac{2a(a-x)}{r_2^2} E + (K - E) \right\} \qquad \dots\dots\dots(22),$$

where $r_1^2 = (a+x)^2 + z^2, \quad r_2^2 = (a-x)^2 + z^2$

and $k^2 = 1 - r_2^2/r_1^2 = 4ax/r_1^2.$

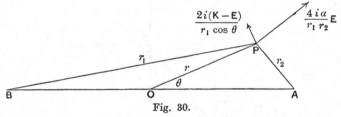

Fig. 30.

If ϕ be the angle APB (Fig. 30), $OP = r$, the angle $POA = \theta$, and $OA = OB = a$, we have $AP = r_2$ and $PB = r_1$. Hence in bipolar coordinates

$$Z = -\frac{2i}{r_2} E \cos \phi + \frac{2i}{r_1} K \qquad \dots\dots\dots\dots\dots(23),$$

and $$X = \frac{2i}{r_2} \cdot \frac{r_1^2 + r_2^2}{r_1^2 - r_2^2} E \sin \phi - \frac{2i}{r_1} \cdot \frac{2r_1 r_2}{r_1^2 - r_2^2} K \sin \phi \dots\dots(24),$$

where $\cos \phi = (r_1^2 + r_2^2 - 4a^2)/(2r_1 r_2)$ and $k^2 = 1 - r_2^2/r_1^2.$

A simple and useful way of considering the force at any point P due to a circular current is to consider that it is the resultant (Fig. 30) of two forces at right angles to AP and OP respectively. The component at right angles to AP equals $(4ia/r_1 r_2) E$, and the component at right angles to OP equals $2i(K - E)/(r_1 \cos \theta)$, the modulus of E and K being $(1 - r_2^2/r_1^2)^{1/2}$.

Let us now find the force at any point in the plane of the circular current and inside the circle. In this case $X = 0$, and by (23)

$$Z = \frac{2i}{r_2} E + \frac{2i}{r_1} K$$

$$= \frac{2i}{a-r} E + \frac{2i}{a+r} K \quad \dots\dots\dots\dots(25),$$

where $k = 2 (ar)^{1/2}/(a + r)$. The same formula holds at points outside the circle, the first term being negative when r is greater than a.

Transforming elliptic functions. The formulae given above for the magnetic force at points in the plane of a circular current enable us to prove the Landen-Legendre formulae for transforming the modulus of elliptic functions. Comparing the formula on p. 47 of Chapter I with (25) we see that, when r is less than a,

$$\frac{4a}{a^2 - r^2} E_1 = \frac{2}{a-r} E + \frac{2}{a+r} K,$$

the modulus k_1 of E_1 being r/a and the modulus k of E and K being $2 (ar)^{1/2}/(a + r)$, which equals $2\sqrt{k_1}/(1 + k_1)$. Hence we see that

$$E_1 = \frac{1 + k_1}{2} E + \frac{1 - k_1}{2} K.$$

Similarly by comparing the formula on p. 48 of Chapter I with (25) we get

$$(1 - k_1^2) K_1 - E_1 = -\frac{1 + k_1}{2} E + \frac{1 - k_1}{2} K.$$

Hence

$$K = (1 + k_1) K_1 \quad \dots\dots\dots\dots\dots\dots(26),$$

and

$$E = \frac{2}{1 + k_1} E_1 - (1 - k_1) K_1 \quad \dots\dots\dots\dots(27),$$

where

$$k_1 = \left(\frac{k}{1 + \sqrt{1 - k^2}}\right)^2.$$

It is to be noticed that k_1 is less than k, and thus by means of (26) and (27) we can substitute for elliptic integrals other elliptic integrals having smaller arguments. As a rule the transformed integrals are easier to compute, and the transformation sometimes simplifies the formulae.

We shall suppose that we have two circular currents of radii

Mutual in-
ductance of
two circles. a and b, the line joining their centres being perpendicular to both and of length c.   The magnetic force X at a point on the circumference of the circle of radius b parallel to the planes of the circles and due to unit current in the other circle is given by

$$X = 2a \int_0^\pi \frac{c \cos \phi \partial \phi}{(a^2 + b^2 + c^2 - 2ab \cos \phi)^{3/2}}.$$

This follows from (16) since $c = r \sin \theta$, and $b = r \cos \theta$.   If we suppose that c becomes $c + \partial c$ then the number of lines of induction linked with the second circle will be diminished by $X \cdot 2\pi b \partial c$. Hence if M be the mutual inductance

$$\frac{\partial M}{\partial c} = -4\pi ab \int_0^\pi \frac{c \cos \phi \partial \phi}{(a^2 + b^2 + c^2 - 2ab \cos \phi)^{3/2}}.$$

Thus integrating and noticing that M vanishes when c is infinite we get

$$M = \int_0^\pi \frac{4\pi ab \cos \phi \partial \phi}{(a^2 + b^2 + c^2 - 2ab \cos \phi)^{1/2}} \quad \ldots\ldots\ldots\ldots(28).$$

If we have another pair of coaxial circles the corresponding radii of which are given by $a_1 = an$, $b_1 = b/n$, and if the distance c_1 between their planes is given by

$$c_1^2 = (1 - n^2)(a^2 - b^2/n^2) + c^2,$$

we see by substituting in (28) that this pair of circles has the same mutual inductance as the first pair.   Since c_1^2 cannot be less than zero we see that all the possible values of n must lie between

$$\{(a + b)^2 + c^2\}^{1/2}/(2a) + \{(a - b)^2 + c^2\}^{1/2}/(2a)$$

and $\qquad \{(a + b)^2 + c^2\}^{1/2}/(2a) - \{(a - b)^2 + c^2\}^{1/2}/(2a).$

Since these numbers are always real there are an infinite number of pairs of coaxial circles all of which have the same M.

If we put $\phi = \pi - 2\theta$ in (28) we get

$$M = -\frac{8\pi ab}{r_1} \int_0^{\pi/2} \frac{\cos 2\theta}{\Delta} \partial \theta$$

and hence, by (4),

$$M = 4\pi \sqrt{ab} \{(2/k - k) K - (2/k) E\} \quad \ldots\ldots\ldots(29),$$

where $r_1^2 = (a + b)^2 + c^2$; $k^2 = 4ab/r_1^2$ and $\Delta^2 = 1 - k^2 \sin^2 \theta.$

If we transform (29) by means of (26) and (27) we get

$$M = \frac{8\pi\sqrt{ab}}{\sqrt{k_1}}\,(K_1 - E_1)\quad\dots\dots\dots\dots(30),$$

where $k_1 = \frac{r_1 - r_2}{r_1 + r_2}$; $r_1{}^2 = (a+b)^2 + c^2$; $r_2{}^2 = (a-b)^2 + c^2$.

Transforming, once more, we get

$$M = 4\pi(\sqrt{r_1} + \sqrt{r_2})^2\,(K_2 + k_2 K_2 - E_2)\quad\dots\dots(31),$$

where $k_2 = \{(\sqrt{r_1} - \sqrt{r_2})/(\sqrt{r_1} + \sqrt{r_2})\}^2$.

By means of formulae (9) and (10) we deduce from (30) and (31) respectively the following formulae for M:

$$M = 2\pi^2 k_1{}^{3/2}\sqrt{ab}\left\{1 + \frac{3}{8}k_1{}^2 + \frac{15}{64}k_1{}^4 + \frac{175}{1024}k_1{}^6 + \dots\right\}\dots(32),$$

and

$$M = 2\pi^2(\sqrt{r_1} - \sqrt{r_2})^2\left\{1 + \frac{1}{2}k_2 + \frac{1}{4}k_2{}^2 + \frac{3}{16}k_2{}^3 + \frac{9}{64}k_2{}^4 + \frac{15}{128}k_2{}^5\right.$$
$$\left. + \frac{25}{256}k_2{}^6 + \frac{175}{2048}k_2{}^7 + \dots\right\}\dots(33).$$

For example, if the circles be concentric and coplanar and their radii be 37 and 12 cms. long respectively, we get

$$r_1{}^2 = (a+b)^2 + c^2 = 49^2,$$

and

$$r_2{}^2 = (a-b)^2 + c^2 = 25^2.$$

Hence $k = \left(1 - \frac{r_2{}^2}{r_1{}^2}\right)^{\frac{1}{2}} = 0\cdot860,$

$$k_1 = \frac{r_1 - r_2}{r_1 + r_2} = \frac{12}{37} = 0\cdot324\dots,$$

and $k_2 = \left(\frac{\sqrt{r_1} - \sqrt{r_2}}{\sqrt{r_1} + \sqrt{r_2}}\right)^2 = \frac{1}{36} = 0\cdot028\dots.$

If we use (33) we get

$$M = 8\pi^2\{1 + 0\cdot0138889 + 0\cdot0001929$$
$$+ 0\cdot0000040 + 0\cdot0000001\dots\}$$
$$= 8\cdot1126872\pi^2$$
$$= 80\cdot06902 \text{ cms. approximately.}$$

When k is not greater than $0\cdot8$ the following approximate formula will be found useful:

$$M = \frac{4\pi^2 (r_1 - r_2)^2}{r_1 + r_2 + 6\sqrt{r_1 r_2}} \dots\dots\dots\dots\dots(34).$$

For example, when $a = 37$, $b = 12$, $c = 0$, we have

$$M = 8\cdot11268\pi^2.$$

The error is therefore less than 1 in 800,000.

When k is nearly unity we can use the following formula obtained from (29) by means of series (13) and (14):

$$M = 4\pi \sqrt{ab} \left\{ \left(1 + \frac{3}{4} k_0^2 + \frac{33}{64} k_0^4 + \frac{107}{256} k_0^6 + \dots \right) \left(\log \frac{4}{k_0} - 1 \right) \right.$$
$$\left. - \left(1 + \frac{15}{128} k_0^4 + \frac{185}{1536} k_0^6 + \dots \right) \right\} \dots(35),$$

where $k_0^2 = 1 - k^2 = r_2^2 / r_1^2$.

For example, when $a = 25$, $b = 20$, $c = 10$; $k_0^2 = 1/17$, and we find that $M = 248\cdot79$ cms. approximately. When $a = b = 25$ and $c = 1$, we find that $k_0^2 = 1/2501$. Hence we find $M = 1036\cdot7$ cms. approximately.

If the circular currents be coaxial and their radii be a and b The attraction between circular currents. respectively, the attractive force F between them when the currents flow in the same sense and have magnitudes i_1 and i_2 respectively is given by

$$F = - i_1 i_2 \frac{\partial M}{\partial c},$$

where c is the distance between their centres. Hence

$$F = 4\pi abc\, i_1 i_2 \int_0^\pi \frac{\cos\phi\, \partial\phi}{(a^2 + b^2 + c^2 - 2ab\cos\phi)^{3/2}} \dots\dots(36).$$

By means of (6) and (7) this may be written

$$F = \frac{8\pi abc\, i_1 i_2}{\{(a + b)^2 + c^2\}^{3/2}} \left[\frac{E}{1 - k^2} - \frac{2}{k^2} (K - E) \right] \dots\dots(37),$$

where $k^2 = 4ab / \{(a + b)^2 + c^2\}$.

When k^2 is small we get by (9) and (10)

$$F = \frac{6 (\pi a^2)\, (\pi b^2)\, c i_1 i_2}{\{(a + b)^2 + c^2\}^{5/2}} \left[1 + \frac{5}{4} k^2 + \dots \right] \dots\dots\dots(38).$$

When c is very great compared with $a + b$, this becomes

$$F = \frac{6\,(\pi a^2)\,(\pi b^2)\,i_1 i_2}{c^4} \text{ approximately } \quad \ldots\ldots\ldots(39).$$

When k^2 is nearly unity we get, by (13) and (14),

$$F = \frac{8\pi abc\,i_1 i_2}{\{(a + b)^2 + c^2\}^{1/2}\{(a - b)^2 + c^2\}} \quad \ldots\ldots\ldots\ldots(40),$$

approximately. If a and b remain constant F attains its maximum value when

$$4c^2 = \{9\,(a^4 + b^4) + 4ab\,(a^2 + b^2) - 10a^2b^2\}^{1/2} - (a + b)^2.$$

This solution applies when a and b are nearly equal to one another and c is very small compared with $a + b$.

We shall suppose that the smaller circle (Fig. 31) passes

Coplanar circles. through O, the centre of the larger one, and is inside it. Let a be the radius of the larger circle and d the diameter of the smaller one. Then if Z denote the magnetic force at the point $P\,(r, \theta)$ due to unit current in the larger circle and M the mutual inductance between the circles, we have

$$M = 2\int_0^{\pi/2}\int_0^{d\cos\theta} Zr\,\partial\theta\,\partial r,$$

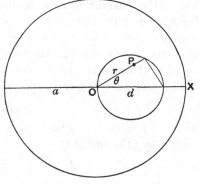

Fig. 31.

and hence, noticing that (p. 47)

$$Z = \frac{2\pi}{a}\left[1 + 3\left(\frac{1}{2}\right)^2\frac{r^2}{a^2} + 5\left(\frac{1.3}{2.4}\right)^2\frac{r^4}{a^4} + \ldots\right],$$

we easily find that

$$M = \frac{2\pi}{a}\cdot\frac{\pi d^2}{4}\left[1 + \frac{2}{3}\left(\frac{3}{4}\right)^3\left(\frac{d}{a}\right)^2 + \frac{3}{5}\left(\frac{3.5}{4.6}\right)^3\left(\frac{d}{a}\right)^4 + \ldots\right.$$

$$\left. + \frac{n}{2n-1}\left(\frac{3.5\ldots 2n-1}{4.6\ldots 2n}\right)^3\left(\frac{d}{a}\right)^{2n-2} + \ldots\right]\ldots(41).$$

If the circles are concentric

$$M = \frac{2\pi}{a} \cdot \frac{\pi d^2}{4} \left[1 + \frac{2}{3} \left(\frac{3}{4}\right)^2 \left(\frac{d}{2a}\right)^2 + \ldots \right.$$
$$\left. + \frac{n+1}{2n+1} \left(\frac{3 \cdot 5 \ldots 2n+1}{4 \cdot 6 \ldots 2n+2}\right)^2 \left(\frac{d}{2a}\right)^{2n} + \ldots \right] \ldots(42).$$

Let us now consider the self inductance of a circular current.
The inductance of a circular current. We shall suppose that the radius of the ring is a and that the radius of the circular cross section of the ring is r. In order to simplify the problem we shall suppose that r/a is very small and that the current is uniformly distributed over the cross section. The construction for the magnetic force given in Fig. 30 shows that the lines of force near a circular filament are very approximately circles with their centres on the circular axis of the filament. The field is therefore similar to that round a straight cylindrical conducting tube in which the current flow is parallel to the axis. The number of linkages due to unit current inside the wire is therefore approximately equal to $2\pi a/2$, i.e. to πa (p. 77). The total linkages of the flux inside the aperture with the current is approximately equal to the mutual inductance between two circular filaments of radii a and $a - r$ respectively. Hence, using (30), we get

$$L = 8\pi a \, (K - E) + \pi a,$$

where the modulus is $(a - r)/a$. Thus finally, since r/a is small, we get by (13) and (14)

$$L = 4\pi a \left\{ \log \frac{8a}{r} - 1 \cdot 75 \right\} \ldots\ldots\ldots\ldots\ldots(43),$$

approximately.

This formula can also be written in the form

$$L = 28 \cdot 935 a \log_{10} \{1 \cdot 3902 \, (a/r)\} \ldots\ldots\ldots\ldots(44).$$

We notice that the smaller the value of r the greater the value of L. In the limiting case when r is infinitely small, L is infinitely great. But in this case the electric resistance of the wire would be infinitely great also, and so if a finite electromotive force was applied to the circuit the current i produced and the electromagnetic energy $Li^2/2$ stored up in the field would both be infinitely small.

If a circular coil of circular cross section be made up of thin **Maximum inductance.** insulated wire and if there be N complete turns of wire, then

$$L = 4\pi N^2 a \{\log (8a/r) - 1\cdot 75\} \text{ approximately.}$$

If l be the length of the wire, $l = 2\pi a N$, and if n be the number of wires per unit area of cross section, $N = n \cdot \pi r^2$. For a given length of wire, l is approximately constant, and for a given thickness n is very approximately constant. The formula for L may be written

$$L = 2Nl [\log \{4l (n/\pi)^{1/2}\} - (3/2) \log N - 1\cdot 75].$$

If we regard N as a variable and l and n as constants, L has a maximum value when $\partial L/\partial N$ is zero. In this case $\log (8a/r) = 13/4$ and hence $a = 3\cdot 22r$ approximately, and

$$L_{\text{max.}} = 6\pi N^2 a = 3Nl = 0\cdot 148 l^2/r = 0\cdot 59 n^{1/3} l^{5/3}.$$

The method we used to prove (43) however, does not warrant the assumption that the formula is accurate when r is as great as $a/(3\cdot 22)$. We shall therefore give a more accurate value for L.

Let us take the origin O at the centre of a cross section of the **Rayleigh's formula.** ring and divide the section into elementary areas $\rho_1 \partial \theta_1 \partial \rho_1$. We shall also suppose that the current density $i/\pi r^2$ is uniform over the cross section so that the current in the elementary ring whose cross sectional area is $\rho_1 \partial \theta_1 \partial \rho_1$ is $(i/\pi r^2) \rho_1 \partial \theta_1 \partial \rho_1$. We see that the linkages of the flux due to this current filament with the current filament whose cross sectional area is $\rho \partial \theta \partial \rho$ equals $M (i/\pi r^2)^2 \rho_1 \rho \partial \theta_1 \partial \rho_1 \partial \theta \partial \rho$. Making the assumption that all the circular filaments are close together so that we can use the simple approximate formula for M suitable for this case we integrate over the cross section keeping ρ_1 and θ_1 constant. We finally integrate this expression with respect to ρ_1 and θ_1 and so get the total linkages of flux and current. Equating this to $Li^2/2$ we get

$$L = 4\pi a \left[\log \frac{8a}{r} - 1\cdot 75 + \frac{r^2}{8a^2} \left(\log \frac{8a}{r} + \frac{1}{3} \right) \right] \quad \ldots \ldots (45).$$

This formula, which is due to Rayleigh, gives us the value of L to a second approximation. A complete proof of it will be found in the *Proc. Roy. Soc.*, Vol. 86, p. 562, 1912.

Let us now suppose that we have a coil of thin insulated wire,
Maximum inductance. the length of the wire being l and the number of turns N. As formerly, we have $l = 2\pi a N$ and $N = n \cdot \pi r^2$. Hence by (45)

$$L = 2Nl\left[\left(1 + \frac{\pi N^3}{2nl^2}\right)\log\frac{4ln^{1/2}}{\pi^{1/2}N^{3/2}} + \frac{\pi N^3}{6nl^2} - \frac{7}{4}\right].$$

For a maximum value $\partial L/\partial N = 0$, and writing

$$x = 8a/r = 4ln^{1/2}/(\pi N^3)^{1/2}$$

in this equation we get

$$12(32 + x^2)\log x - (39x^2 + 16) = 0.$$

It is easy to verify that $20\cdot60$ is an approximate value of the root of this equation. We see, therefore, that when $8a/r = 20\cdot6$, or $a/r = 2\cdot575$, L has its maximum value. Hence

$$L_{\max.} = 5\cdot35\pi N^2 a = 0\cdot165l^2/r = 0\cdot61n^{1/3}\,l^{5/3}.$$

In the last section but one, using a less accurate formula, we found that L was a maximum when $r = a/(3\cdot22)$. Using this value of r/a in (45) we find that $L = 6\cdot17\pi N^2 a = 0\cdot61n^{1/3}l^{5/3}$. We see therefore that our former result was about $2\cdot8$ per cent. too small.

When a rough accuracy only is desired we may use the formula

$$L = L_{\max.} = 0\cdot61n^{1/3}l^{5/3} \quad\dots\dots\dots\dots\dots(46),$$

for all values of r/a ranging from $0\cdot3$ to $0\cdot4$. This formula will be found useful in the practical design of inductive coils. Since the resistance R of a coil of wire varies as its length l and the maximum inductance varies as $l^{5/3}$, it follows that the time constant L/R of the coil increases the more wire we use.

For a circular coil of square section we can use the following formula:

Weinstein's formula.
$$L = 4\pi a\left\{\left(1 + \frac{b^2}{24a^2}\right)\log\frac{8a}{b}\right.$$
$$\left. + 0\cdot03657\,\frac{b^2}{a^2} - 1\cdot194914\right\} \dots(47),$$

where a is the radius of the circular axis and b is the length of the side of the square cross section. It is easy to see that for a given area of cross section the inductance of the ring is greater for a circular than for a square cross section.

Let L_1 be the value of the inductance of a ring computed by

Numerical examples. (43) and L_2 its inductance computed by (45). When r/a is 0·01 or less, L_1 and L_2 are practically identical but when r/a is greater than 0·2 the discrepancy between them exceeds one per cent. The following table illustrates how the inductance diminishes as r increases and also shows how L_1 compares with L_2.

r/a	0·001	0·01	0·1	0·2	0·4
$L_1/(2\pi a)$	14·4744	9·8692	5·2641	3·878	2·491
$L_2/(2\pi a)$	14·4744	9·8694	5·2759	3·918	2·625

Let us suppose that the ring shown in Fig. 29 consists of a thin conductor having a charge q per unit length.

Electrostatic capacity of a circular ring. The potential V at any point P is given by

$$V = 2 \int_0^\pi \frac{qa\partial\phi}{PR}$$

$$= 2qa \int_0^\pi \frac{\partial\phi}{(a^2 + r^2 - 2ar\cos\theta\cos\phi)^{1/2}}$$

$$= (4qa/r_1)\,K,$$

where $r_1^2 = a^2 + r^2 + 2ar\cos\theta$, and the modulus of K is given by $k^2 = 1 - r_2^2/r_1^2 = 4ar\cos\theta/r_1^2$. If ρ be the radius of the cross section of the ring then since by hypothesis ρ/a is a very small fraction, we see that at points on the surface of the ring $r_2 = \rho$ and $r_1 = 2a$ nearly. Thus $k_0^2 = 1 - k^2 = (\rho/2a)^2$ approximately. Hence k is very nearly equal to unity and we may write $\log(4/k_0)$, that is, $\log(8a/\rho)$, for K in this case. If Q_1 be the total charge on the ring $Q_1 = 2\pi aq$, and if V_1 be its potential

$$V_1 = 2qK = 2q\log(8a/\rho) \text{ approximately.}$$

Hence $Q_1/V_1 = \pi a/\{\log(8a/\rho)\}$ approximately,

and this ratio gives the electrostatic capacity of the ring. If S be the surface of the ring $S = 2\pi\rho \cdot 2\pi a$, and hence the capacity equals

$$\frac{S}{4\pi\rho\log(2S/\pi^2\rho^2)}.$$

If S remains constant it is easy to see that the capacity continually increases as ρ diminishes.

Let us now consider the mutual inductance between two single
The mutual inductance of two coaxial coils. layer coaxial cylindrical coils, the radii of their cross sections being a and b respectively, and their axial lengths being $2h_1$ and $2h_2$. Let d be the distance between their centres and N_1 and N_2 the total number of turns in each. We shall suppose that the two coils are closely wound with fine wire so that we may regard a turn of either coil as being practically in a plane perpendicular to the axis. We shall also neglect the thickness of the wires so that we have merely to consider two cylindrical current sheets.

If M_1 be the mutual inductance between two parallel circles at a distance x apart, we have, by (28),

$$M_1 = 4\pi ab \int_0^\pi \frac{\cos\phi\,\partial\phi}{(a^2 + b^2 + x^2 - 2ab\cos\phi)^{1/2}}.$$

If we integrate $M_1(\partial x/2h_2)\,N_2$ from $x = y - h_2$ to $x = y + h_2$, we get M_2 the mutual inductance between a cylindrical coil of radius b and axial length $2h_2$ and a coaxial circle of radius a at a distance y from its centre. Hence

$$\frac{M_2 . 2h_2}{4\pi ab N_2} = \int_0^\pi \int_{y-h_2}^{y+h_2} \frac{\cos\phi\,\partial\phi\,\partial x}{(a^2 + b^2 + x^2 - 2ab\cos\phi)^{1/2}}$$

$$= \int_0^\pi \cos\phi \log\frac{y + h_2 + \{\Delta^2 + (y + h_2)^2\}^{1/2}}{y - h_2 + \{\Delta^2 + (y - h_2)^2\}^{1/2}}\,\partial\phi,$$

where $\Delta^2 = a^2 + b^2 - 2ab\cos\phi$.

Integrating by parts, and noting that $\sin\phi$ vanishes at both limits, we have

$$\frac{M_2 . 2h_2}{4\pi ab N_2}$$

$$= -\int_0^\pi \sin\phi \left[\frac{1}{y + h_2 + \{\Delta^2 + (y + h_2)^2\}^{1/2}} \cdot \frac{ab\sin\phi}{\{\Delta^2 + (y + h_2)^2\}^{1/2}} - \cdots\right]\partial\phi$$

$$= \int_0^\pi \frac{ab\sin^2\phi}{a^2 + b^2 - 2ab\cos\phi} \left[\frac{y + h_2}{\{\Delta^2 + (y + h_2)^2\}^{1/2}} - \frac{y - h_2}{\{\Delta^2 + (y - h_2)^2\}^{1/2}}\right]\partial\phi.$$

If we now integrate $M_2\,(\partial y/2h_1)\,N_1$ from $y = d - h_1$ to $d + h_1$ we obtain the value of the mutual inductance M between the two coils.   Hence

$$\frac{M\,.\,h_1 h_2}{\pi ab\, N_1 N_2} = ab \int_0^\pi \frac{\sin^2 \phi}{a^2 + b^2 - 2ab \cos \phi} \left[\{\Delta^2 + (y + h_2)^2\}^{1/2} \right.$$
$$\left. - \{\Delta^2 + (y - h_2)^2\}^{1/2} \right]_{d-h_1}^{d+h_1} \partial\phi$$
$$= ab \int_0^\pi \frac{\sin^2 \phi}{(a + b)^2 - 4ab \cos^2 (\phi/2)} \left[\{\Delta^2 + (d + h_1 + h_2)^2\}^{1/2} \right.$$
$$- \{\Delta^2 + (d + h_1 - h_2)^2\}^{1/2} - \{\Delta^2 + (d - h_1 + h_2)^2\}^{1/2}$$
$$\left. + \{\Delta^2 + (d - h_1 - h_2)^2\}^{1/2} \right] \partial\phi.$$

This solution may be expressed in terms of elliptic integrals, but the following solution is convenient, in practice.

Put $\phi = 2\theta$, $c^2 = 4ab/(a + b)^2$,
$$R_1^2 = (a + b)^2 + (d + h_1 + h_2)^2; \quad k_1^2 = 4ab/R_1^2,$$
$$R_2^2 = (a + b)^2 + (d + h_1 - h_2)^2; \quad k_2^2 = 4ab/R_2^2,$$
$$R_3^2 = (a + b)^2 + (d - h_1 - h_2)^2; \quad k_3^2 = 4ab/R_3^2,$$
$$R_4^2 = (a + b)^2 + (d - h_1 + h_2)^2; \quad k_4^2 = 4ab/R_4^2.$$

Substituting we get

$$M = \frac{2\pi abc^2}{h_1 h_2} N_1 N_2 \int_0^{\pi/2} \frac{\sin^2 \theta \cos^2 \theta}{1 - c^2 \cos^2 \theta} [R_1\,(1 - k_1^2 \cos^2 \theta)^{1/2}$$
$$- R_2\,(1 - k_2^2 \cos^2 \theta)^{1/2} + R_3\,(1 - k_3^2 \cos^2 \theta)^{1/2} - R_4\,(1 - k_4^2 \cos^2 \theta)^{1/2}] \partial\theta.$$

By the binomial theorem, we have

$$\int_0^{\pi/2} \frac{\sin^2 \theta \cos^2 \theta}{1 - c^2 \cos^2 \theta} (1 - k_1^2 \cos^2 \theta)^{1/2} \partial\theta$$
$$= P_1 - \frac{1}{2} P_2 k_1^2 - \frac{1.1}{2.4} P_3 k_1^4 - \frac{1.1.3}{2.4.6} P_4 k_1^6 - \dots$$

where
$$P_n = \int_0^{\pi/2} \frac{\cos^{2n} \theta \sin^2 \theta}{1 - c^2 \cos^2 \theta} \partial\theta$$
$$= \frac{P_{n-1}}{c^2} - \frac{1}{c^2} \int_0^{\pi/2} \cos^{2n-2} \theta \sin^2 \theta \partial\theta$$
$$= \frac{P_{n-1}}{c^2} - \frac{1}{2n} \cdot \frac{1.3 \dots 2n - 3}{2.4 \dots 2n - 2} \cdot \frac{\pi}{2} \cdot \frac{1}{c^2}.$$

When $n = 1$, we have

$$P_1 = \int_0^{\pi/2} \frac{\cos^2 \theta \sin^2 \theta}{1 - c^2 \cos^2 \theta} \partial\theta$$

$$= \int_0^{\pi/2} \{\cos^2 \theta \sin^2 \theta + c^2 \cos^4 \theta \sin^2 \theta + c^2 \cos^6 \theta \sin^2 \theta + \ldots\} \partial\theta.$$

Writing $\sin^2 \theta = 1 - \cos^2 \theta$, and integrating, we get

$$P_1 = \frac{\pi}{2} \left\{ \frac{1}{2} + \frac{1.3}{2.4} c^2 + \frac{1.3.5}{2.4.6} c^4 + \ldots \right.$$
$$\left. - \frac{1.3}{2.4} - \frac{1.3.5}{2.4.6} c^2 - \ldots \right\}$$

$$= \frac{\pi}{2} \left\{ \frac{(1 - c^2)^{-1/2} - 1}{c^2} - \frac{(1 - c^2)^{-1/2} - 1 - c^2/2}{c^4} \right\}$$

$$= \frac{\pi}{4 \{1 + (1 - c^2)^{1/2}\}^2}.$$

It has to be carefully noticed that $(1 - c^2)^{1/2}$ must be a positive quantity. It therefore equals $(a - b)/(a + b)$ when a is greater than b and $(b - a)/(a + b)$ when a is less than b. We shall now make the supposition that a is greater than b.

Hence $$P_1 = \frac{\pi}{16} \cdot \frac{(a + b)^2}{a^2}.$$

Denoting P_n/P_1 by q_n, we get

$$q_n = \frac{(a + b)^2}{4ab} q_{n-1} - \frac{1}{n} \cdot \frac{1.3 \ldots 2n - 3}{2.4 \ldots 2n - 2} \cdot \frac{a}{b} \ldots\ldots\ldots(48).$$

We find that

$$q_1 = 1, \quad q_2 = (a + b)^2/4ab - a/4b = 1/2 + b/4a,$$

$$q_3 = q_2^2 + 1/16, \quad q_4 = q_2^3 + b/(32a) + 3/32, \text{ etc.}$$

Hence finally, we have

$$M = \frac{\pi^2 b^2}{2h_1 h_2} N_1 N_2 \left[R_1 \left\{ 1 - \frac{1}{2} q_2 k_1^2 - \frac{1.1}{2.4} q_3 k_1^4 - \frac{1.1.3}{2.4.6} q_4 k_1^6 - \ldots \right\} \right.$$

$$- R_2 \left\{ 1 - \frac{1}{2} q_2 k_2^2 - \ldots \right\} + R_3 \left\{ 1 - \frac{1}{2} q_2 k_3^2 - \ldots \right\}$$

$$\left. - R_4 \left\{ 1 - \frac{1}{2} q_2 k_4^2 - \ldots \right\} \right] \ldots(49).$$

By means of this formula the mutual inductance between any two coaxial cylindrical coils can be computed, the series being convergent in all cases.

When d^2 is great compared with ab, so that we can neglect the fourth and higher powers of k_1, k_2, ... we may use the formula

$$M = \frac{\pi^2 b^2}{2 h_1 h_2} N_1 N_2 \left[(R_1 - R_2) \left(1 + \frac{b^2 + 2ab}{2 R_1 R_2} \right) \right.$$
$$\left. + (R_3 - R_4) \left(1 + \frac{b^2 + 2ab}{2 R_3 R_4} \right) \right] \dots(50).$$

If, in addition, $d - h_1 - h_2$ is great compared with $a + b$, this becomes

$$M = \frac{2 (\pi a^2) (\pi b^2) N_1 N_2 d}{(d^2 - h_1^2 - h_2^2)^2 - 4 h_1^2 h_2^2} \dots\dots\dots\dots\dots(51).$$

When d is very great compared with h_1 and h_2, this simplifies to

$$M = \frac{2 (\pi a^2) (\pi b^2) N_1 N_2}{d^3} \dots\dots\dots\dots\dots(52).$$

When the coils are concentric, $d = 0$, and

$$M = \pi^2 b^2 \frac{N_1 N_2}{h_1 h_2} \left[R_1 \left\{ 1 - \frac{1}{2} q_2 k_1^2 - \frac{1 \cdot 1}{2 \cdot 4} q_3 k_1^4 - \dots \right\} \right.$$
$$\left. - R_2 \left\{ 1 - \frac{1}{2} q_2 k_2^2 - \frac{1 \cdot 1}{2 \cdot 4} q_3 k_2^4 - \dots \right\} \right] \dots(53),$$

where $\qquad R_1^2 = (a + b)^2 + (h_1 + h_2)^2; \quad k_1^2 = 4ab/R_1^2,$

and $\qquad R_2^2 = (a + b)^2 + (h_1 - h_2)^2; \quad k_2^2 = 4ab/R_2^2.$

As a numerical example let us take the case of two coils for which $a = 5$, $2h_1 = 30$, $N_1 = 300$ and $b = 4$, $2h_2 = 5$, $N_2 = 200$. By (48) we easily obtain

$$q_2 = 0 \cdot 7, \quad q_3 = 0 \cdot 5525, \quad q_4 = 0 \cdot 4618, \quad q_5 = 0 \cdot 3992,$$
$$q_6 = 0 \cdot 3527, \quad q_7 = 0 \cdot 3170, \dots.$$

We also have

$$R_1^2 = (a + b)^2 + (h_1 + h_2)^2 = 387 \cdot 25; \quad R_1 = 19 \cdot 6787,$$
$$R_2^2 = (a + b)^2 + (h_1 - h_2)^2 = 237 \cdot 25; \quad R_2 = 15 \cdot 4029,$$

and $\qquad k_1^2 = 80/387 \cdot 25, \; k_2^2 = 80/237 \cdot 25.$

Thus substituting in (53), we have

$$M = \frac{(4\pi)^2 . 6 . 10^4}{37\cdot5} [19\cdot6787 (1 - 0\cdot07230 - 0\cdot00295$$
$$- 0\cdot00026 - 0\cdot00003 - \ldots)$$
$$- 15\cdot4029 (1 - 0\cdot11802 - 0\cdot00785 - 0\cdot00111$$
$$- 0\cdot00020 - 0\cdot00004 - 0\cdot00001 - \ldots)]$$
$$= 1199900 \text{ cms.}$$
$$= 0\cdot0011999 \text{ henry.}$$

When $h_1 = h_2 = h$ the series of terms with the suffix 2 in (53) may converge slowly. It is better, therefore, to express the second half of the formula on the right hand side of (53) in terms of elliptic integrals. The formula now becomes

$$M = \pi^2 b^2 \frac{N_1 N_2}{h^2} \left[R_1 \left(1 - \frac{1}{2} q_2 k_1^2 - \frac{1\cdot1}{2\cdot4} q_3 k_1^4 - \frac{1\cdot1\cdot3}{2\cdot4\cdot6} q_4 k_1^6 - \ldots \right) \right]$$
$$- \frac{2\pi (a + b)}{3} . \frac{N_1 N_2}{h^2} \{2abK - (a^2 + b^2)(K - E)\} \ldots(54),$$

where k the modulus of the elliptic functions is $2 (ab)^{1/2}/(a + b)$.

Let us consider the case when the radii of the coils are equal to one another. Both of the coils for instance may be wound on the same insulating cylinder.

Formula (53) now becomes

$$M = 4\pi a^2 \frac{N_1 N_2}{h_1 h_2} \int_0^{\pi/2} \sin^2 \theta [R_1 (1 - k_1^2 \sin^2 \theta)^{1/2}$$
$$- R_2 (1 - k_2^2 \sin^2 \theta)^{1/2}] \partial\theta.$$

Hence from (8)

$$M = 4\pi a^2 \frac{N_1 N_2}{h_1 h_2} \left[R_1 \frac{2k_1^2 - 1}{3k_1^2} E_1 + R_1 \frac{1 - k_1^2}{3k_1^2} K_1 \right.$$
$$\left. - R_2 \frac{2k_2^2 - 1}{3k_2^2} E_2 - R_2 \frac{1 - k_2^2}{3k_2^2} K_2 \right] \ldots(55),$$

where $R_1^2 = 4a^2 + (h_1 + h_2)^2; \quad k_1^2 = 4a^2/R_1^2,$

and $R_2^2 = 4a^2 + (h_1 - h_2)^2; \quad k_2^2 = 4a^2/R_2^2.$

If we have in addition $h_1 = h_2 = a$, (55) becomes

$$M = 13\cdot589 N_1 N_2 a \ldots\ldots\ldots\ldots\ldots\ldots(56).$$

The above formulae (49) to (56) are only strictly true for coaxial cylindrical current sheets of infinitesimal thickness. It can be shown, however, that if one of the current sheets be replaced by a helical coil of very fine wire having the same effective number of turns, the above formula can still be applied. When both the coils are helical, modifications are necessary, but if the pitch of one of the helices be a very small quantity the correcting factor will be small. It is advisable, however, when actually measuring the mutual inductance to investigate whether rotating one of the helices round the axis through various angles alters the mutual inductance.

If we have a wire closely wound on a smooth insulating cylinder, the diameter of the wire being small compared with that of the cylinder, the mutual inductance between it and a cylindrical current sheet of the same radius as the helix formed by the axis of the wire and having the same axial length, will be approximately equal to the self inductance L of the coil. Hence by putting $h_1 = h_2 = h$ in (55) we find that

The self inductance of a single layer cylindrical coil.

$$L = \frac{8\pi a^2}{3h} N^2 \left[\left(1 + \frac{a^2}{h^2}\right)^{1/2} \left\{ E_1 + \frac{h^2}{a^2}(K_1 - E_1) - \frac{a}{h} \right\} \right] \dots(57),$$

where $k = a/(a^2 + h^2)^{1/2}$ is the modulus of the elliptic functions.

This is a most important formula given by Lorenz in 1879.

When h is less than a, we get by the binomial theorem and (13) and (14),

$$L = 4\pi a N^2 \left[\log \frac{4a}{h} - \frac{1}{2} + \frac{h^2}{8a^2} \left(\log \frac{4a}{h} + \frac{1}{4} \right) \right] \dots(58).$$

When $h/a = 3/4$ the number found by this formula is too small by about the quarter of one per cent. For smaller values of h/a the values given by (57) and (58) are in practical agreement. When h is small, a small percentage error in determining it introduces a large percentage error into the calculated value of L. Hence, in making standards, it is advisable to make h not less than $2a$. In this case we can use the following remarkably simple formula,

$$L = \frac{2\pi^2 a^2}{h} N^2 \left[1 - \frac{4a}{3\pi h} + \frac{a^2}{8h^2} - \frac{a^4}{64h^4} + \dots \right.$$
$$\left. + \frac{(-)^n (2n)! (2n+2)!}{n!(n+2)! \{(n+1)!\}^2 2^{4n+3}} \left(\frac{a}{h}\right)^{2n+2} + \dots \right] \dots(59).$$

When $h = 2a$, the inaccuracy of this formula taking into account the first four terms only of the series inside the bracket, as compared with the more accurate elliptic integral formula, is less than 1 in 8000. We may deduce (59) from (57) by the binomial theorem and (9) and (10).

It is convenient to write

$$L = \beta a N^2 \dots\dots\dots\dots\dots\dots\dots(60),$$

and find β for various values of a/h from the following table.

a/h	β	a/h	β
0·20	3·6324	1·80	19·5794
0·30	5·2337	2·00	20·7463
0·40	6·7102	2·20	21·8205
0·50	8·0747	2·40	22·8150
0·60	9·3389	2·60	23·7401
0·70	10·5135	2·80	24·6048
0·80	11·6079	3·00	25·4161
0·90	12·6306	3·20	26·1801
1·00	13·5889	3·40	26·9018
1·20	15·3380	3·60	27·5855
1·40	16·8984	3·80	28·2349
1·60	18·3035	4·00	28·8534

For example, the self inductance of a single layer coil of 1000 turns wound on a cylinder the radius of which is 10 cms. and the length 80 cms. is 0·288534 henry.

If in Lorenz's formula (57) we write $a = h \tan \alpha$, we get

Coil of maximum inductance.

$$L = \frac{8\pi}{3} a N^2 \left[\frac{\sin \alpha}{\cos^2 \alpha} E + \frac{K - E}{\sin \alpha} - \tan^2 \alpha \right] \dots(61).$$

Let l be the length of the wire forming the coil, so that $l = 2\pi a N = 2\pi h \tan \alpha N$. If b be the breadth of the wire including the insulation, we also have

$$2\pi a b N = 2\pi a . 2h, \text{ and } bN = 2h.$$

Hence $l = \pi b N^2 \tan \alpha$. Substituting x for $\tan \alpha$ in (61), we find that

$$L = \frac{4 l^{3/2}}{3 (\pi b)^{1/2}} \left[x^{1/2} (1 + x^2)^{1/2} E + \frac{(1 + x^2)^{1/2}}{x^{3/2}} (K - E) - x^{3/2} \right].$$

Hence assuming that l and b are constant, L has a maximum value when $\partial L/\partial x = 0$. Noticing that

$$\frac{\partial K}{\partial x} = (E - K \cos^2 \alpha) \cot \alpha,$$

$$\frac{\partial E}{\partial x} = (E - K) \cos^2 \alpha \cot \alpha,$$

and $$\frac{\partial \sin \alpha}{\partial x} = \cos^3 \alpha,$$

we find that L is a maximum when

$$E - \frac{K - E}{\tan^2 \alpha} = \sin \alpha.$$

It will be found by trial that $\tan \alpha = 2\cdot4525$ satisfies this equation. Hence when $a/h = 2\cdot4525$ the self inductance of the coil is a maximum. Noticing that $\alpha = 67°49'$, we get

$$L_{max.} = 1\cdot17198 \times \frac{2}{\sqrt{\pi}} \times \frac{l^{3/2}}{b^{1/2}}$$

$$= 1\cdot32244\,(l^{3/2}/b^{1/2}) \quad \dots\dots\dots\dots\dots(62).$$

If this wire were made into a circular coil of circular cross section, we find by the formula given on p. 106 that

$$L_{max.} = 0\cdot33\,(l^2/b).$$

As l is very great compared with b, we see that the ring coil has a much higher inductance than the cylindrical coil.

The above formulae can also be used to calculate mutual

Mutual inductance from self inductance. inductance. For example, let us suppose that we have two coaxial coils A and B each having the same radius and the same number of turns per unit length. Let us suppose that the space between A and B is exactly filled up by a third coil X having the same number of turns per unit length. Then if L_A, L_{AX}, L_{AXB} denote the inductance of the coil A, of the coils A and X in series and of the coils A, X and B in series, and M_{AX} denotes the mutual inductance between the coils A and X, we have

$$L_{AXB} = L_A + L_X + L_B + 2M_{XB} + 2M_{BA} + 2M_{AX},$$

$$L_{AX} = L_A + L_X + 2M_{AX},$$

$$L_{BX} = L_B + L_X + 2M_{BX},$$

and, therefore,

$$2M_{AB} = (L_{AXB} + L_X) - (L_{AX} + L_{BX}).$$

The values of the inductances on the right-hand side can in many cases be written down very easily.

The kinetic energy of the electrons. If we adopt the electron theory of electricity, we must suppose that the kinetic energy of a current is made up not only of magnetic energy but also of the kinetic energy of the moving electrons themselves which compose the current. Let us consider the case of a current filament, the number of electrons in a length ∂s being $n\partial s$. Let m be the mass of an electron and v its average forward velocity in the direction of the current. Let q be the charge on each electron so that $i = nqv$. If the actual velocity at any instant be $u + v$, so that the mean value of u is zero, the kinetic energy will equal $(1/2) \Sigma m (u + v)^2 = (1/2) \Sigma m u^2 + (1/2) \Sigma m v^2$. The first term in this expression depends on the temperature of the body and is independent of i. Hence the total kinetic energy due to the current is $L i^2/2 + (1/2) \Sigma m v^2$, and since $v = i/nq$ this can be written in the form

$$\tfrac{1}{2} \left(L + \int \frac{m}{nq^2} \, \partial s \right) i^2.$$

In most practical cases the correcting term $\int m/(nq^2) \, \partial s$ which has to be added to L when finding the total kinetic energy due to the current is negligibly small.

Values of the Elliptic Integrals K and E from k = sin 0° to k = sin 90°.

sin⁻¹ k	K	E	sin⁻¹ k	K	E	sin⁻¹ k	K	E
0°	1·570 796	1·570 796	30°	1·685 750	1·467 462	60°	2·156 516	1·211 056
1	1·570 916	1·570 677	31	1·694 114	1·460 774	61	2·184 213	1·201 538
2	1·571 275	1·570 318	32	1·702 836	1·453 908	62	2·213 195	1·192 046
3	1·571 874	1·569 720	33	1·711 925	1·446 869	63	2·243 549	1·182 589
4	1·572 712	1·568 884	34	1·721 391	1·439 662	64	2·275 376	1·173 180
5	1·573 792	1·567 809	35	1·731 245	1·432 291	65	2·308 787	1·163 828
6	1·575 114	1·566 497	36	1·741 499	1·424 760	66	2·343 905	1·154 547
7	1·576 678	1·564 948	37	1·752 165	1·417 075	67	2·380 870	1·145 348
8	1·578 486	1·563 162	38	1·763 256	1·409 240	68	2·419 842	1·136 244
9	1·580 541	1·561 142	39	1·774 786	1·401 260	69	2·460 999	1·127 250
10	1·582 843	1·558 887	40	1·786 770	1·393 140	70	2·504 550	1·118 378
11	1·585 394	1·556 400	41	1·799 222	1·384 886	71	2·550 731	1·109 643
12	1·588 197	1·553 681	42	1·812 160	1·376 504	72	2·599 820	1·101 062
13 ·	1·591 254	1·550 732	43	1·825 602	1·367 999	73	2·652 138	1·092 650
14	1·594 568	1·547 554	44	1·839 567	1·359 377	74	2·708 068	1·084 425
15	1·598 142	1·544 150	45	1·854 075	1·350 644	75	2·768 063	1·076 405
16	1·601 978	1·540 521	46	1·869 148	1·341 806	76	2·832 673	1·068 610
17	1·606 081	1·536 670	47	1·884 809	1·332 870	77	2·902 565	1·061 059
18	1·610 454	1·532 597	48	1·901 083	1·323 842	78	2·978 569	1·053 777
19	1·615 101	1·528 306	49	1·917 997	1·314 729	79	3·061 729	1·046 786
20	1·620 026	1·523 799	50	1·935 581	1·305 539	80	3·153 385	1·040 114
21	1·625 234	1·519 079	51	1·953 865	1·296 278	81	3·255 303	1·033 789
22	1·630 729	1·514 147	52	1·972 882	1·286 954	82	3·369 868	1·027 844
23	1·636 517	1·509 007	53	1·992 670	1·277 574	83	3·500 422	1·022 313
24	1·642 604	1·503 662	54	2·013 266	1·268 147	84	3·651 856	1·017 237
25	1·648 995	1·498 115	55	2·034 715	1·258 680	85	3·831 742	1·012 664
26	1·655 697	1·492 368	56	2·057 062	1·249 182	86	4·052 758	1·008 648
27	1·662 716	1·486 427	57	2·080 358	1·239 661	87	4·338 654	1·005 259
28	1·670 059	1·480 293	58	2·104 658	1·230 127	88	4·742 717	1·002 584
29	1·677 735	1·473 970	59	2·130 021	1·220 589	89	5·434 910	1·000 752
30	1·685 750	1·467 462	60	2·156 516	1·211 056	90	infinity	1·000 000

Values of $log_{10} K$ *and* $log_{10} E$ *for values of* k *from*
$sin\ 80°\ to\ 1.\ (Legendre.)$

$sin^{-1} k$	Log K	Δ_1	Δ_2	Log E	Δ_1	Δ_2
80°·0	0·4987 7703	13 3336	1018	0·0170 8111	2 7091	129
80·1	0·5001 1040	13 4354	1036	0·0168 1020	2 6962	131
80·2	0·5014 5394	13 5390	1054	0·0165 4058	2 6831	132
80·3	0·5028 0783	13 6444	1073	0·0162 7227	2 6698·	134
80·4	0·5041 7227	13 7517	1093	0·0160 0529	2 6564	136
80·5	0·5055 4744	13 8610	1113	0·0157 3965	2 6429	137
80·6	0·5069 3354	13 9724	1134	0·0154 7536	2 6291	139
80·7	0·5083 3078	14 0858	1156	0·0152 1245	2 6153	140
80·8	0·5097 3936	14 2014	1178	0·0149 5092	2 6012	142
80·9	0·5111 5949	14 3192	1201	0·0146 9080	2 5870	144
81·0	0·5125 9141	14 4393	1225	0·0144 3210	2 5726	145
81·1	0·5140 3534	14 5618	1249	0·0141 7484	2 5581	147
81·2	0·5154 9151	14 6867	1274	0·0139 1903	2 5433	149
81·3	0·5169 6018	14 8141	1300	0·0136 6470	2 5285	151
81·4	0·5184 4159	14 9441	1327	0·0134 1185	2 5134	152
81·5	0·5199 3600	15 0769	1355	0·0131 6052	2 4981	154
81·6	0·5214 4369	15 2124	1384	0·0129 1070	2 4827	156
81·7	0·5229 6493	15 3508	1414	0·0126 6243	2 4671	158
81·8	0·5245 0001	15 4922	1445	0·0124 1572	2 4513	160
81·9	0·5260 4923	15 6366	1477	0·0121 7058	2 4354	162
82·0	0·5276 1289	15 7843	1510	0·0119 2704	2 4192	163
82·1	0·5291 9132	15 9352	1544	0·0116 8512	2 4029	165
82·2	0·5307 8485	16 0896	1579	0·0114 4483	2 3863	167
82·3	0·5323 9381	16 2476	1616	0·0112 0620	2 3696	169
82·4	0·5340 1857	16 4092	1655	0·0109 6924	2 3527	171
82·5	0·5356 5949	16 5747	1694	0·0107 3397	2 3356	173
82·6	0·5373 1696	16 7441	1736	0·0105 0041	2 3183	175
82·7	0·5389 9137	16 9177	1779	0·0102 6859	2 3007	177
82·8	0·5406 8313	17 0955	1823	0·0100 3851	2 2830	179
82·9	0·5423 9268	17 2778	1870	0·0098 1021	2 2651	182
83·0	0·5441 2047	17 4648	1918	0·0095 8371	2 2469	184
83·1	0·5458 6695	17 6566	1968	0·0093 5902	2 2285	186
83·2	0·5476 3260	17 8534	2021	0·0091 3616	2 2100	188
83·3	0·5494 1795	18 0555	2076	0·0089 1517	2 1912	190
83·4	0·5512 2350	18 2631	2133	0·0086 9605	2 1721	193
83·5	0·5530 4980	18 4764	2193	0·0084 7884	2 1529	195
83·6	0·5548 9744	18 6956	2255	0·0082 6355	2 1334	197
83·7	0·5567 6700	18 9211	2320	0·0080 5021	2 1137	199
83·8	0·5586 5912	19 1532	2389	0·0078 3884	2 0937	202
83·9	0·5605 7443	19 3921	2460	0·0076 2947	2 0735	204

Values of $log_{10} K$ and $log_{10} E$ for values of k from
sin 80° to 1. (*Legendre.*)—Continued

sin⁻¹ k	Log K	Δ₁	Δ₂	Log E	Δ₁	Δ₂
84°·0	0·5625 1364	19 6381	2535	0·0074 2211	2 0531	207
84·1	0·5644 7745	19 8916	2614	0·0072 1680	2 0324	209
84·2	0·5664 6661	20 1531	2697	0·0070 1356	2 0115	212
84·3	0·5684 8192	20 4228	2784	0·0068 1241	1 9903	214
84·4	0·5705 2420	20 7012	2875	0·0066 1338	1 9689	217
84·5	0·5725 9431	20 9887	2972	0·0064 1649	1 9472	220
84·6	0·5746 9318	21 2859	3073	0·0062 2177	1 9252	222
84·7	0·5768 2177	21 5932	3180	0·0060 2925	1 9029	225
84·8	0·5789 8109	21 9112	3293	0·0058 3896	1 8804	228
84·9	0·5811 7221	22 2405	3413	0·0056 5092	1 8576	231
85·0	0·5833 9626	22 5818	3539	0·0054 6516	1 8345	234
85·1	0·5856 5444	22 9357	3673	0·0052 8171	1 8111	237
85·2	0·5879 4801	23 3031	3816	0·0051 0060	1 7874	240
85·3	0·5902 7832	23 6846	3967	0·0049 2185	1 7634	243
85·4	0·5926 4679	24 0813	4127	0·0047 4551	1 7391	246
85·5	0·5950 5492	24 4940	4299	0·0045 7160	1 7145	249
85·6	0·5975 0432	24 9239	4481	0·0044 0015	1 6896	253
85·7	0·5999 9671	25 3720	4676	0·0042 3119	1 6643	256
85·8	0·6025 3391	25 8396	4885	0·0040 6476	1 6387	260
85·9	0·6051 1788	26 3281	5109	0·0039 0089	1 6127	263
86·0	0·6077 5069	26 8390	5349	0·0037 3962	1 5864	267
86·1	0·6104 3459	27 3739	5607	0·0035 8097	1 5598	270
86·2	0·6131 7198	27 9346	5886	0·0034 2499	1 5327	274
86·3	0·6159 6543	28 5231	6186	0·0032 7172	1 5053	278
86·4	0·6188 1775	29 1418	6512	0·0031 2118	1 4775	282
86·5	0·6217 3193	29 7929	6865	0·0029 7343	1 4493	286
86·6	0·6247 1122	30 4794	7248	0·0028 2850	1 4207	290
86·7	0·6277 5916	31 2042	7667	0·0026 8642	1 3917	295
86·8	0·6308 7958	31 9709	8124	0·0025 4725	1 3622	299
86·9	0·6340 7668	32 7834	8626	0·0024 1103	1 3323	304
87·0	0·6373 5501	33 6459	9177	0·0022 7779	1 3020	308
87·1	0·6407 1961	34 5636	9785	0·0021 4759	1 2712	313
87·2	0·6441 7597	35 5422	10459	0·0020 2048	1 2398	318
87·3	0·6477 3019	36 5881	11208	0·0018 9649	1 2080	324
87·4	0·6513 8900	37 7089	12043	0·0017 7569	1 1757	329
87·5	0·6551 5989	38 9132	12980	0·0016 5813	1 1428	335
87·6	0·6590 5121	40 2112	14035	0·0015 4385	1 1093	340
87·7	0·6630 7233	41 6147	15230	0·0014 3292	1 0753	347
87·8	0·6672 3380	43 1377	16590	0·0013 2540	1 0406	353
87·9	0·6715 4757	44 7967	18149	0·0012 2134	1 0053	360

Values of $log_{10} K$ and $log_{10} E$ for values of k from $sin\ 80°$ to 1. (Legendre.)—Continued

$sin^{-1} k$	Log K	Δ_1	Δ_2	Log E	Δ_1	Δ_2
88°·0	0·6760 2724	46 6116	19948	0·0011 2081	9693	367
88·1	0·6806 8840	48 6064	22040	0·0010 2387	9327	374
88·2	0·6855 4904	50 8104	24492	0·0009 3060	8953	382
88·3	0·6906 3009	53 2597	27396	0·0008 4107	8571	390
88·4	0·6959 5605	55 9993	30870	0·0007 5536	8181	399
88·5	0·7015 5598	59 0862	35077	0·0006 7355	7782	408
88·6	0·7074 6460	62 5940	40245	0·0005 9573	7374	418
88·7	0·7137 2400	66 6184	46693	0·0005 2199	6956	429
88·8	0·7203 8584	71 2878	54895	0·0004 5242	6527	441
88·9	0·7275 1462	76 7773	65561	0·0003 8715	6087	453
89·0	0·7351 9234	83 3334	79812	0·0003 2628	5633	468
89·1	0·7435 2568	91 3146	99496	0·0002 6995	5166	483
89·2	0·7526 5714	101 2642	127847	0·0002 1829	4683	501
89·3	0·7627 8356	114 0489	170975	0·0001 7146	4181	522
89·4	0·7741 8844	131 1464	241655	0·0001 2965	3660	546
89·5	0·7873 0308	155 3119	370693	0·0000 9305	3114	576
89·6	0·8028 3427	192 3813	650756	0·0000 6192	2538	615
89·7	0·8220 7240	257 4569	1501510	0·0000 3654	1923	670
89·8	0·8478 1809	407 6079		0·0000 1731	1253	774
89·9	0·8885 7889			0·0000 0479	479	
90·0	infinity			0·0000 0000		

Table of the Neperian Logarithms of the Prime Numbers from 2 to 101.

2	0·693	147	181	43	3·761 200 116	
3	1·098	612	289	47	3·850 147 602	
5	1·609	437	912	53	3·970 291 914	
7	1·945	910	149	59	4·077 537 444	
11	2·397	895	273	61	4·110 873 864	
13	2·564	949	357	67	4·204 692 619	
17	2·833	213	344	71	4·262 679 877	
19	2·944	438	979	73	4·290 459 441	
23	3·135	494	216	79	4·369 447 852	
29	3·367	295	830	83	4·418 840 608	
31	3·433	987	204	89	4·488 636 370	
37	3·610	917	913	97	4·574 710 979	
41	3·713	572	067	101	4·615 120 517	

Example. $\log 9·4 = \log 2 + \log 47 - \log 10$
$$= \log 47 - \log 5.$$

REFERENCES

A. Russell, *Proc. Phys. Soc. London*, Vol. 20, p. 476, 1907, 'The Magnetic Field and Inductance Coefficients of Circular, Cylindrical, and Helical Currents.'

E. B. Rosa and F. W. Grover, *Bull. of the Bureau of Standards*, Washington, Vol. 8, p. 1, 1912, 'Formulas and Tables for the Calculation of Mutual and Self-Inductance.'

Lord Rayleigh, *Proc. Roy. Soc.* Ser. A, Vol. 86, p. 562, 1912, 'On the Self-Inductance of Electric Currents in a Thin Anchor-Ring.'

M. Brooks and H. M. Turner, Univ. of Illinois, *Bulletin*, 9, No. 10, p. 3, 1912, 'Inductance of Coils.'

F. Emde, *Elektrotechnik und Maschinenbau*, Band 30, p. 221, 1912, 'Die Berechnung eisenfreier Drosselspulen für Starkstrom.'

A. M. Legendre, *Traité des Fonctions Elliptiques*, Tome 2, p. 240.

A. Cayley, *Elliptic Functions*.

CHAPTER IV

THE frequency of the alternating currents employed in practice varies between 10 and 100 cycles per second. For lighting purposes 50, and for power transmission purposes 25, have become the standard frequencies. Now the instruments used to measure alternating amperes and volts are comparatively speaking sluggish in their action, and so their indications do not give the instantaneous values. Consider for example a Siemens electro-dynamometer. The effect of a current i through the fixed coil acting on the movable coil which is traversed by the same current is to produce a couple of which the instantaneous values are proportional to i^2. The couple will therefore have the same sign in whichever direction the current passes through the dynamometer, and hence a deflection will be produced by an alternating current. If the periodic time of the fluctuations be small compared with the periodic time of the moving coil, then the coil is insensible to the fluctuations of i^2, and the mean torque on the moving coil, when its plane is perpendicular to the plane of the fixed coil, can be measured by the torsion of the spiral spring required to bring the coil back to this position. This torsion is measured by the angle through which a pointer has to be turned. Hence if R be the reading,

$$k^2 R = \text{the mean value of } i^2,$$

where k^2 is a constant, determined by experiments with direct current. When i is constant, $k\sqrt{R}$ gives us its value in amperes; when i is variable, $k\sqrt{R}$ gives us the square root of its mean square in the same units. The instrument therefore tells us the square root of the mean square (which is sometimes called the R.M.S.), or the effective value, of the current.

When i varies harmonically with the time, it is easy to find the effective value of the current in terms of its maximum value. For example, suppose that

$$i = I \sin \omega t,$$

where $\omega = 2\pi f$ and f is the frequency, then

$$i^2 = I^2 \sin^2 \omega t$$

$$= \frac{I^2}{2} - \frac{I^2}{2} \cos 2\omega t.$$

Now the mean value of $\cos 2\omega t$ over a whole period is zero, for if we plot out the cosine curve, a glance will show that for every positive ordinate in the first and fourth quarter periods there is an equal negative ordinate in the second and third quarter periods, and hence, if we add them all together and divide by the number of them in order to get their mean value, the result will be zero. Therefore the mean value of i^2 is $\dfrac{I^2}{2}$. If we call the effective value of the current A, then

$$A = \frac{I}{\sqrt{2}} = 0{\cdot}7071 I.$$

Therefore the effective value in this case is about 71 per cent. of the maximum value.

In like manner we can show that those alternating current voltmeters, the readings of which depend on the expansion of a heated wire, or on the electromagnetic attractions and repulsions of moving coils, or on the electrostatic repulsions and attractions of movable segments, give us the square root of the mean square values of the voltages which we measure with them. If the instantaneous value e of the potential difference between the voltmeter terminals be given by

$$e = E \sin \omega t,$$

then, as before, its effective value V is given by

$$V = \frac{1}{\sqrt{2}} E = 0.7071 E.$$

Effective values of complex currents and pressures. If i do not vary harmonically, then by Fourier's Theorem it may be written as follows:

$$i = I_1 \sin(\omega t - \alpha_1) + I_3 \sin(3\omega t - \alpha_3) + \dots \quad \dots(1).$$

Hence squaring and taking mean values, we find that the effective value A of the current is given by

$$A^2 = \tfrac{1}{2} I_1{}^2 + \tfrac{1}{2} I_3{}^2 + \tfrac{1}{2} I_5{}^2 + \dots.$$

If R be the ohmic resistance of the conductor in which i is flowing, then

$$RA^2 = RA_1{}^2 + RA_3{}^2 + RA_5{}^2 + \dots,$$

where A_1, A_3, A_5, ... are the effective values of the harmonics of the current. We thus see that each harmonic component produces its own heating effect on the conductor in exactly the same way as if all the other harmonics were absent.

If the alternating current i have a direct current component C, then

$$i = C + I_1 \sin(\omega t - \alpha_1) + I_3 \sin(3\omega t - \alpha_3) + \dots,$$

and therefore $\qquad A^2 = C^2 + A_1{}^2 + A_3{}^2 + \dots \quad \dots\dots\dots\dots(2).$

Hence, to find the heating effect on a conductor produced by a combined direct and alternating current, we calculate the heating effect produced by each separately and add up the results. For example a complex current consisting of 10 amperes of direct current and 10 effective amperes of alternating current would only heat a conductor as much as a direct or an alternating current of 14·14 amperes, and an alternating current ammeter placed in the circuit would only read 14·14 amperes. We shall see later the importance of this result in the theory of rotary converters, and it is utilised in systems of polycyclic distribution.

If the potential difference applied to the voltmeter terminals be given by

$$e = P + E_1 \sin (\omega t - \alpha_1) + E_3 \sin (3\omega t - \alpha_3) + \ldots,$$

then $$V^2 = P^2 + V_1^2 + V_3^2 + \ldots \quad \ldots \ldots \ldots \ldots \ldots (3),$$

where $$V_1^2 = \tfrac{1}{2} E_1^2, \quad V_3^2 = \tfrac{1}{2} E_3^2 \ldots.$$

Hence if we put 100 volts direct current pressure in series with 100 volts alternating current pressure, the reading on a voltmeter would be $100 \sqrt{2}$, *i.e.* 141·4 volts, no matter what was the shape of the wave of alternating pressure.

Although in a few cases the negative half of a wave of alter-

Mean value of an alternating current or pressure. nating current or pressure is not similar to the positive half, as for example when we have aluminium electrodes or electric arcs between metals in the circuit, yet in general they are exactly alike, so that their mean value over a whole period is zero. In a few cases it is useful to know their mean value over half a period. The mean value of $E \sin \omega t$, for example, over half a period, starting from t equal to zero, is

$$\frac{2}{T} \int_0^{\frac{T}{2}} E \sin \omega t \, \partial t = \frac{2}{\pi} E = 0\!\cdot\!6366E \, ;$$

this is less than $0\!\cdot\!7071E$, which is the effective value. The root mean square value of a variable quantity is always greater than its mean value. This follows at once from the algebraical theorem that

$$\left\{ \frac{y_1^2 + y_2^2 + y_3^2 + \ldots + y_n^2}{n} \right\}^{\frac{1}{2}} \text{ is greater than } \frac{y_1 + y_2 + \ldots + y_n}{n},$$

except when all the n quantities are equal to one another—when the two expressions are equal to each other.

In practice the readings on electrostatic, hot wire and moving

Graphical methods of finding mean square values. coil instruments give us at once the square root of the mean square of the alternating currents and pressures. We can also by oscillographs, ondographs, etc. find the shape of the current and pressure waves. It is important to test whether the root mean square pressure or current got from these waves agrees with the voltmeter or ammeter reading. There are several graphical methods of doing this; the following method is particularly convenient when the curve is drawn on sectional paper.

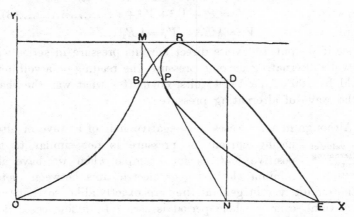

Fig. 32. Graphical construction for finding effective values.

$$Y^2 = \frac{2h}{l} \text{ (area } RPED\text{)}.$$

In Fig. 32 suppose that $OBRE$ is the positive part of any wave curve and that the negative half is exactly similar. Draw a line through R, the highest point of the curve, parallel to OX, and draw any chord BD parallel to it. Draw BM and DN perpendicular to RM and OX respectively and join MN, cutting BD in P. Construct RPE the curve locus of P. Then the volume of the solid generated by the revolution of ORE about OX

$$= \int_0^l \pi y^2 \partial x.$$

Also $$= \int_0^h 2\pi y \cdot BD \partial y,$$

where l is the breadth and h the height of the curve.

Therefore $$\int_0^l y^2 \partial x = 2 \int_0^h y \cdot BD \partial y.$$

But $$y \cdot BD = h \cdot PD.$$

Hence, $$\int_0^l y^2 \partial x = 2 \int_0^h h \cdot PD \partial y = 2hA',$$

where A' is the area of the curve $RPED$. If Y^2 be the mean square value of y, then

$$Y^2 = \frac{1}{l} \int_0^l y^2 \partial x = \frac{2h}{l} A'.$$

Now as the curve RPE can easily be constructed, especially when the given curve is drawn on sectional paper, this is a rapid and accurate graphical method of finding Y.

When the curve has several maxima and minima values the above method becomes inconvenient; for we have to divide the curve into several portions, and this entails drawing several loci curves. In this case it is better to draw n lines from a point O making angles $2\pi/n$ with one another, and measure along these lines lengths OP_1, OP_2, ... equal in length to the value of n equidistant ordinates of the given curve. Then the area A of the curve drawn through P_1, P_2, ... will be given by

$$A = \tfrac{1}{2} \int_0^{2\pi} r^2 \partial\theta = \pi Y^2.$$

Therefore $Y = \sqrt{(A/\pi)}.$

A useful formula for Y can be found by equating the two well-known expressions for the volume of the solid of revolution formed by rotating the given curve round its time axis,

$$\pi \int_0^l y^2 \partial x = 2\pi \bar{y} A,$$

whence $Y^2 = 2y_m \bar{y}$ ....(4).

Y is the R.M.S. value of y, while y_m is its mean value, which equals $\frac{A}{l}$, and $\bar{y}$ is the height of the centre of gravity of the plane figure ORE above OX, which can easily be found by the ordinary methods. When there is only one maximum point, the first method of finding Y is the most accurate in practice, but formula (4) is a useful one.

It is instructive to draw curves the mean square values of the
Equivolt curves.   ordinates of which are all constant. In practice for example, the shape of the wave producing an effective voltage of 50 might be any of the curves drawn in Figs. 33, 34, 35, 36, and 37. We shall see later on that the shape of the

wave has a considerable bearing on the economical working of transformers and motors—and the following equations to families of waves all having the same effective voltage will repay study.

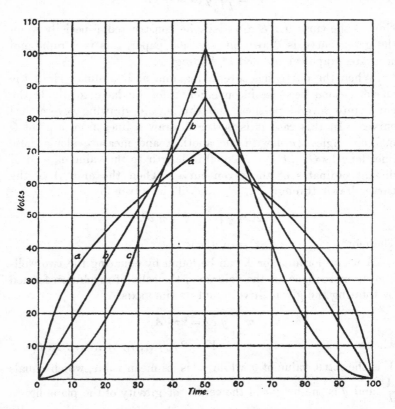

Fig. 33.   Equivolt curves.   The effective voltage of each curve is 50.

(1)   The equations to the curves shown in Fig. 33 are

$$e = E\left(\frac{4}{T}\right)^n t^n \qquad \text{from } t = 0 \ \text{ to } t = \frac{T}{4}$$

$$e = E\left(\frac{4}{T}\right)^n \left(\frac{T}{2} - t\right)^n \text{ from } t = \frac{T}{4} \text{ to } t = \frac{T}{2}$$

$$\dots\dots\dots(a),$$

and the negative half of the wave is supposed to be similar. In these equations e is the instantaneous value of the E.M.F.,

t is the time in seconds and T is the period of the alternating current. It follows that

$$E = V \sqrt{2n + 1},$$

where E is the maximum value of e and V is its effective value. Hence for a given V the maximum value E increases with n.

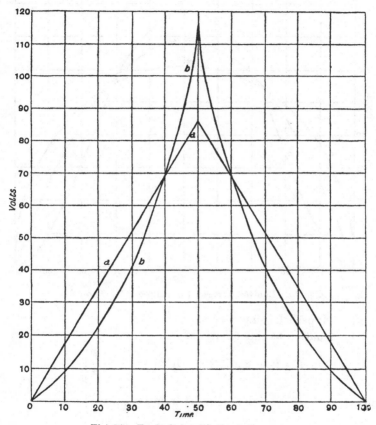

Fig. 34. Equivolt hyperbolic sine curves.

Now in Fig. 33 three of the curves are drawn, namely those corresponding to n equal to 1/2, 1 and 2. If n equals zero we get a rectangle. When n lies between 0 and 1 we get curves similar to (a) in Fig. 33, namely two curves meeting at an angle and concave to the time axis. When n equals 1 we get the triangle (b),

R. I.   9

and when n is greater than 1 we get two curves (c) meeting at an angle and convex to the axis. We see that a knowledge of the value of V only fixes an inferior limit to the value of E.

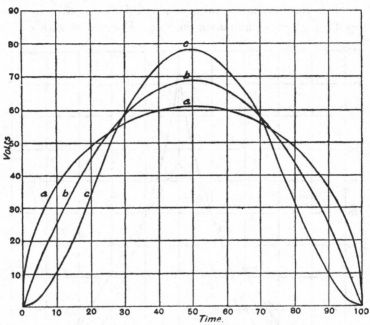

Fig. 35.   Equivolt curves.   (a) Ellipse.   (b) Parabola.

(2)   Hyperbolic sine curves.
The equation to this family is

$$e = B \sinh n \, \frac{t}{T} \qquad \text{from } t = 0 \text{ to } \frac{T}{4}$$
$$e = B \sinh n \left(\frac{1}{2} - \frac{t}{T}\right) \text{ from } t = \frac{T}{4} \text{ to } \frac{T}{2} \qquad \dots\dots\dots(b),$$

where

$$B = V \sqrt{\frac{n}{\sinh\left[(n/2)\right] - n/2}}.$$

In this case

$$E = V \sqrt{\frac{n \cosh\left[(n/2)\right] - n}{2 \sinh\left[(n/2)\right] - n}}.$$

We get the triangle (a) shown in Fig. 34 when n is 0 and the curve (b) when n is equal to 10.

The values of $\sinh \theta$ and $\cosh \theta$ are given in *Mathematical Tables*.

(3) Ellipse and Parabola.

The equation in this case is

$$e = B \left\{ t \left(\frac{T}{2} - t \right) \right\}^n \text{ from } t = 0 \text{ to } \frac{T}{2} \quad \ldots\ldots\ldots\ldots(c),$$

where

$$B = \left(\frac{2}{T} \right)^{2n} \frac{\sqrt{\Gamma(4n+2)}}{\Gamma(2n+1)} V,$$

and

$$E = \frac{\sqrt{\Gamma(4n+2)}}{4^n \Gamma(2n+1)} V.$$

Note that $\Gamma(n+1) = n\Gamma(n)$, $\Gamma(1) = 1$ and $\Gamma(1/2) = \sqrt{\pi}$.

The curves in Fig. 35 are (a) $n = \frac{1}{2}$ an ellipse, (b) $n = 1$ a parabola, and (c) $n = 2$ a biquadratic.

Tables of $\log \Gamma(n)$ are given in Dale's *Mathematical Tables*.

(4) Sine curves.

The equation to the E.M.F. is

$$e = E \sin^n \frac{2\pi t}{T} \text{ from } t = 0 \text{ to } \frac{T}{2} \quad \ldots\ldots\ldots\ldots\ldots(d).$$

Therefore

$$E = \sqrt{\frac{\pi^{\frac{1}{2}} \Gamma(n+1)}{\Gamma\{(2n+1)/2\}}} V.$$

When $n = 0$ we get a rectangle, $n = \frac{1}{2}$ we get the curve (a) in Fig. 36, when $n = 1$ the sine curve (b), and $n = 2$ the curve (c).

It will be noted that all the waves figured above are symmetrical waves, *i.e.* those in which

$$f(t) = f(T/2 - t).$$

(5) Distorted waves.

If $e = f(t)$ be the equation of a symmetrical wave, then

$$\left.\begin{array}{l} e = f\left(\dfrac{T}{4} \cdot \dfrac{t}{\tau} \right) \qquad \text{from } t = 0 \text{ to } \tau \\[2mm] e = f\left(\dfrac{T}{4} \cdot \dfrac{T/2 - t}{T/2 - \tau} \right) \text{ from } t = \tau \text{ to } \dfrac{T}{2} \end{array}\right\} \quad \ldots\ldots\ldots\ldots(e)$$

represents a distorted wave of E.M.F. which has the same maximum height, the same R.M.S. height, the same breadth, and the same area as the original wave.

This is easily proved.   For example, if V be the R.M.S. of the values of e given by the above equations, then

$$\frac{T}{2} V^2 = \int_0^\tau \left\{ f\left(\frac{T}{4} \cdot \frac{t}{\tau}\right) \right\}^2 \partial t + \int_\tau^{\frac{T}{2}} \left\{ f\left(\frac{T}{4} \cdot \frac{T/2 - t}{T/2 - \tau}\right) \right\}^2 \partial t$$

$$= \frac{4\tau}{T} \int_0^{\frac{T}{4}} \{f(t)\}^2 \partial t + \frac{4(T/2 - \tau)}{T} \int_0^{\frac{T}{4}} \{f(t)\}^2 \partial t$$

$$= \int_0^{\frac{T}{2}} \{f(t)\}^2 \partial t.$$

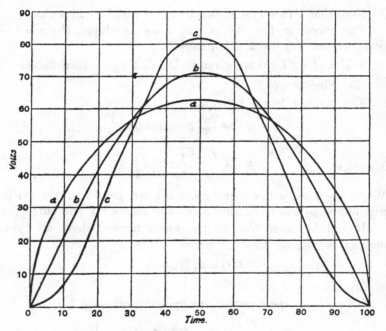

Fig. 36.   Equivolt sine curves.

Hence V is independent of the values of τ.   Similarly we can prove that its area etc. are the same as that of the curve $e = f(t)$. Hence since τ may have any value between 0 and $T/2$ there are an infinite number of waves which have the same maximum height, the same area and the same R.M.S. height as the original symmetrical wave. In Fig. 37 the middle curve shown is the sine curve and the others are distorted members of the same family.

We will refer to the family of waves given by the equations (e) above as a family of waves of equal height. It must be borne in mind, however, that there is an infinite number of families of waves of equal height. For example (a) in Fig. 33 and the sine wave (b) in Fig. 36 are the symmetrical members of two families of waves whose maximum heights are equal.

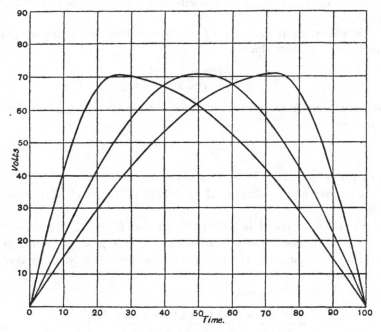

Fig. 37. Equivolt sine curves of equal height.

We have seen (page 65) that the equation connecting the applied P.D. and the current in an inductive coil is

Choking coil currents.

$$e = Ri + L\frac{\partial i}{\partial t}.$$

If the equation to the wave of the applied E.M.F. is $e = E\sin \omega t$, and R is absolutely zero, then (page 66)

$$i = (E/\omega L)(\cos \omega t_1 - \cos \omega t)$$

if the switch be closed at the time t_1. If t_1 be a multiple of π/ω, it is easy to see that the current is always flowing in one direction. Hence, in these cases, the 'choking coil' acts like a valve, only allowing the current to flow in one direction. A piece of apparatus of this nature is called a rectifier. In practice, however, the resistance of choking coils is never absolutely zero and so the exponential factor ultimately becomes zero. In this case the solution is $i = -(E/\omega L)\cos\omega t$, and therefore $A = 0.1591\,(V/fL)$ approximately, where A is the effective value of i and f is the frequency.

Similarly when the wave of P.D. is given by the equations (a) above, we can show that

$$A = \left\{ \frac{2(2n+1)}{(n+2)(2n+3)} \right\}^{1/2} \frac{V}{4fL}.$$

As n increases from zero, A increases until it attains its maximum value, for

$$n = \frac{\sqrt{6}-1}{2} = 0.7247 ;$$

it then diminishes for greater values of n.

The maximum value of A is $0.1589\,(V/fL)$, and E is then $1.565V$.

For a sine curve A is $0.1591\,(V/fL)$, and E equals $1.414V$.

It will be seen that the sine wave of P.D. produces a larger choking coil current than any of the first family of waves considered.

For the hyperbolic sine curves,

$$A = \frac{V}{fL} \left\{ \frac{2n - 6\sinh(n/2) + n\cosh(n/2)}{n^2[2\sinh(n/2) - n]} \right\}^{\frac{1}{2}}.$$

Condenser currents. If K be the capacity in farads (see Chapter V) of a condenser of which the terminals have a P.D. of v volts, then

$$q = Kv,$$

where q is the number of coulombs in the condenser.

Hence if i be the current flowing into the condenser,

$$i = \frac{\partial q}{\partial t} = K\frac{\partial v}{\partial t}.$$

If A be the effective value of the condenser current when waves of P.D. similar to those illustrated above are applied to the terminals of the condenser, then

Minimum value

(a) $A = 4n \sqrt{\dfrac{2n+1}{2n-1}} fKV;$ $6\cdot661 fKV.$

(b) $A = n \left\{\dfrac{\sinh (n/2) + n/2}{\sinh (n/2) - n/2}\right\}^{\frac{1}{2}} fKV;$ $6\cdot928 fKV.$

(c) $A = 2 \sqrt{\dfrac{2n(4n+1)}{2n-1}} fKV;$ $6\cdot293 fKV.$

(d) $A = \dfrac{2\pi n}{\sqrt{2n-1}} fKV;$ $6\cdot283 fKV$ (sine curve).

It will be seen that the sine wave produces the least effective value of the condenser current of any of the waves we have considered. It is also not difficult to show that of all E.M.F. waves of equal height (Fig. 37), applied to choking coils and condensers, the symmetrical wave produces the maximum effective current in the choking coil, and the minimum effective current in the condenser.

We have seen (page 69) that the solution for the current in an inductive coil is

Effect of altering the resistance of a circuit on the form of the current wave.

$$i = \frac{E_1 \sin(\omega t + \alpha_1 - \beta_1)}{\sqrt{R^2 + L^2\omega^2}} + \frac{E_3 \sin(3\omega t + \alpha_3 - \beta_3)}{\sqrt{R^2 + L^2(3\omega)^2}} + \cdots$$
$$\cdots\cdots(5)$$
$$= \frac{1}{\sqrt{R^2 + L^2\omega^2}} \left\{ E_1 \sin(\omega t + \alpha_1 - \beta_1) + \sqrt{\frac{R^2 + L^2\omega^2}{R^2 + L^2(3\omega)^2}} E_3 \sin(3\omega t + \alpha_3 - \beta_3) + \cdots \right\}.$$

When R is very large $\beta_1, \beta_3 \ldots$ are all very small, and the curve of i is practically the same as the curve of e on a diminished scale. The more we diminish R, the smaller do the amplitudes of the higher harmonics become as compared with the amplitude of the fundamental harmonic. Hence by diminishing R in an inductive circuit we make the curve of i more like a simple sine curve.

If we have resistance in series with a condenser in a circuit, then

$$e = Ri + \frac{\int i \partial t}{K},$$

and

$$i = \frac{E_1 \sin(\omega t + \alpha_1 + \gamma_1)}{\sqrt{R^2 + 1/(K^2 \omega^2)}} + \frac{E_3 \sin(3\omega t + \alpha_3 + \gamma_3)}{\sqrt{R^2 + 1/\{K^2 (3\omega)^2\}}} + \dots$$

Hence, proceeding as before, we deduce the following working rule: diminishing the resistance in a condenser circuit makes the current wave less like a sine wave.

The equation of the current in an inductive circuit is

A simple sine wave of E.M.F. produces the maximum current in an inductive coil and the minimum current in a condenser.

$$e = Ri + L \frac{\partial i}{\partial t}.$$

Thus

$$V^2 = R^2 A^2 + \frac{1}{T} \int_0^T \left(L \frac{\partial i}{\partial t} \right)^2 \partial t,$$

since $\int_0^T 2RLi \frac{\partial i}{\partial t} \partial t$ vanishes.

Now from pages 69 and 124,

$$A^2 = \frac{V_1^2}{R^2 + L^2 \omega^2} + \frac{V_3^2}{R^2 + L^2 (3\omega)^2} + \dots,$$

and

$$\frac{1}{T} \int_0^T \left(L \frac{\partial i}{\partial t} \right)^2 \partial t = L^2 \omega^2 \left\{ \frac{V_1^2}{R^2 + L^2 \omega^2} + \frac{3^2 V_3^2}{R^2 + L^2 (3\omega)^2} + \dots \right\}$$

$$= \alpha^2 \omega^2 L^2 A^2,$$

where

$$\alpha^2 = \frac{\dfrac{V_1^2}{R^2 + L^2 \omega^2} + \dfrac{3^2 V_3^2}{R^2 + L^2 (3\omega)^2} + \dots}{\dfrac{V_1^2}{R^2 + L^2 \omega^2} + \dfrac{V_3^2}{R^2 + L^2 (3\omega)^2} + \dots}.$$

Now the numerator of this fraction is greater than the denominator except when

$$V_3 = V_5 = \dots = 0.$$

Hence α has its minimum value unity when the curve of the applied E.M.F. is sine-shaped.

Also

$$A^2 = \frac{V^2}{R^2 + \alpha^2 L^2 \omega^2} \dots \dots \dots \dots \dots \dots (6),$$

and the denominator has its minimum value, and therefore A has its maximum value, when α is unity.

Therefore the sine wave produces the maximum effective current in an inductive coil.

Similarly for a condenser circuit,

$$A^2 = \frac{V^2}{R^2 + \dfrac{1}{\beta^2 K^2 \omega^2}} \quad \dotfill (7),$$

and β has its minimum value unity when the wave is sine-shaped.

Therefore the sine-shaped wave produces the minimum effective current in a condenser circuit.

When we have both inductance and capacity in the circuit **Resonance.** the problem becomes of great practical importance, owing to the high pressures and large currents that are produced in distributing systems through the effects of resonance by comparatively low E.M.F.'s. Suppose that we have an

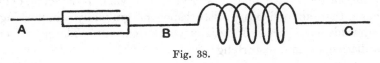

Fig. 38.

inductive coil (Fig. 38) in series with a condenser, and that an alternating P.D. is applied to the terminals A and C. Then in order to find the current we have to solve the equation

$$Ri + L\frac{\partial i}{\partial t} + \frac{\int i \partial t}{K} = E_1 \sin(\omega t + \alpha_1) + E_3 \sin(3\omega t + \alpha_3) + \dots \dots (8).$$

The solution of this equation consists of two parts.

There is first the particular integral

$$i = \sum \frac{E_{2n+1} \sin(\overline{2n+1}\,\omega t + \alpha_{2n+1} - \beta_{2n+1})}{\sqrt{R^2 + \left\{L - \dfrac{1}{K(2n+1)^2\omega^2}\right\}^2 (2n+1)^2 \omega^2}} \quad \dots (9),$$

where
$$\tan\beta_{2n+1} = \frac{(2n+1)\,\omega\left\{L - \dfrac{1}{K(2n+1)^2\omega^2}\right\}}{R}.$$

The other part of the solution—the complementary function—is found by solving the equation

$$Ri + L\frac{\partial i}{\partial t} + \frac{\int i \partial t}{K} = 0.$$

When R^2 is greater than $4L/K$ the solution is of the form

$$i = A\epsilon^{-m_1 t} + B\epsilon^{-m_2 t} \dots\dots\dots\dots(10),$$

where A and B are constants, and

$$m_1 = \frac{R}{2L} + \sqrt{\frac{R^2}{4L^2} - \frac{1}{LK}}$$

and

$$m_2 = \frac{R}{2L} - \sqrt{\frac{R^2}{4L^2} - \frac{1}{LK}}.$$

When $4L/K$ is greater than R^2 the solution is of the form

$$i = A\epsilon^{-\frac{Rt}{2L}} \cos\left\{\sqrt{\frac{1}{LK} - \frac{R^2}{4L^2}}\, t + \alpha\right\} \dots\dots(11),$$

where A and α are constants. The complete solution of (8) is thus given by adding (9) and (10), or (9) and (11) together. When R is large (10) shows that the initial non-oscillatory disturbance of the current wave rapidly subsides, and when R is small, (11) shows that an oscillatory disturbance of gradually diminishing amplitude is superposed initially on the current. The period of oscillation of this disturbing effect is

$$\frac{2\pi}{\sqrt{\frac{1}{LK} - \frac{R^2}{4L^2}}} \dots\dots\dots\dots(12).$$

If the applied P.D. wave were absolutely constant in shape, then, in practice, after a second or two (9) alone would give us the value of the current.

Now if
$$LK(2n+1)^2 \omega^2 = 1 \dots\dots\dots(13),$$
we see from (9) that $\beta_{2n+1} = 0$, and hence that the $(2n+1)$th harmonic of i is in phase with the $(2n+1)$th harmonic of the applied P.D. Its amplitude also is simply $\frac{E_{2n+1}}{R}$, and if R is small compared to $L\omega$, this term practically swamps all the other terms; thus the frequency of the alternating current is practically $(2n+1)\omega/2\pi$, and its effective value only slightly greater than V_{2n+1}/R. Also in this case the effective value of the P.D. across the choking coil will be nearly equal to

$$V_{2n+1}\left\{1 + \frac{(2n+1)^2 \omega^2 L^2}{R^2}\right\}^{\frac{1}{2}},$$

and the effective value of the P.D. across the condenser is approximately equal to

$$(2n + 1)\,\omega L\,\frac{V_{2n+1}}{R}\ \text{ or }\ \frac{V_{2n+1}}{(2n + 1)\,\omega K R}.$$

It will be seen that if R be very small, or if the frequency $(2n + 1)\,\omega/2\pi$ be very high, then these voltages may attain enormous values. The phase of the choking coil P.D. will be practically 90 degrees in advance, and that of the condenser F.D 90 degrees behind the current. Hence the two P.D.'s are nearly in opposition to one another, and their resultant or the applied P.D. between A and C (Fig. 38) will be very small compared with either of them.

We see from (13) that the lowest value of the frequency at which resonance can occur is $1/(2\pi\sqrt{LK})$; hence if the highest harmonic in the applied P.D. have a frequency less than this there will be no danger of resonance.

It is also to be noted that the above investigation shows that it is quite possible to obtain resonance effects in direct current circuits. The voltage curve of a direct current dynamo is never an exact straight line. The voltage has an alternating component due to the facts that the number of bars round the commutator is finite, and that the slots in the armature gave rise to pulsations of the magnetic field. By suitably adjusting the value of the inductance of a choking coil put in series with a condenser between the mains, it is possible to get resonance, so that a large alternating current component starts in the circuit, and the pressure across the condenser attains a very high value. Duddell has shown that if we put a resonant circuit across a direct current arc formed between hard carbons, then resonance often ensues, and a large alternating current flows across the arc, causing it to emit a musical note. This happens even when accumulators are used instead of dynamos, and must be due to constantly recurring irregularities in the burning of the arc which continually renew oscillatory waves of the form given by (11).

In Fig. 39 we have supposed that the curve (I), which is parabolic, represents the shape of the wave of current in a circuit formed by a choking coil and condenser in series, and we have calculated the applied P.D. (E) necessary to produce this wave. It will be seen that it is a peaky wave very different from a sine curve. E_2 is the wave of P.D. at the terminals of the choking coil and is triangular in shape. E_1 gives the shape of the P.D. at the terminals of the condenser, and is very similar to a sine curve. The diagram illustrates the general theorem that except when the applied P.D. wave is a sine curve, the wave of P.D. across the choking coil terminals is much more distorted from the sine shape than the wave of P.D. across the terminals of the condenser.

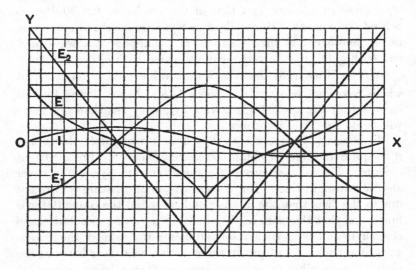

Fig. 39   Choking coil (E_1) and condenser (E_2) P.D.'s when the current wave (I) is part of a parabola.

If we have (Fig. 40) a condenser with capacity K shunted by a choking coil of inductance L and negligible resistance, then in certain cases the current in the main can be very small compared with that either in the choking coil or in the condenser. Let e, i_1 and i_2 be the instantaneous

Resonance of currents.

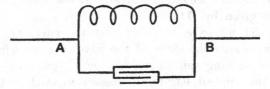

Fig. 40.   Resonance of currents.

values of the P.D., the current in the condenser and the current in the choking coil respectively, then

$$i_1 = K\frac{\partial e}{\partial t}; \quad e = L\frac{\partial i_2}{\partial t}.$$

Now by equations (6) and (7) the effective values of the currents are

$$A_1 = \beta\omega KV \text{ and } A_2 = \frac{V}{\alpha\omega L},$$

where α and β are constants depending on the shape of e; these constants have their minimum value unity when the wave is sine-shaped. If i be the current in the main, then

$$i = i_1 + i_2,$$

and hence, noting that

$$i_1 = KL\frac{\partial^2 i_2}{\partial t^2},$$

we find that

$$\cos\phi = -\frac{\alpha}{\beta},$$

where ϕ is the phase difference (see Chapter X) between A_1 and A_2. If K and V are fixed, then the minimum value of the main current is $A_1\sin\phi$, and in this case the choking coil current is $-A_1\cos\phi$.

Hence

$$LK(\alpha\omega)^2 = 1 \quad \dots\dots\dots\dots\dots(14).$$

If L and V are fixed, then the minimum value of the current in the main is $A_2\sin\phi$, and it has this value when

$$LK(\beta\omega)^2 = 1 \quad \dots\dots\dots\dots\dots(15).$$

1.   For a sine wave ϕ is 180 degrees, and the main current is zero when

Numerical examples.

$$LK\omega^2 = 1.$$

2.   For a parabolic wave ϕ is 173° 46′, and if the condenser current is constant and equal to A_1, the minimum value of the current in the main is $0\cdot1086A_1$, and it has this value when

$$1\cdot001LK\omega^2 = 1.$$

Similarly if the current A_2 in the choking coil be kept constant, the minimum value of the main current is $0\cdot1086A_2$, and it has this value when

$$1\cdot013LK\omega^2 = 1.$$

3.   For a triangular wave ϕ is 155° 54′.   When the condenser current is constant and equal to A_1, then the minimum value of the main current is $0.4083A_1$, and it has this value when

$$1.013LK\omega^2 = 1.$$

When the condenser current varies and A_2 is constant, then the minimum value of the main current is $0.4083A_2$, and we have

$$1.216LK\omega^2 = 1.$$

If a condenser shunted by a choking coil be adjusted so that $LK\omega^2$ is unity, then A_1/A_2 equals $\alpha\beta$.

If A_1 were equal to A_2 the applied P.D. would be sine-shaped, and the greater this ratio the more distorted from the sine shape will be the wave of P.D.   Also if A be the current in the main, the smaller A becomes compared to either A_1 or A_2 the nearer does the shape of the applied wave approximate to that of a sine curve.   Hence we can use this theorem as a rough test for the shape of a wave of P.D.

The inductive effect on a circuit of a coil (R, L) can be neu-
Method of neutralising the inductive effect of a choking coil. tralised by shunting it as in Fig. 41 with a condenser K in series with a resistance R, provided that K equals L/R^2.   Let e be the applied E.M.F. and let i_1, i_2 be the currents in the inductive coil and capacity respectively, also let

$$q = \int i_2 \partial t.$$

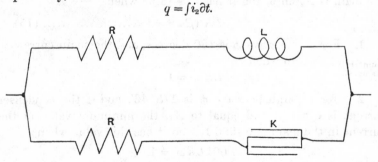

Fig. 41.   Circuits in parallel which act exactly like a non-inductive resistance R when $L = KR^2$.

Then $e = Ri_1 + L\dfrac{\partial i_1}{\partial t}$ ....(16),

and
$$e = \frac{q}{K} + Ri_2 = \frac{q}{K} + R\frac{\partial q}{\partial t} \dotsc\dotsc\dotsc\dotsc(17)$$

$$= \frac{1}{KR}\left\{Rq + \frac{\partial q}{\partial t}\right\} \dotsc\dotsc\dotsc(18).$$

Subtracting (18) from (16), and solving the equation for $i_1 - \frac{q}{KR}$, we get

$$i_1 = \frac{q}{KR} + A\epsilon^{-\frac{R}{L}t}.$$

Therefore, since i_1 and q are zero at the instant of closing the switch in the main, A is zero, and thus

$$i_1 = \frac{q}{KR}.$$

Hence from (17) $e = R\,(i_1 + i_2).$

But $i_1 + i_2$ is the current in the main, and therefore the combination acts like a non-inductive coil of resistance R from the instant the switch is closed.

If L be not equal to KR^2 the combination still acts like a non-inductive resistance R when the frequency is infinite. In this case the apparent resistance or impedance of the choking coil would be infinite, and the impedance of the condenser by itself would be zero.

The capacity effect of a condenser K shunted by a resistance

Method of neutralising the capacity effect of a shunted condenser.
R can be neutralised by putting in series with the combination a purely inductive coil L shunted by a resistance R where L equals KR^2.

If e_1 and e_2 be the potential differences across the condenser and the choking coil respectively, we have

$$i = \frac{e_1}{R} + K\frac{\partial e_1}{\partial t}$$

and
$$i = \frac{e_2}{R} + \frac{\int e_2 \partial t}{L} = K\frac{R}{L}e_2 + \frac{\int \frac{R}{L}e_2 \partial t}{R}.$$

Hence, proceeding as before,

$$e_1 = \int \frac{R}{L}e_2 \partial t + A\epsilon^{-\frac{t}{KR}}.$$

Thus since e_1 and $\int (R/L)\, e_2 \partial t$ are zero at the moment of closing the switch in the main, A is zero, and hence

$$i = \frac{e_1 + e_2}{R}.$$

The combination therefore acts exactly like a non-inductive resistance R from the moment of closing the switch.

Suppose that we have two coils ab and cd with self-inductances L and N respectively; let M be their mutual inductance. If we send an alternating current whose value is i through the coil ab, then the P.D. at its terminals will be given by

Comparison of inductances by means of a voltmeter.

$$e = Ri + L\frac{\partial i}{\partial t},$$

and an electrostatic voltmeter placed across ab will read V, where

$$V^2 = R^2 A^2 + L^2 \frac{1}{T}\int_0^T \left(\frac{\partial i}{\partial t}\right)^2 \partial t$$

$$= R^2 A^2 + L^2 X^2.$$

Now if we connect b and c together, and put the voltmeter across a and d, the same current still flowing through ab, we have

$$e_1 = Ri + L\frac{\partial i}{\partial t} - M\frac{\partial i}{\partial t}.$$

Therefore $V_1^2 = R^2 A^2 + (L - M)^2 X^2.$

Similarly if we join b and d and put the voltmeter across a and c, the current in ab remaining unaltered, we find V_2^2, where

$$V_2^2 = R^2 A^2 + (L + M)^2 X^2.$$

Eliminating X^2 and $R^2 A^2$ from these equations, we find that

$$\frac{M}{L} = 2\,\frac{V_1^2 + V_2^2 - 2V^2}{V_2^2 - V_1^2}.$$

Similarly we could find M/N and hence L/N.

Provided the eddy currents are negligible, these equations are theoretically accurate. It is easily seen, however, that in certain cases a small error in reading the voltmeter may introduce a large error into the ratio of the inductances as calculated by the formula.

REFERENCES

J. A. FLEMING, *Journ. of the Inst. of El. Eng.*, Vol. 20, p. 362, 1891, 'On some Effects of Alternating Current Flow in Circuits having Capacity and Self Inductance.'

A. HAY, *The Electrician*, Vol. 33, June, 1894, 'On Impulsive Current Rushes in Inductive Circuits.'

A. RUSSELL, *The Electrician*, Vol. 35, p. 115, 1895, 'Graphical Methods of Finding Mean Square Values.'

A. RUSSELL, *The Electrician*, Vol. 37, p. 502, 1896, 'Combined Alternating and Continuous Currents.'

A. RUSSELL, *Journ. of the Inst. of El. Eng.*, Vol. 29, p. 154, 1899, 'How Condenser and Choking Coil Currents vary with the Shape of the Wave of the Applied E.M.F.'

W. DUDDELL, *Journ. of the Inst. of El. Eng.*, Vol. 30, p. 232, 1900, 'On Rapid Variations in the Current through the Direct Current Arc.'

A. RUSSELL, *Journ. of the Inst. of El. Eng.*, Vol. 30, p. 596, 1901, 'Note on Resonance with Alternating Currents.'

T. MATHER, *The Electrical Review*, Vol. 48, p. 915, 1901, 'Capacity Currents Influenced by Frequency.'

H. ARMAGNAT, *L'Éclairage Électrique*, Vol. 27, p. 465, 1901, 'Résonance dans les Circuits à Courant Continu.'

CHAPTER V

Coefficients of self and mutual induction for electrostatic charges. Motion of electrified bodies. Charges constant. Potentials constant. Capacity of a conductor. Capacity between two conductors. Capacity of a condenser. Capacity of a concentric main. Dielectric currents. Laplace's formula. The energy of a moving charge. Capacity in electromagnetic units. The microfarad. The capacities of a triple concentric main. The capacity of a condenser formed by two long parallel cylinders. The capacity of a condenser formed by two long parallel cylinders, one wholly enclosed by the other. The maximum value of the potential gradient between two parallel cylinders. Practical application. Condenser currents in concentric cables and in two parallel overhead wires. Two core cable. Three phase cables. Numerical example. Condenser currents in three phase working. Two phase cables with four separate conductors. Twin concentric cable. Model of a polyphase cable. Three core cable.

SUPPOSE that we have n conductors with potentials $v_1, v_2, \ldots v_n$
Capacity coefficients. respectively. Let there be electrical equilibrium when the potential of the first conductor is v_1 and all the other conductors are at zero potential. In this case the charge on the first conductor is proportional to v_1, and hence it may be written $K_{1.1} v_1$, where $K_{1.1}$ is a constant which is called the coefficient of self induction of the conductor for electrostatic charges. The charge on the second conductor will be negative and will be proportional to v_1. Let it equal $K_{2.1} v_1$, where $K_{2.1}$ is a constant and is negative. Similarly the charge on the nth conductor will be $K_{n.1} v_1$. Now consider another state of equilibrium, when the potentials of the conductors are $0, v_2, 0, \ldots 0$. In this case the charge on the first conductor will be $K_{1.2} v_2$ and on the others $K_{2.2} v_2$, $K_{3.2} v_2$, $\ldots K_{n.2} v_2$. Similarly we can write down expressions for the charges on the conductors when the pth is at a potential v_p and all the others are at zero potential. Now if we superpose all these systems, we get another system in a

state of equilibrium, in which the charges on the conductors will be given by the linear equations

$$\left.\begin{aligned}
q_1 &= K_{1.1}\, v_1 + K_{1.2}\, v_2 + K_{1.3}\, v_3 + \ldots\ldots + K_{1.n}\, v_n \\
q_2 &= K_{2.1}\, v_1 + K_{2.2}\, v_2 + K_{2.3}\, v_3 + \ldots\ldots + K_{2.n}\, v_n \\
&\ldots\ldots\ldots\ldots\ldots\ldots\ldots\ldots\ldots\ldots\ldots\ldots\ldots\ldots\ldots
\end{aligned}\right\} \ldots\ldots(1).$$

It is obvious that $K_{1.1}, K_{1.2}, \ldots$ depend only on the shapes and the relative positions of the conductors.

Consider two conductors R and S whose coefficients of self induction are $K_{r.r}$ and $K_{s.s}$ respectively. Suppose that all the other conductors are permanently connected to the earth. Put S to earth and charge R to the potential v_r. The work done during this process

$$= \int v_r \partial q_r,$$

but $q_r = K_{r.r} v_r$, and hence the work done is $\frac{1}{2} K_{r.r} v_r{}^2$. Now keep R at potential v_r and raise the potential of S to v_s. The work done during this stage in changing the charge on R so as to keep v_r constant is

$$\int v_r \partial q_r,$$

but $q_r = K_{r.r} v_r + K_{r.s} v_s$, and therefore, since v_r is constant,

$$\partial q_r = K_{r.s}\, \partial v_s.$$

Hence the work done is $K_{r.s} v_r v_s$. Similarly the work done in changing the charge on S so as to raise its potential from zero to v_s is $\int v_s \partial q_s$. But $q_s = K_{s.s} v_s + K_{s.r} v_r$, and therefore, since v_r is constant, $\partial q_s = K_{s.s}\, \partial v_s$, and the work done is $\frac{1}{2} K_{s.s} v_s{}^2$. Therefore the total work done in raising R and S to the potentials v_r and v_s is

$$\tfrac{1}{2} K_{r.r} v_r{}^2 + K_{r.s} v_r v_s + \tfrac{1}{2} K_{s.s} v_s{}^2.$$

Starting with the conductor S we find in the same way that the work is

$$\tfrac{1}{2} K_{r.r} v_r{}^2 + K_{s.r} v_r v_s + \tfrac{1}{2} K_{s.s} v_s{}^2.$$

Hence, equating these two expressions, we find that

$$K_{r.s} = K_{s.r}.$$

We may therefore call $K_{r.s}$ or $K_{s.r}$ the coefficient of mutual induction between R and S for electrostatic charges.

If we have n conductors, the total potential energy of the system is

$$\tfrac{1}{2} \Sigma K_{r.r} v_r{}^2 + \Sigma K_{r.s} v_r v_s.$$

From equations (1) we may write this by ordinary algebra in the form

$$\tfrac{1}{2} \Sigma q_r v_r.$$

Note the similarity of these expressions to the expressions for the electromagnetic energy of a system of currents,

$$\tfrac{1}{2}\Sigma L_{r.r} i_r{}^2 + \Sigma L_{r.s} i_r i_s$$

and

$$\tfrac{1}{2} \Sigma \phi_r i_r.$$

Again we may write the self energy of a single conductor in the form $\tfrac{1}{2} q V$. Now by Gauss's theorem $4\pi q = \Sigma R \partial S$, where we may suppose that ∂S is an element of the surface of the conductor itself. Also if ∂s be an element of the axis of the tube of force starting from ∂S,

$$V = \int R \partial s.$$

Therefore the self energy of the conductor is

$$\tfrac{1}{2} q V = \frac{1}{8\pi} \Sigma R \partial S \int R \partial s.$$

But along ∂s, $R \partial S$ is constant, therefore $R \partial S \int R \partial s = \int R^2 \partial S \partial s$. Now $\partial S \partial s = \partial v =$ the element of volume of a tube of force. Hence $R \partial S \int R \partial s = \int R^2 \partial v$, the integration being taken along the tube of force standing on ∂S. Therefore, integrating for the whole surface of the conductor, we see that

$$\Sigma \frac{R^2}{8\pi} \partial v$$

is an expression for the self energy of an electrified conductor, the integration being taken throughout all the space occupied by tubes of force.

If λ be the dielectric coefficient of the medium in which the conductors are immersed this expression becomes

$$\lambda \Sigma \frac{R^2}{8\pi} \partial v.$$

Compare this with the expression

$$\mu \Sigma \frac{H^2}{8\pi} \partial v$$

for the self energy of an electric current flowing in a medium of constant permeability μ.

We shall now give two useful practical theorems, due to Kelvin, on the motion of electrified bodies under their mutual attractions and repulsions.

Let us consider the case of n charged conductors which form
Motion of electrified bodies. Charges constant.
a self-contained system screened from outside electric influences. To simplify the problem let us suppose only one of them is movable. Let us also suppose, in the first place, that all the conductors are insulated so that the charges on them remain constant. By the Conservation of Energy the movable conductor X cannot, unless acted on by external forces, move under the action of the electric forces into a position where the electrostatic energy of the system is greater, otherwise the total energy of the system would be increased. The electric forces acting on X move it in the direction along which the electrostatic energy diminishes most rapidly as the force acting on X will be greatest in this direction. We see therefore that the motion which ensues, due to the action of the electric forces, must diminish the electrostatic energy of the system and increase the mechanical energy.

Let ∂w denote the work done on X by the electric forces during an infinitely small displacement from P to P', then, with the notation of p. 147, if $v_1 + \partial v_1, v_2 + \partial v_2, \ldots$ be the new values of $v_1, v_2, \ldots$ we have

$$\partial w = \tfrac{1}{2} q_1 v_1 + \tfrac{1}{2} q_2 v_2 + \ldots$$
$$- \tfrac{1}{2} q_1 (v_1 + \partial v_1) - \tfrac{1}{2} q_2 (v_2 + \partial v_2) - \ldots$$
$$= - \tfrac{1}{2} q_1 \partial v_1 - \tfrac{1}{2} q_2 \partial v_2 - \ldots \qquad\qquad (\alpha).$$

Now q_1 is constant and since (p. 147)

$$q_1 = K_{1.1} v_1 + K_{1.2} v_2 + \ldots,$$

we have $\quad 0 = K_{1.1} \partial v_1 + v_1 \partial K_{1.1} + K_{1.2} \partial v_2 + v_2 \partial K_{1.2} + \ldots,$
and therefore

$$K_{1.1} \partial v_1 + K_{1.2} \partial v_2 + \ldots = - v_1 \partial K_{1.1} - v_2 \partial K_{1.2} - \ldots.$$

Hence by substituting in (α) we get

$$\partial w = - \tfrac{1}{2} (K_{1.1} v_1 + K_{1.2} v_2 + \ldots) \partial v_1 - \tfrac{1}{2} (K_{2.1} v_1 + K_{2.2} v_2 + \ldots) \partial v_2 - \ldots$$
$$= - \tfrac{1}{2} v_1 (K_{1.1} \partial v_1 + K_{1.2} \partial v_2 + \ldots) - \tfrac{1}{2} v_2 (K_{2.1} \partial v_1 + K_{2.2} \partial v_2 + \ldots) - \ldots$$
$$= \tfrac{1}{2} v_1 (v_1 \partial K_{1.1} + v_2 \partial K_{1.2} + \ldots) + \tfrac{1}{2} v_2 (v_1 \partial K_{2.1} + v_2 \partial K_{2.2} + \ldots) + \ldots$$
$$= \tfrac{1}{2} v_1^2 \partial K_{1.1} + \ldots + v_1 v_2 \partial K_{1.2} + \ldots.$$

When the potentials of the n conductors are maintained constant

and have the same initial values as in the last case, the difference between the values of the electrostatic energy when X is in the positions P' and P equals

$$\tfrac{1}{2}(K_{1.1} + \partial K_{1.1})\, v_1^2 + \ldots + (K_{1.2} + \partial K_{1.2})\, v_1 v_2 + \ldots$$
$$- \tfrac{1}{2}K_{1.1}v_1^2 - \ldots - K_{1.2}v_1 v_2 - \ldots.$$

By the preceding paragraph this is equal to ∂w and is therefore positive. Hence the electrostatic energy in the position P' is greater than in the position P. Also, since the force acting on X in the position P is exactly the same in the two cases, and P' is infinitely near to P, it follows that, the electrostatic field being only infinitesimally disturbed by the motion of X, the force at P' is practically the same in the two cases. The mechanical work done on X, therefore, during an infinitesimal displacement is the same whether the charges or the potentials are maintained constant.

Thus the work done on X, when the potentials are constant, is also ∂w, and this equals the gain in the electrostatic energy of the system. It follows, by integration, that the total work done on X during a finite displacement equals the gain in the electrostatic energy of the system. In an electrostatic voltmeter, for instance, the energy taken from the mains equals twice the mechanical energy required to displace the moving part.

These theorems can be stated more generally as follows.

(1) When the relative positions of a system of insulated conductors alter owing to their mutual electric actions, the conductors move in such a way that the electrostatic energy is diminished, the diminution being equal to the work done on the conductors.

(2) When the relative positions of a system of conductors, the potentials of which are maintained constant by means of external sources, alter owing to their mutual electric actions, the conductors move in such a way that the electrostatic energy is increased by an amount exactly equal to the work done on the conductors.

Maxwell defined the capacity of a conductor to be its charge

when its own potential is unity and that of all the other conductors is zero. In other words, it is the coefficient of self induction of the conductor for electrostatic charges. It is to be noted that any alteration in the position of any of the conductors generally alters the capacity of the conductor. In those practical cases, however, where we want to know this coefficient, all the conductors are fixed in position.

The capacity of a conductor is found by measuring the charge q which flows into it when it is connected to one terminal of an insulated battery whose electromotive force is v, the other terminal being connected to earth or to any of the surrounding conductors which are all earthed in this case as required by the definition. The ratio of q to v gives us the required capacity.

Electricians generally refer to the capacity of a conductor as the capacity between the conductor and all neighbouring conductors in parallel with the earth. In order to find the coefficients of mutual induction for electrostatic charges between the various conductors we find relations between the coefficients $K_{1.1}$, $K_{1.2}$, ... by measuring what is called the capacity between two of the conductors or between two groups of the conductors. The capacity between two conductors may be defined as follows.

Let there be any number of conductors 1, 2, 3, ... n, and let 1 and 2 be insulated. Then, if, when all the con-

Capacity between two conductors. ductors are initially uncharged a charge q be given to 1 and a charge $- q$ to 2, and if the potential of 1 now exceed that of 2 by $v_1 - v_2$, then the ratio of q to $v_1 - v_2$ is constant and is called the capacity between the two conductors. It is to be noticed that any alteration in the position of any of the conductors 3, 4, ... n generally alters the capacity between 1 and 2. Also connecting by fine wires any of the insulated conductors to one another or to the earth in general alters the capacity between 1 and 2, and so it is necessary to specify which of the conductors 3, 4, ... n are insulated from earth and whether any are joined together.

In practice the equal and opposite charges are given to the two conductors by connecting the terminals of an insulated battery to them, and the capacity between them is measured in exactly the same way as the capacity of an ordinary condenser.

Maxwell's equations on page 147 enable us to find in all cases

an expression for the capacity between two conductors in terms of the coefficients $K_{1.1}$, $K_{1.2}$, ... We shall apply them to find the capacity between two conductors, 1 and 2, when all other conductors in the neighbourhood are earthed. Give a charge q to 1 and a charge $-q$ to 2 and let their potentials be v_1 and v_2 respectively, then, since v_3, v_4, ... are all zero,

$$\left. \begin{array}{l} q = K_{1.1}v_1 + K_{1.2}v_2 \\ -q = K_{2.1}v_1 + K_{2.2}v_2 \end{array} \right\} \dots\dots\dots\dots\dots(2).$$

Solving these equations for v_1 and v_2 we find that

$$v_1 - v_2 = q \frac{K_{1.1} + K_{2.2} + 2K_{1.2}}{K_{1.1}K_{2.2} - K^2_{1.2}},$$

and

$$K = \frac{K_{1.1}K_{2.2} - K^2_{1.2}}{K_{1.1} + K_{2.2} + 2K_{1.2}} \dots\dots\dots\dots(3),$$

where K is the capacity between the two conductors. If $K_{1.1} = K_{2.2}$, then

$$K = \tfrac{1}{2}(K_{1.1} - K_{1.2}) \dots\dots\dots\dots(4).$$

An important practical case arises when the conductor 1 completely encloses the conductor 2. In this case when the charge on the conductor 2 is q, the induced charge on the inside of 1 will be $-q$. By definition, the difference of potential between 1 and 2 is the work done against the electric forces when we take a unit of positive electricity from 1 to 2. Hence the difference of potential between 1 and 2 depends only on the charge on the conductor 2, since the space inside 1 is completely screened from electrostatic induction from the outside. Hence, also, Maxwell's equation for q is

$$q = K_{2.1}v_1 + K_{2.2}v_2.$$

If v_1 equals v_2, q must be zero, therefore we must have $K_{2.2}$ equal to $-K_{2.1}$, hence

$$\frac{q}{v_2 - v_1} = K_{2.2} = -K_{2.1}.$$

Since this is true in all cases, it is true when the charge on the outside of the conductor 1 is zero, and hence it follows from our definition that the capacity between the conductors 1 and 2 is $K_{2.2}$ or $-K_{1.2}$.

If the conductor 2 be a metal sphere the centre of which coincides with the centre of a spherical cavity in the conductor 1, then if K be the capacity between 1 and 2,

$$K = K_{2.2} = - K_{1.2} = \frac{r_1 r_2}{r_1 - r_2},$$

where r_1 is the radius of the spherical cavity and r_2 is the radius of the metal sphere. When r_1 is infinitely great K equals r_2.

If the conductors instead of being separated from one another by air were separated by insulating materials like india-rubber, paper, oil, etc., then the capacities and coefficients of induction of the conductors would be altered. If they were embedded in a homogeneous insulating mass whose dielectric coefficient (specific inductive capacity) was λ, then the new constants would be $\lambda K_{1.1}$, $\lambda K_{1.2}$, etc. In electric lighting cables the conductors are as a rule separated by various materials whose dielectric coefficients are different. This considerably increases the difficulty of calculating the capacities between the various conductors, but as they are generally arranged in a symmetrical manner inside a metal sheath various useful formulae can be found giving all the capacities in terms of two or three constants. We will first however find an expression for the capacity of a condenser and calculate the capacities of concentric mains and of two parallel cylinders.

A condenser consists of two equal insulated conductors 1 and 2, whose coefficients $K_{1.1}$, $K_{2.2}$, $K_{1.2}$ are very large com-

Capacity of a condenser. pared with the mutual coefficients between 1 or 2 and the earth or other conductors. For a condenser we have, therefore, approximately, since $K_{1.1}$ is taken to be equal to $K_{2.2}$,

$$q = K_{1.1} v_1 + K_{1.2} v_2,$$
$$- q = K_{1.2} v_1 + K_{1.1} v_2.$$

Now these equations must be true when v_2 is zero and thus $K_{1.1} = - K_{1.2}$ approximately. Using this relation, we at once find that the capacity between the two conductors, or briefly the capacity of the condenser, is approximately $K_{1.1}$ or $- K_{1.2}$.

In the theoretical condenser we suppose that $K_{1.2}$ is infinitely great compared with the other mutual coefficients, and hence $K_{1.1}$ or $- K_{1.2}$ is the capacity of the theoretical condenser.

Let a be the outer radius of the inner cylindrical conductor and b the inner radius of the outer conductor. Let also V_1 and V_2 be their potentials, $+q$ and $-q$ be their charges per unit length, and let k be the capacity per unit length, then

Capacity of a concentric main.

$$q = k(V_1 - V_2) \quad \dots\dots\dots\dots\dots(5).$$

Now, from symmetry, the equipotential surfaces between the two cylinders are coaxial cylinders no matter how small a may be. Also the potential inside due to the charge on the outer cylinder is a constant and hence, if V be the potential at a point P distant x from the axis, where x lies between a and b, the force $\partial V/\partial x$ on unit of positive electricity placed at P will be the same, by Green's theorem (see page 9), as if the charge q were concentrated along an infinitely thin conductor coincident with the axis. Hence if ∂z be an element of the axis,

$$-\frac{\partial V}{\partial x} = \int_{-\infty}^{+\infty} \frac{qx\partial z}{(z^2 + x^2)^{\frac{3}{2}}} = \frac{2q}{x}.$$

This could also have been proved easily from Laplace's equation, which in this case gives us (page 6) $V = A + B \log x$; therefore (page 8)

$$\sigma = -\frac{1}{4\pi}\frac{\partial V}{\partial x} = -\frac{B}{4\pi a}$$

on the inner cylinder. Now $q = 2\pi a\sigma$ and hence $B = -2q$.

If λ be the dielectric coefficient of the insulating medium,

$$-\frac{\partial V}{\partial x} = \frac{2q}{\lambda x};$$

and therefore
$$V_1 - V_2 = \frac{2q}{\lambda}\int_a^b \frac{\partial x}{x}$$

$$= \frac{2q}{\lambda}\log\frac{b}{a}.$$

Comparing this result with (5), we see that

$$k = \frac{\lambda}{2\log\dfrac{b}{a}},$$

or if K be the capacity, in electrostatic units, of a length l centimetres of concentric main, then

$$K = kl$$
$$= \frac{\lambda l}{2 \log_\epsilon \frac{b}{a}} \quad \text{............................(6).}$$

We have hitherto used λ for the dielectric coefficient relative
Dielectric currents. to air in connection with the electrostatic system of units. Similarly μ has always appeared as the ratio of B to H measured on the electromagnetic system and thus denotes the permeability relative to that of air. For the purposes of this section it is necessary to generalize the meanings of λ and μ. We may state that in whatever system of units q is measured the force between two charges can be expressed by $qq'/\lambda r^2$, where q and q' are the charges and r is the distance between them. It will be noticed that λ now depends not only upon the medium but also upon the unit of charge. Similarly we can generalize μ. It has to be remembered that both the electrostatic and the electromagnetic systems are absolute systems and the connections between the units of the different electrical and magnetic quantities are the same in each. The following investigation holds whichever system of units is used.

Let us consider the phenomena which take place in a homogeneous dielectric when the Faraday tubes are in motion. In order to simplify the theory as much as possible we shall only consider the case of a parallel plate condenser, the positive electrode (Fig. 42) being the YZ plane and the negative electrode being a parallel plane through O'. We suppose that the lines of force are all parallel to OX, and that the electric density σ of the electricity on the plates is independent of the value of y. Hence any alteration in σ will cause the tubes to move at right angles to their length and parallel to OZ. If σ be the density

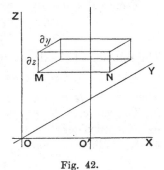

Fig. 42.

of the electricity at M, the end of the tube of induction MN (Fig. 42) the section of which is $\partial y \partial z$, and if P be the induction in the tube we have $P = \sigma = (\lambda/4\pi) R$, where R is the electric force. If a current flows in the electrode, σ alters and so the induction in the tube alters also. The stream lines of current flow will be continuous with the lines of induction in the dielectric. The ordinary current contained in the tube of flow which is continuous with the tube of induction MN stops at M. The magnitude of this current is

$$(\partial\sigma/\partial t)\,\partial y\,\partial z.$$

Following Maxwell, we shall suppose that the varying induction inside the tube MN produces exactly the same magnetic effects as if the ordinary current flowed along it. This hypothetical current in the tube MN we shall call the dielectric current. Its magnitude may be written $(\lambda/4\pi)(\partial R/\partial t)\,\partial y\partial z$. Since the line integral of the magnetic force round this tube must be equal to 4π times the dielectric current, we have, since the contributions to the line integral of H along the sides parallel to OZ are zero,

$$H\partial y - \left(H + \frac{\partial H}{\partial z}\partial z\right)\partial y = \lambda\left(\frac{\partial R}{\partial t}\right)\partial y\,\partial z,$$

and so
$$\frac{\partial H}{\partial z} = -\lambda\frac{\partial R}{\partial t}\dots\dots\dots\dots(a).$$

Again if μ be the permeability of the medium, the magnetic induction through $\partial x\partial z = \mu H\partial x\partial z$, and hence by Faraday's law

$$R\partial x - \left(R + \frac{\partial R}{\partial z}\partial z\right)\partial x = \frac{\partial}{\partial t}\left(\mu H\right)\partial x\partial z,$$

and therefore
$$\frac{\partial R}{\partial z} = -\mu\frac{\partial H}{\partial t}\dots\dots\dots\dots(b).$$

From (a) and (b) we find that

$$\frac{\partial^2 H}{\partial z^2} = \lambda\mu\frac{\partial^2 H}{\partial t^2}\text{ and }\frac{\partial^2 R}{\partial z^2} = \lambda\mu\frac{\partial^2 R}{\partial t^2}.$$

The solution of the differential equation for H is

$$H = f_1\{z - t/(\lambda\mu)^{1/2}\} + f_2\{z + t/(\lambda\mu)^{1/2}\}\dots\dots(c),$$

where f_1 and f_2 are any two arbitrary functions.

The value of R is given by an exactly similar equation. Let us now consider the meaning of the first term of the solution for H. When t becomes $t + t_1$, it becomes

$$f_1[\{z - t_1/(\lambda\mu)^{1/2}\} - t/(\lambda\mu)^{1/2}],$$

and hence its value at all points on the plane whose height is z above the plane of XY at the time $t + t_1$ is the same as its value at all points on the plane whose height above the plane of XY is $z - t_1/(\lambda\mu)^{1/2}$ at the time t. It therefore represents a wave travelling upwards with velocity $1/(\lambda\mu)^1$. Similarly the second term in the solution represents a wave travelling downwards with equal velocity. Hence any disturbance of the initial conditions sets up two waves in the medium. According to this theory the Faraday tubes move at right angles to their length with velocity v and are accompanied by magnetic forces in a direction at right angles to them and to the direction in which the tubes are moving.

From (a), (b) and (c) we get $H = \lambda v R$ and $R = \mu v H$. Hence $\mu H^2/8\pi = \lambda R^2/8\pi = HR/(8\pi v)$. The total energy per unit volume being the sum of the electric and magnetic energies is therefore equal to $\mu H^2/8\pi + \lambda R^2/8\pi$, that is, $HR/(4\pi v)$. Hence the rate at which energy crosses unit area of the wave front equals $HR/(4\pi)$. This is a particular case of a more general theorem due to Poynting.

It follows that the velocity v of electromagnetic waves in air is $1/\lambda^{1/2}$ if λ is the dielectric coefficient of air in the electromagnetic system of units. Similarly in the electrostatic system of units v equals $1/\mu^{1/2}$. We shall now show that v is the ratio of the electromagnetic unit of electric charge to the electrostatic unit.

Let q_e be the measure of a charge on the electrostatic system and let q_m be its measure on the electromagnetic system. Then the force between two equal charges q at a distance r apart in air is q_e^2/r^2 and in the electromagnetic system this force is $q_m^2/(\lambda r^2)$, where λ is the dielectric coefficient of air when the electromagnetic system is used. Hence

$$\frac{q_e^2}{r^2} = \frac{q_m^2}{\lambda r^2}, \text{ and therefore } \frac{q_e}{q_m} = \frac{1}{\lambda^{1/2}} = v.$$

If K_e be the capacity of a condenser and q_e the charge in it, both being measured in the electrostatic system of units, the energy stored in it is $q_e^2/(2K_e)$. Similarly in the electromagnetic system the energy stored in the condenser will be given by $q_m^2/(2K_m)$. Hence

$$\frac{q_e^2}{2K_e} = \frac{q_m^2}{2K_m}, \text{ and thus } K_m = \frac{K_e}{v^2}.$$

According to Maxwell's theory, light is an electromagnetic phenomenon. Its velocity in air should therefore be equal to the ratio of the electromagnetic unit of charge to the electrostatic unit. Determinations of this ratio by electrical experiments agree to a high degree of accuracy with the values obtained for the velocity of light. This agreement is strong evidence of the soundness of the electromagnetic theory.

Let us suppose that a point charge q at O is moving with a
Laplace's
formula. velocity v in the direction OX. After a time ∂t let its position be O', so that $OO' = v\partial t$. Let us consider the change in the flux linked with a circle having its centre at A, and its plane perpendicular to the direction of the motion of the charge. If the radius of the circle be a, the flux linked with it is $(q/4\pi)\omega$, where ω is the solid angle subtended at O by the circle. Hence, since (p. 25), $\omega = 2\pi(1 - \cos\theta)$, where θ is the angle POA, we have

$$\phi = (q/2)(1 - \cos\theta),$$

and so, if we neglect the effect of the forces due to electromagnetic induction,

$$\frac{\partial\phi}{\partial t} = \frac{q\sin\theta}{2} \cdot \frac{\partial\theta}{\partial t}.$$

Draw $O'B$ perpendicular to OP, and let $OP = r$, then $BO' = r\,\partial\theta$ approximately, and so

$$\sin\theta = \frac{r\partial\theta}{v\partial t} = \frac{a}{r}, \text{ and } \frac{\partial\theta}{\partial t} = \frac{av}{r^2}.$$

Hence
$$\frac{\partial\phi}{\partial t} = \frac{qav\sin\theta}{2r^2}.$$

We also have (p. 39) $\int H\partial s = 4\pi i$, and therefore

$$H.2\pi a = 4\pi(q/2)\sin\theta\,(av/r^2),$$

and hence $H = qv\sin\theta/r^2$.

Let us now suppose that a current i is flowing along OX and that q is the charge on an element ∂s at O, so that $i = (q/\partial s)v$, and therefore $qv = i\partial s$. Substituting we get

$$H = \frac{i\sin\theta\,\partial s}{r^2}$$

for the magnetic force at P, due to the current in the element ∂s, which is Laplace's formula (see p. 43).

In proving the formula above we have neglected the displacement of the Faraday tubes produced by the forces due to electromagnetic induction. On the hypothesis, however, that the motion is slow the displacement current through the circle is small and therefore the magnetic force is small. Hence the electric force due to the slow changes of the small magnetic force can be neglected. The above formula is therefore very approximately true when v is small.

In this case the force at the point P due to electromagnetic induction is $\mu H v$, that is, $\mu\,(qv\sin\theta/r^2)v$ and the electrostatic force is $q/(\lambda r^2)$. The ratio therefore of these forces is $\mu\lambda v^2\sin\theta$, that is, $\sin\theta\,(v/V)^2$, where V is the velocity of light. It will be seen that the forces due to electromagnetic induction tend to make the Faraday tubes spread out towards the plane which is at right angles to their direction of motion. If the charge were moving with the velocity of light all the tubes would be forced into this plane.

Let us now consider a sphere of radius a possessing a charge q and moving with a velocity v, where v^2/V^2 **The energy of a moving charge.** can be neglected compared with unity. The magnetic force at a point P distant r from the centre of the sphere is $qv\sin\theta/r^2$, where θ is the angle r makes with the direction of motion. Hence since $\mu H^2/8\pi$ is the kinetic energy per unit volume we see that the energy in the element

$r^2\sin\theta\,\partial r\partial\theta\partial\phi$ is $\{\mu q^2v^2\sin^2\theta/(8\pi r^4)\}\,r^2\sin\theta\,\partial r\partial\theta\partial\phi.$

Hence the total kinetic energy outside the sphere

$$= \frac{\mu q^2v^2}{8\pi}\int_a^\infty \frac{\partial r}{r^2}\int_0^\pi \sin^3\theta\partial\theta\int_0^{2\pi}\partial\phi = \frac{\mu q^2v^2}{3a}.$$

Thus if m be the mass of the uncharged sphere, the total kinetic energy is $(1/2)\,(m+2\mu q^2/3a)\,v^2$. Hence the effect of the charge is to increase the mass by a quantity $2\mu q^2/3a$. It follows, therefore, from Maxwell's theory that a charge of electricity possesses inertia and has the properties of mass.

From the theory of hydrodynamics we know that if a sphere

of mass m be moving in a liquid with velocity v its kinetic energy is $(1/2)(m + m'/2)v^2$, where m' is the mass of the liquid displaced by the sphere. A probable conjecture therefore is that the increased mass is due to the Faraday tubes setting the ether in motion.

In a theory due to J. J. Thomson, material atoms are supposed to be made up of very small particles having negative charges and embedded in a sphere of positive electricity, the sum of the negative charges being exactly equal to the positive charge. In this theory the inertia of matter is explained by the motion given to the ether by the Faraday tubes.

We have already seen (p. 157) that if K_m be the capacity of a condenser measured in electromagnetic units and K_e be its capacity in electrostatic units, we have $K_m = K_e/v^2$, where $v = 3 \times 10^{10}$. The C.G.S. unit of electromagnetic capacity is too great for practical use. We therefore use two smaller units, the farad and the microfarad, the latter being the millionth part of the farad. In practice the microfarad is always used.

Capacity in electromagnetic units. The microfarad.

The farad is the capacity of a condenser which has a P.D. of one volt between its terminals when charged with one coulomb. Since the coulomb is the tenth of the C.G.S. unit of quantity, and the volt equals 10^8 C.G.S. units, we get

$$1 \text{ farad} = \frac{1}{10^9} \text{ C.G.S. unit,}$$

and

$$1 \text{ microfarad} = \frac{1}{10^{15}} \text{ C.G.S. unit.}$$

Hence if K be the capacity of a condenser in microfarads, and K_e be its capacity in electrostatic units,

$$K = \frac{10^{15}}{v^2} K_e$$

$$= \tfrac{1}{900000} K_e.$$

For example, the capacity of a concentric main of length l miles (160,900 l centimetres) in microfarads is given by (p. 155)

$$K = 0.0776 \frac{\lambda l}{2 \log \frac{b}{a}} \quad \dots\dots\dots\dots\dots(7),$$

where the logarithm is to the base 10.

Let a be the outside radius of the inner main, b and c the radii of the middle main, and d the inside radius of the outer main. Then, denoting the capacity between the inner and the middle main by $C_{1.2}$, etc., we get

The capacities of a triple concentric main.

$$
\left.
\begin{aligned}
C_{1.2} &= \frac{\lambda l}{2 \log_\epsilon (b/a)} \\[2mm]
C_{2.3} &= \frac{\lambda l}{2 \log_\epsilon (d/c)} \\[2mm]
C_{1.3} &= \frac{\lambda l}{2 \log_\epsilon (bd/ca)}
\end{aligned}
\right\} \quad \dots\dots\dots\dots(8).
$$

In the case of $C_{1.3}$ the middle main is insulated. The presence of the inner cylinder thus increases the capacity between the two outers.

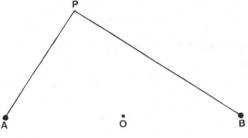

Fig. 43.

Consider first the equipotential lines round two thin cylinders which are placed with their axes A and B (Fig. 43) perpendicular to the plane of the paper. Suppose that they are so thin that they may be regarded as lines, and that the cylinder A is charged with a quantity of electricity $+q$ per unit length and that the B cylinder has a charge $-q$. Now join AB and bisect it in O; let $AO = d/2$. Let v_A be the potential at any point P (Fig. 43) due to the action of the A wire alone and let v_A' be the potential at O. Then if the medium be air,

The capacity of a condenser formed by two long parallel cylinders.

$$
\frac{\partial v_A}{\partial r} = \frac{2q}{r}.
$$

Thus
$$v_A - v_A' = \int_{r_1}^{\frac{d}{2}} \frac{2q}{r}\, \partial r$$
$$= 2q \log \frac{d}{2r_1}$$

where $AP = r_1$.

Similarly if v_B and v_B' be potentials at P and O due to the action of the wire B,
$$v_B - v_B' = -2q \log \frac{d}{2r_2}$$

where $BP = r_2$.

Now by the principle of superposition, if v be the potential at P due to both, then
$$v = v_A + v_B.$$

Also by symmetry the potential at O will be zero,

and therefore $$0 = v_A' + v_B'.$$

Hence $$v = 2q \log \frac{r_2}{r_1}.$$

Therefore the equation to the equipotential surface the potential of which is V is
$$V = 2q \log \frac{r_2}{r_1} \quad \dots\dots\dots\dots\dots\dots(1),$$

or $$\frac{r_1}{r_2} = \text{constant}.$$

Now by a well-known geometrical theorem the locus of a point P which moves so that the ratio of its distances from two fixed points A and B is constant, is a circle, and if C be its centre and a its radius, $CA \cdot CB = a^2$. These points are called inverse points with regard to the circle.

It will be seen that the equipotential surfaces are a series of cylinders surrounding A and B, all of which have A and B for inverse points. A particular case is the plane bisecting AB at right angles and passing through O. Now it follows from Green's theorem that, if we distribute electricity over one of the equipotential cylinders surrounding the wire A and if the surface density of the distribution at any point be $R/4\pi$, where R is the normal force at that point, then this distribution will be in equilibrium, the potential at external points will be unaltered, and the potential at all points inside this cylinder will be constant. Similarly we can replace the wire B by one of the equipotential cylinders surrounding it.

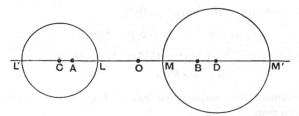

Fig. 44. *A* and *B* are the inverse points of the circles.
$CA . CB = CL^2$; $DA . DB = DM^2$.
$CL = a$, $DM = b$, $CD = c$, $AB = d = 2r$.

Suppose then that we have a solid cylinder LL' (Fig. 44) surrounding A and another MM' surrounding B. Let their potentials be V_1 and V_2 respectively, and their charges $+q$ and $-q$ per unit length as before, then from (1)

$$V_1 = 2q \log \frac{BL}{AL},$$

and

$$V_2 = 2q \log \frac{BM}{AM}.$$

Therefore

$$V_1 - V_2 = 2q \log \frac{BL . AM}{AL . BM} = \frac{q}{k},$$

where k is the capacity per unit length.

Hence,

$$k = \frac{1}{2 \log \dfrac{BL}{AL} \cdot \dfrac{AM}{BM}} \qquad\qquad\qquad (2).$$

Now from Fig. 44,

$$\frac{BL}{AL} = \frac{BL'}{AL'} = \frac{2 . BC}{2 . CL} = \frac{BC}{CL} \left.\right\}$$

Also

$$\frac{AM}{BM} = \frac{AM'}{BM'} = \frac{2 . AD}{2 . MD} = \frac{AD}{MD} \left.\right\} \qquad (3).$$

Let the radii of the two cylinders be a and b respectively, and let c be the distance CD between their centres. Since $CA . CB = a^2$,

11—2

and $DB.DA = b^2$, the circle described on AB as diameter will intersect both of the circles at right angles. If $AB = 2r$, we have

$$c = CO + OD = (a^2 + r^2)^{1/2} + (b^2 + r^2)^{1/2} \quad \dots\dots\dots\dots(4),$$

and hence, $c^2 + a^2 - b^2 = 2c\,(a^2 + r^2)^{1/2}$,

$$4c^2 r^2 = (c^2 + a^2 - b^2)^2 - 4c^2 a^2$$

and

$$c^2 r^2 = 4s\,(s-a)\,(s-b)\,(c-s) \quad \dots\dots\dots\dots(5),$$

where

$$2s = a + b + c.$$

It is convenient to introduce symbols α, β and ω which are defined by the equations,

$$\sinh \alpha = r/a, \quad \sinh \beta = r/b \quad \text{and} \quad \sinh \omega = rc/ab \dots\dots(6).$$

From (4), $c = a \cosh \alpha + b \cosh \beta$, and hence, from (6) we easily find that $\omega = \alpha + \beta$.

We also have,

$$\cosh \omega = (c^2 - a^2 - b^2)/(2ab),$$
$$\left.\cosh \alpha = (c^2 + a^2 - b^2)/(2ca), \quad \cosh \beta = (c^2 + b^2 - a^2)/(2bc)\right\} \dots(7).$$

Now from (3),

$$\frac{BL.AM}{AL.BM} = \frac{BC.AD}{ab} = \frac{b.CB}{a.DB},$$

$$CB = (a^2 + r^2)^{1/2} + r = a\,(\cosh \alpha + \sinh \alpha) = a\epsilon^{\alpha},$$

and

$$DB = (b^2 + r^2)^{1/2} - r = b\,(\cosh \beta - \sinh \beta) = b\epsilon^{-\beta},$$

and thus

$$\frac{BL.AM}{AL.BM} = \frac{\epsilon^{\alpha}}{\epsilon^{-\beta}} = \epsilon^{\omega}.$$

Substituting in (2) we get, $k = 1/(2\omega)$.

If the length of each of the cylinders be l, and λ be the dielectric coefficient of the medium in which they are immersed, the capacity K between them is given by

$$K = \lambda lk = \lambda l/(2\omega) \quad \dots\dots\dots\dots\dots(8).$$

When a table of hyperbolic sines is available ω can be readily found by either of the following formulae,

$$\begin{aligned}\sinh(\omega/2) &= \{s\,(c-s)/ab\}^{1/2} \\ \cosh(\omega/2) &= \{(s-a)\,(s-b)/ab\}^{1/2}\end{aligned} \Big\} \quad \dots\dots\dots(9).$$

Again since $\omega = 2 \log_\epsilon \{\sinh(\omega/2) + \cosh(\omega/2)\}$ its value may also be computed from ordinary logarithmic tables.

If c be large compared with $a + b$, then approximately,

$$K = \frac{\lambda l}{2 \log \left\{ \dfrac{c^2}{ab} - \dfrac{a^2 + b^2}{ab} \right\}} \quad\dots\dots\dots\dots(10).$$

If $c = a + b + x$, where x is very small compared with either a or b, then

$$K = \frac{\lambda l (ab)^{1/2}}{2 \{2(a+b)\}^{1/2}} \cdot \frac{1}{\sqrt{x}} \quad\dots\dots\dots\dots(11).$$

When $a = b$; $\alpha = \beta = \omega/2$, and $\cosh \alpha = c/2a$. Hence

$$K = \frac{\lambda l}{4 \log \left[\{c + (c^2 - 4a^2)^{1/2}\}/(2a)\right]} \quad\dots\dots\dots(12),$$

exactly.

Finally if c be large compared with a,

$$K = \frac{\lambda l}{4 \log (c/a - a/c)} \quad\dots\dots\dots\dots(13),$$

approximately.

In these formulae for K, if l be in miles and the logarithms are to the base 10, then to reduce K to microfarads, we must multiply the right hand side of the equations by 0·0776. For example suppose that $a = 7r'$, $b = r'$, $c = 10r'$, and l is one mile. In this case $s = 9r'$, and we readily find that $\sinh(\omega/2) = 3/\sqrt{7}$, $\cosh(\omega/2) = 4/\sqrt{7}$, and thus $\omega = \log_e 7 = 1·946$. Hence if the medium is air, so that λ is unity,

$$K = 0·0776 (\lambda l/2\omega) = 0·0199 \text{ microfarad.}$$

It is to be noticed that the value of K does not depend on the value of r'.

The cross section of the two cylinders is shown in Fig. 45. The inner one is supposed to have a charge $+q$ per unit length, and the outer one a charge $-q$. Let A and B be the inverse points common to the two circles, then, using the same notation as in the preceding example, we have

The capacity of a condenser formed by two long parallel cylinders one wholly enclosed by the other.

$$V_1 - V_2 = 2q \log \frac{b \cdot CB}{a \cdot DB} \quad\dots\dots\dots\dots(14).$$

It is easy to see that the circle described on AB as diameter (Fig. 45) intersects the two circles orthogonally. Let the radius of this circle be r_1. Then we can easily prove that

$$CB = r_1 + (a^2 + r_1^2)^{1/2}; \quad DB = r_1 + (b^2 + r_1^2)^{1/2},$$

and hence $\qquad c = (b^2 + r_1^2)^{1/2} - (a^2 + r_1^2)^{1/2}$ ....(15),

and $\qquad c^2 r_1^2 = 4s\,(s - a)\,(s - c)\,(b - s)$...............(16).

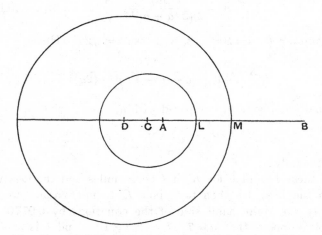

Fig. 45. A and B are the inverse points of the two circles.
$$CA \cdot CB = CL^2; \quad DA \cdot DB = DM^2.$$
$$CL = a, \quad DM = b, \quad CD = c, \quad AB = d = 2r_1.$$

Let α_1, β_1 and ω_1 be defined by the equations,

$$\sinh \alpha_1 = r_1/a, \quad \sinh \beta_1 = r_1/b \quad \text{and} \quad \sinh \omega_1 = r_1 c/ab...(17).$$

It follows from (15) that $\omega_1 = \alpha_1 - \beta_1$.
We also have

$$\left. \begin{array}{c} \cosh \omega_1 = (b^2 + a^2 - c^2)/(2ab) \\ \cosh \alpha_1 = (b^2 - a^2 - c^2)/(2ca), \quad \cosh \beta_1 = (b^2 + c^2 - a^2)/(2cb) \end{array} \right\} \ ...(18)$$

Noticing that

$$\frac{b \cdot CB}{a \cdot DB} = \frac{\sinh \alpha_1 + \cosh \alpha_1}{\sinh \beta_1 + \cosh \beta_1} = \epsilon^{\alpha_1 - \beta_1} = \epsilon^{\omega_1},$$

we get from (14),

$$K = \frac{\lambda l}{2\omega_1} \ ....(19),$$

and ω_1 can be found from either of the following formulae

$$\left.\begin{array}{l} \sinh(\omega_1/2) = \{(s-a)(b-s)/ab\}^{1/2} \\ \cosh(\omega_1/2) = \{s(s-c)/ab\}^{1/2} \end{array}\right\} \dots\dots\dots\dots(20).$$

We also have $\omega_1 = 2 \log \{\sinh(\omega_1/2) + \cosh(\omega_1/2)\}$.

If $c = (b-a) - x$, and x be very small compared with either a or b, we have $\omega_1 = x^{1/2} \{2(b-a)/ab\}^{1/2}$ approximately, and

$$K = \frac{\lambda l (ab)^{1/2}}{2\{2(b-a)\}^{1/2}} \cdot \frac{1}{\sqrt{x}} \dots\dots\dots\dots\dots(21).$$

When c is very small compared with $b - a$, we have the approximate formula,

$$K = \frac{\lambda l}{2 \log[(b/a)\{1 - c^2/(b^2 - a^2)\}]} \dots\dots\dots\dots(22).$$

We see that the capacity is a minimum when the axis of the inner cylinder coincides with the axis of the outer one. The potential energy $q^2/(2K)$ is therefore a maximum and the inner cylinder would be in unstable equilibrium if it were free to move in any direction.

If K be the value of the capacity between the two cylinders when $c = 0$, and $2K$ be the value when $c = c'$, we have $2\omega_1' = \omega_1$,

and therefore, $\cosh \omega_1' = \cosh(\omega_1/2)$,

and $\left(\dfrac{b^2 + a^2 - c'^2}{2ab}\right)^2 = \dfrac{(b+a)^2}{4ab}$,

and so, $c'^2 = a^2 + b^2 - (ab)^{1/2}(a+b)$.

For example, if $b = 10$ and $a = 8.1$, then when $c = 1.646$ the capacity between the cylinders will be twice as great as when c is zero.

Suppose that we have three cables whose sections are shown in Fig. 46 and suppose that they are l miles long and that λ is the

(1)   (2)   (3)

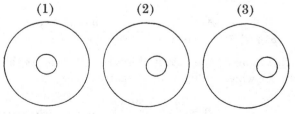

Fig. 46.   Capacities of (1), (2) and (3) are K, $1.05\,K$ and $1.30\,K$.

dielectric coefficient of the insulating material used. Then the following table gives their constants, where $K = 0.0644\lambda l$ microfarads.

Condenser	b	a	c	ω_1 calc. from (20)	Capacity in microfarads
(1)	4	1	0	1·39	K
(2)	4	1	1	1·32	$1.05\ K$
(3)	4	1	2	1·07	$1.30\ K$
(4)	4	1	2·646	0·69	$2\ K$

The maximum value of the potential gradient between two parallel cylinders.

Let us first consider the case when the two cylinders are external to one another and immersed in a medium the dielectric coefficient of which is λ. We have already seen (p. 162) that the potential v at a point P external to both is given by

$$v = (2q/\lambda) \log (BP/AP),$$

where A and B are the inverse points of the circular sections made by a plane through P perpendicular to the axes of the cylinders. In obtaining this formula we supposed that $+ q$ and $- q$ were the charges per unit length on the two cylinders. Hence, if C be the centre of the circle whose radius is a, if $CP = \rho$, and if the angle PCB be θ, we have

$$v = \frac{q}{\lambda} \log \frac{\rho^2 + (CA + 2r)^2 - 2\rho\,(CA + 2r) \cos \theta}{\rho^2 + (CB - 2r)^2 - 2\rho\,(CB - 2r) \cos \theta}$$

$$= \frac{q}{\lambda} \log \frac{\rho^2 + a^2\,\epsilon^{2a} - 2\rho a\epsilon^a \cos \theta}{\rho^2 + a^2\,\epsilon^{-2a} - 2\rho a\epsilon^{-a} \cos \theta}.$$

Hence

$$\frac{\lambda}{2q}\frac{\partial v}{\partial \rho} = \frac{\rho - a\epsilon^a \cos \theta}{\rho^2 + a^2\,\epsilon^{2a} - 2\rho a\epsilon^a \cos \theta} - \frac{\rho - a\epsilon^{-a} \cos \theta}{\rho^2 + a^2\,\epsilon^{-2a} - 2\rho a\epsilon^{-a} \cos \theta}.$$

If R be the value of the potential gradient $- \partial v/\partial \rho$ at the surface of the cylinder whose radius is a, so that $\rho = a$, we have

$$R = \frac{2q}{\lambda a} \cdot \frac{\sinh \alpha}{\cosh \alpha - \cos \theta} \quad \dots\dots\dots\dots(23).$$

If $v_1 - v_2$ be the potential difference between the cylinders, we get by (8), $q = (\lambda/2\omega)(v_1 - v_2)$, and thus

$$R = \frac{v_1 - v_2}{a\omega} \cdot \frac{\sinh \alpha}{\cosh \alpha - \cos \theta} \quad \ldots\ldots\ldots\ldots(24),$$

and α and ω can easily be found by (7). When $\cos \theta = 1$, R has its maximum value $R_{max.}$, and when $\cos \theta = -1$, it has its minimum value $R_{min.}$. We thus have

$$R_{max.} = \frac{v_1 - v_2}{a\omega} \cdot \frac{\sinh \alpha}{\cosh \alpha - 1} = \frac{v_1 - v_2}{a\omega} \cdot \left[\frac{(c+a)^2 - b^2}{(c-a)^2 - b^2}\right]^{1/2} \quad \ldots(25),$$

and $R_{max.}/R_{min.} = \{(c+a)^2 - b^2\}/\{(c-a)^2 - b^2\}$.

Similarly if $R'_{max.}$, $R'_{min.}$, be the maximum and minimum values of the potential gradient on the surface of the cylinder whose radius is b, we have

$$R'_{max.} = -\frac{v_1 - v_2}{b\omega} \cdot \frac{\sinh \beta}{\cosh \beta - 1} = -\frac{v_1 - v_2}{b\omega}\left[\frac{(c+b)^2 - a^2}{(c-b)^2 - a^2}\right]^{1/2} \quad \ldots(26),$$

and $R'_{max.}/R'_{min.} = \{(c+b)^2 - a^2\}/\{(c-b)^2 - a^2\}$.

It is easy to see from (25) and (26) that the maximum value of the potential gradient occurs along the line where the plane through the axes of the two cylinders cuts the smaller one.

As a numerical example, let us suppose that $c = 6$, $a = 1$ and $b = 4$. Then $R_{max.}/R_{min.} = 33/9$ and $R'_{max.}/R'_{min.} = 33$. Hence the surface density is more uniform round the smaller cylinder. By (7) we get $\cosh \omega = 2\cdot375$, and thus $\omega = 1\cdot5105$. Substituting in (25) we get $R_{max.} = (v_1 - v_2)(1\cdot269)$. For example if

$$v_1 - v_2 = 10000 \text{ volts,}$$

$R_{max.}$ will be $12\cdot69$ kilovolts per centimetre.

When $a = b$, (25) becomes

$$R_{max.} = \frac{v_1 - v_2}{2a \log\left[\{c + (c^2 - 4a^2)^{1/2}\}/(2a)\right]} \cdot \left[\frac{c+2a}{c-2a}\right]^{1/2} \quad \ldots(27),$$

and $R_{max.}/R_{min.} = (c+2a)/(c-2a)$.

If the distance $(c - 2a)$ between the two cylinders be very small compared with their diameters

$$R_{max.} = (v_1 - v_2)/(c - 2a) \text{ approximately.}$$

Let us now suppose that the larger cylinder is hollow and that the smaller one is inside it, their axes being parallel and at a distance c apart. If the radius of the inner cylinder be a, then proceeding as before we get

$$R = \frac{2q}{\lambda a} \cdot \frac{\sinh \alpha_1}{\cosh \alpha_1 - \cos \theta},$$

and

$$R_{max.} = \frac{v_1 - v_2}{a\omega_1} \cdot \frac{\sinh \alpha_1}{\cosh \alpha_1 - 1} = \frac{v_1 - v_2}{a\omega_1}\left[\frac{b^2 - (a-c)^2}{b^2 - (a+c)^2}\right]^{1/2} \quad ...(28).$$

We also have $R_{max.}/R_{min.} = \{b^2 - (a-c)^2\}/\{b^2 - (a+c)^2\}$.
Similarly we find that

$$R'_{max.} = -\frac{v_1 - v_2}{b\omega_1}\left[\frac{(b+c)^2 - a^2}{(b-c)^2 - a^2}\right]^{1/2} \quad............(29),$$

and $R'_{max.}/R'_{min.} = \{(b+c)^2 - a^2\}/\{(b-c)^2 - a^2\}$.

When c is zero, $\cosh \omega_1 = b/2a + a/2b$, and therefore

$$\omega_1 = \log(b/a),$$

and (28) becomes

$$R_{max.} = \frac{v_1 - v_2}{a \log(b/a)} \quad....................(30).$$

This well-known formula, which gives the maximum potential gradient in the insulating material between the inner and outer cylinders of a concentric main, can be easily proved directly. For in this case the potential gradient at any point in the dielectric at a distance x from the axis is $2q/x$, and this is a maximum when x equals a.

When power is transmitted by alternating currents in overhead wires, it is found that when the potential of a wire exceeds a definite critical value E_0, a luminous discharge takes place at its surface. At potentials higher than the critical value the wire appears to be surrounded by a halo of light. This appearance is known as the 'corona.' If the wire be cylindrical in shape and perfectly smooth and clean the corona presents a uniform appearance. At a little distance away its diameter appears to be uniform. The slightest surface imperfections however are marked by variations in both the colour and diameter of the corona. In small wires the part of the corona next the wire has a pink tinge shading off to violet, the intensity of which lessens as we approach the outer boundary. On larger

Practical application.

wires the region round the wire is not so bright. There appears to be a constant play of intermittent thread-like sparks from the surface of the wire to the outer boundary of the corona.

It is found by experiment that when the potential of the line is raised above the critical value the power expended on it is considerably increased. If the potential of the line be raised to E, where E is greater than E_0, the power expended on the corona can be expressed by $kf(E - E_0)^2$, where f is the frequency and k is a constant depending on the geometrical and physical conditions of the line. The power loss increases rapidly with E and for this and other reasons the transmission becomes uneconomical. The value of E_0 is thus of great importance in the design of high pressure systems for transmitting power.

Experimental results obtained by J. B. Whitehead and F. W. Peek prove that when cylindrical conductors are at appreciable distances apart the maximum value $R_{max.}$ of the electric force at the surface of a conductor determines the value of the critical voltage. When the corona first appears on a wire the air round it becomes ionised and its conductivity will cause an electroscope to discharge rapidly. By placing wires of various diameters along the axis of a hollow cylinder and then noting the potential difference between the wires and the cylinder when ionisation first ensued at the surface of the inner wire, Whitehead proved that at 21° C. and 76 cms. pressure, the critical value of $R_{max.}$ was given by

$$R_{max.} = 32 + 9 \cdot 5/(a)^{1/2} \quad \ldots\ldots\ldots\ldots\ldots\ldots(31),$$

in kilovolts per centimetre, a, the radius of the wire, being measured in centimetres. Hence

$$E_0 = 32a \log(b/a) + 9 \cdot 5 (a)^{1/2} \log(b/a) \quad \ldots\ldots\ldots\ldots(32),$$

where E_0 is the maximum value of the critical potential difference and b is the inner radius of the outer cylinder. It is to be noticed that $R_{max.}$ is independent of b.

F. W. Peek deduces from his experiments on parallel cylindrical wires of equal radius a that the value of $R_{max.}$ when a disruptive discharge occurs in air at 25° C. and 76 cms. pressure is given by

$$R_{max.} = 30 + 9 \cdot 0/(a)^{1/2} \quad \ldots\ldots\ldots\ldots\ldots\ldots(33).$$

In this case again, $R_{max.}$ is independent of the distance between the wires. The value of $R_{max.}$ being determined by (33), E_0 can be found from (27). Considering the very different methods employed the agreement between (31) and (33) is satisfactory. Peek notices that the value of $R_{max.}$ at which the corona can first be detected by the eye is about three or four per cent. greater than that determined by (33). Any roughness on the surface of the wires lowers the value of the critical voltage. For instance when the wires have been in the open air for a long time the value of $R_{max.}$ may be as much as seven per cent. less than the value calculated from (33). When the conductors are stranded the calculation of $R_{max.}$ is difficult as the individual strands in the outside layer are spiralled round the central strand. We should expect however that the value of $R_{max.}$ for a stranded conductor would be less than for a solid conductor of equal cross section, and this is found to be the case in practice.

When the distance between the wires is less than a millimetre the formulae given above are not even approximately correct. Experimental evidence seems to point to the existence of a minimum value of the potential difference below which a disruptive discharge cannot take place. At ordinary temperatures and pressures this minimum value is about 300 volts.

To a first approximation we can say that the corona appears just as readily when the potential of the wire is positive as when it is negative. With alternating pressures the corona appears as a bluish white glow having a needle-like fringe and interspersed with reddish tufts of light. Examination with a stroboscope proves that the reddish tufts are on the negative wire and that the positive is surrounded by a bluish white glow.

In this case if K be the capacity of the cable measured in microfarads, then the effective value of the condenser current A is given by (p. 137)

Condenser currents in concentric cables and in two parallel overhead wires.

$$A = \alpha f K V\, 10^{-6} \text{ amperes,}$$

where f is the frequency, V the effective voltage between the wires, and α a constant which has its minimum value 2π when the wave of the applied voltage is sine-shaped. We have assumed that the potential drop due to the resistance of the cable is negligible.

The two conductors are embedded in insulating material
Two core cable. (Fig. 47) and are enclosed
in a metal sheath which
is connected to earth and is therefore
at zero potential. Let $K_{1.1}$ be the
capacity per mile of No. 1 conductor
when No. 2 and the sheath are earthed,
as defined on p. 150, let v_1 be its
potential at any instant, and let $K_{1.2}$
be the coefficient of mutual induction
per mile between the two conductors.
We will assume that v_1 is constant

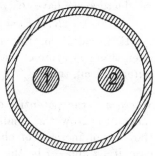

Fig. 47. Two core cable.

throughout the whole length of the cable at any instant. Then if
q_1 be the quantity of electricity on a mile of the No. 1 conductor,

$$q_1 = K_{1.1}v_1 + K_{1.2}v_2,$$

where v_2 is the instantaneous value of the potential of No. 2.

Similarly $\qquad q_2 = K_{2.2}v_2 + K_{2.1}v_1.$

These equations may be written:

$$q_1 = (K_{1.1} + K_{1.2})(v_1 - 0) - K_{1.2}(v_1 - v_2),$$
$$q_2 = (K_{2.2} + K_{1.2})(v_2 - 0) - K_{1.2}(v_2 - v_1).$$

Now if we have two small bodies 1 and 2 (Fig. 48) which are
connected with each other
and with an earthed con-
ductor through condensers of
capacities K_0, K_1 and K_2 as
in the figure, and if the
potentials of 1 and 2 are v_1
and v_2, then the charges on
the conductors connected
with 1 and 2 will be equal
to q_1 and q_2 provided that
$K_1 = K_{1.1} + K_{1.2}$, $K_0 = -K_{1.2}$
and $\qquad K_2 = K_{2.2} + K_{1.2}.$

We thus obtain an exact
electrical model of a two core
cable.

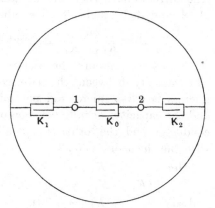

Fig. 48. Equivalent condensers.

When everything is symmetrical with respect to the two conductors we have $K_{1.1} = K_{2.2}$, and thus

$$K_1 = K_2 = K_{1.1} + K_{1.2}, \quad K_0 = -K_{1.2}.$$

If, in addition, the alternator be insulated from earth and everything is symmetrical with regard to the middle point of the external load, then

$$v_1 = -v_2 = \tfrac{1}{2}v,$$

where v is the potential difference between the terminals of the machine. Now the condenser current for any main is defined as the rate of increase of charge upon the main, and hence, in this case, if $+i$ and $-i$ be the condenser currents for the two mains,

$$i = \frac{\partial q_1}{\partial t} = \tfrac{1}{2}(K_{1.1} - K_{1.2})\frac{\partial v}{\partial t}.$$

Leaving the symmetrical case for the moment, let No. 2 main be earthed at the alternator terminal, so that v_2 is zero. Then since v_1 equals v, the condenser currents are given by

$$i_1 = K_{1.1}\frac{\partial v}{\partial t}, \text{ and } i_2 = K_{1.2}\frac{\partial v}{\partial t}.$$

If v is the same in magnitude and wave form in the two cases, then at corresponding instants

$$\frac{i_1}{i} = \frac{2K_{1.1}}{K_{1.1} - K_{1.2}}, \text{ and } \frac{i_2}{i} = \frac{2K_{1.2}}{K_{1.1} - K_{1.2}}.$$

The currents have therefore constant ratios to one another, and if A, A_1 and A_2 be their effective values we have

$$A_1 = \frac{2K_{1.1}}{K_{1.1} - K_{1.2}}A, \text{ and } A_2 = \frac{-2K_{1.2}}{K_{1.1} - K_{1.2}}A.$$

Fig. 49 represents the section of a lead covered twin cable. The capacity between the two conductors is 0·345 microfarad per mile, and the capacity per mile between one conductor and the other in parallel with the sheathing is 0·53.

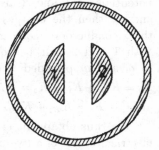

Hence　　　$K_{1.1} = $　0·53,

　　$\tfrac{1}{2}(K_{1.1} - K_{1.2}) = $　0·345,

and so　　　$K_{1.2} = -0·16.$

Therefore if A be the condenser current when both mains are insu-

Fig. 49.　Two core cable.

lated, then when No. 2 main is earthed A_1 will equal $1\cdot54A$ and A_2 will equal $0\cdot46A$ and the current in the sheathing will be $(1\cdot54 - 0\cdot46)A$, *i.e.* $1\cdot08A$.

We will suppose that three conductors are symmetrically

Three phase cables. embedded in a dielectric and surrounded by a metal sheath (Fig. 50). Then if v_0, v_1, v_2 and v_3 be the potentials of the sheath and of the three conductors respectively we have, with our usual notation,

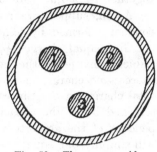

Fig. 50.   Three core cable.

$$q_0 = K_{0.0}v_0 + K_{0.1}v_1 + K_{0.2}v_2 + K_{0.3}v_3,$$
$$q_1 = K_{1.0}v_0 + K_{1.1}v_1 + K_{1.2}v_2 + K_{1.3}v_3,$$
$$q_2 = K_{2.0}v_0 + K_{2.1}v_1 + K_{2.2}v_2 + K_{2.3}v_3,$$
$$q_3 = K_{3.0}v_0 + K_{3.1}v_1 + K_{3.2}v_2 + K_{3.3}v_3.$$

If we make $v_1 = v_2 = v_3 = v_0$, then there will be no charge on any of the internal conductors, since in practice they are completely screened by the sheath from electrostatic induction from the outside. Hence we obtain the three equations,

$$0 = K_{1.0} + K_{1.1} + K_{1.2} + K_{1.3},$$
$$0 = K_{2.0} + K_{2.1} + K_{2.2} + K_{2.3},$$
$$0 = K_{3.0} + K_{3.1} + K_{3.2} + K_{3.3}.$$

Now from symmetry,

$$K_{1.1} = K_{2.2} = K_{3.3}, \quad K_{1.2} = K_{2.3} = K_{3.1}, \quad K_{0.1} = K_{0.2} = K_{0.3}.$$

Using these values we reduce the three equations to the single equation

$$K_{0.1} + K_{1.1} + 2K_{1.2} = 0 \quad \ldots\ldots\ldots\ldots\ldots\ldots(a).$$

Hence if we know $K_{1.1}$ and $K_{1.2}$ we can find $K_{0.1}$, and then we shall be able to calculate the capacities which can be obtained with various combinations of the three conductors and the sheath.

For example, suppose that we wish to find the capacity between the conductors 1 and 2 when 3 and the sheath S are insulated. Putting $q_1 = q$, $q_2 = -q$ in the general equations, we have

$$q = K_{1.0}v_0 + K_{1.1}v_1 + K_{1.2}v_2 + K_{1.2}v_3,$$
$$-q = K_{1.0}v_0 + K_{1.2}v_1 + K_{1.1}v_2 + K_{1.2}v_3.$$

Hence $\qquad 2q = (K_{1.1} - K_{1.2})(v_1 - v_2)$,
and therefore by definition the capacity in question is
$$K = \tfrac{1}{2}(K_{1.1} - K_{1.2}).$$
We obviously get the same result when S and 3 are joined together by a fine wire or are put to earth.

Again, suppose we require the capacity between S and the conductor formed by joining 1 and 2, when 3 is insulated.

Give equal charges $\tfrac{1}{2}q$ to each of the mains 1 and 2. The induced charge on the inside of the sheath will be $-q$, and if there is no charge on the outside of the sheath this will be the total charge on S. Since the conductors 1 and 2 are practically screened by S, the capacity between them and S will be independent of the absolute value of the potential of S, and hence it will simplify our equations to put v_0 zero. Since v_1 equals v_2, the first and fourth equations now become

$$- q = 2K_{0.1}v_1 + K_{0.1}v_3,$$
$$0 = 2K_{1.2}v_1 + K_{1.1}v_3,$$

and hence $\qquad q = -2v_1 K_{0.1}\left(1 - \dfrac{K_{1.2}}{K_{1.1}}\right).$

Employing (a) we obtain for the capacity

$$K = 2\,\frac{(K_{1.1} + 2K_{1.2})(K_{1.1} - K_{1.2})}{K_{1.1}}.$$

The following is the complete list of the capacities that can be got from a three core cable.

(1) Capacity between 1 and 2
$$= \tfrac{1}{2}(K_{1.1} - K_{1.2}).$$

(2) Capacity between 1 and 2, 3
$$= \tfrac{2}{3}(K_{1.1} - K_{1.2}).$$

(3) Capacity between 1 and S (2 and 3 insulated)
$$= \frac{(K_{1.1} - K_{1.2})(K_{1.1} + 2K_{1.2})}{K_{1.1} + K_{1.2}}.$$

(4) Capacity between 1 and S, 2 (3 insulated)
$$= \frac{(K_{1.1} - K_{1.2})(K_{1.1} + K_{1.2})}{K_{1.1}}.$$

(5) Capacity between 1 and S, 2, 3
$$= K_{1.1}.$$

(6) Capacity between S and 1, 2 (3 insulated)

$$= 2 \frac{(K_{1.1} - K_{1.2})(K_{1.1} + 2K_{1.2})}{K_{1.1}}.$$

(7) Capacity between 1, S and 2, 3

$$= 2 (K_{1.1} + K_{1.2}).$$

(8) Capacity between S and 1, 2, 3

$$= 3 (K_{1.1} + 2K_{1.2}).$$

If we measure (5) in the ordinary way, by reading the throw on a mirror galvanometer and comparing with the throw given by a standard condenser, we get $K_{1.1}$. A further measurement of (7) or (8) will give us a simple equation to find $K_{1.2}$.

Let us take as an example the three phase 'clover leaf' extra high tension cable (Fig. 51) supplied to the Manchester Corporation by the British Insulated Wire Co.

Numerical example.

The working pressure between the conductors = 6500 volts.

Working pressure between any conductor and the sheathing = 3750 volts.

Section of a conductor = 0·15 square inch = 0·97 sq. cm.

Minimum distance between conductor and sheathing $\Big\}$ = 0·86 cm.
Minimum distance between any two conductors

Insulating material, specially prepared paper. Mean dielectric coefficient $\lambda = 2\cdot8$.

By measurement, (7) was found to be 0·436 microfarad per mile, and (8) was 0·488 microfarad per mile.

Therefore

$$2 (K_{1.1} + K_{1.2}) = 0\cdot436$$
$$3K_{1.1} + 6K_{1.2} = 0\cdot488 \Big\};$$

and hence $K_{1.1} = 0\cdot273$,

and $K_{1.2} = -0\cdot0553$.

We deduce the other capacities by the formulae given above. The results are expressed in microfarads per mile.

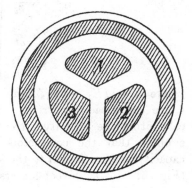

Fig. 51. 'Clover leaf' cable.

(1)   Capacity between 1 and 2 $= 0\cdot164$.

(2)   „   „   1 and 2, 3 $= 0\cdot219$.

(3)   „   „ S and 1 $= 0\cdot245$.

(4)   „   „ S, 1 and 2 $= 0\cdot262$.

(5)   „   „ S, 1, 2 and 3 $= 0\cdot273$.

(6)   „   „ S and 1, 2 $= 0\cdot391$.

(7)   „   „ S, 1 and 2, 3 $= 0\cdot436$.

(8)   „   „ S and 1, 2, 3 $= 0\cdot488$.

We also see that $K_{0.1}$, the coefficient of electrostatic induction between the sheath and a conductor, is $-0\cdot163$.

In practical work v_0 is zero, and when the load is balanced (see

Condenser currents in three phase working.

Chapter XV) we have

$$v_1 + v_2 + v_3 = 0.$$

In this case our equations become

$$q_1 = K_{1.1}v_1 + K_{1.2}(v_2 + v_3)$$
$$= K_{1.1}v_1 - K_{1.2}v_1$$
$$= (K_{1.1} - K_{1.2})v_1.$$

Similarly $q_2 = (K_{1.1} - K_{1.2})v_2,$

and $q_3 = (K_{1.1} - K_{1.2})v_3.$

Hence since $i_1 = \dfrac{\partial q_1}{\partial t},$

we get $i_1 = 2K\dfrac{\partial v_1}{\partial t}$

$$i_2 = 2K\dfrac{\partial v_2}{\partial t}$$

$$i_3 = 2K\dfrac{\partial v_3}{\partial t}$$

where $K = \frac{1}{2}(K_{1.1} - K_{1.2})$, and i_1, i_2 and i_3 are the capacity currents. Hence in calculating the capacity currents we can suppose that the conductors have no capacity, and are joined to the sheathing (Fig. 52) by three condensers each of capacity $2K$.

For example, in the Manchester cable

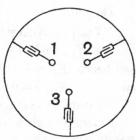

Fig. 52. Equivalent condensers.

described above, $2K$ is double the capacity between any two conductors; it therefore equals 0.328 microfarad per mile. If the working pressure between a conductor and the sheathing be 3750 volts, and the frequency be 50, then the minimum value of the condenser current in each conductor is $2\pi f 2KlV10^{-6}$, *i.e.* $0.386l$ ampere, where l is the length of the cable in miles.

Using the same notation as before, we have

Two phase cables with four separate conductors. $$q_1 = K_{1.1}v_1 + K_{1.2}v_2 + K_{1.3}v_3 + K_{1.4}v_4$$

when the sheath is earthed. Similarly we can write down the three other equations.

From symmetry (see Fig. 53)

$$K_{1.1} = K_{2.2} = K_{3.3} = K_{4.4},$$
$$K_{1.2} = K_{1.4} = K_{2.3} = K_{3.4},$$

and $K_{1.3} = K_{2.4}$. Hence

$$q_1 = K_{1.1}v_1 + K_{1.3}v_3 + K_{1.2}(v_2 + v_4).$$

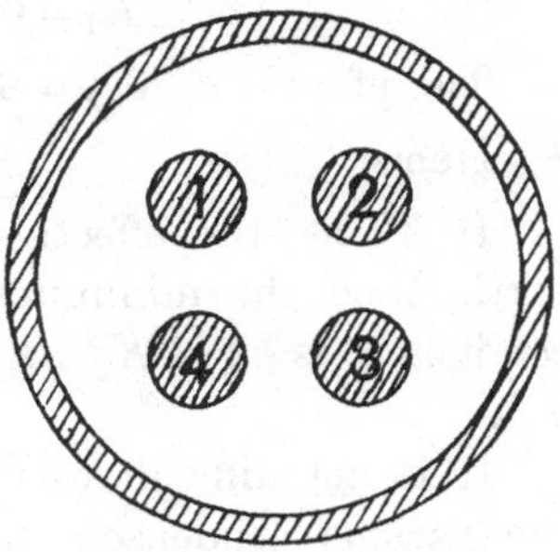

Now if the system is balanced (see Chapter XVI),

$$v_1 + v_3 = 0,$$
$$v_2 + v_4 = 0.$$

Fig. 53. Four core cable.

Hence
$$i_1 = \frac{\partial q_1}{\partial t}$$
$$= (K_{1.1} - K_{1.3})\frac{\partial v_1}{\partial t},$$
$$i_2 = (K_{1.1} - K_{1.3})\frac{\partial v_2}{\partial t},$$
$$i_3 = (K_{1.1} - K_{1.3})\frac{\partial v_3}{\partial t},$$

and
$$i_4 = (K_{1.1} - K_{1.3})\frac{\partial v_4}{\partial t}.$$

Therefore when we neglect the resistance of the conductors, the effect of capacity can be shown by imagining that the conductors have no capacity, but are joined by four condensers each of capacity $K_{1.1} - K_{1.3}$ connected star-wise between the conductors

12—2

(Fig. 54). This capacity is double the capacity between two opposite conductors.

If we measure the capacity K_1 between 1 and 2, 3, 4, S joined in parallel, then

$$K_{1.1} = K_1.$$

Similarly, if we measure the capacity K_2 between 1, 3 and 2, 4, S, then

$$2(K_{1.1} + K_{1.3}) = K_2.$$

Hence $K_{1.3} = -(K_1 - \tfrac{1}{2}K_2).$

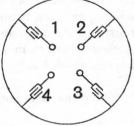

Fig. 54.
Equivalent condensers.

For example, in a lead-covered four core cable,

$$K_1 = 0.234 \text{ and } K_2 = 0.454.$$

Therefore $K_{1.1} = 0.234$ and $K_{1.3} = -0.007.$

Hence $K_{1.1} - K_{1.3} = 0.241.$

If V be the effective pressure between any conductor and earth, then the minimum value of the capacity current in a conductor is $2\pi V K f 10^{-6}$, where K is 0.241 microfarad per mile.

It is not difficult to find expressions for all the capacities of the various condensers that can be made out of the four conductors and the sheath. For example, suppose we wish to find the capacity between 1 and 3 (Fig. 53).

We have, on giving charges $+q$ and $-q$ to 1 and 3 respectively,

$$q = K_{0.1} v_0 + K_{1.1} v_1 + K_{1.2} v_2 + K_{1.3} v_3 + K_{1.2} v_4 \Big\}$$
$$\text{and} \quad -q = K_{0.1} v_0 + K_{1.3} v_1 + K_{1.2} v_2 + K_{1.1} v_3 + K_{1.2} v_4 \Big\} .$$

Thus $2q = (K_{1.1} - K_{1.3})(v_1 - v_3),$

and therefore $q = \tfrac{1}{2}(K_{1.1} - K_{1.3})(v_1 - v_3).$

Hence the capacity between a pair of opposite conductors is $\tfrac{1}{2}(K_{1.1} - K_{1.3})$, which is half the capacities of the condensers shown in Fig. 54.

In the twin concentric cable, a section of which is shown in Fig.
Twin concen-   55, 1 and 4 are copper conductors and X is a third
tric cable.   cylindrical copper conductor enclosing the other two and itself enclosed by a lead sheath. This cable is used for two

phase working; 1 and 4 are what are ordinarily called the two outside conductors, and 2, 3 or X is their common return. The copper used in X is 1·414 times the copper used in either 1 or 4. When the system is balanced, the P.D. between 1 and X is equal to the P.D. between 4 and X as regards effective value but differs in phase from it by 90 degrees. The effective value of the P.D. between 1 and 4 is 1·414 times the effective value of the P.D. between 1 and X or between 4 and X. We also

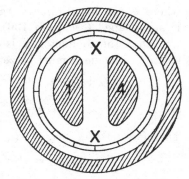

Fig. 55.   Twin concentric cable.

know that its phase difference from either of them is 135 degrees (Chapter XVI).

Let v_1, v_x and v_4 be the potentials from earth of 1, X and 4 respectively, then as before

$$q_1 = K_{1.1} v_1 + K_{1.x} v_x + K_{1.4} v_4,$$
and
$$q_4 = K_{4.1} v_1 + K_{4.x} v_x + K_{4.4} v_4.$$

From symmetry $K_{1.1} = K_{4.4}$ and $K_{1.x} = K_{4.x}$.

Since one conductor surrounds the other two, we must have q_1 and q_4 equal to zero when v_1, v_x and v_4 are all equal, therefore

$$K_{1.1} + K_{x.1} + K_{4.1} = 0.$$

Hence
$$q_1 = K_{1.1}(v_1 - v_x) + K_{1.4}(v_4 - v_x),$$
and
$$i_1 = K_{1.1} \frac{\partial v'}{\partial t} + K_{1.4} \frac{\partial v''}{\partial t},$$

where
$$v' = v_1 - v_x \text{ and } v'' = v_4 - v_x.$$

If the waves of v' and v'' are similar and their effective values differ in phase by 90 degrees, then the effective values of the curves representing $\frac{\partial v'}{\partial t}$ and $\frac{\partial v''}{\partial t}$ will also differ in phase by 90 degrees; if in addition the effective values of v' and v'' are equal, the effective value of the capacity current in either 1 or 4 is given by

$$A_1{}^2 = (K^2{}_{1.1} + K^2{}_{1.4}) \left\{ \text{mean value of} \left(\frac{\partial v'}{\partial t} \right)^2 \right\}.$$

Therefore $A_1 = 2\pi\alpha V_{1.x} (K^2_{1.1} + K^2_{1.4})\ f,$

where α has its minimum value unity when the wave of P.D. is sine-shaped (see page 136).

When the conductor X is earthed, since the total quantity of electricity inside it must be zero at every instant, we have

$$q_x + q_1 + q_4 = 0,$$

and $i_x = \dfrac{dq_x}{dt}.$

Therefore $i_x = -(i_1 + i_4)$

$$= -(K_{1.1} + K_{1.4})\left(\frac{\partial v'}{\partial t} + \frac{\partial v''}{\partial t}\right)$$

$$= K_{1.x}\left(\frac{\partial v'}{\partial t} + \frac{\partial v''}{\partial t}\right).$$

Thus $A_x = -2\pi\alpha V_{1.x}\sqrt{2}K_{1.x}f$

$$= 2\pi\alpha V_{1.x}\sqrt{2}(K_{1.1} + K_{1.4})f$$

$$= 2\pi\alpha V_{1.4}(K_{1.1} + K_{1.4})f.$$

It follows from these formulae that A_x is always less than $\sqrt{2}A_1$.

As before we can show that

(1)   The capacity between 1 and 4 $= \frac{1}{2}(K_{1.1} - K_{1.4}).$

(2)   The capacity between 1 and X when 4 is insulated
$$= K_{1.1} - K^2_{1.4}/K_{1.1}.$$

(3)   The capacity between 1 and X, $4 = K_{1.1}.$

(4)   The capacity between 1, 4 and $X = 2(K_{1.1} + K_{1.4}).$

The following are some of the data for a twin concentric cable made by the British Insulated Wire Co. for two phase work:

Working pressure between inner conductors = 2700 volts.

Working pressure between either inner and the other conductor = 1900 volts.

Section of inner conductor = 0·025 sq. inch = 0·161 sq. cm.

Section of outer ring conductor = 1·414 × 0·161 sq. cm.
$$= 0·228 \text{ sq. cm.}$$

Minimum distance between inner conductors = 0·56 cm.

„   „   between inner and outer = 0·63 cm.

The capacity between 1 and 4, X

$$=: K_{1.1} = 0.233 \text{ microfarad per mile.}$$

The capacity between 1, 4 and X

$$= 2(K_{1.1} + K_{1.4}) = 0.370 \text{ microfarad per mile.}$$

Hence $K_{1.1} = 0.233$ and $K_{1.4} = -0.048$.

The capacity between 1 and 4

$$= \tfrac{1}{2}(K_{1.1} - K_{1.4}) = 0.141 \text{ microfarad per mile.}$$

The capacity between 1 and X

$$= K_{1.1} - K^2_{1.4}/K_{1.1} = 0.223 \text{ microfarad per mile.}$$

In calculating the formulae for cables it is to be noted that we have supposed that the conductors are arranged symmetrically. This is the case in practice, and it is exceptional for example to find appreciable discrepancies in the values of the capacities between any two conductors of a three phase cable. When such discrepancies occur the above formulae have to be modified. The 'spiral,' that is, the twist of the cores round the central axis, in two and three core cables does not affect their symmetry. For cables with cores 0·1 of a square inch in section the spiral is in general one turn for about every eight feet of length. In cables made by some makers, however, the spiral is one turn for every four feet.

Let us consider the case of a cable with a metal sheath enclosing *Model of a polyphase cable.* n conductors 1, 2, ..., n. If the diameter of the sheath be small compared with its length, as it always is in practice, then the inside conductors are practically screened from electrostatic induction from the outside, and so the coefficients of mutual induction between these conductors and outside bodies are quite negligible when compared with the mutual coefficients between any two conductors or between the conductors and the sheath. The equations for the charges on 1, 2, ..., n are therefore

$$\left.\begin{aligned}
q_1 &= K_{1.0}v_0 + K_{1.1}v_1 + \ldots + K_{1.n}v_n \\
q_2 &= K_{2.0}v_0 + K_{2.1}v_1 + \ldots + K_{2.n}v_n \\
&\cdots\cdots\cdots\cdots\cdots\cdots\cdots\cdots\cdots\cdots\cdots\cdots
\end{aligned}\right\} \quad \ldots\ldots\ldots\ldots(a).$$

When the potential of each of the n conductors is equal to that of the sheath,

$$q_1 = q_2 = \ldots = q_n = 0,$$

for all values of v_0. We thus obtain the n relations

$$\left.\begin{array}{l} K_{1.0} + K_{1.1} + \ldots + K_{1.n} = 0 \\ K_{2.0} + K_{2.1} + \ldots + K_{2.n} = 0 \\ \ldots\ldots\ldots\ldots\ldots\ldots\ldots\ldots\ldots\ldots \end{array}\right\} \quad\ldots\ldots\ldots\ldots\ldots(b).$$

Substituting in (a) the values of $K_{1.1}$, $K_{2.2}$, ..., $K_{n.n}$ given by (b), we have

$$q_1 = -K_{1.0}(v_1 - v_0) - K_{1.2}(v_1 - v_2) - \ldots - K_{1.n}(v_1 - v_n),$$
$$q_2 = -K_{2.0}(v_2 - v_0) - K_{2.1}(v_2 - v_1) - \ldots - K_{2.n}(v_2 - v_n),$$
$$\ldots\ldots\ldots\ldots\ldots\ldots\ldots\ldots\ldots\ldots\ldots\ldots\ldots\ldots\ldots\ldots\ldots\ldots\ldots$$

These equations show us that we may suppose the conductors to have no capacity, but that any one of them is connected to any of the others or the sheath by a condenser whose capacity equals the coefficient of electrostatic induction between the two with its sign changed. For example, the condenser connecting the conductor p to a conductor q will have a capacity equal to $-K_{p.q}$, and the condenser connecting it to the sheath will have a capacity $-K_{p.0}$.

Hence we can construct a model to illustrate the capacity effects of a polyphase cable as follows. Take n small conductors $1, 2, \ldots, n$ and join them to a conductor S by condensers of capacities $-K_{0.1}, -K_{0.2}, \ldots$. Now join any two of them p and q by a condenser of capacity $-K_{p.q}$ and do this for every pair of conductors. The model thus constructed would act so far as capacity is concerned in a manner similar to the polyphase cable which has $K_{0.1}, K_{0.2}, \ldots, K_{p.q}, \ldots$ for its coefficients of mutual induction for electrostatic charges. The number of condensers required to construct the model would be

$$n + \tfrac{1}{2}n(n-1), \text{ that is, } \tfrac{1}{2}n(n+1).$$

In the particular case of a three core cable, the capacities **Three core cable.** can be represented as in Fig. 56. The capacities of the condensers joining the conductors to the sheath will be $-K_{1.0}$, $-K_{2.0}$ and $-K_{3.0}$, and the capacities of the condensers joining the conductors will be $-K_{1.2}$, $-K_{2.3}$ and $-K_{3.1}$ respectively.

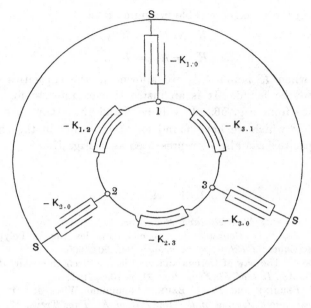

Fig. 56. Model of a three core cable.

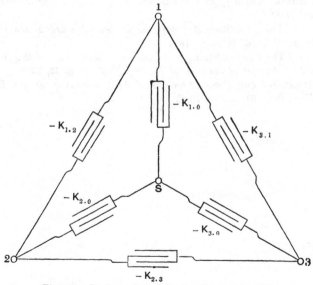

Fig. 57. Reciprocal model of a three core cable.

When the three core cable is symmetrical,

$$K_{1.0} = K_{2.0} = K_{3.0}$$

and $$K_{1.2} = K_{2.3} = K_{3.1}.$$

Hence when $K_{1.0}$ and $K_{1.2}$ are known, all the capacities of the cable can be found. It is an instructive exercise to find these capacities from Fig. 56. It will be found that they agree with the values which we have found for them earlier in the chapter. The capacities can also be represented as in Fig. 57.

REFERENCES

CLERK MAXWELL, *Electricity and Magnetism*, Vol. 1.

C. E. GUYE, 'Les Courants de Capacité dans les Lignes Polyphasées Symétriques,' *L'Éclairage Électrique*, Jan. 20, 1900.

F. W. PEEK, 'The Law of Corona and the Electric Strength of Air,' *Journal of the Am. Inst. of El. Eng.*, Proc. 31, p. 1085, 1912.

F. A. C. PERRINE and F. G. BAUM, 'Aluminium Wires and Polyphase Transmission,' *Journal of the Am. Inst. of El. Eng.*, Trans. 17, p. 391, 1900.

A. DELLA RICCIA, 'Capacity of Polyphase Cables,' *Soc. Belge Élect. Bull.*, 19, p. 318, 1902.

A. RUSSELL, 'The Capacities of Polyphase Cables,' *Journal of the Inst. of El. Eng.*, Vol. 70, p. 1022, 1901.

A. RUSSELL, 'The Electric Strength of Insulating Materials and the Grading of Cables,' *Journal of the Inst. of El. Eng.*, Vol. 40, p. 33, 1908.

J. B. WHITEHEAD, 'The Electric Strength of Air,' *Journal of the Am. Inst. of El. Eng.*, Proc. 31, p. 839, 1912.

CHAPTER VI

Formulae for a three core cable. Formula for a four core cable. Cable with n cores. The capacity of a cylinder parallel to the earth. The capacity coefficients of two horizontal parallel wires near the earth. The capacity coefficients of horizontal antennae. The capacity between two parallel horizontal wires one vertically over the other. Relation between the electrostatic and the electromagnetic coefficients of parallel wires. The capacity coefficients of three phase overhead wires.

IN Chapter V we have considered the mutual relations between the capacities of the cores and the sheathing in polyphase cables In this chapter we shall investigate formulae for these capacities. These formulae are in some cases only approximate, but the approximations are sufficiently close to be practically useful, and the simple method employed, combined with the method of electrical images due to Kelvin, is so powerful that it is deserving of attentive study.

Formulae for a three core cable.

We shall suppose that the copper conductors or cores as we shall call them are three parallel cylinders, and we shall first consider the case when each has a charge $+ q/3$ per unit length. The equipotential surface whose potential is v is given by

$$v = C - 2\frac{q}{3}\log r_1 - 2\frac{q}{3}\log r_2 - 2\frac{q}{3}\log r_3,$$

where C is a constant and r_1, r_2 and r_3 are the distances of a point on the surface from the axes of the three cores respectively. If A be a constant, this equation may be written in the form

$$r_1{}^2 r_2{}^2 r_3{}^2 = A^6.$$

Now, if the axes of the cores are at the angular points of an equilateral triangle whose centre is O, and if OP equals r, then it is easy to prove directly or by means of De Moivre's property of the circle (Loney's *Trigonometry*) that the last equation may be written

$$r^6 - 2a^3r^3 \cos 3\theta + a^6 = r_1{}^2r_2{}^2r_3{}^2 = A^6 \ldots\ldots\ldots\ldots(1),$$

where θ is the angle which OP makes with a line passing through O and one of the angular points of the equilateral triangle, and a is the radius of the circle circumscribing it.

When the constant in (1) is zero, the curves are simply the three points where the axes of the cores cut the plane of the paper. When the constant is small, the equipotential curves are ovals which are nearly circular in shape and enclose the three points. When the constant equals a^6, the curves are given by

$$r^3 = 3a^3 \cos 3\theta,$$

which represents three loops, each enclosing a core. Each of these loops has a double point and two tangents at the origin, the angle between the tangents being 60 degrees.

When the constant is greater than a^6, we get a single curve enclosing the three cores and having three elevations and three depressions on it. For the curve passing through the point $r = 2a$, $\theta = 0$ the constant equals $(7a^3)^2$, and the maximum value of r for this curve is $2a$ and the minimum value is $1\cdot82a$. Hence the radii of this curve differ from the radius of the circle $r = 1\cdot91a$ by less than five per cent.

Now, by Green's theorem, we can replace any conductor by another surrounding it provided that the surface of the outer conductor is an equipotential surface of the system of distribution. Suppose then that the three core cable has the section shown in Fig. 58.

If b is the minimum distance of a core from the centre of the cable, then the equation to the boundaries of the cross sections of the three cores is

$$r^6 - 2a^3r^3 \cos 3\theta + a^6 = (a^3 - b^3)^2 \ldots\ldots\ldots\ldots(2).$$

This equation has equal roots when

$$\cos^2 3\theta = \frac{b^3 (2a^3 - b^3)}{a^6}.$$

If θ_1 is the positive solution of this equation, $2\theta_1$ is the angular breadth of the core as seen from the centre of the cable, and

$120° - 2\theta_1$ is the angular breadth of the space between the cores as seen from the centre.

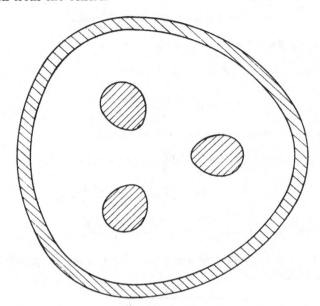

Fig. 58. Section of a three core cable, the equipotential surfaces being given by

$$r^5 - 2a^3r^3 \cos 3\theta + a^6 = \text{constant},$$

when the three cores are at the same potential.

If $b = \frac{1}{2}a$, as in Fig. 58, then $\theta_1 = 20°\cdot3$ nearly, and hence $2\theta_1 = 40°\cdot6$ and $120° - 2\theta_1 = 79°\cdot4$.

If c be the maximum distance of a point on a core from the centre of the cable, then from (2)

$$c^3 = 2a^3 - b^3 ;$$

and hence, if b is $\frac{1}{2}a$, c will be $1\cdot23a$.

This is the case illustrated in Fig. 58, the boundary of the lead sheath being supposed to coincide with the curve

$$r^6 - 2a^3r^3 \cos 3\theta + a^6 = (7a^3)^2,$$

so that its maxima and minima radii are $2a$ and $1\cdot82a$ respectively.

If v_1 be the potential of each core, and v_2 be the potential of the lead sheath, then

$$v_1 = C - 2\frac{q}{3}\log r_1 r_2 r_3 = C - 2\frac{q}{3}\log A^3.$$

Putting $\theta = 0$, and $r = b$ we have, by (1),

$$v_1 = C - 2\frac{q}{3}\log(a^3 - b^3)$$

$$= C - 2\frac{q}{3}\log\frac{c^3 - b^3}{2}.$$

If R is a maximum radius of the lead sheath,

$$v_2 = C - 2\frac{q}{3}\log(R^3 - a^3)$$

$$= C - 2\frac{q}{3}\log\left(R^3 - \frac{c^3 + b^3}{2}\right),$$

and hence

$$v_1 - v_2 = \frac{2}{3}q\log\frac{2R^3 - (c^3 + b^3)}{c^3 - b^3}.$$

Therefore the capacity between the three cores in parallel and the lead sheath is

$$\frac{3\lambda l}{2\log\dfrac{2R^3 - (c^3 + b^3)}{c^3 - b^3}},$$

where l is the length of the conductor, λ the dielectric coefficient, R the maximum inner radius of the sheath, and b and c are the minimum and maximum distances of points on the cores from the centre of the cable.

The formula may also be written in the form

$$\frac{\lambda l}{2\log\dfrac{R}{a} + \dfrac{2}{3}\log\dfrac{1 - \dfrac{a^3}{R^3}}{1 - \dfrac{b^3}{a^3}}},$$

since

$$2a^3 = b^3 + c^3.$$

Hence when b/a is greater than a/R, the capacity is less than that of a concentric main whose inner radius is a and outer radius R. When b/a equals a/R, the capacity equals that of this concentric main, and when b/a is less than a/R it is greater than it.

With our usual notation (see page 177)

$$3\left(K_{1.1}+2K_{1.2}\right)=\frac{3\lambda l}{2\log\dfrac{R^3-a^3}{a^3-b^3}},$$

and therefore $K_{1.1}+2K_{1.2}=\dfrac{\lambda l}{2\log\dfrac{R^3-a^3}{a^3-b^3}},$ ....(3).

If the sections of the cores, instead of having the shapes shown in Fig. 58, were true circles, then we should expect that the equipotential surfaces would still be very similar to the curves

$$r_1 r_2 r_3 = \text{constant},$$

and hence that the formulae given above could be used as first approximations. The exact shapes of the equipotential curves in this case could be found by the well-known laboratory method of tracing out the equipotential lines on a circular sheet of tinfoil whose boundary was maintained at zero potential whilst three circular copper electrodes were pressed on it at symmetrical points and maintained at constant equal potentials by a suitable battery.

If, however, we make the supposition that the circular cross sections of the wires are small compared with the cross section of the sheath, we can find by the method of images approximate formulae to give us the values of $K_{1.1}$ and $K_{1.2}$.

Let O (Fig. 59) be the centre of the section of the sheath by a plane perpendicular to its axis. Give charges $+q/3$ to each of the conductors the sections of which we suppose to be almost coincident with the points A, B and C. Let OA, OB and OC be each equal to a and let the angles between them be each equal to $120°$. Let A', B' and C' be the inverse points (page 162) of A, B and C with respect to the circle formed by the section

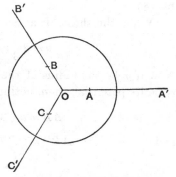

Fig. 59. The images of the three wires A, B and C are at A', B' and C' where

$$OA \cdot OA' = OB \cdot OB' = OC \cdot OC' = R^2.$$

In finding formulae we replace the sheath by these images.

of the inner surface of the sheath.   Then

$$OA \cdot OA' = OB \cdot OB' = OC \cdot OC' = R^2,$$

where R is the inner radius of the sheath.   Let charges $-q/3$ be given to fine wires passing through the points A', B' and C' and parallel to the three original wires.   Then if r_1, r_2, r_3, r_1', r_2' and r_3' are the distances of A, B, C, A', B' and C' from a point P where the potential is v, we should have if the sheath were removed

$$v = C - 2\frac{q}{3}\log\frac{r_1 r_2 r_3}{r_1' r_2' r_3'},\quad\dots\dots\dots\dots\dots\dots\dots(4).$$

Now for all points on the circle, we have (page 163)

$$\frac{r_1}{r_1'} = \frac{r_2}{r_2'} = \frac{r_3}{r_3'} = \frac{a}{R} = \text{constant.}$$

Hence, substituting these values in (4), we see that the inner surface of the sheath is an equipotential surface of the six charged wires.   Therefore, by Green's theorem, we can replace the three outside wires by the sheath without disturbing the equipotential surfaces inside.   Conversely, as we desire in this case, we can replace the sheath by the three outside wires.   It is necessary to suppose that the section of the wires inside is very small, otherwise the bounding surfaces of the three wires cannot be considered as equipotential surfaces which are determined approximately by (4).

When the sheath is at zero potential, (4) becomes

$$v = 2\frac{q}{3}\log\frac{r_1' r_2' r_3'}{r_1 r_2 r_3} \cdot \frac{a^3}{R^3}.$$

Now if r be the radius of the circular cross section of a wire and v_1 be its potential, then, noting that (Fig. 59)

$$AB = a\sqrt{3}, \quad AA' = \frac{R^2}{a} - a$$

and

$$AB'^2 = OB'^2 + OA^2 + OB' \cdot OA$$
$$= \frac{R^4}{a^2} + a^2 + R^2,$$

we get

$$v_1 = 2\frac{q}{3}\log\frac{(R^2/a - a)\,(R^4/a^2 + a^2 + R^2)}{r \cdot a\sqrt{3} \cdot a\sqrt{3}} \cdot \frac{a^3}{R^3}$$
$$= 2\frac{q}{3}\log\frac{R^6 - a^6}{3R^3 a^2 r}.$$

VI] THREE CORE CABLE 193

Hence since the capacity between the three conductors in parallel and the sheath is $3(K_{1.1} + 2K_{1.2})$ we get

$$K_{1.1} + 2K_{1.2} = \frac{\lambda l}{2\log\dfrac{R^6 - a^6}{3R^3 a^2 r}} \quad \dots\dots\dots\dots(5).$$

If we put $a - r$ for b in (3) and suppose that r/a and $(a/R)^3$ are negligibly small, it is easy to see that the two formulae agree.

To find the capacity between the wires A and B we give a charge $+q$ to A and a charge $-q$ to B. We replace the sheath by wires A' and B' having charges $-q$ and $+q$ respectively. Assuming that there is no charge on C, the equipotential surfaces are given by the equation

$$v = C - 2q\log\frac{r_1}{r_1'}\cdot\frac{r_2'}{r_2}.$$

Hence when the sheath is at zero potential,

$$v = 2q\log\frac{r_1'}{r_1}\cdot\frac{r_2}{r_2'}.$$

If v_1 be the potential of A and v_2 be the potential of B, then

$$v_1 = 2q\log\frac{\dfrac{R^2 - a^2}{a}}{r}\cdot\frac{a\sqrt{3}}{\left(\dfrac{R^4 + R^2 a^2 + a^4}{a^2}\right)^{\frac{1}{2}}}$$

$$= -v_2.$$

Hence since the capacity between the two wires is $\frac{1}{2}(K_{1.1} - K_{1.2})$ we get

$$\tfrac{1}{2}(K_{1.1} - K_{1.2}) = \frac{q}{2v_1},$$

and therefore

$$2(K_{1.1} - K_{1.2}) = \frac{\lambda l}{\log\dfrac{a\sqrt{3}}{r}\cdot\dfrac{R^2 - a^2}{(R^4 + R^2 a^2 + a^4)^{\frac{1}{2}}}} \quad \dots\dots(6).$$

From equations (5) and (6), $K_{1.1}$ and $K_{1.2}$ can be readily determined.

By the help of the formulae given in Chapter V we can calculate all the capacities of a three core cable when the cores are of small section and not too close to one another in terms of these approximate values of $K_{1.1}$ and $K_{1.2}$.

R. I   13

Assume that the equation to the boundaries of the sections of the four cores is

Formula for a four core cable.

$$r^8 - 2a^4r^4 \cos 4\theta + a^8 = (a^4 - b^4)^2,$$

where a is the distance of the axes of the cores from the centre of the cable, and b is the minimum distance of a conductor from the centre. Also assume that the equation to the boundary of the lead sheath is

$$r^8 - 2a^4r^4 \cos 4\theta + a^8 = (R^4 - a^4)^2,$$

where R is the greatest value of the radius vector. Then proceeding in the same way as for a three core cable, we find that the equation to the equipotential surfaces is

$$v = C - 2\frac{q}{4} \log r_1 r_2 r_3 r_4.$$

Hence if v_1 be the potential of the four cores and v_2 be the potential of the sheath,

$$v_1 = C - 2\frac{q}{4} \log (a^4 - b^4),$$

and

$$v_2 = C - 2\frac{q}{4} \log (R^4 - a^4).$$

Thus

$$v_1 - v_2 = 2\frac{q}{4} \log \frac{R^4 - a^4}{a^4 - b^4}.$$

The capacity between the four cores in parallel and the sheath is therefore

$$\frac{2\lambda l}{\log \dfrac{R^4 - a^4}{a^4 - b^4}}.$$

If $R = ma$ and $b = a/m$, this becomes

$$\frac{\lambda l}{2 \log m},$$

which is the same formula as for a concentric main whose outer radius is m times its inner one.

If c be the maximum distance of a point on the boundary of the cross section of a core from the centre of the cable, then

$$c^4 = 2a^4 - b^4.$$

And hence the formula becomes

$$\frac{2\lambda l}{\log \dfrac{2R^4 - (c^4 + b^4)}{c^4 - b^4}}.$$

When b and c are nearly equal to a, we can find approximate formulae for $K_{1.1}$, $K_{1.2}$ and $K_{1.3}$ by methods similar to that employed for the three core cable.

Suppose that the n cores are all equal and parallel, and that
Cable with n
cores.
they are arranged symmetrically in the sheath. If q/n be the charge per unit length on each wire, then the potential at any point P inside the cylinder will be given by

$$v = C - 2\frac{q}{n}\log r_1 - 2\frac{q}{n}\log r_2 - \ldots\ldots - 2\frac{q}{n}\log r_n$$

$$= C - 2q \log (r_1 r_2 \ldots\ldots r_n)^{1/n}.$$

If the axes of the cores are arranged on a circle of radius a, this becomes

$$v = C - q \log \{r^{2n} - 2a^n r^n \cos n\theta + a^{2n}\}^{1/n},$$

where θ is the angle which OP makes with OA, where A is one of the points of intersection of the axes of the conductors with a perpendicular plane.

The equation to the equipotential curve passing through the point $r = d$, $\theta = 0$ is

$$r^{2n} - 2a^n r^n \cos n\theta + a^{2n} = (d^n - a^n)^2.$$

This only meets the line $\theta = \pi/n$ when d is greater than $2^{1/n}a$. When d is very little greater than $2^{1/n}a$, the curve is rippled symmetrically and encloses the n cores. The maximum values of the radius vector are when $\theta = 0, 2\pi/n, 4\pi/n, \ldots\ldots$, and the minimum values when $\theta = \pi/n, 3\pi/n, 5\pi/n, \ldots\ldots$. The maximum values are each equal to d and the minimum values to $(d^n - 2a^n)^{1/n}$. For example, when n is 20 and d is $1\cdot 2a$, the curve would differ from the circle $r = 1\cdot 199a$ by less than one part in a thousand. Hence no great error is introduced by the assumption that the equipotential lines near the outer cylinder are circles.

When $d = 2^{1/n}a$ the equipotential lines are n loops, each enclosing a core and each having a double point at the origin. The

13—2

angle between the tangents at the origin to one of these loops is π/n. Hence the length of the loops is much greater than their breadth.

When d is less than $2^{1/n}a$, there are n oval curves, one round each wire, and the smaller d is, the rounder these curves become. When d equals a, $\cos n\theta$ must equal unity, and therefore θ is $0, 2\pi/n, 4\pi/n, \ldots\ldots$, that is, the curves are reduced to points which coincide with the axes of the cores. Hence, when d is nearly equal to a, the equipotential curves are n small rounded curves, and we may suppose the sections of the n cores to coincide with them.

Let v_1 be the potential of each core whose minimum distance from the centre of the cable is b, then

$$v_1 = C - 2\,(q/n) \log\,(r_1 r_2 \ldots r_n)$$
$$= C - 2\,(q/n) \log\,(a^n - b^n).$$

Similarly, if v_2 be the potential of the sheath whose maximum inner radius is R, then

$$v_2 = C - 2\,(q/n) \log\,(R^n - a^n),$$

and
$$v_1 - v_2 = 2\frac{q}{n} \log \frac{R^n - a^n}{a^n - b^n}.$$

If c be the maximum distance of any point on a conductor from the centre of the cable, we have

$$c^n = 2a^n - b^n.$$

Hence the capacity between the n wires in parallel and the sheath is

$$\frac{n\lambda l}{2 \log \dfrac{2R^n - (c^n + b^n)}{c^n - b^n}}.$$

This may also be written in the form

$$\frac{\lambda l}{2 \log \dfrac{R}{a} + \dfrac{2}{n}\log \dfrac{1 - (a/R)^n}{1 - (b/a)^n}}.$$

Now b/a will in general be greater than a/R, and hence the capacity will be less than that of a concentric main whose inner and outer radii are a and R respectively. If $bR = a^2$, or if n is infinite, the capacity will be the same as that of this concentric main.

The general appearance of the equipotential surfaces for the case of eight cores can be understood from the figure shown in Chapter XIX. When b is greater than 1·5 times a, the equipotential surfaces are practically circular cylinders having the same axis as the cable.

In the case of overhead wires which, unlike the conductors in cables, are not screened from outside influences by an enclosing metallic screen, we have to take into account the effect of the earth. We shall first give the exact solution of the capacity of a single cylindrical wire parallel to the earth, and then give various practical approximate solutions for the case of several wires in parallel.

Suppose that we have two equal cylinders parallel to one **The capacity** another, one charged with a quantity q of electricity **of a cylinder** per unit of length, and the other charged with a **parallel to the** **earth.** quantity $-q$. Then, by pages 10 and 162, the equipotential surfaces are given by

$$v = 2q \log \frac{r_2}{r_1},$$

where r_1 and r_2 are the distances of a point P from the inverse points A and B of the two circular sections made by a plane cutting the cylinders at right angles. Now, at every point on the plane bisecting AB at right angles, r_1 equals r_2, and therefore v equals zero, so that this plane is an equipotential surface. Since the earth is at zero potential, we see by Green's theorem, that the equation

$$v = 2q \log \frac{r_2}{r_1}$$

gives the equipotential surfaces for a single cylinder and the earth.

If a be the radius of the cylinder, v_1 its potential, and h be the height of its axis above the earth, then

$$v_1 = 2q \log \frac{r_2}{r_1}$$

$$= 2q \log \frac{a}{OA},$$

where O is the centre of the cylinder.

But $\qquad OA . OB = a^2,$

and $\qquad OA + OB = 2h.$

Thus $\qquad OA = h - \sqrt{h^2 - a^2},$

$$v_1 = 2q \log \frac{a}{h - \sqrt{h^2 - a^2}},$$

and therefore $\qquad K_{1.1} = \dfrac{l}{2 \log \dfrac{h + \sqrt{h^2 - a^2}}{a}}.$

Let us suppose that the heights of the wires above the earth are h_1 and h_2 respectively, and that d is the distance

The capacity coefficients of two horizontal parallel wires near the earth. between the vertical planes through their axes. Let the charges on the wires per unit length be q_1 and q_2 respectively. If the radii of the wires, a_1 and a_2, be small compared with d and h, we can easily find an approximate solution by the method of images. The electrostatic field in the air is the same as if we suppose the earth removed and replaced by images of the given wires at depths h_1 and h_2, and having charges $-q_1$ and $-q_2$ per unit length respectively. Hence the potential v at any point P in the air is given by

$$v = c - 2q_1 \log r_1 + 2q_1 \log r_1' - 2q_2 \log r_2 + 2q_2 \log r_2',$$

where c is a constant and r_1, r_2 are the distances of P from the wires and r_1', r_2' are the distances of P from their images.

At the surface of the earth v is zero, and thus c is zero. Hence if v_1 be the potential of the first wire, we have

$$v_1 = q_1 \log \frac{4h_1^2}{a_1^2} + q_2 \log \frac{d^2 + (h_1 + h_2)^2}{d^2 + (h_1 - h_2)^2}.$$

Similarly

$$v_2 = q_2 \log \frac{4h_2^2}{a_2^2} + q_1 \log \frac{d^2 + (h_1 + h_2)^2}{d^2 + (h_1 - h_2)^2}.$$

We shall write these equations in the form

$$v_1 = p_{1.1} q_1 + p_{1.2} q_2, \quad \text{and} \quad v_2 = p_{2.2} q_2 + p_{1.2} q_1,$$

where $\qquad p_{1.1} = \log(4h_1^2/a_1^2), \quad p_{2.2} = \log(4h_2^2/a_2^2),$

and $\qquad p_{1.2} = \log \dfrac{d^2 + (h_1 + h_2)^2}{d^2 + (h_1 - h_2)^2}.$

Solving them, we get

$$q_1 = (p_{2.2}/\Delta)\, v_1 - (p_{1.2}/\Delta)\, v_2,$$

and
$$q_2 = (p_{1.1}/\Delta)\, v_2 - (p_{1.2}/\Delta)\, v_1,$$

where $\Delta = p_{1.1}\, p_{2.2} - p^2_{1.2}$.

Hence $k_{1.1} = p_{2.2}/\Delta$, $k_{2.2} = p_{1.1}/\Delta$, and $k_{1.2} = -\,p_{1.2}/\Delta$,

where $k_{1.1}$, $k_{2.2}$ and $k_{1.2}$ are the capacity coefficients per unit length. The capacity between the two wires in parallel and the earth is $k_{1.1} + k_{2.2} + 2k_{1.2}$ which equals

$$(p_{1.1} + p_{2.2} - 2p_{1.2})/(p_{1.1}\, p_{2.2} - p^2_{1.2}).$$

The capacity between the two wires (see p. 152) is

$$(p_{1.1} + p_{2.2} - 2p_{1.2})^{-1}.$$

In radio-telegraphic work the antennae often consist of parallel horizontal wires at the same height h and equally spaced apart. We shall suppose that there are n wires, that the distance between neighbouring wires is d and that the charges on them per unit length are $q_1, q_2, \ldots$ respectively. Making the assumption that the radii of the wires are very small compared with d and h, we get at once by the method of images the following equations:

The capacity coefficients of horizontal antennae.

$$v_1 = q_1 \log \frac{4h^2}{a^2} + q_2 \log \frac{d^2 + 4h^2}{d^2} + \ldots + q_m \log \frac{(m-1)^2\, d^2 + 4h^2}{(m-1)^2\, d^2} + \ldots,$$

$$v_2 = q_1 \log \frac{d^2 + 4h^2}{d^2} + q_2 \log \frac{4h^2}{a^2} + \ldots + q_m \log \frac{(m-2)^2\, d^2 + 4h^2}{(m-2)^2\, d^2} + \ldots,$$

and $n-2$ similar equations.

Writing these equations in the form

$$v_1 = p_{1.1}\, q_1 + p_{1.2}\, q_2 + \ldots + p_{1.m}\, q_m + \ldots,$$

$$v_2 = p_{2.1}\, q_1 + p_{2.2}\, q_2 + \ldots + p_{2.m}\, q_m + \ldots, \text{ etc.,}$$

we get, by determinants,

$$q_1 = k_{1.1}\, v_1 + k_{1.2}\, v_2 + \ldots + k_{1.m}\, v_m + \ldots,$$

$$q_2 = k_{2.1}\, v_1 + k_{2.2}\, v_2 + \ldots + k_{2.m}\, v_m + \ldots, \text{ etc.,}$$

where $k_{l.m}\, \Delta = M_{l.m}$, Δ being the symmetrical determinant $\Sigma \pm p_{1.1}\, p_{2.2} \ldots p_{n.n}$, and $M_{l.m}$ being the coefficient of $p_{l.m}$ in Δ.

All the capacity coefficients of the n wires have thus been determined.

The capacity
between two
parallel hori-
zontal wires
one vertically
over the other.
Let d be the distance between the wires, and h the height of
the lower wire above the ground. Let the radius of
each wire be a, and suppose that it is small compared
with either d or h. Then, if the charge on the lower
wire be q per unit length and that on the upper wire
$-q$ per unit length, the equipotential surfaces are given by

$$v = 2q \log \frac{r_1'}{r_1} - 2q \log \frac{r_2'}{r_2},$$

where r_1 and r_1' are the distances of a point on the surface from
the axis of the lower wire and its image respectively, and r_2 and
r_2' are its distances from the upper wire and its image. If v_1 and
v_2 be the potentials of the lower and upper wires, then

$$v_1 = 2q \log \frac{2h}{a} - 2q \log \frac{d+2h}{d}$$

$$= 2q \log \frac{2hd}{a(d+2h)}.$$

Similarly

$$v_2 = 2q \log \frac{d+2h}{d} - 2q \log \frac{2(d+h)}{a}$$

$$= 2q \log \frac{a(d+2h)}{2d(d+h)}.$$

Thus

$$v_1 - v_2 = 2q \log \frac{4hd^2(d+h)}{a^2(d+2h)^2}.$$

Therefore the capacity between the two wires is approximately

$$\frac{l}{4 \log \dfrac{d}{a} + 2 \log \left\{ 1 - \dfrac{d^2}{(d+2h)^2} \right\}}.$$

If d be small compared with $2h$, this may be written

$$\frac{l}{4 \log \dfrac{d}{a} - \dfrac{2d^2}{(d+2h)^2}}.$$

If the wires had been in the same horizontal plane at a height
h above the ground, then the capacity would be

$$\frac{l}{4 \log \dfrac{d}{a} + 2 \log \left\{ 1 - \dfrac{d^2}{d^2+4h^2} \right\}}.$$

Now since $\dfrac{d^2}{(d+2h)^2}$ is less than $\dfrac{d^2}{d^2+4h^2}$, the capacity between the wires for a given distance between them is a little smaller when they are arranged one over the other than when they are placed side by side, provided that the height of the lower wire in the one case is the same as the height of the two wires in the other. If however the mean height is the same in both cases we see, by writing $h - d/2$ instead of h, that the capacities for the two arrangements are practically equal when d is small compared with $2h$.

Let us first consider a single overhead wire of radius a_1 at a **Relation between the electrostatic and the electromagnetic coefficients of parallel wires.** height h_1 above the ground. If we suppose that a current flows along this wire and returns uniformly along a thin sheet on the surface of the earth, we see by the method of images that the self induction is the same as half that between two wires of radius a_1 at a distance $2h_1$ apart. Hence by p. 88 the self inductance $l_{1.1}$ per unit length is $1/2 + 2\log(2h_1/a_1)$. If we neglect the part of $l_{1.1}$ due to linkages inside the wire itself, we have $l_{1.1} = \log(4h_1^2/a_1^2)$. This would be rigorously true if the conductivity of the wire were infinite. Let us now suppose that there is a second parallel wire at a horizontal distance d from the first. Then $l_{2.2} = \log(4h_2^2/a_2^2)$ and we see by p. 75 that $l_{1.2} = \log\{d^2 + (h_1+h_2)^2\}/\{d^2 + (h_1-h_2)^2\}$. Hence by p. 198 $l_{1.1} = p_{1.1}$, $l_{1.2} = p_{1.2}$ and $l_{2.2} = p_{2.2}$.

From the equations given at the top of p. 199 we see that
$$k_{1.1} = l_{2.2}/\Delta, \ k_{2.2} = l_{1.1}/\Delta \text{ and } k_{1.2} = - l_{1.2}/\Delta,$$
where $\Delta = l_{1.1}\, l_{2.2} - l^2_{1.2}$. We deduce that
$$l_{1.1} = k_{2.2}/\Delta_1, \ l_{2.2} = k_{1.1}/\Delta_1 \text{ and } l_{1.2} = - k_{1.2}/\Delta_1,$$
where $\Delta_1 = k_{1.1} k_{2.2} - k^2_{1.2}$. Hence also
$$(l_{1.1}\, l_{2.2} - l^2_{1.2})\,(k_{1.1}\, k_{2.2} - k^2_{1.2}) = 1.$$

Let us now suppose that the two long parallel wires have negligible resistance and are earthed at both ends. Let a voltage e_1 per unit length be applied to the first wire. The equations to determine the currents i_1 and i_2 in the wires are
$$e_1 = l_{1.1}\,(\partial i_1/\partial t) + l_{1.2}\,(\partial i_2/\partial t),$$
and
$$0 = l_{2.2}\,(\partial i_2/\partial t) + l_{1.2}\,(\partial i_1/\partial t).$$

Hence eliminating $\partial i_2 / \partial t$, we get $e_1 = (\Delta / l_{2.2})(\partial i_1 / \partial t) = l'_{1.1}(\partial i_1 / \partial t)$, where $l'_{1.1}$ is the effective self inductance per unit length of the circuit formed by the first wire and the earth. We see that $k_{1.1}\, l'_{1.1} = 1$.

If k be the self capacity of the first wire in microfarads per mile and l the effective self inductance in microhenrys per mile, we get

$$kl = \frac{k_{1.1}}{9 \times 10^5} \times 1\cdot 61 \times 10^5 \times \frac{l'_{1.1}}{10^6} \times 1\cdot 61 \times 10^5 = \frac{288}{10^4}.$$

This relation between k and l is useful in practical work, as it enables us to find the value of l when k is known and *vice versâ*. The relations between the capacity and inductance coefficients given above can easily be generalised by means of the equations given at the foot of p. 199.

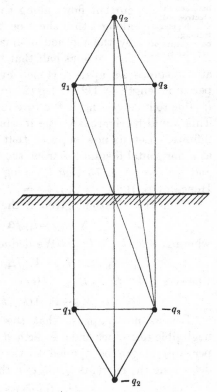

In three phase overhead

The capacity coefficients of three phase overhead wires. systems the mains are usually so arranged that, to a first approximation, we can consider that their axes form the edges of an equilateral prism the base of which is horizontal (Fig. 60). Let d be the distance between the axes of the wires and h the height of the two lower wires above the ground, so that $h + \sqrt{3}d/2$ is the height of the upper wire. Let the radius of each wire equal a and let q_1 and q_3 be the charges per unit length on the lower wires, q_2 being the charge on the upper wire. Then

Fig. 60. Image of overhead three phase mains.

proceeding as above and using the method of images, we get

$$v_1 = p_{1.1}\, q_1 + p_{1.2}\, q_2 + p_{1.3}\, q_3,$$
$$v_2 = p_{2.1}\, q_1 + p_{2.2}\, q_2 + p_{2.3}\, q_3,$$
$$v_3 = p_{3.1}\, q_1 + p_{3.2}\, q_2 + p_{3.3}\, q_3,$$

where $p_{1.1} = p_{3.3} = \log(4h^2/a^2)$, $p_{2.2} = \log\{(2h + d\sqrt{3})^2/a^2\}$,

$p_{1.2} = p_{2.1} = p_{2.3} = p_{3.2} = \log\{(d^2 + 2hd\sqrt{3} + 4h^2)/d^2\}$,

and $p_{1.3} = p_{3.1} = \log\{(d^2 + 4h^2)/d^2\}$.

Solving these equations for q_1, q_2 and q_3, we get

$$q_1 = k_{1.1}\, v_1 + k_{1.2}\, v_2 + k_{1.3}\, v_3,$$
$$q_2 = k_{2.1}\, v_1 + k_{2.2}\, v_2 + k_{2.3}\, v_3,$$

and $$q_3 = k_{3.1}\, v_1 + k_{3.2}\, v_2 + k_{3.3}\, v_3,$$

where $k_{1.1} = k_{3.3} = (p_{1.1} p_{2.2} - p^2_{1.2})/\{(p_{1.1} - p_{1.3})\,\Delta\}$,

$k_{2.2} = (p_{1.1} + p_{1.3})/\Delta$, $k_{1.2} = k_{2.1} = k_{3.2} = k_{2.3} = -p_{1.2}/\Delta$,

and $k_{1.3} = k_{3.1} = -(p_{2.2}\, p_{1.3} - p^2_{1.2})/\{(p_{1.1} - p_{1.3})\,\Delta\}$,

where $\Delta = p_{2.2}(p_{1.1} + p_{1.3}) - 2p^2_{1.2}$.

When h is very great so that we can neglect the effect of the earth, we have

$$k_{1.1} = 1/\{3 \log(d/a)\}, \quad \text{and} \quad k_{1.2} = k_{1.3} = -1/\{6 \log(d/a)\}.$$

In applying these formulae it has to be remembered that we have made the assumption that the earth in the neighbourhood of the wires is a level plane of perfectly conducting matter. This assumption is in many cases not permissible. The presence of trees, rocks, buildings, etc. considerably complicates the problem; moreover the heights of the wires above the earth are generally not constant. In addition the wires are often 'spiralled' relatively to one another. For example if a, b and c be the insulators on the first pole and a', b' and c' be the corresponding insulators on the next, then No. 1 main would be connected between a and b', No. 2 main between b and c' and No. 3 main between c and a'. The mains would then probably be slung parallel to one another for a given number of poles and then another twist of 120° in the same direction would be given to them round the central axis. On long transmission lines one effect of this spiralling is to make the capacity and inductance between any two of the mains the same.

The capacity between an overhead wire and the earth is generally measured by charging the overhead wire to a given potential and then measuring its discharge to a good 'earth,' a water pipe for example, by means of the throw produced on a calibrated ballistic galvanometer. The ratio of the charge found in this manner to the potential is taken to be the capacity of the wire. Now after a prolonged drought the surface of the earth is a very bad conductor, and so it takes a considerable time to charge and discharge the condenser formed by the wire and the surface of the earth. Hence, as we should have expected from theoretical considerations, the capacity found in the above manner varies, sometimes by several hundred per cent., from day to day, depending on whether the ground is damp or dry. The value of the capacity of a main found by noting the charging current it takes when one pole of an alternator is connected to it and the other is put to earth corresponds practically with the capacity found by an instantaneous discharge of the main.

REFERENCES

OLIVER HEAVISIDE, *Electrical Papers*, Vol. 1, p. 42, 'On the Electrostatic Capacity of Suspended Wires,' and also p. 101, 'The Inductances of Suspended Wires.'
LOUIS COHEN, 'Inductance and Capacity of Linear Conductors and the Determination of the Capacity of Horizontal Antennae,' *The Electrician*, Vol. 70, p. 881, 1913.

CHAPTER VII

THE alternating currents supplied for lighting and power

High frequency currents. purposes generally have a frequency of about fifty per second. The frequency of the currents used in other practical applications of electricity, however, is often much higher. In telephony, for instance, the currents which flow in the wires have an average frequency of 800 per second and in radio-telegraphy the frequency may be many millions per second. At these high frequencies the laws of current flow are apparently different from the laws of current flow at low frequencies. As high frequency currents are now being extensively used for various purposes, a careful study of them is necessary. It will be seen from what follows that their apparently anomalous behaviour can be deduced from the laws laid down in the preceding chapters.

There are three methods ordinarily employed for producing high frequency currents. The first is by a special type of alternator, the second is by discharging a Leyden jar through an inductive coil of small resistance, and the third by shunting an electric arc by a condenser and an inductive coil in series.

The high frequency alternators designed by Alexanderson have proved successful in practice. They are of the inductor type (see Vol. II) and some of these machines give currents having a frequency as high as 200,000. In the 100,000 frequency machine the rotating part consists of a steel disc with 300 slots cut round its edge. It has thus 300 steel teeth. Three hundred times during a revolution each of these teeth closes magnetically the air-gap between two of the poles of the stationary part of the machine. Hence, when the rotating part makes 20,000 revolutions per minute, the frequency of the electromotive force generated is 100,000. At this frequency an output of 2 kilowatts can be obtained. Owing to the high speed it is necessary to have a flexible shaft and the wheel is placed on one side of the middle point between the bearings. Otherwise the slight want of balance due to unavoidable imperfections in manufacture would produce excessive stresses at the high speeds. These machines are manufactured for radio-telegraphic purposes.

By the Leyden jar method, frequencies varying between unity and a hundred million can be obtained without difficulty. If we charge a Leyden jar by means of a high voltage battery, the charge will usually be oscillatory. If we now discharge it through an air-gap the discharge will be oscillatory if the resistance in the discharging circuit, including the resistance of the spark-gap, be sufficiently small. In practice, a Leyden jar and spark-gap are sometimes put in series with the high tension winding of an alternating current transformer. When the pressure attains a certain critical value during its first half period, the spark-gap breaks down and oscillations ensue until the pressure falls below the critical value when the spark goes out. Similar oscillations ensue during the spark which occurs during the second half of the period. Hence if f be the frequency the ' oscillation transformer ' will give rise to $2f$ trains of waves per second.

Another method used by the Marconi Company for trans-Atlantic signalling is to employ a high voltage battery and a condenser in series with a spark-gap. A disc fitted with studs round its edges rotates in such a way that the studs pass in rapid succession through the spark-gap. At each passage of a stud an oscillatory discharge takes place and so by driving the disc at high speed, trains of waves can be sent off with very short idle intervals between them. For example, if 600 studs pass per second through the spark-gap, there will be 600 spark discharges per second. As the frequency of the trains of waves used is about 50,000, we see that if about 80 waves occur at each discharge we get practically a continuous emission of waves.

In the arc method, oscillations are set up in a resonating circuit by means of an electric arc. If the arc is in hydrogen and is placed in a strong transverse magnetic field, as in the Poulsen system of radio-telegraphy, the frequency of the alternating currents set up can be as high as a million.

In any alternating current supply system, any sudden change in the load may give rise to high frequency currents being set up in the mains owing to their capacity and inductance. Even in direct current supply systems destructive high frequency oscillations can be set up by an arc occurring at a short circuit. We shall first discuss high frequency currents in a concentric main, as in this case the complete solution can be obtained.

Let the inner conductor of the concentric main be a solid
Concentric   metal cylinder of radius a, and let the outer be a
main.   coaxial hollow cylinder of inner and outer radii
b and c respectively. Let μ be the value of the permeability of the metals forming the conductors and let μ' be the permeability of the insulating material separating them. Let ρ be the volume resistivity of the conducting metal. We shall suppose that μ, μ' and ρ are constants and that both the capacity and leakage currents in the dielectric can be neglected. We can assume, therefore, that the flow of current in the conductors is parallel to their common axis, and hence, that the equipotential surfaces in each conductor are planes perpendicular to this axis.

Let us consider the current in a cylindrical tube of unit length in the inner conductor, the inner and outer radii of which are r and $r + \partial r$ respectively. If e_1 be the potential difference between the ends of this tube, the equation to determine the current density i in it is

$$e_1 = (\rho/2\pi r \partial r)(i \cdot 2\pi r \partial r) + \partial \phi/\partial t$$

$$= \rho i + \partial \phi/\partial t \quad \dots \dots \dots \dots \dots \dots \dots \dots \dots \dots \dots \dots (1),$$

where ϕ is the number of lines of induction, between two planes perpendicular to the axis of the main and one centimetre apart, which are linked with the current in this cylindrical tube.

By hypothesis, the equipotential surfaces in the inner conductor are planes perpendicular to the axis. Hence e_1 is independent of the value of r, and thus

$$0 = \rho \frac{\partial i}{\partial r} + \frac{\partial}{\partial t} \frac{\partial \phi}{\partial r} \quad \dots\dots\dots\dots\dots(2).$$

From the symmetry of a concentric main, we see that the intensity of the current is the same at all points equidistant from the axis. Hence, since the magnetic force outside an infinite cylindrical tube, carrying a current flowing parallel to its axis, is the same as if all the current were concentrated at this axis (p. 55), we have, at all points of the inner conductor,

$$\phi = \mu \int_r^a \frac{2I_x}{x} \, \partial x + 2\mu' I \log \frac{b}{a} + \mu \int_b^c \frac{2 \, (I - I_\xi')}{\xi} \, \partial\xi \dots(3),$$

where I_x is the algebraical sum of the currents flowing through the cross section of a coaxial cylinder whose radius is x, I is the total current flowing in the inner conductor, and I_ξ' is the sum of the currents flowing in the tube of the outer conductor whose inner radius is b and outer radius is ξ.

By differentiating ϕ with respect to r, we get

$$\frac{\partial \phi}{\partial r} = -\frac{2\mu I_r}{r} = -\frac{2\mu}{r} \int_0^r 2\pi i r \partial r \dots\dots\dots\dots(4),$$

and hence, by (2),

$$\rho \frac{\partial i}{\partial r} = \frac{4\pi\mu}{r} \int_0^r r \frac{\partial i}{\partial t} \partial r \quad \dots\dots\dots\dots(5),$$

and finally, by differentiating,

$$\frac{\partial^2 i}{\partial r^2} + \frac{1}{r}\frac{\partial i}{\partial r} = \frac{4\pi\mu}{\rho}\frac{\partial i}{\partial t} \quad \dots\dots\dots\dots(6).$$

We shall assume that i varies according to the harmonic law.

The differential equation.

Hence we can write $i = u\epsilon^{\omega t \iota}$, where u is a function of r but not of t, and ι stands for $(-1)^{1/2}$. Substituting this value of i in equation (6), we get

$$\frac{\partial^2 u}{\partial r^2} + \frac{1}{r}\frac{\partial u}{\partial r} - m^2 \iota u = 0 \quad \dots\dots\dots\dots(7),$$

where $m^2 = 4\pi\mu\omega/\rho = 8\pi^2\mu f/\rho$, f being the frequency of the current.

From the theory of Bessel's functions we know that the solution of (7) may be written

$$u = A \cdot I_0(mr\sqrt{\iota}) + B \cdot K_0(mr\sqrt{\iota}) \dots\dots\dots(8),$$

where I and K are the Bessel's functions of the first and second kinds for unreal values of the argument. The values of the functions $I_0(x)$ and $K_0(x)$ can be computed by means of the following series.

When x is small

$$I_0(x) = 1 + \frac{x^2}{2^2} + \frac{x^4}{2^2 \cdot 4^2} + \dots \quad \dots\dots\dots(9),$$

and $K_0(x) = \alpha \cdot I_0(x) - \log x \cdot I_0(x)$

$$+ \frac{x^2}{2^2} + (1 + \tfrac{1}{2}) \frac{x^4}{2^2 \cdot 4^2} + (1 + \tfrac{1}{2} + \tfrac{1}{3}) \frac{x^6}{2^2 \cdot 4^2 \cdot 6^2} + \dots(10),$$

where $\alpha = \log 2 - \gamma$, and γ is Euler's constant.

Hence $\alpha = 0\cdot1159315\dots$.

When x is large we use the series

$$I_0(x) = \frac{\epsilon^x}{(2\pi x)^{1/2}} \left\{ 1 + \frac{1^2}{8x} + \frac{1^2 \cdot 3^2}{\lfloor 2 \,(8x)^2} + \frac{1^2 \cdot 3^2 \cdot 5^2}{\lfloor 3 \,(8x)^3} + \dots \right\} \dots(11),$$

and

$$K_0(x) = \left(\frac{\pi}{2x}\right)^{1/2} \cdot \epsilon^{-x} \left\{ 1 - \frac{1^2}{8x} + \frac{1^2 \cdot 3^2}{\lfloor 2 \,(8x)^2} - \frac{1^2 \cdot 3^2 \cdot 5^2}{\lfloor 3 \,(8x)^3} + \dots \right\} \dots(12).$$

It is necessary to separate the real and imaginary parts of the terms in (8). The new functions we shall use may be defined as follows:

The ber and bei and the ker and kei functions.

$$I_0(x\sqrt{\iota}) = \text{ber } x + \iota \text{ bei } x \dots\dots(13),$$

and

$$K_0(x\sqrt{\iota}) = \text{ker } x + \iota \text{ kei } x \dots\dots(14).$$

When x is small the functions can be found by the following series:

$$\left. \begin{aligned} \text{ber } x &= 1 - \frac{x^4}{2^2 \cdot 4^2} + \frac{x^8}{2^2 \cdot 4^2 \cdot 6^2 \cdot 8^2} - \dots, \\ \text{bei } x &= \frac{x^2}{2^2} - \frac{x^6}{2^2 \cdot 4^2 \cdot 6^2} + \frac{x^{10}}{2^2 \cdot 4^2 \cdot 6^2 \cdot 8^2 \cdot 10^2} - \dots \end{aligned} \right\} \dots\dots(15),$$

$$\text{ker } x = (\alpha - \log x) \text{ ber } x + (\pi/4) \text{ bei } x$$
$$- \left(1 + \frac{1}{2}\right) \frac{x^4}{2^2 \cdot 4^2} + \left(1 + \frac{1}{2} + \frac{1}{3} + \frac{1}{4}\right) \frac{x^8}{2^2 \cdot 4^2 \cdot 6^2 \cdot 8^2} - \cdots \quad \cdots(16),$$

and kei $x = (\alpha - \log x) \text{ bei } x - (\pi/4) \text{ ber } x$
$$+ \frac{x^2}{2^2} - \left(1 + \frac{1}{2} + \frac{1}{3}\right) \frac{x^6}{2^2 \cdot 4^2 \cdot 6^2} + \cdots \quad \cdots(17).$$

When x is large we can use the formulae

$$\text{ber } x = \frac{\epsilon^\beta}{\sqrt{2\pi x}} \cos \delta, \text{ and bei } x = \frac{\epsilon^\beta}{\sqrt{2\pi x}} \sin \delta \cdots\cdots(18),$$

where
$$\beta = \frac{x}{\sqrt{2}} + \frac{1}{8\sqrt{2}x} - \frac{25}{384\sqrt{2}x^3} - \frac{13}{128x^4} - \cdots \quad \cdots(19),$$

and
$$\delta = \frac{x}{\sqrt{2}} - \frac{\pi}{8} - \frac{1}{8\sqrt{2}x} - \frac{1}{16x^2} - \frac{25}{384\sqrt{2}x^3} + \cdots \quad \cdots(20).$$

Similarly when x is large, we have

$$\text{ker } x = \left(\frac{\pi}{2x}\right)^{1/2} \epsilon^{\beta'} \cos \delta', \text{ and kei } x = \left(\frac{\pi}{2x}\right)^{1/2} \epsilon^{\beta'} \sin \delta' \cdots\cdots(21),$$

where
$$\beta' = - \frac{x}{\sqrt{2}} - \frac{1}{8\sqrt{2}x} + \frac{25}{384\sqrt{2}x^3} - \frac{13}{128x^4} + \cdots \quad \cdots(22),$$

and
$$\delta' = - \frac{x}{\sqrt{2}} - \frac{\pi}{8} + \frac{1}{8\sqrt{2}x} - \frac{1}{16x^2} + \frac{25}{384\sqrt{2}x^3} - \cdots \quad \cdots(23).$$

It will be seen that β' and δ' can be deduced from β and δ by merely changing the sign of x in the latter.

The functions defined in the preceding section generally occur **Special formulae.** in the solutions of physical problems associated together in one or other of the following ways, the expressions for which we shall denote by special symbols:

$$X(x) = \text{ber}^2 x + \text{bei}^2 x, \quad V(x) = \text{ber}'^2 x + \text{bei}'^2 x,$$
$$\dot{Z}(x) = \text{ber } x \text{ ber}'x + \text{bei } x \text{ bei}'x,$$

and
$$W(x) = \text{ber } x \text{ bei}'x - \text{bei } x \text{ ber}'x.$$
$$X_1(x) = \text{ker}^2 x + \text{kei}^2 x, \quad V_1(x) = \text{ker}'^2 x + \text{kei}'^2 x,$$
$$S(x) = \text{ber}'x \text{ ker}'x + \text{bei}'x \text{ kei}'x,$$

and
$$T(x) = \text{bei}'x \text{ ker}'x - \text{ber}'x \text{ kei}'x.$$

In the expressions above ber$'x$, ker$'x$, etc. represent the differential coefficients of ber x and ker x with respect to x. The combinations V/X, Z/X, W/X, Z/V, and W/V also occur in the solutions of several electrical problems. The values of these functions may be easily computed by the following series, provided that x be not too large:

$$X = 1 + \left(\frac{1}{\lfloor 1}\right)^2 \frac{1}{\lfloor 2} \left(\frac{x}{2}\right)^4 + \cdots$$
$$+ \left(\frac{1}{\lfloor n-1}\right)^2 \frac{1}{\lfloor 2n-2} \left(\frac{x}{2}\right)^{4n-4} + \cdots \quad \ldots(24),$$

$$V = \frac{x^2}{4}\left\{1 + \frac{1}{2}\left(\frac{1}{\lfloor 1}\right)^2 \frac{1}{\lfloor 3} \left(\frac{x}{2}\right)^4 + \cdots\right.$$
$$\left. + \frac{1}{n}\left(\frac{1}{\lfloor n-1}\right)^2 \frac{1}{\lfloor 2n-1} \left(\frac{x}{2}\right)^{4n-4} + \cdots\right\} \ldots(25),$$

$$Z = \frac{x^3}{16}\left\{1 + \left(\frac{1}{2}\right)^2 \left(\frac{1}{\lfloor 1}\right)^2 \frac{1}{\lfloor 3} \left(\frac{x}{2}\right)^4 + \cdots\right.$$
$$\left. + \left(\frac{1}{n}\right)^2 \left(\frac{1}{\lfloor n-1}\right)^2 \frac{1}{\lfloor 2n-1} \left(\frac{x}{2}\right)^{4n-4} + \cdots\right\} \ldots(26),$$

$$W = \frac{x}{2}\left\{1 + \left(\frac{1}{\lfloor 1}\right)^2 \frac{1}{\lfloor 3} \left(\frac{x}{2}\right)^4 + \cdots\right.$$
$$\left. + \left(\frac{1}{\lfloor n-1}\right)^2 \frac{1}{\lfloor 2n-1} \left(\frac{x}{2}\right)^{4n-4} + \cdots\right\} \ldots(27).$$

These series show that the four functions X, V, Z, and W continually increase as x increases.

Approximate formulae. When x is not greater than 2 the following approximate formulae can be employed:

$$\frac{V}{X} = \frac{x^2}{4}\left\{1 - \frac{5}{12}\left(\frac{x}{2}\right)^4 + \frac{143}{720}\left(\frac{x}{2}\right)^8 - \frac{7661}{4^2 \lfloor 7}\left(\frac{x}{2}\right)^{12}\right\} \quad \ldots\ldots(28),$$

$$\frac{Z}{X} = \frac{x^3}{16}\left\{1 - \frac{11}{24}\left(\frac{x}{2}\right)^4 + \frac{473}{3 \lfloor 6}\left(\frac{x}{2}\right)^8 - \frac{304107}{4 \cdot 12^2 \lfloor 7}\left(\frac{x}{2}\right)^{12}\right\} \quad \ldots\ldots(29),$$

$$\frac{W}{X} = \frac{x}{2}\left\{1 - \frac{1}{3}\left(\frac{x}{2}\right)^4 + \frac{19}{\lfloor 5}\left(\frac{x}{2}\right)^8 - \frac{687}{7 \cdot 6^4}\left(\frac{x}{2}\right)^{12}\right\} \quad \ldots\ldots\ldots(30),$$

$$\frac{Z}{V} = \frac{x}{4}\left\{1 - \frac{1}{24}\left(\frac{x}{2}\right)^4 + \frac{13}{4320}\left(\frac{x}{2}\right)^8 - \frac{647}{12^2 \cdot 360 \cdot 56}\left(\frac{x}{2}\right)^{12}\right\} \quad (31),$$

and
$$\frac{W}{V} = \frac{2}{x}\left\{1 + \frac{1}{12}\left(\frac{x}{2}\right)^4 - \frac{1}{180}\left(\frac{x}{2}\right)^8 + \frac{11}{12.28.80}\left(\frac{x}{2}\right)^{12}\right\} \quad \ldots(32).$$

When x is large we get by (18)
$$X = \frac{\epsilon^{2\beta}}{2\pi x} \quad \ldots\ldots\ldots\ldots\ldots\ldots\ldots(33),$$

where the value of β is given by (19).

In a similar manner we find that

$$\frac{V}{X} = 1 - \frac{1}{x\sqrt{2}} + \frac{1}{4x^2} + \frac{3}{8\sqrt{2}x^3} \quad \ldots\ldots\ldots\ldots(34),$$

$$\frac{Z}{X} = \frac{1}{\sqrt{2}} - \frac{1}{2x} - \frac{1}{8\sqrt{2}x^2} \quad \ldots\ldots\ldots\ldots\ldots(35),$$

$$\frac{W}{X} = \frac{1}{\sqrt{2}} + \frac{1}{8\sqrt{2}x^2} + \frac{1}{8x^3} \quad \ldots\ldots\ldots\ldots\ldots(36),$$

$$\frac{Z}{V} = \frac{1}{\sqrt{2}} - \frac{3}{8\sqrt{2}x^2} - \frac{3}{8x^3} \quad \ldots\ldots\ldots\ldots\ldots(37),$$

$$\frac{W}{V} = \frac{1}{\sqrt{2}} + \frac{1}{2x} + \frac{3}{8\sqrt{2}x^2} \quad \ldots\ldots\ldots\ldots\ldots(38).$$

Formulae (33) to (38) give a four figure accuracy when x is not less than 8.

When x is small it follows from (16) and (17) that
$$\text{ker}\, x = \alpha - \log x \quad \text{and} \quad \text{kei}\, x = -\pi/4.$$

Hence
$$X_1 = \text{ker}^2 x + \text{kei}^2 x = (\alpha - \log x)^2 + \pi^2/16 \quad \ldots\ldots(39).$$

Similarly, when x is large,
$$X_1 = \pi\epsilon^{2\beta'}/(2x) \quad \ldots\ldots\ldots\ldots\ldots\ldots\ldots(40),$$

where β' is given by (22), and
$$V_1 = X_1\left\{1 + \frac{1}{x\sqrt{2}} + \frac{1}{4x^2} - \frac{3}{8\sqrt{2}x^3}\right\} \quad \ldots\ldots\ldots\ldots(41),$$

approximately.

When x is large we shall use the formulae
$$\text{ker}\, x = \left(\frac{\pi}{2x}\right)^{1/2} \epsilon^{-x/\sqrt{2}} \cos\left(\frac{x}{\sqrt{2}} + \frac{\pi}{8}\right) \quad \ldots\ldots\ldots(42),$$

$$\text{kei } x = -\left(\frac{\pi}{2x}\right)^{1/2} \epsilon^{-x/\sqrt{2}} \sin\left(\frac{x}{\sqrt{2}} + \frac{\pi}{8}\right) \quad \ldots\ldots(43),$$

$$\text{ker}' x = -\left(\frac{\pi}{2x}\right)^{1/2} \epsilon^{-x/\sqrt{2}} \cos\left(\frac{x}{\sqrt{2}} - \frac{\pi}{8}\right) \quad \ldots\ldots(44),$$

$$\text{kei}' x = \left(\frac{\pi}{2x}\right)^{1/2} \epsilon^{-x/\sqrt{2}} \sin\left(\frac{x}{\sqrt{2}} - \frac{\pi}{8}\right) \ldots\ldots\ldots(45).$$

We can also show that, when x is small,

$$S(x) = \text{ber}'x\, \text{ker}'x + \text{bei}'x\, \text{kei}'x$$

$$= \frac{x^2}{4}\left(\alpha + \frac{3}{4} - \log x\right)$$

$$+ \frac{x^6}{12 \cdot 64}(\alpha - \log x) + \frac{37x^6}{2^2 \cdot 4^2 \cdot 6^2 \cdot 8} \ldots(46),$$

$$T(x) = \text{bei}'\, x\, \text{ker}'\, x - \text{ber}'\, x\, \text{kei}'\, x$$

$$= -\frac{1}{2} + \frac{\pi}{4}\cdot\frac{x^2}{4} - \frac{x^4}{48}\ldots\ldots\ldots\ldots\ldots\ldots(47),$$

approximately.

When c is great we may write with high accuracy

$$S(x) = -g\cos\psi \quad \text{and} \quad T(x) = -g\sin\psi \quad \ldots\ldots(48),$$

where $\quad g = 1/(2x) + 27/(128x^5)$

and $\quad \left. \psi = x\sqrt{2} + \frac{3}{4\sqrt{2}x} + \frac{21}{64\sqrt{2}x^3} - \frac{1899}{2560\sqrt{2}x^5} \right\} \quad \ldots\ldots(49).$

With the help of the formulae found above we can now discuss the solution of the equation

Particular solution.

$$\frac{\partial^2 i}{\partial r^2} + \frac{1}{r}\frac{\partial i}{\partial r} = \frac{m^2}{\omega}\frac{\partial i}{\partial t},$$

which determines the current density in the conductors. We have seen above that $i = I_0(mr\sqrt{\iota})\,\epsilon^{\omega t \iota}$ is a particular solution of this equation. Hence $i = (\text{ber } mr + \iota\,\text{bei } mr)(\cos \omega t + \iota \sin \omega t)$ must also satisfy it. Both the real and imaginary parts of this solution must separately satisfy the equation. Hence a particular solution may be written in either of the two following forms

$$i = (A\,\text{ber } mr + B\,\text{bei } mr)\cos \omega t$$

$$+ (-A\,\text{bei } mr + B\,\text{ber } mr)\sin \omega t \ldots(50),$$

or $\quad i = (A^2 + B^2)^{1/2}\{X(mr)\}^{1/2}\cos(\omega t - \epsilon) \ldots\ldots\ldots\ldots(51),$

where A and B are constants, and

$$\tan \epsilon = (- A \text{ bei } mr + B \text{ ber } mr)/(A \text{ ber } mr + B \text{ bei } mr)...(52).$$

For a solid core and a return conductor of infinite conductivity this solution suffices. It will also give the current density in a cylindrical conductor when the return conductor parallel to it is so far away that the action of the magnetic field due to the return current in disturbing the distribution of the current in the conductor can be neglected.

Let us suppose that the current density along the axis of the cylinder is given by $i = i_0 \cos \omega t$. At the axis r is zero, and since ber $0 = 1$ and bei $0 = 0$, we see that $A = i_0$ and $B = 0$. Hence

$$i = i_0 \text{ ber } mr \cos \omega t - i_0 \text{ bei } mr \sin \omega t \quad ....(53).$$

Noticing that

$$\int r \text{ ber } mr \, \partial r = (r/m) \text{ bei}' \, mr \quad ....(54),$$

$$\int r \text{ bei } mr \, \partial r = - (r/m) \text{ ber}' \, mr \quad ....(55),$$

we find that
$$I_x = 2\pi \int_0^x ir \partial r$$
$$= (2\pi/m) \, x \text{ bei}' \, mx \, . \, i_0 \cos \omega t$$
$$+ (2\pi/m) \, x \text{ ber}' \, mx \, . \, i_0 \sin \omega t ....(56).$$

By (3), therefore, we see that

$$\phi = \mu \int_r^a \frac{2I_x}{x} \partial x + 2\mu' I \log \frac{b}{a}$$
$$= (\rho/\omega) \, (\text{bei } ma - \text{bei } mr) \, . \, i_0 \cos \omega t$$
$$+ (\rho/\omega) \, (\text{ber } ma - \text{ber } mr) \, . \, i_0 \sin \omega t + 2\mu' I \log (b/a),$$

and thus by (1)

$$e_1 = \rho i + \partial \phi / \partial t$$
$$= \rho \text{ ber } ma \, . \, i_0 \cos \omega t$$
$$- \rho \text{ bei } ma \, . \, i_0 \sin \omega t + 2\mu' \log (b/a) \, . \, \frac{\partial I}{\partial t} \, ...(57).$$

By (56), we have

$$I = (2\pi/m) \, a \text{ bei}' \, ma \, . \, i_0 \cos \omega t + (2\pi/m) \, a \text{ ber}' \, ma \, . \, i_0 \sin \omega t,$$

and

$$\frac{\partial I}{\partial t} = - (2\pi\omega/m) \, a \text{ bei}' \, ma \, . \, i_0 \sin \omega t + (2\pi\omega/m) \, a \text{ ber}' \, ma \, . \, i_0 \cos \omega t.$$

Solving these equations for $i_0 \cos \omega t$ and $i_0 \sin \omega t$, we get

$$i_0 \cos \omega t = \frac{m \, \text{bei}' \, ma}{2\pi a V (ma)} I + \frac{m \, \text{ber}' \, ma}{2\pi a \omega V (ma)} \frac{\partial I}{\partial t} \quad \dots\dots(58),$$

and

$$i_0 \sin \omega t = \frac{m \, \text{ber}' \, ma}{2\pi a V (ma)} I - \frac{m \, \text{bei}' \, ma}{2\pi a \omega V (ma)} \frac{\partial I}{\partial t} \quad \dots\dots(59).$$

Substituting the values in (57) we get

$$e_1 = \frac{\rho m \, W (ma)}{2\pi a V (ma)} I + \left\{ 2\mu' \log \frac{b}{a} + \frac{2\mu Z (ma)}{ma V (ma)} \right\} \frac{\partial I}{\partial t} \quad \dots(60)$$

$$= RI + L \frac{\partial I}{\partial t},$$

where R and L are the resistance and self inductance per unit length of the inner conductor. Hence

$$R = \frac{\rho m}{2\pi a} \cdot \frac{W (ma)}{V (ma)} \quad \dots\dots\dots\dots\dots\dots(61),$$

and

$$L = 2\mu' \log \frac{b}{a} + \frac{2\mu Z (ma)}{ma V (ma)} \quad \dots\dots\dots\dots(62).$$

When ma is small, we see by (32) and (31) that

$$R = \frac{\rho}{\pi a^2} \left\{ 1 + \frac{1}{12} \left(\frac{ma}{2}\right)^4 - \frac{1}{180} \left(\frac{ma}{2}\right)^8 + \dots \right\} \quad \dots\dots(63),$$

and

$$L = 2\mu' \log \left(\frac{b}{a}\right) + \frac{\mu}{2} \left\{ 1 - \frac{1}{24} \left(\frac{ma}{2}\right)^4 + \frac{13}{4320} \left(\frac{ma}{2}\right)^8 - \dots \right\} \quad \dots(64).$$

When ma is large we see similarly by (38) and (37) that

$$R = \frac{\rho m}{2\pi a} \left\{ \frac{1}{\sqrt{2}} + \frac{1}{2ma} + \frac{3}{8\sqrt{2} \, m^2 a^2} + \dots \right\} \quad \dots\dots\dots(65),$$

and

$$L = 2\mu' \log \frac{b}{a} + \frac{2\mu}{ma} \left(\frac{1}{\sqrt{2}} - \frac{3}{8\sqrt{2} \, m^2 a^2} - \frac{3}{8 m^3 a^3} + \dots \right) \quad \dots(66).$$

If R' and L' be the resistance and inductance per unit length with direct current, we have

$$\frac{R}{R'} = \frac{ma}{2} \cdot \frac{W (ma)}{V (ma)} \quad \dots\dots\dots\dots\dots\dots(67),$$

and

$$\frac{L - 2\mu' \log (b/a)}{L' - 2\mu' \log (b/a)} = \frac{4}{ma} \cdot \frac{Z (ma)}{V (ma)} \quad \dots\dots\dots\dots(68).$$

Tables of the values of these fractions computed to five decimal places are given at the end of this chapter. When ma is zero the value of both fractions is unity, but when ma is infinite the value of the first fraction is infinity and the value of the second is zero, so that $L = 2\mu' \log (b/a)$. It is easy to see that this would be the case if all the current were confined to the surface of the inner conductor when the frequency is infinitely high. We shall now show that this is the case.

If i be the density of the current at a distance r from the axis of the inner conductor, we have by (53)

The distribution of the current in the inner conductor.

$$i = i_0 \{X (mr)\}^{1/2} \cos (\omega t + \epsilon) \dots\dots(69),$$

and $\tan \epsilon = \text{bei } mr/\text{ber } mr.$

Since $X (mr)$ always increases as r increases we see that the amplitude of the current density continually increases as we pass from the axis to the surface of the conductor. We also see that the phases of the various layers of current are different. As the value of ma increases the current gets more and more concentrated on the outer layer. With very high frequencies, therefore, and thick conductors the amplitude of the current is practically zero after penetrating a very little way into the conductor. This phenomenon is known as the skin effect.

When mr is not greater than unity we may write

$$i = i_0(1 + m^4r^4/64) \cos (\omega t + \epsilon) \quad \dots\dots\dots(70),$$

where $\tan \epsilon = m^2r^2/4.$

For example, if $mr = 1$,

$$i = 1{\cdot}02 \, i_0 \cos (\omega t + 14°), \text{ approximately.}$$

When mr is large

$$\frac{i}{i_0} = \frac{\epsilon^{mr/\sqrt{2}}}{\sqrt{2\pi mr}} \cos \left(\omega t + \frac{mr}{\sqrt{2}} - \frac{\pi}{8}\right) \quad \dots\dots\dots(71).$$

For example, if $mr = 10$,

$$i = 149 \, i_0 \cos (\omega t + 383°),$$

and if $mr = 30$,

$$i = 119{\cdot}3 \times 10^6 i_0 \cos (\omega t + 1193°).$$

In this case at a depth equal to one-thirtieth of the radius the current density will have fallen to half its surface value.

Let us now consider the case when the conductivity of the **Complete solution.** return conductor is not infinite. Let its inner radius be b and its outer radius c. Let i' be the current density in this cylinder at a distance r from the axis. Let e' be the potential difference per unit length measured from the distributing station to the generator, i' being considered positive when flowing in this direction.

We have, therefore,

$$e' = (\rho/2\pi r \partial r)(2\pi r \partial r . i') - \partial\phi'/\partial t$$

$$= \rho i'' - \partial\phi'/\partial t \quad\dots\dots\dots\dots\dots\dots\dots\dots\dots\dots(72),$$

where

$$\phi' = \mu \int_r^c \frac{2(I - I_x')}{x} \partial x \quad\dots\dots\dots\dots\dots(73),$$

and thus

$$r\frac{\partial\phi'}{\partial r} = -2\mu\,(I - I_r')$$

$$= -4\pi\mu \int_r^c r i'' \partial r \quad\dots\dots\dots\dots\dots(74).$$

Also, since we have supposed that e' does not vary with r, we get from (72) and (74)

$$\rho\frac{\partial i''}{\partial r} = -\frac{4\pi\mu}{r}\int_r^c r\frac{\partial i''}{\partial t}\partial r \quad\dots\dots\dots\dots\dots(75),$$

and thus

$$\frac{\partial^2 i''}{\partial r^2} + \frac{1}{r}\frac{\partial i''}{\partial r} = \frac{m^2}{\omega}\frac{\partial i''}{\partial t} \quad\dots\dots\dots\dots\dots(76).$$

By (8) it is easy to see that the complete solution of this equation may be written as follows:

$$i' = (A\,\mathrm{ber}\,mr + B\,\mathrm{bei}\,mr + C\,\mathrm{ker}\,mr + D\,\mathrm{kei}\,mr)\,i_0\cos\omega t$$

$$+ (-A\,\mathrm{bei}\,mr + B\,\mathrm{ber}\,mr - C\,\mathrm{kei}\,mr + D\,\mathrm{ker}\,mr)\,i_0\sin\omega t$$

$$\dots(77),$$

where A, B, C and D are constants which have to be determined from the data of the problem.

Noticing that

$$\int r\,\mathrm{ker}\,mr\,\partial r = (r/m)\,\mathrm{kei}'\,mr\dots\dots\dots\dots\dots(78),$$

and

$$\int r\,\mathrm{kei}\,mr\,\partial r = -(r/m)\,\mathrm{ker}'\,mr\dots\dots\dots\dots(79),$$

we find, by substituting the value of i' given by (77) in (75), that

$$A \text{ bei}' \, mc - B \text{ ber}' \, mc + C \text{ kei}' \, mc - D \text{ ker}' \, mc = 0 \; ...(80),$$

and $$A \text{ ber}' \, mc + B \text{ bei}' \, mc + C \text{ ker}' \, mc + D \text{ kei}' \, mc = 0 \; ...(81).$$

Hence using our special notation

$$A \, V(mc) = -CS(mc) + DT(mc) \; ....(82),$$

and $$B V(mc) = -CT(mc) - DS(mc) \; ....(83).$$

By equating the integral value $\int_b^c 2\pi r i' \, \partial r$ of the current in the return conductor to the value of I found from (56) by putting $x = a$, we get

$$A \text{ bei}' \, mb - B \text{ ber}' \, mb + C \text{ kei}' \, mb - D \text{ ker}' \, mb$$
$$= -(a/b) \text{ bei}' \, ma \; ...(84),$$

and $$A \text{ ber}' \, mb + B \text{ bei}' \, mb + C \text{ ker}' \, mb + D \text{ kei}' \, mb$$
$$= -(a/b) \text{ ber}' \, ma \; ...(85).$$

Substituting the values of A and B given by (82) and (83) in (84) and (85), we get

$$C \{V(mc) \, S(mb) - V(mb) \, S(mc)\}$$
$$+ D \{V(mb) \, T(mc) - V(mc) \, T(mb)\}$$
$$= -(a/b) \{V(mc)\} \{\text{ber}' \, ma \, \text{ber}' \, mb + \text{bei}' \, ma \, \text{bei}' \, mb\}...(86),$$

and $$- C \{V(mb) \, T(mc) - V(mc) \, T(mb)\}$$
$$+ D \{V(mc) \, S(mb) - V(mb) \, S(mc)\}$$
$$= -(a/b) \{V(mc)\} \{\text{ber}' \, ma \, \text{bei}' \, mb - \text{bei}' \, ma \, \text{ber}' \, mb\} \; ...(87).$$

From these equations C and D are readily found, and then A and B are found from (82) and (83). Hence the current density in the outer conductor can be found by (77).

From (72) and (73) we can see at once that

$$e' = \rho i_b' - \mu \frac{\partial}{\partial t} \int_b^c \frac{2(I - I_x')}{x} \, \partial x ....(88),$$

where i_b' is the current density on the inner surface of the outer conductor. We also see from (1) and (3) that

$$e = \rho i_a + 2\mu' \log \frac{b}{a} \frac{\partial I}{\partial t} + \mu \frac{\partial}{\partial t} \int_b^c \frac{2(I - I_x')}{x} \, \partial x \; ......(89),$$

where i_a is the current density on the outer surface of the inner conductor. Thus, by addition,

$$e + e' = \rho i_a + \rho i_b' + 2\mu' \log \frac{b}{a} \frac{\partial I}{\partial t} \quad \dots\dots\dots\dots(90).$$

Hence writing for i_a and i_b' their values found from (53) and (77) respectively, and also writing for $i_0 \cos \omega t$ and $i_0 \sin \omega t$ their values in terms of I and $\partial I/\partial t$ from (58) and (59), we get after a little reduction

$$e + e' = RI + L \frac{\partial I}{\partial t} \quad \dots\dots\dots\dots\dots(91),$$

where

$$R = \frac{\rho m}{2\pi a V(ma)} [W(ma) + \text{bei}'\, ma\, (A\, \text{ber}\, mb + B\, \text{bei}\, mb$$

$$+ C\, \text{ker}\, mb + D\, \text{kei}\, mb) - \text{ber}'\, ma\, (A\, \text{bei}\, mb - B\, \text{ber}\, mb$$

$$+ C\, \text{kei}\, mb - D\, \text{ker}\, mb)] \dots(92),$$

and

$$L = 2\mu' \log (b/a) + \frac{2\mu}{ma V(ma)} [Z(ma) + \text{ber}'\, ma\, (A\, \text{ber}\, mb$$

$$+ B\, \text{bei}\, mb + C\, \text{ker}\, mb + D\, \text{kei}\, mb) + \text{bei}'\, ma\, (A\, \text{bei}\, mb$$

$$- B\, \text{ber}\, mb + C\, \text{kei}\, mb - D\, \text{ker}\, mb)] \dots(93).$$

The formulae above give the complete solution of the problem when the applied wave is sine-shaped and the capacity and leakage currents are neglected.

If the applied wave be not sine-shaped, but if it be a periodic function of the time, it may be expanded in a series of sines by Fourier's theorem; and hence we could write down the solution without difficulty, but it would be very cumbrous.

When mc is small we find by (77) that

The density of the current in the outer conductor.
$$i'/i_0 = \{a^2/(c^2 - b^2)\}\, \{1 - (m^2c^2/2) \log mr\} \cos \omega t \dots(94),$$
approximately. Hence the amplitude of the current density in the outer conductor diminishes as r increases. Again when mc is great

$$\frac{i'}{i_0} = \frac{\epsilon^{ma/\sqrt{2}}}{\sqrt{2\pi mr}} \left(\frac{a}{b}\right)^{1/2}$$

$$\times \left[\frac{\cosh m(c-r)\sqrt{2} + \cos m(c-r)\sqrt{2}}{\cosh m(c-b)\sqrt{2} - \cos m(c-b)\sqrt{2}}\right]^{1/2} \cos(\omega t + \theta) \dots (95),$$

where
$$\tan\theta = \frac{(\sin\psi_1)\,\epsilon^{-m(c-r)/\sqrt{2}} + (\sin\psi_2)\,\epsilon^{m(c-r)/\sqrt{2}}}{(\cos\psi_1)\,\epsilon^{-m(c-r)/\sqrt{2}} + (\cos\psi_2)\,\epsilon^{m(c-r)/\sqrt{2}}},$$

and
$$\psi_1 = \gamma - mr/\sqrt{2} + \pi/8, \quad \psi_2 = \delta + mr/\sqrt{2} + \pi/8,$$

$$\tan\gamma = B/A, \quad \tan\delta = D/C, \quad \text{and} \quad \gamma - \delta = mc\sqrt{2}.$$

Since $\{\cosh m(c-r)\sqrt{2} + \cos m(c-r)\sqrt{2}\}/r$ continually diminishes as r increases, we see that the amplitude of the current density in the outer conductor always diminishes as r increases. Hence we see that in a concentric main when the frequency of the current is very high the current is practically confined to thin layers on the surface of the inner conductor and on the inner surface of the outer conductor respectively.

When mc is not greater than 2 we can use the formula

The effective resistance and inductance of a concentric main.

$$R = \frac{\rho}{\pi a^2}\left(1 + \frac{m^4 a^4}{192}\right)$$

$$+ \frac{\rho}{\pi(c^2 - b^2)}\{1 + m^4(\sigma + \sigma_1\,\xi + \sigma_2\,\xi^2)\} \dots (96),$$

where
$$\sigma = \frac{(7c^2 - b^2)(c^2 - b^2)}{192}, \quad \sigma_1 = \frac{b^2 c^4}{8(c^2 - b^2)},$$

$$\sigma_2 = -\frac{b^2 c^6}{4(c^2 - b^2)^2}, \quad \text{and} \quad \xi = \log\frac{c}{b}.$$

It is interesting to notice that the coefficient of m^2 vanishes and that the numerical constants α and $\log m$ have cancelled out. This formula was first given by Oliver Heaviside. It follows after very lengthy algebraical work from the formulae given previously in this chapter.

In power transmission cables $a^2 = c^2 - b^2$. In this case the greater the value of b the smaller the value of $\sigma + \sigma_1\xi + \sigma_2\xi^2$. In these cables, when the frequency is not greater than 50, the increase in the effective resistance of the outer conductor due to the skin effect is negligibly small.

We find in a similar way that

$$L = 2\mu' \log \frac{b}{a} + \frac{\mu}{2} + \frac{2\mu c^4}{(c^2 - b^2)^2} \log \frac{c}{b} - \mu \frac{3c^2 - b^2}{2(c^2 - b^2)}$$

$$- \frac{\mu}{2} \cdot \frac{m^4 a^4}{192} - \mu\, (\tau + \tau_1\, \xi + \tau_2\, \xi^2 + \tau_3\, \xi^3)\, m^4 \ldots (97),$$

where
$$\tau = \frac{19c^6 + 103c^4 b^2 - 41c^2 b^4 + 3b^6}{2^2 \cdot 4^2 \cdot 6^2 (c^2 - b^2)},$$

$$\tau_1 = -\frac{14c^4 b^2 (2c^2 - b^2)}{2^2 \cdot 4^2 \cdot 3 (c^2 - b^2)^2},$$

$$\tau_2 = -\frac{c^6 b^4}{4 (c^2 - b^2)^3},$$

$$\tau_3 = \frac{c^8 b^4}{2 (c^2 - b^2)^4}, \quad \text{and} \quad \xi = \log \frac{c}{b}.$$

Formula (97) shows that at low frequencies the inductance diminishes as the frequency increases, the effect being more pronounced the thicker the shell of the outer conductor.

When ma is greater than 5, we get in a similar way

$$R = \frac{\rho m}{2\pi a} \left\{ \frac{1}{\sqrt{2}} + \frac{1}{2ma} + \frac{3}{8\sqrt{2} m^2 a^2} \right\}$$

$$+ \frac{\rho m}{2\pi b \sqrt{2}} \cdot \frac{\sinh m(c - b)\sqrt{2} + \sin m(c - b)\sqrt{2}}{\cosh m(c - b)\sqrt{2} - \cos m(c - b)\sqrt{2}} \ldots (98),$$

and
$$L = 2\mu' \log \frac{b}{a} + \frac{2\mu}{ma} \left(\frac{1}{\sqrt{2}} - \frac{3}{8\sqrt{2} m^2 a^2} - \frac{3}{8m^3 a^3} \right)$$

$$+ \frac{2\mu}{mb \sqrt{2}} \cdot \frac{\sinh m(c - b)\sqrt{2} - \sin m(c - b)\sqrt{2}}{\cosh m(c - b)\sqrt{2} + \cos m(c - b)\sqrt{2}} \ldots (99).$$

The first term in (98) gives the resistance of the inner conductor and the second that of the outer conductor. In (99) the first term gives the linkages of the magnetic flux in the dielectric with unit current in the inner conductor. The second term gives the linkages of the magnetic flux in the substance of the inner conductor, and the third the linkages with the outer conductor.

When ma is very great provided that c be not nearly equal to b, we may write

$$R = \frac{\rho m}{2\pi \sqrt{2}} \left(\frac{1}{a} + \frac{1}{b}\right) = \sqrt{\mu f \rho} \left(\frac{1}{a} + \frac{1}{b}\right) \quad \ldots\ldots\ldots(100),$$

and

$$L = 2\mu' \log \frac{b}{a} + \frac{\mu}{2\pi} \sqrt{\frac{\rho}{\mu f}} \left(\frac{1}{a} + \frac{1}{b}\right) \quad \ldots\ldots\ldots(101).$$

Hence as f increases, R continually increases, but L diminishes and approaches the value $2\mu' \log (b/a)$ asymptotically. It is easy to see that it has this value when the currents are confined to an infinitely thin skin on the outer and inner surfaces of the inner and outer conductors respectively.

When the inner conductor is hollow the complete solution can still be written down without difficulty but the formulae become more complicated. Let us suppose that a_1 is the inner radius of the outer conductor. The solution of the differential equation for i is

Concentric main with hollow inner conductor.

$$i = (A_1 \operatorname{ber} mr + B_1 \operatorname{bei} mr + C_1 \operatorname{ker} mr + D_1 \operatorname{kei} mr) i_0 \cos \omega t$$
$$+ (- A_1 \operatorname{bei} mr + B_1 \operatorname{ber} mr - C_1 \operatorname{kei} mr + D_1 \operatorname{ker} mr) i_0 \sin \omega t$$
$$\ldots(102).$$

If we suppose that $i = i_0 \cos \omega t$, when $r = a_1$, we must have

$$A_1 \operatorname{ber} ma_1 + B_1 \operatorname{bei} ma_1 + C_1 \operatorname{ker} ma_1 + D_1 \operatorname{kei} ma_1 = 1$$
and $\quad A_1 \operatorname{bei} ma_1 - B_1 \operatorname{ber} ma_1 + C_1 \operatorname{kei} ma_1 - D_1 \operatorname{ker} ma_1 = 0$
$$\ldots(103).$$

The equation corresponding to (75) is

$$r \frac{\partial i}{\partial r} = \frac{m_1^2}{\omega} \int_{a_1}^{r} r \frac{\partial i}{\partial t} \partial r.$$

Substituting for i its value given by (102) in this equation and equating the coefficients of $\cos \omega t$ and $\sin \omega t$ to zero, we get

$$A_1 \operatorname{ber}' ma_1 + B_1 \operatorname{bei}' ma_1 + C_1 \operatorname{ker}' ma_1 + D_1 \operatorname{kei}' ma_1 = 0$$
and $\quad A_1 \operatorname{bei}' ma_1 - B_1 \operatorname{ber}' ma_1 + C_1 \operatorname{kei}' ma_1 - D_1 \operatorname{ker}' ma_1 = 0$
$$\ldots(104).$$

The equations (103) and (104) completely determine the four

constants A_1, B_1, C_1 and D_1. The values of these constants, therefore, are independent of the values of a, b and c.

For the outer cylinder equations (77), (80) and (81) still hold.
Two other equations are obtained by equating the integral value
$\int_b^c 2\pi r i'' \partial r$ of the current in the outer conductor to the value
$\int_{a_1}^a 2\pi r i \partial r$ of the current I in the inner conductor. Hence the
four constants A, B, C and D can be found. As formerly we can
easily find $i_0 \cos \omega t$ and $i_v \sin \omega t$ in terms of I and $\partial I/\partial t$, and hence
$\rho i_a + \rho i_b'$ can be expressed in terms of I and $\partial I/\partial t$. Finally
substituting this value in (90) the coefficient of I in the resulting
equation gives us R and the coefficient of $\partial I/\partial t$ gives us L.

In connexion with the transmission of power by concentric
Numerical mains it is important to know the magnitude of the
examples. skin effect. Let us suppose that the inner core is
solid and that its radius is one centimetre. For a very high
pressure cable a suitable value of b would be 2·4 cms. Hence since
the section of the outer conductor is made equal to that of the
inner, c is 2·6 cms. We shall suppose that mc is not greater than
2 so that we may use formula (96). Substituting these numbers
in the formula, we find that

$$\frac{R}{l} = \frac{\rho}{\pi}\left(1 + \frac{m^4}{192}\right) + \frac{\rho}{\pi}\left(1 + \frac{0\cdot0072 m^4}{192}\right),$$

where l is the length of the main in centimetres.

When the frequency is 25, m is nearly equal to unity. In this
case we see that the skin effect increases the resistance of the
inner conductor by about the half of one per cent., and the
increase in the effective resistance of the outer conductor is less
than the hundredth part of this. For a low voltage cable we might
have $c = 5/3$ and $b = 4/3$. In this case

$$\frac{R}{l} = \frac{\rho}{\pi}\left(1 + \frac{m^4}{192}\right) + \frac{\rho}{\pi}\left(1 + \frac{0\cdot059 m^4}{192}\right),$$

and hence the increase in the loss of the outer conductor due to
the skin effect is only about the twentieth part of the corresponding

quantity for the inner. If the return conductor were a tight fitting tube so that $b = 1$ and $c = 2^{1/2}$, the formula becomes

$$\frac{R}{l} = \frac{\rho}{\pi}\left(1 + \frac{m^4}{192}\right) + \frac{\rho}{\pi}\left(1 + \frac{0\cdot148m^4}{192}\right).$$

Even in this case the increase in the loss of the outer is less than the sixth part of the increase in the loss of the inner, although the losses at high frequencies would be practically the same in each conductor.

Let us first consider the case of two equal parallel cylindrical **Parallel** conductors, the distance between their axes being c. **conductors.** If this distance be sufficiently large compared with the radius a of either conductor the distribution of current may be regarded as made up of cylindrical sheets. We have seen (p. 88) that the linkages of the lines of induction due to the current in a conductor but outside it with the currents in the two conductors equals $2\mu'I^2\log(c/a)$, where I is the current. Hence by (62) we may write

$$L = 4\mu'\log\frac{c}{a} + \frac{4\mu}{ma}\cdot\frac{Z(ma)}{V(ma)} \quad\ldots\ldots\ldots(105).$$

Similarly if the radii of the wires be a and b respectively,

$$L = 2\mu'\log\frac{c^2}{ab} + \frac{2\mu}{ma}\cdot\frac{Z(ma)}{V(ma)} + \frac{2\mu}{mb}\cdot\frac{Z(mb)}{V(mb)} \quad\ldots(106).$$

When the larger radius is not small compared with the distance between the wires, the mutual actions between the currents in the two wires will distort the distribution of the current in both so that the assumption of cylindrical current sheets will not be admissible.

In the case of very high frequencies, for instance, it can be shown that

$$L = 4\mu'\log\left[\{c + (c^2 - 4a^2)^{1/2}\}/2a\right]\ldots\ldots\ldots(107),$$

when the radii are equal. Hence when $c = 2a$, L is zero. In this case the currents simply concentrate themselves along the lines of contact.

A table of the values of $Z(x)/V(x)$ is given at the end of this chapter.

In the case of a ring conductor when we only have a surface

Ring conductor. current of uniform density flowing in circles parallel to the circular axis, it can be shown by Maxwell's method (see p. 105) that

$$L = 4\pi a \{(1 + r^2/4a^2) \log_\epsilon (8a/r) - 2\} \ldots\ldots(108),$$

approximately, where r is the radius of the circular cross section and a is the radius of the circular axis.

For a uniform solid ring when currents are slowly generated in it by variations of the flux parallel to the axis of symmetry, the flux being uniform with regard to space, Rayleigh has shown that

$$L = 4\pi a \{(1 + 3r^2/8a^2) \log_\epsilon (8a/r) - 7/4\} \ldots\ldots(109).$$

When the alternations of the flux are very rapid, the distribution of the current becomes independent of the resistance and is determined by induction alone. In this case the currents are superficial and (108) is the formula to use. It is easy to see that with very high frequencies the magnetic forces produced by the superficial currents must be zero at every point in the substance of the conductor, as otherwise very large eddy currents would be generated.

By comparing equation (6) with the corresponding equation

Analogy with theory of heat. for the diffusion of heat into a solid cylinder, we see that the resulting currents diffuse into the conductors from their surfaces in exactly the same way as heat would diffuse into the conductors if their surfaces were exposed to variable temperatures. There is this difference, however, that high electrical conductivity in the one problem corresponds to low thermal conductivity in the other. When alternating E.M.F.s of very high frequency act upon conductors, the currents are practically confined to very thin layers at the surfaces of the conductors. These surface currents are distributed in such a way that they give rise to no magnetic force in the interior of the conductors.

When the frequency of the oscillations and the electrical con-

Analogy with electrostatics. ductivity of the metal are finite, the problem of finding the distribution of alternating currents can only be solved in the very simplest cases. When, however, the conductivity is supposed infinite, the problem of finding the distribution of the current can be solved easily in certain cases by the aid of results already obtained in the solution of electrostatic problems. When the conductivity is infinite, the currents are entirely confined to the surfaces of the conductors just as the charges are in the case of an electrostatic distribution.

For example we know from electrostatics that when we have any number of parallel cylindrical conductors charged with electricity, the charges distribute themselves on the surface of the conductors in such a way that they produce no electric force in the interior of the conductors. The electric force at a point P at a distance r from an infinite straight filament of the surface of the conductor having a charge of q units of electricity per unit length is $2q/r$, the direction of the force being along r. We also know that if the filament carried a current i the magnetic force at P would be $2i/r$ in a direction at right angles to r. We see therefore by integrating round the surfaces of the cylinders that if i be proportional to q, the resultant magnetic force at P is zero, and this proves the theorem.

We can readily prove by this analogy that if L be the inductance and K the capacity per unit length between two parallel cylinders of infinite conductivity, then $LK = 1$. Since the currents must be wholly on the surface, we can thus easily prove formula (107).

Let (R_1, L_1), (R_2, L_2) be the resistances and inductances of the two branches of a divided circuit, and let M be their mutual inductance. Then, if e be the E.M.F. applied to the branches, we have

The currents in a divided circuit.

$$e = (R_1 + L_1 D) i_1 + MDi_2 = (R_2 + L_2 D) i_2 + MDi_1 \ ...(110),$$

where D stands for $\partial/\partial t$. Hence

$$(R_1 + R_2) i_1 + (L_1 + L_2 - 2M) Di_1 = R_2 i + (L_2 - M) Di \ ...(111),$$

where $i = i_1 + i_2 =$ the current in the main.

In the particular case of high frequency currents the terms $(R_1 + R_2) i_1$ and $R_2 i$ in equation (111) can be neglected in comparison with the other terms; we thus get

$$(L_1 + L_2 - 2M) Di_1 = (L_2 - M) Di,$$

and integrating,

$$(L_1 + L_2 - 2M)\,i_1 = (L_2 - M)\,i + \text{constant}.$$

The constant must vanish, for i_1 and i are periodic functions of the time the mean values of which are zero, and thus

$$i_1 = \{(L_2 - M)/(L_1 + L_2 - 2M)\}\,i \dots\dots\dots(112),$$

and similarly,

$$i_2 = \{(L_1 - M)/(L_1 + L_2 - 2M)\}\,i \dots\dots\dots(113).$$

It is to be noticed that the values of i_1/i and i_2/i are constant and independent of the resistances of the mains. Also since M can be greater than L_1, we see that i_2 can be negative and $i_1 = i - i_2$ can therefore be greater than i.

The value of the energy stored in the field, at the instant when the currents are i_1 and i_2 respectively, is

$$\tfrac{1}{2} L_1 i_1^2 + \tfrac{1}{2} L_2 (i - i_1)^2 + M i_1 (i - i_1),$$

and for a given value of i this has its minimum value when

$$(L_1 + L_2 - 2M)\,i_1 = (L_2 - M)\,i.$$

Since this is the actual value of i_1 we see that, with high frequency currents, the main current splits up in such a way that the energy stored in the field is a minimum.

Let U denote the energy expended in heating the branches from the moment of closing the switch to the time t, **Condition that the energy expended have a stationary value.** together with the energy stored in the field at this instant. Then we have

$$U = \tfrac{1}{2} L_1 i_1^2 + M i_1 (i - i_1) + \tfrac{1}{2} L_2 (i - i_1)^2$$

$$+ \int_0^t \{R_1 i_1^2 + R_2 (i - i_1)^2\}\,dt.$$

Now if δU be the increment of U due to an alteration in the distribution of the current in the two branches, we have

$$\delta U = L_1 i_1 \delta i_1 + M i_2 \delta i_1 - M i_1 \delta i_1 - L_2 i_2 \delta i_1$$

$$+ \int_0^t \{2R_1 i_1 \delta i_1 - 2R_2 i_2 \delta i_1\}\,dt.$$

By the calculus of variations the condition for a maximum or a minimum value of U is that the coefficients of δi_1 inside and

15—2

outside of the integral sign must vanish simultaneously, and hence we get

$$(L_1 - M)\,i_1 = (L_2 - M)\,i_2 \text{ and } R_1 i_1 = R_2 i_2$$

as the required conditions. We see therefore that when

$$\frac{L_1 - M}{R_1} = \frac{L_2 - M}{R_2}$$

U has a maximum or a minimum value.

It is interesting to notice that

$$R_1 i_1 = R_2 i_2$$

is the condition that the heating effect at a given instant on the branch conductors is a minimum and that

$$(L_1 - M)\,i_1 = (L_2 - M)\,i_2$$

is the condition that the energy stored in the field is a minimum.

In practice, when M is less than L_1 and L_2, it is easy to adjust the ratio of the resistances of the arms so that

Important consequences in this case.

$$\frac{R_1}{R_2} = \frac{L_1 - M}{L_2 - M}.$$

We shall now prove that in this case, from the moment of closing the switch, the current waves in the branches are similar to one another and are therefore also similar to the current in the main. From the equation (110) we see that

$$R_1 i_1 + (L_1 - M)\,Di_1 = R_2 i_2 + (L_2 - M)\,Di_2,$$

and therefore $\quad R_1 i_1 - R_2 i_2 = -\dfrac{L_1 - M}{R_1}\dfrac{d}{dt}\,(R_1 i_1 - R_2 i_2).$

Solving this equation we get

$$R_1 i_1 - R_2 i_2 = A\,\epsilon^{-\frac{R_1}{L_1 - M}\,t},$$

where A is a constant. Now at the instant when t is zero both i_1 and i_2 must be zero, otherwise we should have a finite amount of energy stored in the field in an infinitely short time. Thus A is zero and so

$$R_1 i_1 - R_2 i_2 = 0.$$

We see therefore that i_1 and i_2 are similar waves whatever may be

the shape of the applied potential difference wave. We see also
that

$$i_1 = \frac{R_2}{R_1 + R_2} i \text{ and } i_2 = \frac{R_1}{R_1 + R_2} i.$$

It is evident that i_1, i_2 and i vanish simultaneously, but the
applied potential difference e does not vanish at this instant.
It is to be noticed that for a given effective value of the applied
potential difference the values of A_1, A_2 and A depend on the
shape of the wave and the frequency, but their ratios are always
constant. This theorem may be usefully applied in the design
of alternating current ammeters.

. In order to facilitate calculation, tables of the numerical values
of some of the functions we have used are appended to this
chapter. As an illustration of their use let us find the resistance
R and the inductance L per unit length, at frequencies of
1000 and 500,000, of two parallel cylindrical wires each of radius
0·125 cm., the distance between their axes being 1·5 cm. If the
wires are of 'high conductivity annealed copper,' the volume
resistivity ρ is 1721 absolute units at 20° C. and so

$$1/\rho = 5·811 \times 10^{-4}.$$

Hence from the table on p. 234, $m = 6·774$ and therefore
$ma = 0·8468$. We find by interpolation from the table on p. 233
that $(ma/2)\{W(ma)/V(ma)\} = 1·003$
and $(4/ma)\{Z(ma)/V(ma)\} = 0·9987$.
Hence if R' be the resistance of unit length of the copper wire for
direct current, we see by (67) p. 215 and (105) p. 224 that
$R = 1·003R'$, and $L = 9·9396 + 0·9987$. Similarly, when the
frequency is 500,000, we find that $ma = 18·93$ and so $R = 6·950R'$
and $L = 9·9396 + 0·1492$.

If the wires had been of manganin the conductivity of which
is one-thirtieth that of copper $ma = 0·1546$, when f is 1000, and
$ma = 3·457$, when f is 500,000. Similarly for iron wire assuming
that the conductivity is one-seventh that of copper and that the
permeability μ is 100, we get $ma = 3·200$, when f is 1000, and
$ma = 71·56$, when f is 500,000. The resistances and inductances
are then easily calculated. For purposes of comparison all the
results are given below.

	Copper	Manganin	Iron ($\mu=100$)
R/R', when f is 1000,	1·003	1·000	1·385
R/R', when f is 500,000,	6·950	1·476	25·55
$L - 9\cdot9396$, when f is 1000,	0·9987	1·0000	81·39
$L - 9\cdot9396$, when f is 500,000,	0·1492	0·7726	3·953

The increase of the resistance and the diminution of the inductance at the high frequency is very noticeable when the wires are of iron.

Tables of ber x, bei x, ker x and kei x, computed by Harold G. Savidge.

If $J_0(x)$ and $Y_0(x)$ are the Bessel's functions of the first and second kinds of zero order, we have

$$J_0(\iota x\sqrt{\iota}) = I_0(x\sqrt{\iota}) = \text{ber } x + \iota \text{ bei } x,$$

and

$$Y_0(\iota x\sqrt{\iota}) = K_0(x\sqrt{\iota}) = \text{ker } x + \iota \text{ kei } x.$$

x	ber x	bei x	ker x	kei x
0	1	0	∞	$-0\cdot7854$
1	$9\cdot844 \times 10^{-1}$	$2\cdot496 \times 10^{-1}$	$2\cdot867 \times 10^{-1}$	$-4\cdot950 \times 10^{-1}$
2	$7\cdot517 \times 10^{-1}$	$9\cdot723 \times 10^{-1}$	$-4\cdot166 \times 10^{-2}$	$-2\cdot024 \times 10^{-1}$
3	$-2\cdot214 \times 10^{-1}$	$1\cdot938$	$-6\cdot703 \times 10^{-2}$	$-5\cdot112 \times 10^{-2}$
4	$-2\cdot563$	$2\cdot293$	$-3\cdot618 \times 10^{-2}$	$2\cdot198 \times 10^{-3}$
5	$-6\cdot230$	$1\cdot160 \times 10^{-1}$	$-1\cdot151 \times 10^{-2}$	$1\cdot119 \times 10^{-2}$
6	$-8\cdot858$	$-7\cdot335$	$-6\cdot531 \times 10^{-4}$	$7\cdot216 \times 10^{-3}$
7	$-3\cdot633$	$-2\cdot124 \times 10$	$1\cdot922 \times 10^{-3}$	$2\cdot700 \times 10^{-3}$
8	$2\cdot097 \times 10$	$-3\cdot502 \times 10$	$1\cdot486 \times 10^{-3}$	$3\cdot696 \times 10^{-4}$
9	$7\cdot394 \times 10$	$-2\cdot471 \times 10$	$6\cdot372 \times 10^{-4}$	$-3\cdot192 \times 10^{-4}$
10	$1\cdot388 \times 10^{2}$	$5\cdot637 \times 10$	$1\cdot295 \times 10^{-4}$	$-3\cdot075 \times 10^{-4}$
11	$1\cdot330 \times 10^{2}$	$2\cdot572 \times 10^{2}$	$-4\cdot779 \times 10^{-5}$	$-1\cdot495 \times 10^{-4}$
12	$-1\cdot285 \times 10^{2}$	$5\cdot470 \times 10^{2}$	$-6\cdot308 \times 10^{-5}$	$-3\cdot899 \times 10^{-5}$
13	$-8\cdot827 \times 10^{2}$	$6\cdot466 \times 10^{2}$	$-3\cdot474 \times 10^{-5}$	$5\cdot387 \times 10^{-6}$
14	$-2\cdot131 \times 10^{3}$	$-1\cdot609 \times 10^{2}$	$-1\cdot088 \times 10^{-5}$	$1\cdot268 \times 10^{-5}$
15	$-2\cdot967 \times 10^{3}$	$-2\cdot953 \times 10^{3}$	$-1\cdot514 \times 10^{-8}$	$7\cdot963 \times 10^{-6}$
16	$-6\cdot595 \times 10^{2}$	$-8\cdot191 \times 10^{3}$	$2\cdot466 \times 10^{-6}$	$2\cdot895 \times 10^{-6}$
17	$9\cdot484 \times 10^{3}$	$-1\cdot309 \times 10^{4}$	$1\cdot797 \times 10^{-6}$	$2\cdot861 \times 10^{-7}$
18	$3\cdot096 \times 10^{4}$	$-7\cdot454 \times 10^{3}$	$7\cdot438 \times 10^{-7}$	$-4\cdot555 \times 10^{-7}$
19	$5\cdot625 \times 10^{4}$	$2\cdot804 \times 10^{4}$	$1\cdot293 \times 10^{-7}$	$-3\cdot982 \times 10^{-7}$
20	$4\cdot749 \times 10^{4}$	$1\cdot148 \times 10^{5}$	$-7\cdot715 \times 10^{-8}$	$-1\cdot859 \times 10^{-7}$
21	$-7\cdot616 \times 10^{4}$	$2\cdot337 \times 10^{5}$	$-8\cdot636 \times 10^{-8}$	$-4\cdot388 \times 10^{-8}$
22	$-4\cdot155 \times 10^{5}$	$2\cdot539 \times 10^{5}$	$-4\cdot535 \times 10^{-8}$	$1\cdot097 \times 10^{-8}$

Tables of ber x, bei x, ker x and kei x—(continued)

x	ber x	bei x	ker x	kei x
23	$-9{\cdot}536 \times 10^5$	$-1{\cdot}527 \times 10^5$	$-1{\cdot}320 \times 10^{-8}$	$1{\cdot}824 \times 10^{-8}$
24	$-1{\cdot}242 \times 10^6$	$-1{\cdot}460 \times 10^6$	$8{\cdot}786 \times 10^{-10}$	$1{\cdot}083 \times 10^{-8}$
25	$9{\cdot}798 \times 10^3$	$-3{\cdot}809 \times 10^6$	$3{\cdot}723 \times 10^{-9}$	$3{\cdot}703 \times 10^{-9}$
26	$4{\cdot}936 \times 10^6$	$-5{\cdot}744 \times 10^6$	$2{\cdot}532 \times 10^{-9}$	$1{\cdot}912 \times 10^{-10}$
27	$1{\cdot}489 \times 10^7$	$-2{\cdot}308 \times 10^6$	$9{\cdot}915 \times 10^{-10}$	$-7{\cdot}257 \times 10^{-10}$
28	$2{\cdot}553 \times 10^7$	$1{\cdot}578 \times 10^7$	$1{\cdot}367 \times 10^{-10}$	$-5{\cdot}791 \times 10^{-10}$
29	$1{\cdot}825 \times 10^7$	$5{\cdot}695 \times 10^7$	$-1{\cdot}320 \times 10^{-10}$	$-2{\cdot}563 \times 10^{-10}$
30	$-4{\cdot}612 \times 10^7$	$1{\cdot}100 \times 10^8$	$-1{\cdot}294 \times 10^{-10}$	$-5{\cdot}290 \times 10^{-11}$

Tables of ber′ x and bei′ x.

x	ber′ x	bei′ x	x	ber′ x	bei′ x
0	0·0000	0·0000	4·5	−3·754	−2·053
0·5	−0·0078	0·2499	5·0	−3·845	−4·354
1·0	−0·0624	0·4974	5·5	−2·907	−7·373
1·5	−0·2100	0·7303	6·0	−0·2931	−10·85
2·0	−0·4931	0·9170	8·0	38·31	−7·661
2·5	−0·9436	0·9983	10·0	51·20	135·2
3·0	−1·570	0·8805	15·0	91·01	−4088
3·5	−2·336	0·4353	20·0	-4880×10	1119×10^2
4·0	−3·135	−0·4911	30·0	-1096×10^5	4330×10^4

Tables of ber x, bei x, ber′ x and bei′ x computed to nine significant figures are given in the *British Association Report*, Dundee, 1912, p. 57, for values of x from 0·1 to 10, the difference between successive values of x being 0·1.

Tables of $X_1(x)$, $V_1(x)$, $S(x)$ and $T(x)$, computed by Harold G. Savidge.

$$X_1(x) = \text{ker}^2 x + \text{kei}^2 x,$$
$$V_1(x) = \text{ker}'^2 x + \text{kei}'^2 x,$$
$$S(x) = \text{ber}' x \, \text{ker}' x + \text{bei}' x \, \text{kei}' x,$$

and
$$T(x) = \text{bei}' x \, \text{ker}' x - \text{ber}' x \, \text{kei}' x.$$

x	$X_1(x)$	$V_1(x)$	$S(x)$	$T(x)$
0	∞	∞	0	-0.5
1	3.272×10^{-1}	6.066×10^{-1}	2.186×10^{-1}	-3.235×10^{-1}
2	4.270×10^{-2}	5.968×10^{-2}	2.541×10^{-1}	1.063×10^{-2}
3	7.106×10^{-3}	8.933×10^{-3}	4.733×10^{-2}	1.634×10^{-1}
4	1.314×10^{-3}	1.563×10^{-3}	-1.104×10^{-1}	5.949×10^{-2}
5	2.577×10^{-4}	2.962×10^{-4}	-6.254×10^{-2}	-7.801×10^{-2}
6	5.250×10^{-5}	5.901×10^{-5}	5.501×10^{-2}	-6.263×10^{-2}
7	1.099×10^{-5}	1.214×10^{-5}	6.086×10^{-2}	3.741×10^{-2}
8	2.344×10^{-6}	2.560×10^{-6}	-2.347×10^{-2}	5.793×10^{-2}
9	5.078×10^{-7}	5.491×10^{-7}	-5.421×10^{-2}	-1.217×10^{-2}
10	1.113×10^{-7}	1.195×10^{-7}	2.910×10^{-3}	-4.992×10^{-2}
11	2.464×10^{-8}	2.628×10^{-8}	4.521×10^{-2}	-4.684×10^{-3}
12	5.500×10^{-9}	5.833×10^{-9}	1.087×10^{-2}	4.022×10^{-2}
13	1.236×10^{-9}	1.305×10^{-9}	-3.506×10^{-2}	1.582×10^{-2}
14	2.792×10^{-10}	2.936×10^{-10}	-1.967×10^{-2}	-2.981×10^{-2}
15	6.341×10^{-11}	6.646×10^{-11}	2.456×10^{-2}	-2.254×10^{-2}
16	1.446×10^{-11}	1.512×10^{-11}	2.451×10^{-2}	1.939×10^{-2}
17	3.311×10^{-12}	3.452×10^{-12}	-1.438×10^{-2}	2.566×10^{-2}
18	7.608×10^{-13}	7.912×10^{-13}	-2.607×10^{-2}	-9.593×10^{-3}
19	1.753×10^{-13}	1.820×10^{-13}	5.085×10^{-3}	-2.582×10^{-2}
20	4.051×10^{-14}	4.197×10^{-14}	2.498×10^{-2}	9.118×10^{-4}
21	9.383×10^{-15}	9.704×10^{-15}	2.883×10^{-3}	2.363×10^{-2}
22	2.178×10^{-15}	2.250×10^{-15}	-2.185×10^{-2}	6.261×10^{-3}
23	5.068×10^{-16}	5.226×10^{-16}	-9.195×10^{-3}	-1.970×10^{-2}
24	1.181×10^{-16}	1.216×10^{-16}	1.726×10^{-2}	-1.166×10^{-2}
25	2.757×10^{-17}	2.836×10^{-17}	1.366×10^{-2}	1.461×10^{-2}
26	6.447×10^{-18}	6.625×10^{-18}	-1.182×10^{-2}	1.517×10^{-2}
27	1.510×10^{-18}	1.550×10^{-18}	-1.621×10^{-2}	-8.948×10^{-3}
28	3.540×10^{-19}	3.631×10^{-19}	6.073×10^{-3}	-1.679×10^{-2}
29	8.312×10^{-20}	8.517×10^{-20}	1.693×10^{-2}	3.253×10^{-3}
30	1.954×10^{-20}	2.000×10^{-20}	-3.196×10^{-4}	1.666×10^{-2}
∞	0	0	0	0

Functions for calculating the Resistance and Inductance of straight cylindrical wires at various frequencies (see pp. 215 and 224), computed by the Bureau of Standards, Washington.

x	$\dfrac{x}{2}\cdot\dfrac{W(x)}{V(x)}$	$\dfrac{4}{x}\cdot\dfrac{Z(x)}{V(x)}$	x	$\dfrac{x}{2}\cdot\dfrac{W(x)}{V(x)}$	$\dfrac{4}{x}\cdot\dfrac{Z(x)}{V(x)}$
0·0	1·00000	1·00000	4·8	1·97131	0·57852
0·1	1·00000	1·00000	4·9	2·00710	0·56703
0·2	1·00001	1·00000	5·0	2·04272	0·55597
0·3	1·00004	0·99998	5·2	2·11353	0·53506
0·4	1·00013	0·99993	5·4	2·18389	0·51566
0·5	1·00032	0·99984	5·6	2·25393	0·49764
0·6	1·00067	0·99966	5·8	2·32380	0·48086
0·7	1·00124	0·99937	6·0	2·39359	0·46521
0·8	1·00212	0·99894	6·2	2·46338	0·45056
0·9	1·00340	0·99830	6·4	2·53321	0·43682
1·0	1·00519	0·99741	6·6	2·60313	0·42389
1·1	1·00758	0·99621	6·8	2·67312	0·41171
1·2	1·01071	0·99465	7·0	2·74319	0·40021
1·3	1·01470	0·99266	7·2	2·81334	0·38933
1·4	1·01969	0·99017	7·4	2·88355	0·37902
1·5	1·02582	0·98711	7·6	2·95380	0·36923
1·6	1·03323	0·98342	7·8	3·02411	0·35992
1·7	1·04205	0·97904	8·0	3·09445	0·35107
1·8	1·05240	0·97390	8·2	3·16480	0·34263
1·9	1·06440	0·96795	8·4	3·23518	0·33460
2·0	1·07816	0·96113	8·6	3·30557	0·32692
2·1	1·09375	0·95343	8·8	3·37597	0·31958
2·2	1·11126	0·94482	9·0	3·44638	0·31257
2·3	1·13069	0·93527	9·2	3·51680	0·30585
2·4	1·15207	0·92482	9·4	3·58723	0·29941
2·5	1·17538	0·91347	9·6	3·65766	0·29324
2·6	1·20056	0·90126	9·8	3·72812	0·28731
2·7	1·22753	0·88825	10·0	3·79857	0·28162
2·8	1·25620	0·87451	10·5	3·97477	0·26832
2·9	1·28644	0·86012	11·0	4·15100	0·25622
3·0	1·31809	0·84517	11·5	4·32727	0·24516
3·1	1·35102	0·82975	12·0	4·50358	0·23501
3·2	1·38504	0·81397	12·5	4·67993	0·22567
3·3	1·41999	0·79794	13·0	4·85631	0·21703
3·4	1·45570	0·78175	13·5	5·03272	0·20903
3·5	1·49202	0·76550	14·0	5·20915	0·20160
3·6	1·52879	0·74929	14·5	5·38560	0·19468
3·7	1·56587	0·73320	15·0	5·56208	0·18822
3·8	1·60314	0·71729	16·0	5·91509	0·17649
3·9	1·64051	0·70165	17·0	6·26817	0·16614
4·0	1·67787	0·68632	18·0	6·62129	0·15694
4·1	1·71516	0·67135	19·0	6·97446	0·14870
4·2	1·75233	0·65677	20·0	7·32767	0·14128
4·3	1·78933	0·64262	40·0	14·39545	0·07069
4·4	1·82614	0·62890	60·0	21·46541	0·04713
4·5	1·86275	0·61563	80·0	28·53593	0·03535
4·6	1·89914	0·60281	100·0	35·60666	0·02828
4·7	1·93533	0·59044	∞	∞	0

234 ALTERNATING CURRENT THEORY [CH. VII

Values of the argument m for copper wires of conductivity $5\cdot811 \times 10^{-4} c.g.s.$ *units.* (*Bureau of Standards.*)

('High Conductivity Annealed Copper' at 20° C.)

$$m^2 = 4\pi\mu\omega/\rho = 8\pi^2 f \times 5\cdot811 \times 10^{-4}.$$

Wave length in metres $= 3 \times 10^8/f.$

f	m	f	m	f	m
25	1·071	6000	16·59	200,000	95·79
50	1·515	7000	17·92	250,000	107·1
100	2·142	8000	19·16	300,000	117·3
200	3·029	9000	20·32	333,333	123·7
300	3·710	10,000	21·42	375,000	131·2
400	4·284	15,000	26·23	428,570	140·2
500	4·790	20,000	30·29	500,000	151·5
600	5·247	30,000	37·10	600,000	165·9
700	5·667	40,000	42·84	700,000	179·2
800	6·058	50,000	47·90	750,000	185·5
900	6·426	60,000	52·47	800,000	191·6
1000	6·774	70,000	56·67	900,000	203·2
2000	9·579	80,000	60·58	1,000,000	214·2
3000	11·73	90,000	64·26	1,500,000	262·3
4000	13·55	100,000	67·74	3,000,000	371·0
5000	15·15	150,000	82·96	6,000,000	524·7

REFERENCES

CLERK MAXWELL, *Electricity and Magnetism*, Vol. 2, § 690.
OLIVER HEAVISIDE, *Electrical Papers*, Vol. 2, pp. 64 *et seq.*
LORD RAYLEIGH, *Scientific Papers* and *Roy. Soc. Proc.* ser. A. 86, p. 562, June, 1912.
LORD KELVIN, *Physical Papers*, Vol. 3, p. 491.
J. W. NICHOLSON, 'The Inductance of Two Parallel Wires.' *Proc. Phys. Soc.* June, 1908.
A. RUSSELL, 'The Effective Resistance and Inductance of a Concentric Main.' *Phil. Mag.* April, 1909.
HAROLD G. SAVIDGE, 'Tables of the ber and bei and ker and kei Functions.' *Proc. Phys. Soc.* November, 1909.
C. S. WHITEHEAD, 'On a Generalization of the Functions ber x, bei x, ker x, kei x.' *Quart. Journ. of Math.* No. 168, 1911.
E. B. ROSA and F. W. GROVER, 'Formulas and Tables.' *Bulletin of the Bureau of Standards*, Jan. 1912.
S. BUTTERWORTH, 'On the Evaluation of the Ber, Bei and Allied Functions.' *Proc. Phys. Soc.* May, 1913.

CHAPTER VIII

IN this chapter we shall discuss problems in connection with

Problems in connection with spherical electrodes.

electrified spherical electrodes. These problems are of importance in practical work and are of great historical interest. Poisson, Murphy, Kelvin, Clerk Maxwell, and Kirchhoff, all successfully expended a great amount of labour and mathematical ingenuity over them, so that the solutions can now be given quite simply. In certain cases we have altered their formulae so as to make the computation of the numerical values much easier. We have given new formulae which can be applied when the spheres are close together; we have also given new tables and have recomputed and considerably extended Kelvin's and Schuster's tables. The first problem we shall discuss is the electrostatic capacity of spherical electrodes. We have therefore to find first the formulae for the capacity coefficients.

Let us suppose that the radii of the conducting spheres are a

The capacity coefficients of spherical electrodes.

and b respectively and let c be the distance between their centres. Let their potentials also be 1 and 0 respectively. Now we shall show that if the spheres are removed the actual field round the conductors can be exactly reproduced by an infinite series of point charges placed at certain definite points $A, A_1, A_2, ..., B_1, B_2, ...$ inside the two spheres.

Let us suppose that the conducting spheres X and Y (Fig. 61) are removed and that a charge q_0 equal to a is placed at A, the centre of the sphere X. The potential due to this charge will be 1 over the spherical surface which has A for its centre. But the potential over the spherical surface whose centre is B will not be zero. Let us now place a charge $q_1' = - q_0(b/c)$ at B_1 where $c . BB_1 = b^2$. B_1 will be the electric image of q_0 in the sphere Y and these two charges (p. 10) will make the surface of Y at zero potential. But now the potential over X is not unity. If we place a charge $q_1 = - q_1'(a/AB_1)$ at A_1 where $AA_1 . AB = a^2$, the potential over X due to the three point charges at A, A_1 and B_1 will be unity but that over Y will no longer be zero. Proceeding

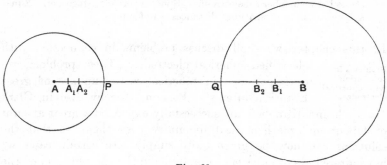

Fig. 61.

$$AP = a\,; \quad BQ = b\,; \quad AA_n = a_n, \quad BB_n = b_n\,; \quad AB = c\,; \quad PQ = d\,;$$
$$(c - b_n)\, a_n = a^2\,; \quad (c - a_{n-1})\, b_n = b^2.$$

in this way and noticing that these point charges are getting rapidly smaller and smaller, we see that we get an infinite number of point charges which produce the same potential at points external to X and Y as the two given spheres would if replaced and maintained at potentials 1 and 0 respectively. By Green's theorem the sum of the point charges inside X will equal the charge on X and the sum of the point charges inside Y will equal the charge on Y.

We have

$$q_0 = a, \qquad\qquad\qquad q_1' = - q_0\, b/c,$$
$$q_1 = - q_1'\, a/(c - BB_1), \qquad q_2' = - q_1\, b/(c - AA_1),$$
$$q_2 = - q_2'\, a/(c - BB_2), \qquad q_3' = - q_2\, b/(c - AA_2),$$

....

Let $AA_n = a_n$ and $BB_n = b_n$, then

$$q_n = -q_n' a/(c - b_n) \text{ and } q_n' = -q_{n-1} b/(c - a_{n-1}) \quad \dots(1).$$

We also have

$$(c - b_1) a_1 = (c - b_2) a_2 = \dots = (c - b_n) a_n = \dots = a^2,$$

and $\quad cb_1 = (c - a_1) b_2 = \dots = (c - a_{n-1}) b_n = \dots = b^2.$

Hence $\quad \dfrac{q_{n+1}}{q_n} = \dfrac{ab}{(c - b_{n+1})(c - a_n)} = \dfrac{a_{n+1} b_{n+1}}{ab}$

and $\quad \dfrac{q_n}{q_{n-1}} = \dfrac{a_n b_n}{ab} = \dfrac{ca_n - a^2}{ab}.$

Hence $\quad \dfrac{q_n}{q_{n+1}} + \dfrac{q_n}{q_{n-1}} = \dfrac{(c - b_{n+1})(c - a_n) + ca_n - a^2}{ab}$

$$= \frac{c^2 - a^2 - b^2}{ab}$$

$$= \epsilon^\omega + \epsilon^{-\omega} \quad \dots\dots\dots\dots\dots\dots\dots\dots\dots(2),$$

using the notation adopted on p. 164. Hence $\sinh \omega = rc/ab$, where r is the radius of the sphere which cuts the two given spheres at right angles. We shall also, as before, introduce α and β where $\sinh \alpha = r/a$, and $\sinh \beta = r/b$.

Writing (2) in the form

$$\frac{1}{q_{n-1}} + \frac{1}{q_{n+1}} = \frac{\epsilon^\omega}{q_n} + \frac{\epsilon^{-\omega}}{q_n},$$

we see, at once, by inspection that the complete solution of this difference equation is

$$\frac{1}{q_n} = C\epsilon^{n\omega} + D\epsilon^{-n\omega},$$

where C and D are constants.

Hence

$$\frac{1}{q_n} = (C + D) \cosh n\omega + (C - D) \sinh n\omega \quad \dots\dots(3).$$

Now we have seen that $q_0 = a$ and $q_1 = a^2b/(c^2 - b^2)$, and thus substituting we get

$$C + D = 1/a \dots\dots\dots\dots\dots\dots\dots(4),$$

and $\quad (\cosh \omega)/a + (C - D) \sinh \omega = (c^2 - b^2)/a^2b$

$$= (\cosh \omega)/a + \{(\coth \alpha)/a\} \sinh \omega,$$

and therefore $\quad C - D = (\coth \alpha)/a \quad \dots\dots\dots\dots(5).$

Substituting these values in (3) we get

$$q_n = r/\sinh(\alpha + n\omega) \quad \dots\dots\dots\dots\dots(6).$$

Again from (1) we get

$$\frac{q'_{n+1}}{q_n'} = \frac{ab}{(c - a_n)(c - b_n)}.$$

Hence proceeding exactly as before and noticing that

$$q_1' = -ab/c \quad \text{and} \quad q_2' = -a^2b^2/\{c(c^2 - a^2 - b^2)\},$$

we get

$$q_n' = -r/\sinh n\omega \dots\dots\dots\dots\dots(7).$$

Now if $k_{1.1}$ and $k_{2.2}$ be the capacity coefficients of the two spheres and v, v' and q, q' be their potentials and charges respectively, we have, by Chapter V,

$$q = k_{1.1}\,v + k_{1.2}\,v'$$

and

$$q' = k_{2.2}\,v' + k_{1.2}\,v.$$

Hence putting $v = 1$, and $v' = 0$, we get

$$k_{1.1} = q = q_0 + q_1 + q_2 + \dots$$

$$= r \sum_{s=0}^{s=\infty} \frac{1}{\sinh(\alpha + s\omega)} \dots\dots\dots\dots(8),$$

$$-k_{1.2} = -q' = -q_1' - q_2' - q_3' - \dots$$

$$= r \sum_{1}^{\infty} \frac{1}{\sinh s\omega} \dots\dots\dots\dots\dots(9),$$

and from symmetry

$$k_{2.2} = r \sum_{0}^{\infty} \frac{1}{\sinh(\beta + s\omega)} \dots\dots\dots\dots(10).$$

The values of r, α, β and ω can be rapidly found by the formulae (5), (6) and (7), p. 164.

If we write $\epsilon^\alpha = 1/x$, $\epsilon^\beta = 1/y$ and $\epsilon^\omega = 1/w$, equations (8), (9) and (10) become

Kirchhoff's formulae.

$$k_{1.1} = 2r\left\{\frac{x}{1 - x^2} + \frac{wx}{1 - w^2x^2} + \frac{w^2x}{1 - w^4x^2} + \dots\right\} \dots(11),$$

$$-k_{1.2} = 2r\left\{\frac{w}{1 - w^2} + \frac{w^2}{1 - w^4} + \frac{w^3}{1 - w^6} + \dots\right\} \dots\dots(12),$$

and

$$k_{2.2} = 2r\left\{\frac{y}{1 - y^2} + \frac{wy}{1 - w^2y^2} + \frac{w^2y}{1 - w^4y^2} + \dots\right\} \dots(13).$$

These formulae are due to Kirchhoff. Since $\omega = \alpha + \beta$, it follows that $w = xy$. We also have

$$x = (1 + r^2/a^2)^{1/2} - r/a \quad \text{and} \quad w = (cx - a)/b.$$

It is to be noticed that x and y are always fractional and that w is smaller than either.

When $c - a - b$ is small compared with a and b, all the formulae given above converge very slowly. Kirchhoff's formulae, however, can very easily be made to converge rapidly. Let $2rR_n$ denote the sum after the first n terms of the series for $k_{1.1}$. Then

$$R_n = \frac{xw^n}{1 - x^2 w^{2n}} + \frac{xw^{n+1}}{1 - x^2 w^{2n+2}} + \cdots.$$

Hence by expanding by the binomial theorem and grouping the terms differently, we get

$$R_n = \frac{xw^n}{1 - w} + \frac{(xw^n)^3}{1 - w^3} + \frac{(xw^n)^5}{1 - w^5} + \ldots \qquad \ldots\ldots\ldots(14).$$

When xw^n is small, this series converges very rapidly.

Similarly if $2rR_n'$ be the remainder after n terms of the series for $-k_{1.2}$, we get

$$R_n' = \frac{w^{n+1}}{1 - w} + \frac{w^{3n+3}}{1 - w^3} + \frac{w^{5n+5}}{1 - w^5} + \ldots \qquad \ldots\ldots\ldots(15).$$

Using this method it will be found that we can readily compute the values of the coefficients even when the distance between the spheres is only the hundredth part of the radius of the smaller sphere.

In the particular case when n is zero, the formula for $k_{1.1}$ becomes

$$k_{1.1} = 2r \left\{ \frac{x}{1 - w} + \frac{x^3}{1 - w^3} + \frac{x^5}{1 - w^5} + \ldots \right\} \qquad \ldots\ldots(16).$$

Equally simple formulae for $-k_{1.2}$ and $k_{2.2}$ are obtained by writing w and y respectively for x in (16). When $n = 1$, the formulae become

$$k_{1.1} = a + 2rx \left\{ \frac{w}{1 - w} + \frac{w^3}{1 - w^3} x^2 + \frac{w^5}{1 - w^5} x^4 + \ldots \right\} \ldots\ldots\ldots(17),$$

and

$$-k_{1.2} = \frac{ab}{c} + 2rw \left\{ \frac{w}{1 - w} + \frac{w^3}{1 - w^3} w^2 + \frac{w^5}{1 - w^5} w^4 + \ldots \right\} \ldots\ldots(18).$$

Let us suppose for example that $a = 7l$, $b = l$, and $c = 10l$. We find by the formulae that

$$r = 2 \cdot 4l, \quad x = 5/7, \quad y = 1/5 \quad \text{and} \quad w = 1/7.$$

Hence, by (17) and (18),

$$k_{1.1} = 7l + (4/7)\, l \,\{1 + 0 \cdot 0089509 + 0 \cdot 0000929 + 0 \cdot 0000009 + \ldots\}$$
$$= 7 \cdot 5765970l,$$
$$-k_{1.2} = 0 \cdot 7\, l + (8/70)l \,\{1 + 0 \cdot 0003580 + 0 \cdot 0000001 + \ldots\}$$
$$= 0 \cdot 8143266\, l.$$

Similarly $k_{2.2} = 1 \cdot 1601124\, l.$

The formulae given above are laborious for computation when the spheres are very close together. In this case the following formulae due to the author may be used:

Spheres close together.

$$\frac{k_{1.1}}{2r} = -\frac{\psi\,(a/\omega) + \log_e (\omega/2)}{2\omega} + \frac{\omega}{24}\left(\frac{1}{6} - \frac{\alpha\beta}{\omega^2}\right)$$
$$+ \frac{7\omega^3}{2880}\left(\frac{1}{30} - \frac{\alpha^2\beta^2}{\omega^4}\right) + \ldots \quad \ldots\ldots\ldots\ldots\ldots(19),$$

$$-\frac{k_{1.2}}{2r} = \frac{\gamma - \log_e (\omega/2)}{2\omega} + \frac{\omega}{144} + \frac{7\omega^3}{86400} + \ldots \quad \ldots\ldots\ldots(20),$$

and

$$\frac{k_{2.2}}{r} = \frac{k_{1.1}}{r} - \frac{\pi}{\omega}\cot\frac{\pi\alpha}{\omega} \quad \ldots\ldots\ldots\ldots\ldots\ldots\ldots\ldots\ldots\ldots\ldots(21),$$

where $\psi\,(x) = \dfrac{\partial}{\partial x}\,\{\log \Gamma\,(x)\}$ and $\gamma = 0 \cdot 577216\ldots = $ Euler's constant.

The function $\psi\,(x)$ is easily computed by the series

$$\psi\,(x) = -\,1/x - \gamma + x/(1 + x) + (S_2 - 1)\,x - \ldots$$
$$+ (-)^m\,(S_m - 1)\,x^{m-1} + \ldots \quad \ldots(22),$$

where $S_m = \overset{\infty}{\underset{1}{\Sigma}}\,(1/n)^m$, and

$S_2 = 1 \cdot 64493\,;$	$S_7 = 1 \cdot 00835\,;$	$S_{12} = 1 \cdot 00025\,;$
$S_3 = 1 \cdot 20206\,;$	$S_8 = 1 \cdot 00408\,;$	$S_{13} = 1 \cdot 00012\,;$
$S_4 = 1 \cdot 08232\,;$	$S_9 = 1 \cdot 00201\,;$	$S_{14} = 1 \cdot 00006\,;$
$S_5 = 1 \cdot 03693\,;$	$S_{10} = 1 \cdot 00099\,;$	$S_{15} = 1 \cdot 00003\,;$
$S_6 = 1 \cdot 01734\,;$	$S_{11} = 1 \cdot 00049\,;$	$S_{16} = 1 \cdot 00002.$

We also have

$$\psi(1+x) - \psi(x) = 1/x \;;\; \psi(1-x) - \psi(x) = \pi \cot \pi x \;;$$

$$\psi(1) = -\gamma \;;\; \psi(3/4) = -\gamma + \pi/2 - 3 \log_\epsilon 2 \;;$$

$$\psi(2/3) = -\gamma + \pi/(2\sqrt{3}) - (3/2)\log_\epsilon 3 \;;\; \psi(1/2) = -\gamma - 2\log_\epsilon 2 \;;$$

$$\psi(1/3) = -\gamma - \pi/(2\sqrt{3}) - (3/2)\log_\epsilon 3 \;;\; \psi(1/4) = -\gamma - \pi/2 - 3 \log_\epsilon 2.$$

Values of $\psi(x)$.

x	0	1	2	3	4	5	6	7	8	9
1·0	0·5772	5609	5448	5289	5133	4978	4826	4676	4528	4382
1·1	0·4238	4095	3955	3816	3679	3543	3410	3277	3147	3018
1·2	0·2890	2764	2640	2517	2395	2275	2155	2038	1921	1806
1·3	0·1692	1579	1467	1357	1248	1139	1032	0926	0821	7169
1·4	0·0614	0512	0411	0311	0211	0113	0016	0081	0176	0271
1·5	0·0365	0458	0550	0642	0732	0822	0911	1000	1087	1174
1·6	0·1260	1346	1431	1515	1598	1681	1763	1845	1926	2006
1·7	0·2085	2165	2243	2321	2398	2475	2551	2626	2701	2776
1·8	0·2850	2923	2996	3069	3141	3212	3283	3353	3423	3493
1·9	0·3562	3630	3699	3766	3833	3900	3967	4033	4098	4163

* All the numbers above this line are negative and all below it are positive. To find the values of $\psi(x)$ when x lies between 0 and 1, we use the formula

$$\psi(x) = \psi(1+x) - 1/x.$$

For example, when $x = 0\cdot1$,

$$\psi(0\cdot1) = \psi(1\cdot1) - 10 = -10\cdot4238,$$

and when $x = 0\cdot5$,

$$\psi(0\cdot5) = \psi(1\cdot5) - 2 = -1\cdot9635.$$

When the distance $d = c - a - b$ between the spheres is very small compared with either radius, we get the following approximate formulae:

$$k_{1.1} = \frac{ab}{a+b}\left\{ -\psi\left(\frac{b}{a+b}\right) + \frac{1}{2}\log_\epsilon \frac{2ab}{(a+b)d} \right\} \quad\ldots\ldots(23),$$

$$-k_{1.2} = \frac{ab}{a+b}\left\{ \gamma + \frac{1}{2}\log_\epsilon \frac{2ab}{(a+b)d} \right\} \quad\ldots\ldots\ldots\ldots(24),$$

and $\quad k_{2.2} = \dfrac{ab}{a+b}\left\{ -\psi\left(\dfrac{a}{a+b}\right) + \dfrac{1}{2}\log_\epsilon \dfrac{2ab}{(a+b)d} \right\} \quad\ldots(25).$

Equal spheres. If the spheres are equal, we may use the formulae

$$k_{1.1} = \frac{a}{2}\left(1 + \frac{d}{6a} - \frac{d^2}{180a^2}\right)\left(1 \cdot 9635 + \frac{1}{2}\log_\epsilon \frac{a}{d} + \frac{d}{72a} - \frac{213d^2}{43200a^2}\right)$$
$$\dots(26),$$

and

$$-k_{1.2} = \frac{a}{2}\left(1 + \frac{d}{6a} - \frac{d^2}{180a^2}\right)\left(0 \cdot 57722 + \frac{1}{2}\log_\epsilon \frac{a}{d} + \frac{7d}{72a} - \frac{251d^2}{43200a^2}\right)$$
$$\dots(27),$$

when they are close together. For example, if d/a does not exceed 1/10, the maximum inaccuracy of either of these formulae is less than 1 in 20,000.

For values of d/a between 0·1 and 1 we may use formulae such as (17) and (18).

Finally for values of $c/a = d/a + 2 = \xi$, not less than three, we have

$$\frac{k_{1.1}}{a} = 1 + \frac{1}{\xi^2} + \frac{2\xi^2 + 1}{(\xi^3 + \xi)^2 - (2\xi^2 + 1)^2} - \frac{1}{\xi^8} \dots\dots(28),$$

and

$$-\frac{k_{1.2}}{a} = \frac{1}{\xi} + \frac{\xi^3 + \xi}{(\xi^3 + \xi)^2 - (2\xi^2 + 1)^2} \dots\dots(29),$$

approximately, the maximum inaccuracy occurring when ξ is three.

The following table giving the values of $k_{1.1}/a$ and $-k_{1.2}/a$ when the spheres are close together will be found useful.

d/a	$k_{1.1}/a$	$-k_{1.2}/a$	d/a	$k_{1.1}/a$	$-k_{1.2}/a$
0·000001	4·43563	3·74249	0·08	1·63519	0·93619
0·00001	3·85998	3·16684	0·09	1·60805	0·90833
0·0001	3·28439	2·59124	0·1	1·58396	0·88352
0·001	2·70920	2·01598	0·15	1·49328	0·78927
0·01	2·13667	1·44246	0·2	1·43131	0·72378
0·02	1·96643	1·27181	0·25	1·38491	0·67321
0·03	1·86788	1·17253	0·3	1·34828	0·63384
0·04	1·79864	1·10256	0·35	1·31830	0·60049
0·05	1·74543	1·04862	0·4	1·29316	0·57202
0·06	1·70234	1·00480	0·45	1·27173	0·54728
0·07	1·66622	0·96795	0·5	1·25302	0·52537

We have already (p. 152) defined the capacity K between the electrodes to be the ratio of q to $v_1 - v_2$, when

The capacities of the electrodes.

$q_1 = -q_2 = q$. Therefore

$$K = \frac{k_{1.1}\,k_{2.2} - k_{1.2}^2}{k_{1.1} + k_{2.2} + 2k_{1.2}}.$$

Hence K can be found by means of the formulae for the capacity coefficients given above. It is also convenient in practice to define the component capacity K' of a conductor, when it and all neighbouring conductors are at potential v, as q/v, where q is the charge on the conductor.

Hence the component capacity K_1' of the sphere whose radius is a equals $k_{1.1} + k_{1.2}$.

The combined capacity of the spheres

$$K_1' + K_2' = k_{1.1} + k_{2.2} + 2k_{1.2},$$

and thus, when they touch one another, we see by formulae (23), (24) and (25) that

$$K_1' + K_2' = -\frac{ab}{a+b}\left\{\psi\left(\frac{a}{a+b}\right) + \psi\left(\frac{b}{a+b}\right) + 2\gamma\right\} \ \ ...(30).$$

Hence, when $a = b$, this capacity is $2a \log_e 2$; when $a = 2b$, it equals $a \log_e 3$; when $a = 3b$, it equals $(3/4)\, a \log_e 4$, etc.

When the spheres are equal and an accuracy of the hundredth part of one per cent. suffices, we have

$$K = \frac{a}{2}\left(1 + \frac{d}{6a}\right)\left(1{\cdot}2704 + \frac{1}{2}\log_e \frac{a}{d} + \frac{d}{18a}\right) \ \ ......(31),$$

and

$$K' = \frac{a}{2}\left(1 + \frac{d}{6a}\right)\left(1{\cdot}3863 - \frac{d}{12a}\right) \ \ ....(32),$$

when d/a is not greater than a tenth.

When $\xi = c/a$ is not less than three, we can use the formulae

$$2K/a = \frac{\xi(\xi-1)^2}{\xi(\xi-1)^2 - 1} + \frac{\xi+1}{\xi^2} - \frac{1}{\xi^3} ....(33),$$

and

$$K'/a = \frac{\xi(\xi+1)^2}{\xi(\xi+1)^2 + 1} - \frac{\xi-1}{\xi^2} - \frac{1}{\xi^3} ....(34),$$

the maximum inaccuracy being less than the hundredth part of one per cent.

The values of K and K' for equal spheres are given in the following table:

c/a	K/a	K'/a	c/a	K/a	K'/a
$2 \cdot 0_6 1$	4·6647	0·6931	2·6	0·8552	0·7340
$2 \cdot 0_5 1$	4·0891	0·6931	2·7	0·8275	0·7401
$2 \cdot 0_4 1$	3·5134	0·6931	2·8	0·8044	0·7460
$2 \cdot 0_3 1$	2·9378	0·6931	2·9	0·7847	0·7517
$2 \cdot 0_2 1$	2·3626	0·6932	3·0	0·7677	0·7572
2·01	1·7896	0·6942	3·1	0·7528	0·7625
2·02	1·6191	0·6946	3·2	0·7396	0·7677
2·03	1·5202	0·6954	3·3	0·7278	0·7726
2·04	1·4506	0·6961	3·4	0·7172	0·7774
2·05	1·3970	0·6968	3·5	0·7076	0·7820
2·06	1·3536	0·6975	3·6	0·6989	0·7864
2·07	1·3171	0·6983	3·7	0·6909	0·7907
2·08	1·2857	0·6990	3·8	0·6836	0·7948
2·09	1·2582	0·6997	3·9	0·6768	0·7988
2·10	1·2337	0·7004	4	0·6705	0·8026
2·15	1·1413	0·7040	5	0·6263	0·8345
2·20	1·0775	0·7075	6	0·6006	0·8577
2·25	1·0291	0·7117	7	0·5836	0·8753
2·30	0·9911	0·7144	8	0·5712	0·8891
2·35	0·9594	0·7178	9	0·5626	0·9001
2·40	0·9326	0·7211	10	0·5556	0·9092
2·45	0·9095	0·7245	100	0·5051	0·9901
2·50	0·8892	0·7277	1000	0·5005	0·9990

When the charges on the spheres are given, we get, by solving

The potentials of the spheres for given charges. the fundamental equations,

$$v_1 = (k_{2.2}/\Delta) q_1 - (k_{1.2}/\Delta) q_2$$
$$\text{and} \qquad v_2 = (k_{1.1}/\Delta) q_2 - (k_{1.2}/\Delta) q_1 \qquad \right\} \dots (35),$$

where $\Delta = k_{1.1} k_{2.2} - k^2_{1.2}$.

When the charges are equal and opposite,

$$- \frac{v_1}{v_2} = \frac{k_{2.2} + k_{1.2}}{k_{1.1} + k_{1.2}} \quad \dots \dots \dots \dots (36).$$

When the distance between the spheres is infinite

$$- \frac{v_1}{v_2} = \frac{b}{a},$$

and when they are very close together,

$$- \frac{v_1}{v_2} = \frac{\psi \{a/(a+b)\} + \gamma}{\psi \{b/(a+b)\} + \gamma} \dots \dots \dots (37).$$

For example, when $a = 1$ and $b = 3$, then at infinity $-v_1/v_2 = 3$, and when they are almost touching,

$$-\frac{v_1}{v_2} = \frac{\psi(1/4) + \gamma}{\psi(3/4) + \gamma} = \frac{6 \log_\epsilon 2 + \pi}{6 \log_\epsilon 2 - \pi} = 7\cdot 178.$$

When the spheres are at the same potential,

$$\frac{q_1}{q_2} = \frac{k_{1.1} + k_{1.2}}{k_{2.2} + k_{1.2}}.$$

When for instance the spheres whose radii are 1 and 9 respectively are at a great distance apart and are at the same potential, $q_1/q_2 = 1/9 = 0\cdot 1111\ldots$, but when they are touching

$$q_1/q_2 = 0\cdot 07667\ldots.$$

Finally, when the charges on the spheres are equal,

$$\frac{v_1}{v_2} = \frac{k_{2.2} - k_{1.2}}{k_{1.1} - k_{1.2}}.$$

In this case, when the spheres are at a great distance apart, $v_1/v_2 = b/a$; but when the distance between the spheres is very small compared with either radius, v_1/v_2 is unity.

If i_1 and i_2 denote the instantaneous values of the currents in the wires connected with the electrodes, we have

The capacity currents to spherical electrodes.

$$i_1 = \frac{\partial q_1}{\partial t} = k_{1.1} \frac{\partial v_1}{\partial t} + k_{1.2} \frac{\partial v_2}{\partial t}$$

and

$$i_2 = \frac{\partial q_2}{\partial t} = k_{2.2} \frac{\partial v_2}{\partial t} + k_{1.2} \frac{\partial v_1}{\partial t}.$$

Hence when the laws according to which the potentials of the terminals vary are known the capacity currents can be found. In practice, we generally have either $q_1 = -q_2$ or $v_2 = 0$ at every instant.

In the former case

$$i_1 = -i_2 = K \frac{\partial}{\partial t}(v_1 - v_2),$$

where K is the capacity between the electrodes, and in the latter

$$i_1 = (k_{1.1}/k_{1.2}) i_2 = k_{1.1} \frac{\partial v_1}{\partial t}.$$

In the first case—when the charges are equal and opposite—we have

$$A_1 = A_2 = \omega K V \times 10^{-6},$$

where A_1 and A_2 are the effective values of the currents in the leads in amperes, $\omega = 2\pi f$, V the effective value in volts of the potential difference $E \sin \omega t$, and K the capacity in microfarads, between the electrodes. Similarly when $v_2 = 0$ and $v_1 = E \sin \omega t$ we have

$$A_1 = \omega k_{1.1} \, V \times 10^{-6}; \quad A_2 = -\omega k_{1.2} \, V \times 10^{-6}.$$

For example, let us suppose that the frequency of the alternating current is 50,000 and that the pressure is 100 kilovolts. If the radius a of each electrode be 10 cms., and the distance c between their centres be 30 cms., we get by (28), (29) and (33)

$$k_{1.1} = 1\cdot146 \times 10/(900,000) \text{ microfarads},$$

$$- k_{1.2} = 0\cdot389 \times 10/(900,000) \text{ microfarads},$$

and $\qquad K = 0\cdot768 \times 10/(900,000) \text{ microfarads}.$

Hence, when the charges are always equal and opposite,

$$A_1 = A_2 = 0\cdot268 \text{ ampere},$$

and when the second electrode is earthed,

$$A_1 = 0\cdot400 \text{ ampere} \quad \text{and} \quad A_2 = 0\cdot136 \text{ ampere}.$$

We have seen in Chapter v that when the potentials of the spheres are maintained constant, and they alter their positions owing to their mutual electric actions, they move in such a way that the electrostatic energy of the system is increased by an amount exactly equal to the work done on the spheres by the electric forces. If W be the electrostatic energy,

The attraction and repulsion between spheres at given potentials.

$$W = \tfrac{1}{2} k_{1.1} v_1^2 + \tfrac{1}{2} k_{2.2} v_2^2 + k_{1.2} v_1 v_2,$$

and therefore

$$\frac{\partial W}{\partial c} = \tfrac{1}{2} v_1^2 \frac{\partial k_{1.1}}{\partial c} + \tfrac{1}{2} v_2^2 \frac{\partial k_{2.2}}{\partial c} + v_1 v_2 \frac{\partial k_{1.2}}{\partial c} = F,$$

where F is the force between the spheres. If F is negative, W increases as c diminishes, and therefore the force is attractive, but if F be positive the force is repulsive. The values of $\partial k_{1.1}/\partial c$, $\partial k_{2.2}/\partial c$, and $\partial k_{1.2}/\partial c$ can be readily found from the preceding formulae. For example, when the spheres are very close together, we get by (23), (24) and (25)

$$\frac{\partial k_{1.1}}{\partial c} = \frac{\partial k_{2.2}}{\partial c} = -\frac{\partial k_{1.2}}{\partial c} = -\frac{ab}{2(a+b)d}.$$

approximately, where $d = c - a - b$, and hence

$$F = - \frac{ab\,(v_1 - v_2)^2}{4\,(a + b)\,d} \quad \text{...................(38)},$$

except when $(v_1 - v_2)^2$ is very small compared with $v_1{}^2 + v_2{}^2$. We see that when v_1 and v_2 are of opposite sign, the force is always attractive. In order to simplify the problem we shall now assume that the electrodes are of equal size.

In this case

$$F = - A\,(v_1{}^2 + v_2{}^2)/2 + B\,v_1 v_2 \quad \text{...............(39)},$$

where $A = - \partial k_{1.1}/\partial c$, and $B = \partial k_{1.2}/\partial c$. The quantities A and B are always positive.

Hence when v_1 and v_2 are of opposite sign the value of F must be negative and the force is therefore always attractive. When, however, v_1 and v_2 are of the same sign, the force can be attractive, zero or repulsive according to the distance of the spheres apart. If v_1 be the smaller potential, then the force is zero when:

$$\frac{v_1}{v_2} = \frac{B}{A} - \left(\frac{B^2}{A^2} - 1\right)^{1/2} \quad \text{.................(40)}.$$

The values of $A/2$ and $B/2$, when the distance between the spheres is small, are given in the following table:

d/a	$A/2$	$B/2$	d/a	$A/2$	$B/2$
$0 \cdot 0_5 1$	125000	125000	0·08	1·4461	1·4823
$0 \cdot 0_4 1$	12500	12500	0·09	1·2751	1·3111
$0 \cdot 0_3 1$	1249·8	1249·8	0·10	1·13844	1·17439
$0 \cdot 0_2 1$	124·82	124·86	0·15	0·73056	0·76603
0·01	12·340	12·377	0·20	0·52853	0·56352
0·02	6·1043	6·1410	0·25	0·40848	0·44299
0·03	4·0294	4·0661	0·30	0·32924	0·36325
0·04	2·9938	3·0304	0·35	0·27319	0·30673
0·05	2·3736	2·4101	0·40	0·23158	0·26464
0·06	1·9608	1·9972	0·45	0·19961	0·23216
0·07	1·6665	1·7028	0·50	0·17424	0·20630

In practice the three most important cases are, (1) when the potentials of the spheres are equal and opposite, (2) when they are equal and (3) when one of them is zero. In the first case the force is an attraction F and is given by

$$F = (B + A)\,v^2 \quad \text{......................(41)},$$

where v and $-v$ are the potentials of the spheres. In the second case it is repulsive and is given by

$$F' = (B - A)\,v^2 \dots\dots\dots\dots\dots\dots(42),$$

where the potential of each sphere is v. In the third case we get an attraction

$$F'' = (A/2)\,v_1^2 \dots\dots\dots\dots\dots\dots(43),$$

where v_1 and 0 are the potentials of the spheres.

From (28) and (29) we get

$$A = -\frac{\partial k_{1.1}}{\partial c} = \frac{2}{\xi^3 - 4\xi} - \frac{2}{\xi\,(\xi^3 - 2\xi)^2} \dots\dots\dots(44),$$

and

$$B = \frac{\partial k_{1.2}}{\partial c} = \frac{1}{\xi^2 - 4} - \frac{\xi^2 + 1}{\xi^6} - \frac{1}{\xi^3 - 4\xi^6}\dots\dots\dots(45).$$

Hence when d is not less than a the values of F, F' and F''' can be readily found. The values when d lies between $0.5a$ and a can be found from Kelvin's tables (W. Thomson, *Reprint*, p. 83). When c/a is large we find that

$$F = \frac{v^2}{\xi^2 - 2\xi} \dots\dots\dots\dots\dots\dots(46),$$

a formula found experimentally by Snow Harris in 1834. When c is less than $2.001a$, then, to a four figure accuracy, we have

$$F = \frac{av^2}{2\,(c - 2a)} \dots\dots\dots\dots\dots\dots(47),$$

and

$$F' = 0.07386v^2 \dots\dots\dots\dots\dots\dots(48).$$

If we write $F = \alpha\,(a^2/c^2)\,v^2$ and $F' = \alpha'\,(a^2/c^2)\,v^2$,

the following table shows how the values of α and α' vary with the distance between the spheres.

c/a	α	α'	c/a	α	α'
2.0_61	20×10^6	0.295	7	1.379	0.757
2.0_51	20×10^5	0.295	8	1.318	0.784
2.0_41	20×10^4	0.295	9	1.273	0.805
2.0_31	20×10^3	0.295	10	1.240	0.824
2.0_21	20×10^2	0.296	20	1.111	0.909
2.01	19.9×10	0.298	30	1.071	0.937
2.1	20.399	0.317	40	1.053	0.952
3.0	2.860	0.487	50	1.042	0.962
4.0	1.931	0.603	100	1.020	0.980
5.0	1.624	0.673	1000	1.002	0.998
6.0	1.472	0.722	10000	1.000	1.000

For values of c/a greater than 10, the maximum inaccuracy of the formulae

$$F = \frac{a^2}{c^2} \cdot \frac{1}{1 - 2a/c} \, v^2 \quad \dots (49)$$

and

$$F' = \frac{a^2}{c^2} \cdot \frac{1}{1 + 2a/c} \, v^2 \quad \dots (50)$$

is less than 1 in 10,000.

If S be the surface of each of the opposing hemispheres, and the distance between them is sufficiently great, we have

$$F = \frac{a^2 v^2}{c^2} = \frac{S(2v)^2}{8\pi c^2} \quad \dots (51),$$

the same formula as for the attraction between two parallel plates each of area S at a very small distance c apart, and the difference of potential between which is $2v$.

Let us now suppose that the charges q_1 and q_2 on the spheres have constant values. We find by (35) and (39)

The attraction and repulsion between equal spheres with given charges. that

$$F = C q_1 q_2 - D(q_1^2 + q_2^2)/2 \quad \dots (52),$$

where

$$C = [B(k^2_{1.1} + k^2_{1.2}) + 2A\, k_{1.1} k_{1.2}]/(k^2_{1.1} - k^2_{1.2})^2$$

and

$$D = [A(k^2_{1.1} + k^2_{1.2}) + 2B\, k_{1.1} k_{1.2}]/(k^2_{1.1} - k^2_{1.2})^2.$$

When F is negative the force is attractive, and when it is positive the force is repulsive. When q_1 and q_2 are of opposite signs the force is always attractive. But when they are of the same sign it may be attractive, zero or repulsive. If q_1 be the smaller charge, F' vanishes when

$$\frac{q_1}{q_2} = \frac{C}{D} - \left(\frac{C^2}{D^2} - 1\right)^{1/2} \dots (53).$$

When $q_1 = -q_2 = q$, formula (52) gives for the attraction

$$F = q^2 \frac{B + A}{(k_{1.1} - k_{1.2})^2} \quad \dots (54),$$

and when $q_1 = q_2 = q$, the repulsion F' is given by

$$F' = q^2 \frac{B - A}{(k_{1.1} + k_{1.2})^2} \quad \dots (55).$$

Writing $\quad F = \beta \, (q^2/c^2) \quad$ and $\quad F' = \beta' \, (q^2/c^2)$,

the manner in which β and β' alter as c increases can be seen from the following table:

c/a	β	β'	c/a	β	β'
2·0	∞	0·615	2·01	$1\cdot56 \times 10$	0·618
$2\cdot0_61$	$5\cdot77 \times 10^5$	0·615	2·10	3·355	0·643
$2\cdot0_51$	$2\cdot99 \times 10^4$	0·615	3·00	1·213	0·850
$2\cdot0_41$	$4\cdot05 \times 10^3$	0·615	4·00	1·074	0·935
$2\cdot0_31$	$5\cdot79 \times 10^2$	0·615	10·00	1·004	0·996
$2\cdot0_21$	$9\cdot11 \times 10$	0·615	20·00	1·001	1·000

It will be seen that β and β' approach their asymptotic value of unity much more rapidly than α and α' do.

In using the formulae given above it has to be remembered Numerical that the electrostatic units of pressure, quantity and illustration. capacity equal 300, $1/(3 \times 10^9)$ and $1/(9 \times 10^{11})$ times the volt, the coulomb and the farad respectively.

To illustrate these formulae let us consider the case of two conducting spheres, each of radius 1 cm. and at a distance apart of 1 cm. We shall find their charges, their potentials, and their capacities (a) when they attract one another with a force of 1 dyne, and (b) when they repel one another with a force of 1 dyne.

(a) *Attraction.* Let $q, -q$, and $v, -v$ be the charges and potentials of the spheres. From the tables given above we get, when the attractive force equals 1 dyne,

$$1\cdot213 \, \frac{q^2}{9} = 1, \quad \text{and} \quad 2\cdot860 \, \frac{v^2}{9} = 1.$$

Hence $\qquad\qquad q = 2\cdot724$ electrostatic units

$\qquad\qquad\qquad = 0\cdot908 \times 10^{-9}$ coulombs,

and $\qquad\qquad v = 1\cdot774$ electrostatic units

$\qquad\qquad\qquad = 532$ volts.

In this case the capacity K between the spheres is given by

$$K = q/(2v) = 0\cdot908 \times 10^{-9}/1064 \text{ farads}$$

$$= 0\cdot768 \text{ electrostatic units,}$$

which agrees with the number given in the table on p. 244.

(b) *Repulsion.* Let q_1, q_1, and v_1, v_1 be the charges and potentials of the spheres in this case. We have

$$0\cdot850\,\frac{q_1^2}{9}=1,\quad\text{and}\quad 0\cdot487\,\frac{v_1^2}{9}=1.$$

Hence $q_1 = 3\cdot255$ electrostatic units
                    $= 1\cdot085 \times 10^{-9}$ coulombs,

and $v_1 = 4\cdot225$ electrostatic units
                    $= 1290$ volts.

Therefore $K' = q_1/v_1 = 1\cdot085 \times 10^{-9}/1290$ farads
                    $= 0\cdot757$ electrostatic units.

It has to be noticed that the electric strength of the dielectric fixes a maximum possible limit to the attractive force between two electrified conductors. In the above case if the air were at 0° C. and 76 cm. pressure a disruptive discharge would take place when the potential difference between the spheres was a little less than 32,000 volts. The attraction between them when the potential difference has this value is only about 0·92 of the weight of a gramme.

When discussing experiments on brush discharges from electrodes or on sparking distances between them it is essential to know the maximum value of the electric stress on the dielectric in which they are immersed. We shall now, therefore, obtain formulae and give tables by means of which this quantity can be computed without difficulty. The formulae show that it is in general necessary to know the potential of each electrode. Except in the case when one of the electrodes is earthed the apparatus is practically always so arranged that the charges on the two electrodes are equal and of opposite sign at the instant of the discharge. We therefore have

$$q = k_{1.1}\,v_1 + k_{1.2}\,v_2,\quad\text{and}\quad -q = k_{2.2}\,v_2 + k_{1.2}\,v_1.$$

Hence we easily deduce that

$$v_1 = \frac{k_{2.2} + k_{1.2}}{k_{1.1} + k_{2.2} + 2k_{1.2}}\,V \quad\ldots\ldots\ldots\ldots\ldots(56),$$

and

$$v_2 = -\frac{k_{1.1} + k_{1.2}}{k_{1.1} + k_{2.2} + 2k_{1.2}}\,V \quad\ldots\ldots\ldots\ldots(57),$$

The maximum electric stress between spherical electrodes.

where $V = v_1 - v_2 =$ the potential difference between the spheres. Hence when we know the value of V the formulae given above enable us to find v_1 and v_2.

To find the maximum electric stress $R_{max.}$ between the spheres we replace (Fig. 61) the spheres by the series of point charges A, A_1, A_2, ... B_1, B_2, which produce the same external field. The potential gradient is a maximum at that point P of the smaller sphere where the line joining their centres cuts its surface. Let us first suppose that the potential of the sphere whose radius is a is unity, and that the potential of the other is zero. We obviously have

$$R_{max.} = \frac{q_0}{a^2} + \frac{q_1}{(a - a_1)^2} + \frac{q_2}{(a - a_2)^2} + \frac{q_3}{(a - a_3)^2} + \dots$$

$$- \frac{q_1'}{(c - b_1 + a)^2} - \frac{q_2'}{(c - b_2 + a)^2} - \frac{q_3'}{(c - b_3 + a)^2} - \dots$$

Hence by addition and using (1) etc. we get

$$R_{max.} = \sum_0^\infty \frac{a + a_n}{(a - a_n)^2} \cdot \frac{q_n}{a},$$

and

$$R'_{max.} = - \sum_0^\infty \frac{b + b_{n+1}}{(b - b_{n+1})^2} \cdot \frac{q'_{n+1}}{a},$$

where $R'_{max.}$ is the electric stress at the point Q where the line joining the centres cuts the larger sphere. It follows by the principle of superposition that when the potentials of the spheres are v_1 and v_2 respectively, the maximum value of the electric stress, which is at the point P where the central line cuts the smaller sphere, is given by

$$R_{max.} = v_1 \sum \frac{a + a_n}{(a - a_n)^2} \cdot \frac{q_n}{a} + v_2 \sum \frac{a + a'_{n+1}}{(a - a'_{n+1})^2} \cdot \frac{q''_{n+1}}{a} \dots (58),$$

where q_n'' is the charge of the nth image in the smaller sphere, when its potential is zero and the potential of the larger is unity.

Now with the notation of Fig. 61, we see that

$$b_n (c - a_{n-1}) = b^2 \quad \text{and} \quad a_n (c - b_n) = a^2.$$

Hence we easily prove that

$$a_n a_{n-1} - \{(c^2 - b^2)/c\} a_n - (a^2/c) a_{n-1} = - a^2.$$

Writing $a_n = u_{n+1}/u_n + (c^2 - b^2)/c$, this becomes

$$u_{n+1} + (ab/c)(w + 1/w) u_n + (a^2b^2/c^2) u_{n-1} = 0,$$

where $\qquad w + 1/w = 2 \cosh \omega = (c^2 - a^2 - b^2)/(ab).$

Assuming that $u_{n+1} = Eu_n = E^2 u_{n-1} = \dots$, we get

$$\{E + (ab/c) w\} \{E + (ab/c)(1/w)\} = 0,$$

and hence, by the theory of finite differences, we have

$$u_n = A \{-(ab/c) w\}^n + B \{-(ab/c)(1/w)\}^n,$$

where A and B are constants.

Since $a_0 = 0$, we get $u_1/u_0 = -(c^2 - b^2)/c.$

Hence noticing that $(a + bw)(b + aw) = c^2 w$ and that $x = (a + bw)/c$, we get $A = -Bx^2$, and

$$a_n = \frac{ax(1 - w^{2n})}{1 - x^2 w^{2n}} = \frac{a \sinh n\omega}{\sinh(n\omega + \alpha)},$$

$$b_n = c - \frac{a^2}{a_n} = b\frac{w(1 - x^2 w^{2n-2})}{x(1 - w^{2n})} = b\frac{\sinh\{(n-1)\omega + \alpha\}}{\sinh n\omega},$$

and

$$a'_{n+1} = a\frac{w(1 - y^2 w^{2n})}{y(1 - w^{2n+2})} = a\frac{\sinh(n\omega + \beta)}{\sinh(n + 1)\omega}.$$

By (6) and (7)

$$q_n = a\frac{\sinh \alpha}{\sinh(n\omega + \alpha)}; \quad q_n' = -a\frac{\sinh \alpha}{\sinh n\omega};$$

and therefore

$$q''_{n+1} = -b\frac{\sinh \beta}{\sinh(n+1)\omega} = -a\frac{\sinh \alpha}{\sinh(n+1)\omega}.$$

Substituting in (58) we get

$$R_{\max.} = \frac{\cosh^2(\alpha/2)}{a \sinh(\alpha/2)} \left[v_1 \left\{ \frac{\sinh(\alpha/2)}{\cosh^2(\alpha/2)} + \frac{\sinh(\alpha/2 + \omega)}{\cosh^2(\alpha/2 + \omega)} \right. \right.$$

$$\left. + \frac{\sinh(\alpha/2 + 2\omega)}{\cosh^2(\alpha/2 + 2\omega)} + \dots \right\}$$

$$\left. - v_2 \left\{ \frac{\sinh(\omega - \alpha/2)}{\cosh^2(\omega - \alpha/2)} + \frac{\sinh(2\omega - \alpha/2)}{\cosh^2(2\omega - \alpha/2)} + \dots \right\} \right] \dots(59).$$

It can also be written in the form

$$R_{\text{max.}} = \frac{(1+x)^2}{a\,(1-x)} \left[v_1 \left\{ \frac{1-x}{(1+x)^2} + w\,\frac{1-xw^2}{(1+xw^2)^2} + w^2\,\frac{1-xw^4}{(1+xw^4)^2} + \ldots \right\} \right.$$

$$\left. - v_2 \left\{ y\,\frac{1-yw}{(1+yw)^2} + wy\,\frac{1-yw^3}{(1+yw^3)^2} + w^2 y\,\frac{1-yw^5}{(1+yw^5)^2} + \ldots \right\} \right] \ldots(60).$$

Formulae (59) and (60) only converge rapidly when the spheres are far apart. We can modify them, however, so as to make them converge more rapidly, in practically the same way as we modified Kirchhoff's formulae for the capacity coefficients. For example, if we expand all the terms above the second in the coefficients of v_1 and v_2 in (60) and use the theorem

$$\frac{1-z}{(1+z)^2} = 1 - 3z + 5z^2 - 7z^3 + \ldots,$$

we get

$$R_{\text{max.}} = \frac{v_1}{a} \left[1 + \frac{ab\,\{c\,(c+a)-b^2\}}{\{c\,(c-a)-b^2\}^2} + w\,\frac{(1+x)^2}{1+x}\{A_1 - 3A_3\,(xw^2) \right.$$

$$\left. + 5A_5\,(xw^2)^2 - \ldots\} \right]$$

$$- \frac{v_2}{a} \left[\frac{b\,(c+a)}{(c-a)^2} + \frac{ab^2\,\{(c+a)^2(c-a)-cb^2\}}{\{(c-a)^2\,(c+a)-cb^2\}^2} \right.$$

$$\left. + yw\,\frac{(1+x)^2}{1-x}\{A_1 - 3A_3\,(yw^3) + 5A_5\,(yw^3)^2 - \ldots\} \right] \ldots(61),$$

where $A_n = w^n/(1-w^n)$.

As a numerical example, let us suppose that $a = 2\cdot1l$, $b = 9\cdot6l$ and $c = 12\cdot5l$, so that $d = 0\cdot8l$, $x = 7/15$, $y = 5/6$ and $w = 7/18$. Hence, by the formulae given above,

$$k_{1.1} = (3\cdot1684147\ldots)\,l, \quad k_{2.2} = (11\cdot6081338\ldots)\,l$$

and $\quad -k_{1.2} = (2\cdot4971357\ldots)\,l.$

By (56) and (57) therefore

$$v_1 = (0\cdot9313782\ldots)V, \quad \text{and} \quad v_2 = -(0\cdot0686218\ldots)V.$$

Hence by (61)

$$R_{\text{max.}} = \frac{v_1}{2\cdot1l}\{1 + 1\cdot271945 + 0\cdot998148 - 0\cdot020754$$

$$+ 0\cdot000351 - 0\cdot000004 + \ldots\}$$

$$-\frac{v_2}{2\cdot1l}\{1\cdot295758 + 1\cdot129597 + 0\cdot831790 - 0\cdot012010$$
$$+ 0\cdot000141 - 0\cdot000002\}$$
$$=\frac{V}{0\cdot8l}\times 1\cdot237870....$$

As a further example, let us suppose that $a = l$, $b = 7l$ and $c = 10l$, so that $d = 2l =$ twice the radius of the smaller sphere. We find easily that $x = 1/5$, $y = 5/7$ and $w = 1/7$, and hence we get

$$k_{1.1} = (1\cdot1601124...)\,l, \quad k_{2.2} = (7\cdot5765970...)\,l$$

and $\qquad -k_{1.2} = (0\cdot814326...)\,l.$

Hence $\qquad v_1 = (0\cdot9513530...)V, \quad v_2 = -(0\cdot0486470...)V$

and thus by (61)

$$R_{\text{max.}} = \frac{V}{2l}\times 2\cdot5807756....$$

These examples show that if the spheres be close together the calculation would be very laborious. In this case we shall show how a much more suitable formula can be obtained.

By the Euler-Maclaurin sum formula, we have

Spheres very close together.
$$\frac{\sinh x}{\cosh^2 x} + \frac{\sinh (x + \omega)}{\cosh^2 (x + \omega)} + \frac{\sinh (x + 2\omega)}{\cosh^2 (x + 2\omega)} + \ldots$$

$$= \frac{\operatorname{sech} x}{\omega} - \frac{1}{2}\frac{\partial}{\partial x}\operatorname{sech} x + \frac{B_1}{\lfloor 2}\omega\frac{\partial^2}{\partial x^2}\operatorname{sech} x - \frac{B_3}{\lfloor 4}\omega^3\frac{\partial^4}{\partial x^4}\operatorname{sech} x + \ldots,$$

where $B_1, B_3, B_5, \ldots$ are Bernoulli's numbers.

Using this theorem in (59), and developing approximate formulae for the various functions on the assumption that d is small, we find after heavy algebraical work that

$$R_{\text{max.}} = \frac{v_1 - v_2}{d}\left\{1 + \frac{2b - a}{3ab}d + \frac{4(a - b)^2 + ab}{45a^2b^2}d^2 + \ldots\right\}\ldots(62),$$

where $d = c - a - b =$ the distance between the spheres.

In this case, therefore, it is unnecessary to compute the values of v_1 and v_2 as their difference only is required. Applying this formula to the first of the numerical examples in the preceding section we get, at once, that

$$R_{\text{max.}} = \frac{v_1 - v_2}{0\cdot8l}\times 1\cdot235 \text{ approximately};$$

the error, therefore, is about 3 in a thousand.

Even in the second case when the distance between the spheres equals the diameter of the smaller one, the error introduced by using (62) is less than three per cent. When the distance between the spheres is very small, formula (62) gives $R_{max.}$ to a high degree of accuracy.

When the spheres are equal, so that $a = b$, and $v_1 = -v_2 = V/2$, **Equal spheres.** formula (59) becomes

$$R_{max.} = \frac{V}{2a} \cdot \frac{\cosh^2(\alpha/2)}{\sinh(\alpha/2)} \left\{ \frac{\sinh(\alpha/2)}{\cosh^2(\alpha/2)} + \frac{\sinh(3\alpha/2)}{\cosh^2(3\alpha/2)} \right.$$
$$\left. + \frac{\sinh(5\alpha/2)}{\cosh^2(5\alpha/2)} + \dots \right\} \dots (63),$$

where $\cosh \alpha = c/(2a)$.

The formula (61) becomes

$$R_{max.} = \frac{V}{2a} \left[1 + \frac{a(c+a)}{(c-a)^2} + \frac{a^2\{c(c+a)-a^2\}}{\{c(c-a)-a^2\}^2} \right.$$
$$+ \frac{a^3\{(c+a)^2(c-a)-ca^2\}}{\{(c-a)^2(c+a)-ca^2\}^2} + \frac{(x^2+x^3)^2}{1-x}\left\{ \frac{1}{1-x} - \frac{3x^9}{1-x^3} \right.$$
$$\left. \left. + \frac{5x^{18}}{1-x^5} - \dots \right\} \right] \dots (64),$$

where $x = \{c - (c^2 - 4a^2)^{1/2}\}/(2a) = \epsilon^{-\alpha}$.

Applying the Euler-Maclaurin sum formula to (63), we get

$$R_{max.} = \frac{V}{d} \left\{ 1 + \frac{1}{3}\frac{d}{a} + \frac{1}{45} \cdot \frac{d^2}{a^2} + \frac{2}{945} \cdot \frac{d^3}{a^3} + \frac{17}{14175} \cdot \frac{d^4}{a^4} + \dots \right\} \dots (65).$$

The first three terms of this series may be deduced by putting $a = b$ in (62).

As a numerical example let us suppose that $c = 2\cdot 5a$, so that $d = a/2$ and $x = 1/2$.

Substituting in (64), we get

$$R_{max.} = \frac{V}{2a} \left\{ 1 + \frac{14}{9} + \frac{124}{121} + \frac{15\cdot 875}{(5\cdot 375)^2} \right.$$
$$\left. + \frac{9}{16}\left(1 - \frac{3}{7} \cdot \frac{1}{128} + \frac{5}{31} \cdot \frac{1}{128^2} - \frac{7}{127} \cdot \frac{1}{128^3} + \dots \right) \right\}$$
$$= \frac{V}{a/2} \times 1\cdot 1726148.$$

By formula (65)

$$R_{\text{max.}} = \frac{V}{a/2} \times 1 \cdot 17256.$$

Hence when the distance between the spheres is not greater than half the radius of either, the maximum inaccuracy of (65) is less than 1 in 20,000.

The two most important cases in practice are, first when the potentials of the spheres are equal and opposite at every instant as, for example, when the middle point of the secondary winding of the high tension transformer is in metallic connection with the

Values of f.

d/a	f	d/a	f
0·0	1·0000	2·0	1·7704
0·1	1·0336	3·0	2·2149
0·2	1·0676	4·0	2·6777
0·3	1·1021	5·0	3·1513
0·4	1·1371	6·0	3·6317
0·5	1·1726	7·0	4·1165
0·6	1·2086	8·0	4·6044
0·7	1·2453	9·0	5·0946
0·8	1·2826	10·0	5·5865
0·9	1·3205	100·0	50·5098
1·0	1·3593	1000·0	500·5010
1·5	1·5594	10000·0	5000·5

Values of f_1.

d/a	f_1	d/a	f_1
0·0	1·000	2·0	2·339
0·1	1·034	3·0	3·252
0·2	1·068	4·0	4·201
0·3	1·106	5·0	5·167
0·4	1·150	6·0	6·143
0·5	1·199	7·0	7·125
0·6	1·253	8·0	8·111
0·7	1·313	9·0	9·100
0·8	1·378	10·0	10·091
0·9	1·446	100·0	100·0
1·0	1·517	1000·0	1000·0
1·5	1·909	10000·0	10000·0

earth and secondly when one of the spheres is earthed. In the first case we may write

$$R_{max.} = (V/d)f,$$

and in the second $\quad R_{max.} = (V/d)f_1,$

where f and f_1 are numbers which depend on the value of d/a. In general, if V_1 and V_2 are the potentials of the two electrodes, we have

$$R_{max.} = \{(V_1 - V_2)/d\}\, f_1 + 2\,(V_2/d)\,(f_1 - f).$$

In the following table an analysis is given of a test on sparking distances between 5 cm. spheres made by J. Algermissen. The experimental figures are taken from Zenneck's *Elektromagnetische Schwingungen und Drahtlose Telegraphie.*

J. Algermissen. 5 cm. spheres ($a = 2\cdot5$).

d is measured in cms. and V in kilovolts.

d	d/a	f (calc.)	V (obs.)	$R_{max.}$ (calc.)
1·5	0·60	1·209	46·2	36·6
1·6	0·64	1·223	48·6	36·5
1·7	0·68	1·238	51·0	36·6
1·8	0·72	1·253	53·4	36·6
1·9	0·76	1·268	55·8	36·7
2·0	0·80	1·283	58·2	36·8
2·1	0·84	1·298	60·6	37·0
2·2	0·88	1·312	62·8	36·9
2·3	0·92	1·326	65·0	37·0
2·4	0·96	1·342	67·0	37·0
2·5	1·00	1·359	69·0	37·1
2·6	1·04	1·374	70·8	37·0
2·7	1·08	1·390	72·6	37·0
2·8	1·12	1·406	74·4	37·0
2·9	1·16	1·421	76·2	37·0
3·0	1·20	1·437	78·0	37·0
3·1	1·24	1·452	79·7	37·0
3·2	1·28	1·469	81·3	37·0
3·3	1·32	1·484	83·0	37·0
3·4	1·36	1·500	84·7	37·0
3·5	1·40	1·515	86·4	37·1
3·6	1·44	1·533	88·0	37·1
3·7	1·48	1·549	89·6	37·2
3·8	1·52	1·566	91·2	37·3
3·9	1·56	1·583	92·7	37·3
4·0	1·60	1·599	94·2	37·4
4·1	1·64	1·616	95·7	37·4
4·2	1·68	1·632	97·2	37·4

We have assumed that the potentials of the electrodes were equal and opposite at the instant of the discharge. As the values of R_{max}. are all approximately the same, our assumption is justified.

By analysing in a similar way experimental results on sparking distances made with 2 inch spheres ($a = 2·54$) in air and using alternating pressures of frequency 125, the author finds that the mean value of R_{max}. from $d = 0·3$ to 14 cms. is 37·8 kilovolts per centimetre. The discharge takes place at the instant when the voltage is a maximum.

When the distance between the spheres is less than about a millimetre the values of R_{max}. increase rapidly. J. E. Almy has given a direct experimental proof of the existence of a minimum sparking potential of about 350 volts. Hence R_{max}. is approximately constant only for distances between the spheres greater than about a millimetre.

REFERENCES

S. D. Poisson, 'Sur la Distribution de l'Electricité à la Surface des Corps Conducteurs.' *Mémoires de l'Institut Impérial de France*, May and August, 1812.

W. Snow Harris, ' On some Elementary Laws of Electricity.' *Phil. Trans. Roy. Soc.* p. 213, 1834.

W. Thomson, 'On the Mutual Attractions or Repulsions between two Electrified Spherical Conductors.' *Phil. Mag.* April and August 1853 or *Reprint*, p. 83.

G. R. Kirchhoff, ' Ueber die Vertheilung der Elektricität auf zwei leitenden Kugeln.' *Crelle's Journal*, p. 89, 1860 or *Gesammelte Abhandlungen*, p. 78 ; also *Annalen der Physik*, Vol. 27, p. 673, 1886.

A. Schuster, ' The Disruptive Discharge of Electricity through Gases.' *Phil. Mag.* Vol. 29, p. 192, 1890.

A. Russell, ' The Dielectric Strength of Air.' *Phil. Mag.* [6] Vol. 11, p. 258, 1906.

—— ' Coefficients of Capacity.' *Proc. Roy. Soc.* A. Vol. 82, p. 524, 1909.

—— ' Electric Stress at which Ionisation begins in Air.' *Proc. Phys. Soc.* Vol. 23, p. 86, 1911.

—— ' Capacity Coefficients.' *Proc. Phys. Soc.* Vol. 23, p. 352, 1911.

—— ' Maximum Value of the Electric Stress.' *Proc. Phys. Soc.* Vol. 24, p. 22, 1911.

—— ' The Mutual Attractions or Repulsions of Two Electrified Spherical Conductors.' *Proc. of the Inst. of El. Eng.* Vol. 48, p. 257, 1912.

J. E. Almy, ' Minimum Spark Potentials.' *Phil. Mag.* [6] Vol. 16, p. 456, 1908.

CHAPTER IX

Current oscillations. Oscillatory discharge. The effect of shunting the inductive coil. Inductively coupled electric circuits. No damping. Two equal circuits. No condenser in the secondary circuit. Complete criterion for the roots of a biquadratic. Two sets of oscillations. Forced oscillations. A damped E.M.F. applied to an oscillatory circuit. References.

WE saw in Chapter IV that when a condenser in series with an

Current oscillations. inductive coil was connected across the alternating current supply mains, abnormal oscillations of the current occurred, in certain cases, at the moment of closing the switch. When the resistance of the inductive coil is small the current can be decomposed into two trains of waves. The frequency of one of these trains of waves is the same as that of the applied electromotive force, but the frequency of the other depends only on the constants of the circuit. In general this latter train is rapidly damped out.

We shall now show that when a condenser is discharged through a circuit containing capacity and inductance, then in certain cases oscillatory currents are set up in this circuit and in circuits linked inductively with it. These oscillations have sometimes a very high frequency and are particularly useful in radio-telegraphy.

Let us first consider a condenser K in series with an inductive

Oscillatory discharge. coil (R, L) and let q_0 be the initial charge in it. The potential difference between its terminals at the time t is $\{q_0 - \int i \partial t\}/K$, and thus

$$\frac{q_0 - \int i \partial t}{K} = Ri + L\frac{\partial i}{\partial t} \quad\dots\dots\dots\dots\dots(1).$$

Hence differentiating, we have

$$\frac{\partial^2 i}{\partial t^2} + \frac{R}{L}\frac{\partial i}{\partial t} + \frac{i}{KL} = 0 \quad\dots\dots\dots\dots\dots(2).$$

If $i = A\epsilon^{mt}$ be a solution of this equation, we must have

$$m^2 + (R/L)\,m + 1/KL = 0.$$

Let m_1 and m_2 be the roots of this equation, then we have

$$m_1 = -\alpha + \beta_1 = -\alpha + \beta_2\iota,$$

and $$m_2 = -\alpha - \beta_1 = -\alpha - \beta_2\iota,$$

where $\alpha = R/(2L)$ and $\beta_1 = (R^2/4L^2 - 1/LK)^{1/2} = \beta_2\iota$.

There are three cases to be considered. In the first case KR^2 is greater than $4L$, and so β_1 is a real quantity. Hence the solution of (2) is

$$i = A\epsilon^{-(\alpha-\beta_1)t} + B\epsilon^{-(\alpha+\beta_1)t},$$

where A and B are constants. Since i is zero when t is zero, we see that $B = -A$, and thus

$$i = 2A\epsilon^{-\alpha t}\sinh\beta_1 t.$$

Now $$\int i\partial t = (1/D)\,2A\epsilon^{-\alpha t}\sinh\beta_1 t$$
$$= 2A\epsilon^{-\alpha t}\{1/(D-\alpha)\}\sinh\beta_1 t.$$

Thus multiplying the numerator and denominator of the operator by $D + \alpha$, and writing β_1^2 for D^2, we easily find that

$$\int i\partial t = -2A\,(LK)^{1/2}\,\epsilon^{-\alpha t}\sinh(\beta_1 t + \gamma_1) + C,$$

where $\tanh\gamma_1 = \beta_1/\alpha$, $\sinh\gamma_1 = \beta_1/(\alpha^2 - \beta_1^2)^{1/2} = \beta_1\,(LK)^{1/2}$,

$\cosh\gamma_1 = \alpha\,(LK)^{1/2}$, and C is a constant.

When t is infinite $\int i\partial t = q_0$, and thus $C = q_0$.

Also when t is zero, $\int i\partial t = 0$, and therefore

$$2A = q_0/\{(LK)^{1/2}\sinh\gamma_1\} = q_0/(\beta_1 LK),$$

and hence $$i = \{q_0/(\beta_1 LK)\}\,\epsilon^{-\alpha t}\sinh\beta_1 t\dots\dots\dots\dots(3),$$

and $$q = q_0\,\epsilon^{-\alpha t}\{\sinh(\beta_1 t + \gamma_1)/\sinh\gamma_1\}\dots\dots\dots\dots(4),$$

where q is the charge in the condenser at the time t.

In the second case when $KR^2 = 4L$, β_1 is zero, and (3) and (4) become

$$i = \{q_0/(LK)\}\,t\epsilon^{-\alpha t}, \quad\text{and}\quad q = q_0\,\epsilon^{-\alpha t}(1 + \alpha t)\dots\dots(5).$$

Hence the current attains its maximum value $2q_0/(\epsilon RK)$ when $t = 1/\alpha = 2L/R$.

In the third case when KR^2 is less than $4L$, β_2 is real, and proceeding in the same way we find that

$$i = \{q_0/(LK\beta_2)\} \, \epsilon^{-at} \sin \beta_2 t \dots\dots\dots\dots\dots(6),$$

and $$q = q_0 \, \epsilon^{-at} \{\sin (\beta_2 t + \gamma_2)/\sin \gamma_2\} \dots\dots\dots(7),$$

where $\tan \gamma_2 = \beta_2/\alpha$, $\sin \gamma_2 = \beta_2 (LK)^{1/2}$ and $\cos \gamma_2 = \alpha (LK)^{1/2}$.

These equations show that in this case the discharge is oscillatory, the frequency being $\beta_2/(2\pi)$, that is

$$(1/LK - R^2/4L^2)^{1/2}/(2\pi).$$

The maximum positive values of i occur at the times

$$(\gamma_2/\beta_2 + 2n\pi/\beta_2)$$

and the maximum negative values at the times

$$\{\gamma_2/\beta_2 + (2n+1)\pi/\beta_2\}.$$

Hence the amplitude of the oscillations diminishes in geometrical progression, the common ratio for the maximum positive or negative values being $\epsilon^{-2\pi\alpha/\beta_2}$.

When the frequency is high the solution given above is only approximate as we have neglected the increase in the value of R due to the skin effect. In several practical cases this increase has to be taken into account.

Let us suppose that the coil (R, L) which is in series with the condenser K is shunted by a non-inductive resistance x.

The effect of shunting the inductive coil. If q_0 be the initial charge in the condenser before the circuit is closed, we have

$$\frac{q_0 - \int i \partial t}{K} = Ri_1 + L \frac{\partial i_1}{\partial t} = xi_2,$$

where i_1 is the current in the inductive coil, i_2 the current through x and $i = i_1 + i_2$.

Eliminating i and i_2 from these equations we readily find that

$$L \frac{\partial^2 i_1}{\partial t^2} + \left(R + \frac{L}{Kx}\right) \frac{\partial i_1}{\partial t} + \frac{1}{K} \left(1 + \frac{R}{x}\right) i_1 = 0.$$

Assuming that $i_1 = A\epsilon^{mt}$, we see that the quadratic to determine m has imaginary roots when

$$4 (L/K)(1 + R/x) \text{ is greater than } (R + L/Kx)^2,$$

that is, when $4L/K$ is greater than $(R - L/Kx)^2$. Hence, if x is greater than $L/\{KR + 2\,(KL)^{1/2}\}$ but less than $L/\{KR - 2\,(KL)^{1/2}\}$, the roots of the quadratic equation are imaginary and an oscillatory discharge will therefore ensue.

Inductively coupled electric circuits. Let us suppose that we have two separate circuits each consisting of a condenser in series with an inductive coil. Let (K_1, R_1, L_1) and (K_2, R_2, L_2) be the constants of the two circuits and let M be the mutual inductance between the two coils. We shall suppose that the condenser K_1 has a charge q_0 at the time $t = 0$, but that the condenser K_2 is uncharged and that the currents i_1 and i_2 in the two circuits are zero at this instant.

The equations to determine the currents are

$$R_1 i_1 + L_1 \frac{\partial i_1}{\partial t} + M \frac{\partial i_2}{\partial t} + \frac{\int i_1 \partial t}{K_1} = \frac{q_0}{K_1} \quad\text{......}(8),$$

and

$$R_2 i_2 + L_2 \frac{\partial i_2}{\partial t} + M \frac{\partial i_1}{\partial t} + \frac{\int i_2 \partial t}{K_2} = 0 \quad\text{......}(9).$$

Eliminating i_2 between the two equations, we get

$$\{(L_1 D^2 + R_1 D + 1/K_1)(L_2 D^2 + R_2 D + 1/K_2) - M^2 D^4\}\, i_1 = 0,$$

where D stands for $\partial/\partial t$.

Now writing

$$t_1 = L_1/R_1, \quad \tau_1 = K_1 R_1, \quad t_2 = L_2/R_2, \quad \tau_2 = K_2 R_2, \quad \sigma = 1 - M^2/(L_1 L_2),$$

and $i_1 = A \epsilon^{mt}$, we get

$$am^4 + bm^3 + cm^2 + dm + e = 0 \quad\text{......}(10),$$

where

$$a = \sigma, \quad b = 1/t_1 + 1/t_2, \quad c = 1/(t_1 t_2) + 1/(t_1 \tau_1) + 1/(t_2 \tau_2),$$
$$d = (1/t_1 t_2)(1/\tau_1 + 1/\tau_2) \quad\text{and}\quad e = 1/(t_1 t_2 \tau_1 \tau_2).$$

In general this equation has four roots, and since all the coefficients are positive, the real roots—if any—must all be negative. Again we can show that

$$bcd - ad^2 - eb^2 = \{1/(t_1 t_2 \tau_1 \tau_2)\}^3 \{\tau_1 \tau_2 (t_1 + t_2)(\tau_1 + \tau_2)$$
$$+ t_1 t_2 (\tau_1 + \tau_2)^2 (1 - \sigma) + (t_1 \tau_1 - t_2 \tau_2)^2\}$$
$$= \text{a positive quantity.}$$

Thus by Vol. II, Chap. VI, the real parts of the imaginary roots are negative and the oscillations are therefore stable. It is to be noticed that a lies in value between 0 and 1, so that c^2 is always greater than $4ae$.

The relations between the coefficients which show whether the roots are all real, two real and two imaginary, or all four imaginary are complicated, and we shall delay giving them for the present. We can prove as follows, however, that if $8ac$ is greater than $3b^2$ or if $8ec$ is greater than $3d^2$, there will be one pair of imaginary roots at least and so oscillations will ensue.

Let x_1, x_2, x_3 and x_4 be the roots of the biquadratic equation (10), then $\Sigma x_1 x_2 = c/a$, and by Newton's theorem $\Sigma x_1^2 = b^2/a^2 - 2c/a$. If the roots are real $\Sigma (x_1 - x_2)^2$ cannot be negative, and hence, by expanding, $3\Sigma x_1^2 - 2\Sigma x_1 x_2$ cannot be negative, i.e. $3(b^2/a^2 - 2c/a)$ cannot be less than $2c/a$ or $3b^2$ less than $8ca$. If, therefore, $8ca$ is greater than $3b^2$, two at least of the roots of the equation must be imaginary. Similarly by considering the biquadratic for $1/m$ we see that oscillations will also ensue if $8ce$ is greater than $3d^2$. Hence substituting for a, b, c, d and e their values we find that oscillations will ensue if

$$8\left(\frac{1}{t_1 \tau_1} + \frac{1}{t_2 \tau_2}\right) \text{ is greater than } \frac{3}{\sigma}\left(\frac{1}{t_1} + \frac{1}{t_2}\right)^2 - \frac{8}{t_1 t_2} \quad ...(11),$$

or if it is greater than

$$3\frac{\tau_1 \tau_2}{t_1 t_2}\left(\frac{1}{\tau_1} + \frac{1}{\tau_2}\right)^2 - \frac{8}{t_1 t_2} \quad ....(12).$$

When $\tau_1 = \tau_2 = \tau$ these equations show us that oscillations will ensue if $1/\tau$ is greater than

$$(3/8\sigma)(1/t_1 + 1/t_2) - 1/(t_1 + t_2) \text{ or } (1/2)/(t_1 + t_2).$$

It must be remembered that although (11) and (12) may not be satisfied we cannot conclude that the motion is non-oscillatory. The rigorous conditions are much more complicated.

Let us now suppose that (10) has two pairs of imaginary roots. We have already shown that the real parts of the imaginary roots must be negative. Let the roots be $-x_1 \pm \iota y_1$ and $-x_2 \pm \iota y_2$. Then

$$x_1 + x_2 = b/2a, \text{ and } c/a - b^2/4a^2 = y_1^2 + y_2^2 + 2x_1 x_2$$

$$= \text{a positive quantity.}$$

Hence, if b^2 be greater than $4ac$, there cannot be two pairs of imaginary roots. Similarly there cannot be two pairs of imaginary roots if d^2 be greater than $4ce$.

In practice the damping is sometimes very small and the frequency is high. In this case if $x + \iota y$ is a root of equation (10) x is very small compared with y. Now if we write equation (10) in the form $f(m) = 0$, we must have $f(x + \iota y) = 0$. Expanding by Taylor's theorem,

$$f(x) + \iota y f'(x) - \frac{y^2}{\lfloor 2} f''(x) - \ldots = 0 \qquad \ldots\ldots\ldots(13).$$

Noticing that $f^v(x) = 0$ and equating the coefficient of ι in this equation to zero, we get

$$y^2 = 6f'(x)/f'''(x)$$
$$= (2cx + d)/(4ax + b),$$

when we neglect squares and higher powers of x.

Solving this equation, we get

$$x = -\frac{by^2 - d}{2(2ay^2 - c)} \qquad \ldots\ldots\ldots\ldots\ldots\ldots(14).$$

Hence when we can determine the frequency $(y/2\pi)$ of the oscillation, we can find the damping coefficient.

Equating the real part of (13) to zero, we get

$$ay^4 - (y^2/2)(12ax^2 + 6bx + 2c) + f(x) = 0,$$

and hence, when x is very small,

$$y^2 = c/2a \pm \{(c/2a)^2 - e/a\}^{1/2}.$$

When there is absolutely no damping R_1 and R_2 are zero and equation (10) becomes

No damping.

$$am^4 + cm^2 + e = 0.$$

If the roots of this equation are $\pm \iota y_1$ and $\pm \iota y_2$, we have

$$2\sigma y_1^2 = \frac{1}{K_1 L_1} + \frac{1}{K_2 L_2} + \left\{\left(\frac{1}{K_1 L_1} + \frac{1}{K_2 L_2}\right)^2 - \frac{4\sigma}{K_1 K_2 L_1 L_2}\right\}^{1/2}$$

and

$$2\sigma y_2^2 = \frac{1}{K_1 L_1} + \frac{1}{K_2 L_2} - \left\{\left(\frac{1}{K_1 L_1} + \frac{1}{K_2 L_2}\right)^2 - \frac{4\sigma}{K_1 K_2 L_1 L_2}\right\}^{1/2}.$$

The quantity $\sigma = 1 - M^2/L_1 L_2$ is called the leakage factor. When the coils are so arranged that σ is nearly zero and there is therefore little magnetic leakage, the coupling is said to be tight. In this case

$$y_1 = (1/\sigma^{1/2})(1/K_1L_1 + 1/K_2L_2)^{1/2}$$

and $$y_2 = 1/(K_1L_1 + K_2L_2)^{1/2}.$$

In this case therefore the frequency of one of the oscillations is very high and the frequency of the other is $1/\{2\pi(K_1L_1 + K_2L_2)^{1/2}\}$. With a very loose coupling, that is, when σ is nearly unity, $y_1^2 = 1/K_1L_1$ and $y_2^2 = 1/K_2L_2$, and hence the frequencies are $1/\{2\pi(K_1L_1)^{1/2}\}$ and $1/\{2\pi(K_2L_2)^{1/2}\}$ respectively.

When $K_1L_1 = K_2L_2$ the circuits are said to be tuned to one another. In this case, writing $1/\omega^2 = K_1L_1 = K_2L_2$, we get

$$y_1^2 = \omega^2/\{1 - M/(L_1L_2)^{1/2}\} \quad \text{and} \quad y_2^2 = \omega^2/\{1 + M/(L_1L_2)^{1/2}\}.$$

Two equal circuits. Let us now suppose that the time constants of the two circuits are the same, that is, that $t_1 = t_2 = t$ and $\tau_1 = \tau_2 = \tau$. In this case the auxiliary biquadratic may be written

$$(M^2/L_1L_2)\,m^4 - (m^2 + m/t + 1/t\tau)^2 = 0.$$

The roots are therefore determined by the equations

$$\{1 \pm M/(L_1L_2)^{1/2}\}\,m^2 + m/t + 1/t\tau = 0 \quad \ldots\ldots\ldots(15).$$

Hence, if there are two sets of oscillations,

$$4\{1 - M/(L_1L_2)^{1/2}\}/(t\tau) \text{ is greater than } 1/t^2,$$

or M is less than $(L_1L_2)^{1/2}\{1 - \tau/4t\}$.

If there is only one set of oscillations

$$4t \text{ must be less than } \tau/\{1 - M/(L_1L_2)^{1/2}\}$$

but greater than $\tau/\{1 + M/(L_1L_2)^{1/2}\}$.

Hence M must be greater than $\pm(L_1L_2)^{1/2}\{1 - \tau/4t\}$.

Finally there are no oscillations at all, when

$$M \text{ is less than } (L_1L_2)^{1/2}\{\tau/4t - 1\}.$$

No condenser in the secondary circuit. Let us now suppose that there is no condenser in the secondary circuit. In this case we can write $\tau_2 = K_2R_2 = $ infinity, as an infinitely great condenser in the circuit is simply equivalent to a short circuit. Thus the auxiliary equation (10) becomes

$$am^3 + bm^2 + cm + d = 0.$$

IX] TWO SETS OF OSCILLATIONS 267

Writing $m = z - b/3a$ and substituting, this equation becomes

$$z^3 + qz + r = 0,$$

where $q = c/a - b^2/(3a^2)$ and $r = d/a - bc/(3a^2) + 2b^3/(27a^3)$.

This equation can be solved by Cardan's method and the result shows that there are two imaginary roots when $q^3/27 + r^2/4$ is greater than zero.

We find, on substituting, that this is equivalent to the inequality,

$$(bc - 9ad)^2 \text{ greater than } 4(b^2 - 3ac)(c^2 - 3bd).$$

In this equation $a = \sigma$, $b = 1/t_1 + 1/t_2$, $c = 1/t_1 t_2 + 1/t_1 \tau_1$, and $d = 1/(t_1 t_2 \tau_1)$. The inequality above is the complete criterion. It is also, however, useful to remember that oscillations ensue when $3ac$ is greater than b^2 and when $3bd$ is greater than c^2. Substituting for a, b, c and d their values, we see that oscillations ensue when

$$\sigma \text{ is greater than } (t_1 + t_2)^2 / \{3t_1 t_2 (1 + t_2/\tau_1)\},$$

and when $3(t_1 + t_2)$ is greater than $\tau_1 (1 + t_2/\tau_1)^2$.

A proof of the following rules for determining the nature of
Complete criterion for the roots of a biquadratic. the roots of a biquadratic equation will be found in treatises on the *Theory of Equations*. For the sake of brevity the following symbols are usually employed :

$$H = ac/6 - b^2/16, \quad I = ae - bd/4 + c^2/12, \quad \Delta = I^3 - 27J^2,$$

where $\quad J = ace/6 + bcd/48 - (ad^2 + eb^2)/16 - c^3/216.$

I. If Δ is negative, equation (10) has two real and two imaginary roots.

II. If Δ is zero, the biquadratic has two equal roots. If in addition both I and J are zero, it has three equal roots.

III. When Δ is positive, the roots are either all real or all imaginary. In this case if H and $2HI - 3aJ$ are both negative the roots are all real. but if either of them be positive the roots are all imaginary.

Let us now suppose that we have found that Δ is positive and
Two sets of oscillations. that either H or $2HI - 3aJ$ is also positive. In this case the roots of the equation are all imaginary and

so they are of the forms $x_1 \pm \iota y_1$ and $x_2 \pm \iota y_2$, and hence

$$x_1 + x_2 = - b/2a,$$

and

$$x_1^2 + y_1^2 + x_2^2 + y_2^2 + 4x_1x_2 = c/a.$$

Since we have shown that x_1 and x_2 have the same sign, we see that the greatest possible value of $4x_1x_2$ is $b^2/4a^2$. When therefore $b^2/4a^2$ can be neglected compared with c/a we may write

$$x_1^2 + y_1^2 + x_2^2 + y_2^2 = c/a.$$

We also have

$$(x_1^2 + y_1^2)(x_2^2 + y_2^2) = e/a,$$

and hence solving these equations we get

$$x_1^2 + y_1^2 = c/2a + (c^2 - 4ae)^{1/2}/2a \quad \ldots\ldots\ldots\ldots(16),$$

and

$$x_2^2 + y_2^2 = c/2a - (c^2 - 4ae)^{1/2}/2a \quad \ldots\ldots\ldots\ldots(17).$$

Again, since

$$2x_1(x_1^2 + y_1^2) + 2x_2(x_2^2 + y_2^2) = - d/a,$$

and

$$2x_1 \qquad\qquad + 2x_2 \qquad\qquad = - b/a,$$

we get the following approximate formulae

$$x_1 = - b/4a + (bc - 2ad)/\{4a(c^2 - 4ae)^{1/2}\}$$

and

$$x_2 = - b/4a - (bc - 2ad)/\{4a(c^2 - 4ae)^{1/2}\},$$

and hence, knowing x_1 and x_2, y_1 and y_2 are determined by (16) and (17).

In some practical cases the circuits are adjusted so that $K_1L_1 = K_2L_2$. In these cases $d = be^{1/2}$ and $c = 2e^{1/2} + 1/t_1t_2$. If in addition $2L_2/K_1$ be much greater than R_1R_2, we have $c = 2e^{1/2}$ approximately. Hence substituting in the formulae given above, we easily get that

$$x_1 = - \frac{1}{4}\left(\frac{R_1}{L_1} + \frac{R_2}{L_2}\right)\Big/\left\{1 + \frac{M}{(L_1L_2)^{1/2}}\right\}$$

and

$$x_2 = - \frac{1}{4}\left(\frac{R_1}{L_1} + \frac{R_2}{L_2}\right)\Big/\left\{1 - \frac{M}{(L_1L_2)^{1/2}}\right\}.$$

These formulae are sometimes used in practical work. It must be remembered however that they can only be used when $b^2/4a^2$ can be neglected compared with c/a, when R_1R_2 is negligible compared with $2L_2/K_1$, and when in addition the criterion for four imaginary roots is satisfied. We cannot assume, for example, that a is practically zero, as the first of these conditions is obviously not true in this case.

Let us now suppose that an alternating electromotive force,
Forced oscillations. $E \sin \omega t$, is impressed on the primary circuit. Our equations are easily expressed as follows:

$$(L_1 D^2 + R_1 D + 1/K_1)\, i_1 + M D^2 i_2 = \omega E \cos \omega t$$

and

$$(L_2 D^2 + R_2 D + 1/K_2)\, i_2 + M D^2 i_1 = 0.$$

Eliminating i_1 between these equations, we get

$$i_2 = \frac{(M/L_1 L_2)\, \omega^3 E \cos \omega t}{a D^4 + c D^2 + e + D\,(b D^2 + d)},$$

where a, b, c, d and e have the same meanings as before. Noticing that $D^2 = -\omega^2$, and proceeding in the ordinary way, we get

$$i_2 = -\frac{(M/L_1 L_2)\, \omega^3 E \sin(\omega t - \alpha)}{\{(a\omega^4 - c\omega^2 + e)^2 + (b\omega^2 - d)^2\, \omega^2\}^{1/2}} \dots \dots (18),$$

where

$$\tan \alpha = \frac{a\omega^4 - c\omega^2 + e}{\omega\,(b\omega^2 - d)}\,.$$

Hence, if the frequency vary, the amplitude of the forced oscillation has a maximum or a minimum value when

$$(a\omega^4 - c\omega^2 + e)\,(a\omega^4 + c\omega^2 - 3e) + 2d\omega^2\,(b\omega^2 - d) = 0.$$

Let us first suppose that there is no magnetic leakage between the two coils. In this case $a = 0$, and the amplitude attains a minimum value when

$$\omega^2 = \frac{2ce - d^2}{c^2 - 2bd} + \frac{\{c^2 e^2 - 2de\,(2cd - 3be) + d^4\}^{1/2}}{c^2 - 2bd},$$

and has a maximum value when

$$\omega^2 = \frac{2ce - d^2}{c^2 - 2bd} - \frac{\{c^2 e^2 - 2de\,(2cd - 3be) + d^4\}^{1/2}}{c^2 - 2bd}.$$

If in addition the resistance of the coils can be neglected, so that b and d may be put equal to zero, the amplitude becomes very large when $\omega^2 = e/c$, and has a minimum value when $\omega^2 = 3e/c$.

Let us next suppose that there is magnetic leakage but that the resistances of the coils are negligible. In this case the amplitude attains maximum values when $a\omega^4 - c\omega^2 + e = 0$, that is, when $\omega^2 = c/2a \pm (c^2 - 4ae)^{1/2}/2a$. It has a minimum value when $a\omega^4 + c\omega^2 - 3e = 0$. In this case $\omega^2 = (c^2 + 12ae)^{1/2}/2a - c/2a$.

In order to obtain the complete expression for i_2 we have to add on to the particular integral given by (18) the general solution $\Sigma A \epsilon^{mt}$, where m is a root of equation (10). Similarly we can find the value of i_1. The constants are then determined from the initial conditions. In practice the resistances are finite and so the terms corresponding to the general solution are rapidly damped out, so that after a fraction of a second the value of i_2 is given accurately by (18).

A damped E.M.F. applied to an oscillatory circuit. In certain systems used in radio-telegraphy we have at the receiving station a long vertical or horizontal conducting wire called the antenna. This is connected with one terminal of the primary of an air-core transformer the other terminal of which is connected with the earth through a condenser. The secondary circuit of the air-core transformer is shunted by a variable condenser. This circuit is adjusted until resonance ensues. In parallel with this variable condenser is a crystal or other detector and another condenser across the terminals of which there is a telephone. The secondary circuit is adjusted by means of the variable condenser until the sounds heard in the telephone are loudest. The damped train of waves coming through the ether from the signalling station sets up oscillations in the primary and hence also in the secondary of the transformer. Owing to the unilateral conductivity of the crystal a pulse of electricity sufficiently strong to cause a sound passes through the telephone. In the case of long distance signalling, sounds are only heard when the 'tuning' is good, that is, when the secondary circuit is adjusted so that resonance is nearly complete.

As a study of this problem would take up too much of our space, we shall merely discuss the case when the secondary circuit consists of a constant resistance in series with a condenser. We shall also suppose that the E.M.F. applied to the secondary circuit can be represented by $E\epsilon^{-\lambda t}\sin \omega t$.

Making these assumptions, we have

$$Ri + L\frac{\partial i}{\partial t} + \frac{\int i \partial t}{K} = E\epsilon^{-\lambda t}\sin \omega t \quad \ldots\ldots\ldots\ldots(19).$$

We have already seen that when $4L$ is greater than KR^2 the complementary function of this differential equation is

$$A\epsilon^{-at} \sin \omega_1 t + B\epsilon^{-at} \cos \omega_1 t,$$

where $a = R/2L$ and $\omega_1 = (1/LK - R^2/4L^2)^{1/2}$. Let $I\epsilon^{-\lambda t} \sin(\omega t - \gamma)$ be a particular integral. Substituting this value of i in (19) and dividing out by $I\epsilon^{-\lambda t}$, we get

$$R \sin(\omega t - \gamma) - L(\omega^2 + \lambda^2)^{1/2} \sin(\omega t - \gamma - \beta) - \frac{\sin(\omega t - \gamma + \beta)}{K(\omega^2 + \lambda^2)^{1/2}}$$
$$= (E/I) \sin \omega t,$$

where $\tan \beta = \omega/\lambda$. As this equation must be true for all values of t, it is true when $\omega t = \gamma$, and when $\omega t = \gamma + \pi/2$.

Hence substituting, we get

$$\frac{E}{I} \sin \gamma = \left\{L - \frac{1}{K(\omega^2 + \lambda^2)}\right\} \omega \quad\dots\dots\dots\dots\dots\dots(20),$$

and $$\frac{E}{I} \cos \gamma = \left\{L - \frac{1}{K(\omega^2 + \lambda^2)}\right\} \lambda + 2L(a - \lambda) \quad\dots\dots\dots(21).$$

Therefore

$$\tan \gamma = \frac{[L - 1/\{K(\omega^2 + \lambda^2)\}]\,\omega}{[L - 1/\{K(\omega^2 + \lambda^2)\}]\lambda + 2L(a - \lambda)} \quad\dots\dots(22),$$

and $(E/I)^2 = L^2(\omega^2 + \lambda^2) + \{1 + 4LK\lambda(\lambda - a)\}/\{K^2(\omega^2 + \lambda^2)\}$
$$- 2L/K - 4L^2 a(\lambda - a)$$
$$= [L(\omega^2 + \lambda^2)^{1/2} - \{1 + 4LK\lambda(\lambda - a)\}^{1/2}/\{K(\omega^2 + \lambda^2)^{1/2}\}]^2$$
$$+ (2L/K)[\{1 + 4LK\lambda(\lambda - a)\}^{1/2} - 1] - 4L^2 a(\lambda - a)$$
$$\dots(23).$$

We see that if ω vary and if a be less than

$$\lambda + (1 - L^2 K^2 \lambda^4)/(4LK\lambda),$$

E/I has a minimum, and therefore I has a maximum, value when

$$\omega^2 + \lambda^2 = (1/LK)\{1 + 4LK\lambda(\lambda - a)\}^{1/2} \quad\dots\dots\dots(24).$$

But if a be equal to or greater than $\lambda + (1 - L^2 K^2 \lambda^4)/(4LK\lambda)$, I continually diminishes as ω increases.

The complete solution of (19) is

$$i = I\epsilon^{-\lambda t} \sin(\omega t - \gamma) + A\epsilon^{-at} \sin \omega_1 t + B\epsilon^{-at} \cos \omega_1 t,$$

where A and B are constants to be determined by the initial conditions. Let us suppose, for instance, that when t is zero both

i and $\int i\partial t$ are zero. In this case we see from (19) that $\partial i/\partial t$ must also be zero initially.   Hence

$$0 = -I \sin \gamma + B$$

and $$0 = I (\omega \cos \gamma + \lambda \sin \gamma) + A\omega_1 - B\alpha.$$

Therefore $$B = I \sin \gamma$$

and $$A = (I/\omega_1)\{(\alpha - \lambda) \sin \gamma - \omega \cos \gamma\}.$$

Hence the complete solution in this case is

$$i = I\epsilon^{-\lambda t} \sin (\omega t - \gamma) + (I/\omega_1) \{(\alpha - \lambda) \sin \gamma - \omega \cos \gamma\} \, \epsilon^{-at} \sin \omega_1 t$$
$$+ I \sin \gamma \epsilon^{-at} \cos \omega_1 t \ldots\ldots(25).$$

We shall now compute the heat H developed in the circuit when $\omega^2 + \lambda^2 = \omega_1^2 + \alpha^2 = 1/LK$, as this is an important practical case and the final formula is simple. We see by (20) that γ is now zero, and by (21) that $I = E/\{2L(\alpha - \lambda)\}$.

Hence by (25)

$$i = \frac{E}{2L(\alpha - \lambda)} \left\{ \epsilon^{-\lambda t} \sin \omega t - \frac{\omega}{\omega_1} \epsilon^{-at} \sin \omega_1 t \right\} \ldots\ldots(26).$$

Noticing that $\int_0^\infty \epsilon^{-\lambda t} \cos \omega t \partial t = \lambda/(\lambda^2 + \omega^2)$,

we get $H = R \displaystyle\int_0^\infty i^2 \partial t$

$$= \frac{E^2 R \omega^2 (\alpha + \lambda)}{16 L^2 (\alpha - \lambda)^2 (\lambda^2 + \omega^2) \lambda \alpha} \cdot \frac{(\omega^2 + \lambda^2 - \lambda \alpha)^2 - \omega^2 \omega_1^2}{(\omega^2 + \lambda^2 + \lambda \alpha)^2 - \omega^2 \omega_1^2}$$

$$= \frac{E^2 R K \omega^2}{16 L \, \alpha \lambda (\alpha + \lambda)}$$

$$= \frac{E^2 R (1 - LK\lambda^2)}{16 L^2 \, \alpha \lambda (\alpha + \lambda)}.$$

We see that the smaller the values of α and λ the greater the value of H. In the particular case when $\lambda = \alpha = R/(2L)$, we have

$$H = \frac{E^2 (4L - KR^2)}{16 R^2}.$$

To find the form that equation (26) assumes when $\lambda = \alpha$, we notice that

$$(\omega/\omega_1) \epsilon^{-at} \sin (\omega_1 t)$$
$$= \{1 + (\lambda^2 - \alpha^2)/\omega^2\}^{-1/2} \epsilon^{-at} \sin [\{1 + (\lambda^2 - \alpha^2)/\omega^2\}^{1/2} \, \omega t].$$

When the squares and higher powers of $(\lambda^2 - \alpha^2)/\omega^2$ can be neglected compared with unity, we may write

$$\{1 + (\lambda^2 - \alpha^2)/\omega^2\}^{-1/2} = 1 - (\lambda^2 - \alpha^2)/(2\omega^2), \text{ approx.},$$

and $\sin[\{1 + (\lambda^2 + \alpha^2)/\omega^2\}^{1/2} \omega t]$

$$= \sin \omega t + \{(\lambda^2 - \alpha^2) t/(2\omega)\} \cos \omega t, \text{ approx.}$$

Noticing also that

$$(\epsilon^{-\lambda t} - \epsilon^{-\alpha t})/(\alpha - \lambda) = t\epsilon^{-\alpha t},$$

when $\lambda = \alpha$, we get, by substituting in (26),

$$i = \{E/(2L\omega)\} \{(\alpha \cos \omega t + \omega \sin \omega t) t - (\alpha/\omega) \sin \omega t\} \epsilon^{-\alpha t}.$$

REFERENCES

W. S. BURNSIDE and A. W. PANTON, *Theory of Equations.*

P. DRUDE, *Physik des Aethers.*

J. A. FLEMING, *The Principles of Electric Wave Telegraphy.*

J. ZENNECK, *Elektromagnetische Schwingungen und Drahtlose Telegraphie.*

L. COHEN, *Calculation of Alternating Current Problems.*

P. DRUDE, 'Über induktive Erregung zweier elektrischer Schwingungskreise, etc.' *Ann. der Physik*, Vol. 13, p. 512, 1904.

G. BENISCHKE, 'Der Resonanztransformotor.' *Elektrotechnische Zeitschrift*, p. 25, 1907.

J. BETHENOD, 'Über den Resonanztransformotor.' *Jahrbuch der Drahtlosen Telegraphie*, Vol. 1, p. 534, 1907.

L. COHEN, 'Theory of Coupled Circuits having Distributed Inductance and Capacity.' *Bulletin of the Bureau of Standards*, Vol. 5, p. 511, 1909.

J. A. FLEMING, 'Some Oscillograms of Condenser Discharges and a Simple Theory of Coupled Oscillatory Circuits.' *Proc. Phys. Soc.* Vol. 25, p. 217, 1913.

CHAPTER X

The power factor. When the power factor is unity, the volt and ampere waves are similar. The maximum value of the power factor is unity. Geometrical interpretation of the power factor. Definition of phase difference. Time lag. Numerical examples. Zero power factor. Watt E.M.F. and wattless E.M.F. Impedance. Reactance. Watt current and wattless current. References.

IF e represent the instantaneous value of the P.D. across a circuit and i the instantaneous value of the current in it,

The power factor. then ei gives the instantaneous value of the watts expended in the circuit. The mean value of ei over a whole period gives us the rate at which work is being done in the circuit, or the power (W) being expended in it. The value of ei is sometimes negative for a fraction of a period and hence its mean value can be very small. Now the reading V of the voltmeter gives us the R.M.S. value of e, and the reading A of the ammeter gives us the R.M.S. value of i, but VA will not in general give us the mean value of ei. It is found convenient in practice to call the ratio of W to VA the power factor of the circuit, and we shall show that the maximum possible value of the power factor is unity. It may therefore be denoted by $\cos\phi$, where ϕ is a certain auxiliary angle of great use in graphical calculations. In mathematical symbols the power factor is defined by the equation

$$\cos\phi = \frac{\dfrac{1}{T}\displaystyle\int_0^T ei\,\partial t}{\left\{\dfrac{1}{T}\displaystyle\int_0^T e^2\partial t \cdot \dfrac{1}{T}\int_0^T i^2\partial t\right\}^{\frac{1}{2}}} = \frac{\displaystyle\int_0^T ei\,\partial t}{\left\{\displaystyle\int_0^T e^2\partial t \cdot \int_0^T i^2\partial t\right\}^{\frac{1}{2}}} \quad \ldots\ldots(1).$$

To prove this we notice, from the meaning of the integral sign, that

When the power factor is unity, the volt and ampere waves are similar.

$$\frac{1}{T}\int_0^T e^2 \partial t = \mathop{L}_{n=\infty} \frac{e_1^2 + e_2^2 + e_3^2 + \ldots + e_n^2}{n},$$

where $e_1, e_2, \ldots, e_n$ are equidistant ordinates of the P.D. wave.

Dividing up the current wave into the same number of ordinates, we see that, if the power factor equals unity, then from (1)

$$\frac{(e_1 i_1 + e_2 i_2 + \ldots + e_n i_n)^2}{(e_1^2 + e_2^2 + \ldots + e_n^2)(i_1^2 + i_2^2 + \ldots + i_n^2)} = 1;$$

and thus $(e_1 i_1 + e_2 i_2 + \ldots)^2 - (e_1^2 + e_2^2 + \ldots)(i_1^2 + i_2^2 + \ldots) = 0,$

and $(e_1 i_2 - e_2 i_1)^2 + (e_1 i_3 - e_3 i_1)^2 + \ldots = 0.$

Now since the square of a number is always positive, every term on the left-hand side is positive; and since the sum of them is zero, every term must be zero.

Hence $\quad e_1 i_2 - e_2 i_1 = 0 \; ; \quad e_1 i_3 - e_3 i_1 = 0 \; ; \quad$ etc.

Thus $\qquad \dfrac{e_1}{i_1} = \dfrac{e_2}{i_2} = \dfrac{e_3}{i_3} = \ldots = \dfrac{e_n}{i_n}.$

Therefore, at every instant, the ratio of the volts to the amperes is constant, which proves the theorem.

Again since $(e_1 i_2 - e_2 i_1)^2 + (e_1 i_3 - e_3 i_1)^2 + \ldots$ is always greater than zero except when each term equals zero, it easily follows by going through the above proof backwards that the power factor is less than unity

The maximum value of the power factor is unity.

except in the very particular case when the current and P.D. waves are the same curve drawn on different scales. It is convenient to call these waves similar waves. Hence if the power factor of a circuit is unity, the volt and ampere waves are similar, vanishing at the same instant and attaining all their maximum and minimum values at the same instants.

We may show in a similar manner that the power factor can not be less than -1. When the power factor is -1, e is negative when i is positive and *vice versâ*, but, as before, the ratio e to i is constant. Hence in this case also the waves are similar waves. They are drawn however on opposite sides of the axis.

18—2

It will be seen that $\cos \phi$ can only be equal to its limiting values, $+1$ and -1, in very special cases. As the angle ϕ is mainly useful in the application of graphical methods to alternating current problems, it is convenient to make the limitation that ϕ lies between $0°$ and $180°$.

If a simple circuit is absolutely non-inductive and has no capacity, then

$$e = Ri,$$

where R is the resistance of the circuit. Substituting this value of e in (1) we see that the power factor equals unity. This could be proved directly as follows.

We have $\qquad\qquad e^2 = R^2 i^2,$

and thus $\qquad\qquad V^2 = R^2 A^2,$

and $\qquad\qquad\qquad V = RA.$

Also $\qquad\qquad\qquad ei = Ri^2.$

Therefore the mean value of $ei =$ the mean value of Ri^2,

and hence $\qquad\qquad W = RA^2$

$$= VA,$$

and the power factor $= W/VA = 1$.

By defining the power factor as the cosine of a certain angle ϕ, **Geometrical interpretation of the power factor.** we are able in many cases to give a geometrical interpretation to the quantities involved, and can easily prove many algebraical relations between them by the help of known theorems in algebra and trigonometry.

We will consider a simple case.

Fig. 62.   Inductive resistance in series with non-inductive resistances.

Suppose that B_2C (Fig. 62) is part of an alternating current circuit. Let the resistances between BB_1 and B_1B_2 be R_1 and R_2 respectively, and suppose them non-inductive. Let e_1, e' be the instantaneous values of the P.D. between B_1 and C and between B and C respectively, and let the instantaneous value of the current be i.

Then evidently we have always

$$e' = e_1 - R_1 i,$$

thus
$$e'^2 = e_1{}^2 + R_1{}^2 i^2 - 2R_1 e_1 i.$$

Hence, by taking mean values for a whole period,

$$V'^2 = V_1{}^2 + R_1{}^2 A^2 - 2R_1 W \dots\dots\dots\dots(2),$$

where V', V_1 and A are the effective values of the volts and amperes, and W is the mean value of $e_1 i$, that is, the power being expended in the circuit $B_1 C$. Denoting the power factor of this circuit by $\cos \phi_1$, we have

$$W = V_1 A \cos \phi_1,$$

and substituting in (2) we get

$$V'^2 = V_1{}^2 + R_1{}^2 A^2 - 2R_1 V_1 A \cos \phi_1.$$

If we now construct a triangle CBB_1 (Fig. 63) whose sides CB, BB_1 and $B_1 C$ are V', $R_1 A$, and V_1 respectively, we see by trigonometry that the angle $CB_1 B$ is ϕ_1. The cosine of $CB_1 B$ is the power factor of the circuit CB_1 (Fig. 62).

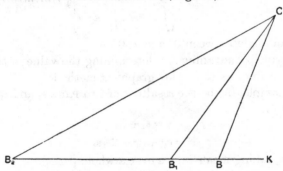

Fig. 63. This diagram shows graphically the magnitudes and phases of the potential differences in the circuit shown in Fig. 62.

Similarly producing BB_1 to B_2 (Fig. 63) and making $B_1 B_2$ equal to $R_2 A$, it is easy to show that the cosine of the angle $CB_2 B$ is the power factor of the circuit CB_2 (Fig. 62) and that CB_2 (Fig. 63) is the effective value of the P.D. between C and B_2.

If V_2 denote this P.D. and ϕ_2 denote the angle $CB_2 B$, then by trigonometry

$$V_2 \sin \phi_2 = V_1 \sin \phi_1.$$

If we denote the angle CBK by ϕ', then

$$V_1 \cos \phi_1 - V' \cos \phi' = R_1 A,$$

and thus $$V_1 A \cos \phi_1 - V' A \cos \phi' = R_1 A^2.$$

Since $V_1 A \cos \phi_1$ is the work done between B_1 and C, it follows from this equation that $V' A \cos \phi'$ is the work done in the circuit BC (Fig. 62). Hence $\cos \phi'$ is the power factor of the circuit BC. Fig. 63, then, can be used to prove many relations between the volts, amperes and watts in the circuit. Thus to define the power factor as the cosine of an angle is a real help in understanding the relations of the various quantities involved.

It is convenient to call the angle whose cosine is the power factor the phase difference between the waves of P.D. and current. More generally, if e_1 and e_2 be two periodic functions of the same frequency, and ϕ be their phase difference, then

Definition of phase difference.

$$\cos \phi = \frac{\int_0^T e_1 e_2 \partial t}{\left\{ \int_0^T e_1^2 \partial t \cdot \int_0^T e_2^2 \partial t \right\}^{\frac{1}{2}}} \quad \ldots\ldots\ldots\ldots\ldots(3),$$

ϕ being an angle between $0°$ and $180°$.

The principal advantage of determining the value of this angle is that it enables us to employ graphical methods.

For example, if e be the resultant of two P.D.'s e_1 and e_2, then at every instant

$$e = e_1 + e_2,$$

and $$e^2 = e_1^2 + e_2^2 + 2e_1 e_2.$$

Hence, taking mean values over a whole period,

$$V^2 = V_1^2 + V_2^2 + 2 V_1 V_2 \cos \phi \quad \ldots\ldots\ldots\ldots(4),$$

where ϕ is given by equation (3). We see then that V, the effective P.D. of the resultant, is the diagonal of the parallelogram constructed on V_1 and V_2 as adjacent sides when the angle between them is ϕ.

By the time lag of two periodic functions of the same frequency we mean the interval that elapses between the instants when they pass through their zero values in the positive direction. The angle of time lag may be defined

Time lag.

as the angle described in an interval equal to the time lag by a uniformly rotating radius which makes one revolution in a time equal to the period of the alternating current.

If t_1, t_2 be the epochs at which e_1 and e_2 pass through zero in the positive direction, we can write

$$\omega t_1 = \alpha_1 \text{ and } \omega t_2 = \alpha_2.$$

The time lag between e_1 and e_2 is $t_1 - t_2$, and the angle of time lag is $\omega (t_1 - t_2)$, that is $\alpha_1 - \alpha_2$.

If $e_1 = E_1 \sin (\omega t - \alpha_1)$ and $e_2 = E_2 \sin (\omega t - \alpha_2)$, by substituting in (3) we find that

$$\cos \phi = \cos (\alpha_1 - \alpha_2),$$

and therefore $$\pm \phi = \alpha_1 - \alpha_2.$$

In this case, then, the phase difference is numerically equal to the angle of time lag between e_1 and e_2, and, when e_1 and e_2 vanish at the same instant, then ϕ is zero or 180°. When e_1 and e_2 are not similar curves, then for no value of the time lag is the phase difference zero or 180°. This will be best understood by solving a few numerical examples.

In order to simplify the calculations we will suppose that one Numerical examples. of the curves is a sine curve, and will find the phase difference between it and the curves of which the positive halves are shown in Fig. 64.

All the curves drawn give an effective voltage of 50, but the absolute value of the voltage has nothing to do with the phase difference, which depends only on the class of curve and its position relatively to the sine curve. If one of the curves represents a F.D. curve and the other the current curve to which it gives rise, then the cosine of the angle of phase difference between them will give the power factor.

The equations to the first halves of the curves shown in Fig. 64 are

     (a) *Rectangle*, $e = V,$

     (b) *Parabola*, $e = (4\sqrt{30}/T^2)\, V\, \{(T/2)\, t - t^2\},$

     (c) *Sine Curve*, $e = \sqrt{2}\, V \sin (2\pi/T)\, t,$

     (d) *Triangle*, $e = 4\sqrt{3}\, V\, (t/T),$

     (e) *Inverted Parabolas*, $e = 16\sqrt{5}\, V\, (t/T)^2,$

     (f) *Inverted Cubics*, $e = 64\sqrt{7}\, V\, (t/T)^3.$

All these curves have the same effective voltage V.
We shall find their phase differences with the curves
$$i = I \sin (2\pi/T)\, t \quad \text{and} \quad i = I \sin \{(2\pi/T)\, t - \alpha\}.$$

Curve	Maximum value of e	Height of c.g.	$\cos \phi$ with $I \sin (2\pi/T)\, t$	ϕ in degrees	$\cos \phi$ with $I \sin \{(2\pi/T)\, t - \alpha\}$
(a)	V	$0\cdot5\,V$	$0\cdot9003$	$25\cdot8$	$0\cdot9003 \cos \alpha$
(b)	$1\cdot370\,V$	$0\cdot5476\,V$	$0\cdot9995$	$2\cdot2$	$0\cdot9995 \cos \alpha$
(c)	$1\cdot414\,V$	$0\cdot5552\,V$	$1\cdot0000$	0	$\cos \alpha$
(d)	$1\cdot732\,V$	$0\cdot5773\,V$	$0\cdot9928$	$6\cdot75$	$0\cdot9928 \cos \alpha$
(e)	$2\cdot236\,V$	$0\cdot6708\,V$	$0\cdot9322$	$21\cdot2$	$0\cdot9322 \cos \alpha$
(f)	$2\cdot646\,V$	$0\cdot7560\,V$	$0\cdot8628$	$30\cdot4$	$0\cdot8628 \cos \alpha$

We might also have calculated the phase difference between any two of the curves shown in the figure. For example, the phase difference ϕ between the rectangle (a) and the very peaky curve (f) is $48\cdot6$ degrees, and the power factor ($\cos \phi$) is $0\cdot6613$.

We have proved at the beginning of this chapter that the power factor can only be unity, and consequently the phase difference can only be zero, when the ratio of e to i is constant throughout the whole wave. A first essential condition for zero phase difference is, thus, that e and i should both vanish at the same instant. This makes possible the second essential condition, namely, that the ratio of the ordinates of the two waves should be constant.

Each of the curves shown in Fig. 64 is the first half of a symmetrical alternating curve. In such a curve the curve from $t = 0$ to $T/2$ is symmetrical about the ordinate corresponding to $t = T/4$, and the curve from $t = T/2$ to T is the exact counterpart of the curve from $t = 0$ to $T/2$, except that it lies on the opposite side of the axis. We may express these conditions by the equations
$$f(t) = f(T/2 - t) = -f(T/2 + t) \dots\dots\dots(c).$$
Since the curve is a continuous one it follows that $f(t) = 0$ when $t = 0$ and when $t = T$ and also when $t = T/2$.

We can now show that, if ϕ_0 be the phase difference when the time lag is zero, and if α be the angle of lag between a symmetrical alternating curve and a sine curve, then
$$\cos \phi = \cos \phi_0 \cos \alpha \dots\dots\dots\dots(5).$$

To prove this, we have

$$\int_0^{T/2} f(t) \sin\{(2\pi/T)t - \alpha\}\, \partial t = \int_0^{T/2} f(t) \sin\{(2\pi/T)t\}\, \partial t \cos\alpha$$

$$-\int_0^{T/2} f(t) \cos\{(2\pi/T)t\}\, \partial t \sin\alpha.$$

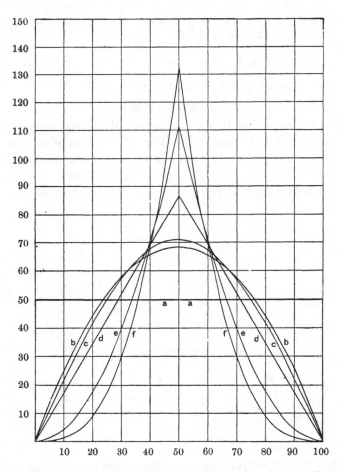

Fig. 64. Voltage curves, each of which has an effective value of 50. (*a*) Rectangle. (*b*) Parabola. (*c*) Sine curve. (*d*) Triangle. (*e*) Inverted parabolas. (*f*) Inverted cubic curves.

Putting $t = T/2 - x$, we get, from (c),

$$\int_0^{T/2} f(t) \cos \{(2\pi/T)\, t\}\, \partial t = \int_{T/2}^0 f(T/2 - x) \cos \{(2\pi/T)x\}\, \partial x$$

$$= -\int_0^{T/2} f(t) \cos \{(2\pi/T)\, t\}\, \partial t.$$

The last integral therefore vanishes. If the limits in the last integral had been $T/2$ to T, the integral would have vanished on account of the symmetry of $f(t)$ about the ordinate corresponding to $t = (3/4)\, T$. We thus obtain

$$\int_0^T f(t) \sin \{(2\pi/T)\, t - \alpha\}\, \partial t = \cos \alpha \int_0^T f(t) \sin \{(2\pi/T)\, t\}\, \partial t \ \ldots(d).$$

Hence from (d) with the aid of (1) or (3)

$$\cos \phi = \cos \phi_0 \cos \alpha.$$

It is instructive to give a graphical interpretation to this formula. Let OAB and OBC (Fig. 65) be two planes at right angles to one another, and let the angle AOB equal ϕ_0, and the angle BOC equal α, then the angle AOC will be ϕ.

Draw BC (Fig. 65) perpendicular to OC and join AC.

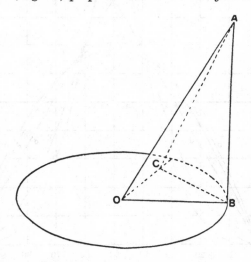

Fig. 65. Time lag and phase difference. For all positions of OC, the angle BOC gives the angle of time lag and the angle AOC gives the angle of phase difference between the two periodic quantities. The minimum value of the angle of phase difference is the angle AOB.

Then $\qquad OA^2 = OB^2 + AB^2$

$\qquad\qquad\qquad = OC^2 + CB^2 + AB^2$

$\qquad\qquad\qquad = OC^2 + AC^2.$

Therefore the angle OCA is a right angle.

Now $\qquad\qquad OC = OA \cos AOC,$

and $\qquad\qquad OC = OB \cos \alpha$

$\qquad\qquad\qquad = OA \cos \phi_0 \cos \alpha.$

Thus $\qquad \cos AOC = \cos \phi_0 \cos \alpha ;$

and therefore by (5) the angle AOC is the phase difference.

This gives us a graphical construction for the phase difference between a symmetrical wave and any sine wave. It is to be noticed that this construction is in three dimensions. We shall return to this method of representing phase differences in Chapter xII.

In practice it is the exception to have both curves symmetrical. Suppose that the current is a sine curve, and that there is no time lag between it and the P.D. curve. Then if we consider a family of P.D. curves all of the same height (see Fig. 37, Chap. IV), a little consideration will show that the power factor is a maximum when the peak of the wave occurs at the quarter period. Suppose now that the current is lagging, and that the P.D. wave has its maximum value before the quarter period. Then it is evident that the power factor will be less than it would be if the P.D. wave were symmetrical. The maximum value of the power factor in this case occurs when the peak of the P.D. wave is in the second quarter.

The curves considered above have only one maximum ordinate during the half wave; they might however have several maxima and minima ordinates. Suppose for example the P.D. and current were given by the curves in Fig. 66. In this case the power factor would be 0·75 and the phase difference 41·4 degrees. It is fairly obvious that if the P.D. have a minimum value when the current has its maximum value, then the power factor will be low and the phase difference large, even if there be no time lag between the current and the P.D.

We shall now consider what happens when the power factor

Zero power factor. vanishes. We see from equation (1) that we have in this case

$$\int_0^T ei\, \partial t = 0.$$

Since in practice when the time is increased by $T/2$ the values e_1 and i_1 become $-e_1$ and $-i_1$ respectively, we see that

$$\int_0^{T/2} ei\, \partial t = \int_{T/2}^T ei\, \partial t;$$

and therefore

$$\int_0^T ei\, \partial t = 2\int_0^{T/2} ei\, \partial t.$$

Let us suppose that the curve e only cuts the axis at points which are at distances $T/2$ apart, and let us also make the same supposition about i.

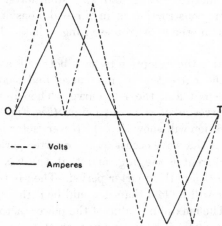

Fig. 66.   Phase difference 41·4°.   Power factor 0·75.

When the time lag between e and i is zero, then ei is positive over the half period, and therefore the integral $\int_0^T ei\, \partial t$ is positive. When the time lag between e and i is $T/2$, ei is negative over the half period and hence the integral is negative. When the time lag τ is less than $T/2$, then from 0 to τ, ei is negative and from τ to $T/2$, ei is positive. We see then that, as τ increases from zero to $T/2$, $\int_0^T ei\partial t$ gradually changes from negative to positive.

Hence $\int_0^T ei\partial t$ is a continuous function of τ, and therefore for some value of τ the integral will vanish. It is therefore always possible to get a power factor of zero with voltage and current waves of any shape by giving to the time lag between them a proper value.

In the particular case when both waves are symmetrical, the value of this time lag is a quarter of a period. This may be proved graphically. An analytical proof can be given as follows.

Let $f(t)$ and $F(t + T/4)$ be the functions which represent the voltage and current waves respectively. Then, since they are symmetrical, alternating curves, both $f(t)$ and $F(t)$ satisfy (c), and thus

$$\int_0^{T/2} f(t) F(t + T/4)\, \partial t = \int_0^{T/2} f(T/2 - t) F(T/4 - t)\, \partial t;$$

putting $t = T - x$, this

$$= -\int_T^{T/2} f(x - T/2) F\{x - (3/4)\,T\}\, \partial x$$

$$= \int_T^{T/2} f(x) F(x + T/4)\, \partial x$$

$$= -\int_{T/2}^T f(t) F(t + T/4)\, \partial t.$$

Thus

$$\int_0^T f(t) F(t + T/4)\, \partial t = 0.$$

This proves that, when both the current and voltage waves are symmetrical, the power factor vanishes when the time lag is a quarter of a period.

We have seen (page 134) that the charging current of a condenser is

$$i = K\frac{\partial e}{\partial t},$$

where e is the potential difference at the condenser terminals. In this case

$$\int_0^T ei\, \partial t = K\int_0^T e\frac{\partial e}{\partial t}\, \partial t$$

$$= \left[(1/2)\,Ke^2\right]_0^T$$

$$= 0,$$

since e has the same value after an interval equal to the period. We see also that when i is zero $\partial e/\partial t$ is zero, and therefore e has a maximum or a minimum value. Assuming that e has only one maximum value in a period, we see that the time lag between e and i equals the time e takes to increase from zero to its maximum value.

Now e can have its maximum value at any time during the half period when it is positive. Therefore the time lag between e and i can have any value between 0 and $T/2$. In all cases however the power factor is zero. This proves that we can infer nothing concerning the value of the time lag from a mere knowledge that the power factor is zero.

If W be the wattmeter reading in an alternating current circuit, V and A the voltmeter and ammeter readings respectively, and ϕ the angle of phase difference between the volt and ampere waves, a quantity which we have seen depends only on the shapes of these waves and their relative positions, then

Watt E.M.F. and wattless E.M.F.

$$W = VA \cos \phi \quad \dots\dots\dots\dots\dots(6).$$

This follows from the definitions (1) and (3) above. Now if we suppose that V is resolved into two components whose values are $V \cos \phi$ and $V \sin \phi$ respectively, then these components are called the watt E.M.F. and the wattless E.M.F. respectively. We see from (6) that $V\cos\phi$ or the watt E.M.F. multiplied by the effective value of the current gives us the true mean power expended in the circuit.

If we have a simple inductive coil subjected to an alternating P.D., then

$$e = Ri + L\frac{\partial i}{\partial t},$$

and therefore

$$VA \cos \phi = RA^2$$

and

$$V^2 = R^2 A^2 + \frac{L^2}{T} \int_0^T \left(\frac{\partial i}{\partial t}\right)^2 \partial t.$$

Thus the effective value of Ri is $V \cos \phi$, and the effective value of $L(\partial i/\partial t)$ is $V \sin \phi$. Hence we can suppose the applied P.D. split up into two components Ri and $L(\partial i/\partial t)$, which have a phase difference of 90 degrees, and it is convenient to give names to the effective values of the two components. In the general case, however, when iron is present, we need to be careful when reasoning about the watt and wattless components of the E.M.F., as they do not seem to have much physical significance.

The impedance of a circuit is the ratio of the applied effective voltage to the effective value of the current produced.

Impedance. If V be the reading of a voltmeter placed across the circuit and A the reading of an ammeter in the circuit, then

$$Z = V/A,$$

where Z denotes the impedance.

When direct current is used, Z is simply the resistance R of the circuit. With alternating currents, Z may be a very complicated function, as it depends on the shape and frequency of the applied potential difference wave, on eddy currents, the position of neighbouring circuits, capacity, inductance and magnetic permeability. In the case of a simple coil whose self inductance is constant, we have (page 68), when the potential difference wave is sine shaped,

$$Z^2 = R^2 + \omega^2 L^2.$$

If the wave be not sine shaped, we can write (page 136)

$$Z^2 = R^2 + \alpha^2 \omega^2 L^2,$$

where α has its minimum value unity when the applied wave is sine shaped.

If V be the applied P.D., A the current and $\cos \phi$ the power factor of a circuit, then $(V \sin \phi)/A$ is called the reactance of the circuit. If there is no iron near the circuit and there are no eddy currents, then the reactance equals $\{(V/A)^2 - R^2\}^{\frac{1}{2}}$ or $\alpha\omega L$, and if in addition the applied P.D. be sine shaped, then the reactance is simply ωL.

The reactance may also be defined as the ratio of the wattless P.D. to the current.

Instead of supposing that the E.M.F. is resolved into two components, we may suppose that the current is so resolved. In this case $A \cos \phi$ is the watt current and $A \sin \phi$ is the wattless current.

Watt current and wattless current.

For an inductive coil

$$e = Ri + L\,\frac{\partial i}{\partial t}.$$

When e is of the form $E \sin \omega t$, then

$$i = \frac{R}{R^2 + \omega^2 L^2}\,E \sin \omega t - \frac{\omega L}{R^2 + \omega^2 L^2}\,E \cos \omega t.$$

The first term on the right-hand side, being in phase with e, may be considered as the watt component of the current, and the other term may be considered as the wattless component of the current.

In this case we have

$$\text{the watt current} = A \cos \phi = \frac{RV}{R^2 + \omega^2 L^2} = \frac{RA^2}{V},$$

and

$$\text{the wattless current} = A \sin \phi = \frac{\omega L V}{R^2 + \omega^2 L^2} = \frac{\omega L A^2}{V}.$$

If e were the parabolic wave whose equation (b) is given above (p. 279), then we can show that

$$A \cos \phi = \frac{V}{R}\left\{ 1 - 40\left(\frac{Lf}{R}\right)^2 + 1920\left(\frac{Lf}{R}\right)^4 - 7680\left(\frac{Lf}{R}\right)^5 \frac{\epsilon^{\frac{R}{2Lf}} - 1}{\epsilon^{\frac{R}{2Lf}} + 1}\right\}$$

$$= \frac{RA^2}{V}.$$

This proves that in general $A \cos \phi$ is a very complicated function of the quantities involved, and reasoning based on it has to be closely examined.

REFERENCES

A. Russell, 'The Theory of the Power Factor.' *The Electrician*, Vol. 44, p. 49, 1899.
—— 'The Current produced in an Inductive Coil by a Parabolic Wave of e.m.f.' *The Electrical Review*, Vol. 45, p. 744, 1899.

CHAPTER XI

Argand's method of representing a complex variable. Vectors. Addition of vectors. Polygon law for compounding vectors. Multiplication of a vector by a complex number. Division of a vector by a complex number. Application to the theory of alternating currents. The currents in a divided circuit. Inductive coil in series with a choking coil shunted by a non-inductive resistance. The apparent resistance and inductance of branched circuits. The currents in a branched circuit when mutual inductance is taken into account. Graphical solution. References.

IT is convenient in Algebra and Trigonometry to introduce the conception of the square root of negative unity, and to make the convention that it must obey all the ordinary algebraical laws. An expression of the form $x + y\iota$, where $\iota = (-1)^{1/2}$, is called a complex number, and has many properties which make it a great aid in calculation. The expression $x - y\iota$ is said to be conjugate to $x + y\iota$, and $\sqrt{x^2 + y^2}$, the square root of the product of the two, is called the modulus of either. The following three fundamental theorems are proved in text-books on algebra.

1. If a, b, c and d are real quantities and

$$a + b\iota = c + d\iota,$$

then $\qquad a = c,$ and $b = d.$

2. $\dfrac{1}{a + b\iota} = \dfrac{a}{a^2 + b^2} - \dfrac{b}{a^2 + b^2}\,\iota.$

3. $\dfrac{c + d\iota}{a + b\iota} = \dfrac{ac + bd}{a^2 + b^2} + \dfrac{ad - bc}{a^2 + b^2}\,\iota.$

R. I.

19

If we agree that the abscissa represents the real part of the

Argand's
method of
representing
a complex
variable.
complex variable and the ordinate the imaginary part, then a line OP (Fig. 67) can be represented by $x + y\iota$. The length of OP, $\sqrt{x^2 + y^2}$, is the modulus of the complex variable, and its inclination to the axis of x is $\tan^{-1}\frac{y}{x}$. If r and θ be the polar coordinates of P, then

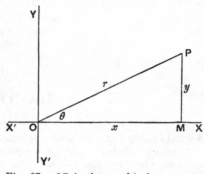

Fig. 67. OP is the graphical representation of the complex variable $x + y\iota$.

$$r = \sqrt{x^2 + y^2},$$

and
$$\tan \theta = \frac{y}{x}.$$

If OP rotate about the point P with uniform angular velocity
Vectors.
ω and if OX be the initial position of OP, then

$$\theta = \omega t,$$

$$OM = r \cos \omega t,$$

and
$$PM = r \sin \omega t,$$

where t is the time in seconds since OP coincided with OX. The line OP is called a vector and may be represented by

$$r \cos \omega t + \iota r \sin \omega t.$$

By trigonometry this expression may be written in the form $r\epsilon^{\omega\iota t}$ where ϵ is the base of Neperian logarithms.

Two vectors OP and OQ are compounded by the paral-
Addition
of vectors.
lelogram law. To prove this, suppose that OP represents $x_1 + y_1\iota$, and that OQ represents $x_2 + y_2\iota$.

Now, if OR (Fig. 68) represents $X + Y\iota$, we see at once by projections that

$$X = x_1 + x_2, \text{ and } Y = y_1 + y_2.$$

Hence OR represents

$$x_1 + x_2 + (y_1 + y_2)\,\iota,$$

and may be called the resultant of the addition of the vectors represented by OP and OQ. If R be the magnitude of the resultant and θ its inclination to the axis of X, then

$$R = \{(x_1 + x_2)^2 + (y_1 + y_2)^2\}^{\frac{1}{2}},$$

and
$$\tan \theta = \frac{y_1 + y_2}{x_1 + x_2}.$$

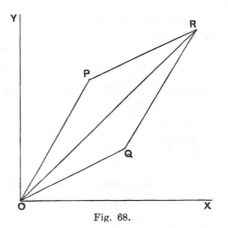

Fig. 68.

Proceeding in exactly the same way as in the last paragraph,

Polygon law for compounding vectors.

we can show that the equations

$$R = \{(\Sigma x)^2 + (\Sigma y)^2\}^{\frac{1}{2}}$$

and
$$\tan \theta = \frac{\Sigma y}{\Sigma x}$$

give the magnitude and the inclination of the resultant of any number of vectors.

Suppose that the vector is represented by $x + y\iota$ and that the

Multiplication of a vector by a complex number.

number is $a + b\iota$. Then by ordinary multiplication we find that the result of the operation is a vector whose length is

$$\sqrt{a^2 + b^2}\,\sqrt{x^2 + y^2}$$

and its inclination to the axis

$$\tan^{-1} \frac{bx + ay}{ax - by} = \theta + \alpha,$$

where
$$\tan \theta = \frac{y}{x} \quad \text{and} \quad \tan \alpha = \frac{b}{a}.$$

The result of the operation is therefore to multiply the length of the vector by $\sqrt{a^2 + b^2}$ and to advance the vector through an angle α, where $\tan \alpha = \dfrac{b}{a}$.

This theorem can be proved more simply trigonometrically as follows:

$$(a + b\iota)(x + y\iota) = A\epsilon^{a\iota}\, r\epsilon^{\theta\iota},$$
$$= A r\epsilon^{(a+\theta)\iota},$$

where A is $\sqrt{a^2 + b^2}$. This obviously proves the theorem.

Division of a vector by a complex number.

We have

$$\frac{x + y\iota}{a + b\iota} = \frac{r\epsilon^{\theta\iota}}{A\epsilon^{a\iota}}$$
$$= \frac{r}{A}\epsilon^{(\theta - a)\iota}.$$

Hence the length of the vector is diminished in the ratio of 1 to $\sqrt{a^2 + b^2}$, and the vector is turned back through an angle $\tan^{-1}\dfrac{b}{a}$.

Application to the theory of alternating currents.

We have seen (page 65) that the equation for the current in a simple inductive circuit is

$$e = Ri + L\frac{\partial i}{\partial t}.$$

This may be written in the form

$$e = (R + LD)i,$$

where D is an operator. If i be a harmonic function $I\sin\omega t$, then

$$D \,.\, I\sin\omega t = \omega\, I\cos\omega t,$$

and

$$D^2 \,.\, I\sin\omega t = -\,\omega^2 I\sin\omega t.$$

Thus

$$D^2 = -\,\omega^2,$$

and

$$D = \omega\iota.$$

Similarly, if i were $I\cos\omega t$, D would have the same value. Now if we suppose that both e and i are complex quantities and can be represented by vectors, then by the preceding theorems the operations of multiplying or dividing by the complex number $R + L\omega\iota$ can easily be performed and the real part of the result will give the required solution. Although at first sight it may appear clumsy to introduce an imaginary term in the expression for i, seeing that we ultimately reject the coefficient of ι in our result, yet it will be found that this artifice introduces trigonometrical symmetry, and in some cases greatly simplifies the calculation. We will use square brackets to denote a vector quantity, so that if $i = I \cos \omega t$, then

$$[i] = I \cos \omega t + \iota I \sin \omega t.$$

Denoting the operator $R + L\omega\iota$ by $[\rho]$, the equation in alternating current theory corresponding to Ohm's law in direct current theory will be

$$[e] = [\rho][i] \quad\text{..........................}(1).$$

To prove this it is sufficient to notice that since, by hypothesis i and therefore also e is a harmonic function of the time, we have

$$E \cos(\omega t + \alpha) = [\rho] I \cos(\omega t + \beta),$$

and hence, by differentiating and multiplying by $-(\iota/\omega)$, we get

$$\iota E \sin(\omega t + \alpha) = [\rho] \iota I \sin(\omega t + \beta).$$

By adding these equations we arrive at (1). We can use this equation to find e.

We have

$$[e] = (R + L\omega\iota)(I \cos \omega t + \iota I \sin \omega t)$$

$$= \sqrt{R^2 + L^2\omega^2}\, \epsilon^{a\iota}\, I \epsilon^{\omega t \iota}$$

$$= \sqrt{R^2 + L^2\omega^2}\, I \epsilon^{(\omega t + a)\iota}$$

$$= \sqrt{R^2 + L^2\omega^2}\, I \{\cos(\omega t + \alpha) + \iota \sin(\omega t + \alpha)\},$$

and therefore $e = \sqrt{R^2 + L^2\omega^2}\, I \cos(\omega t + \alpha)$.

Similarly if

$$[e] = E \cos \omega t + \iota E \sin \omega t,$$

then

$$[i] = \frac{[e]}{[\rho]}$$

$$= \frac{E}{\sqrt{R^2 + L^2\omega^2}}\, \epsilon^{(\omega t - a)\iota}.$$

Thus

$$i = \frac{E \cos(\omega t - \alpha)}{\sqrt{R^2 + L^2\omega^2}},$$

where

$$\tan \alpha = \frac{L\omega}{R}.$$

It will be noticed that this method only gives us the particular integral of the equation

$$e = Ri + L\frac{\partial i}{\partial t},$$

in the special case when e is a simple harmonic function. To get the complete integral we have to add on the term $A\epsilon^{-(R/L)t}$. This term however is as a rule only important for a fraction of a second after switching on (see page 66).

If the currents in the branches be i_1 and i_2 (Fig. 69), and the resistances be r_1 and r_2, then for direct currents,

The currents in a divided circuit.

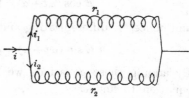

$$i_1 = \frac{r_2}{r_1 + r_2}i,$$

$$i_2 = \frac{r_1}{r_1 + r_2}i,$$

and $$R = \frac{r_1 r_2}{r_1 + r_2},$$

Fig. 69.   Calculation of harmonic currents in a divided circuit.

where i is the current in the main and R the resistance of the two branches in parallel.

With alternating currents, if we denote the impedance of the branches by Z_1, Z_2, and if Z_1 equals $\sqrt{R_1^2 + L_1^2\omega^2}$ and $\tan \alpha_1$ equals $\frac{L_1\omega}{R_1}$, then, if we neglect mutual inductance,

$$[i_1] = \frac{[\rho_2]}{[\rho_1] + [\rho_2]}\cdot[i]$$

$$= \frac{Z_2\epsilon^{\alpha_2 t}}{Z'\epsilon^{\alpha' t}}I\epsilon^{\omega t},$$

where $Z' = \{(R_1 + R_2)^2 + \omega^2(L_1 + L_2)^2\}^{\frac{1}{2}}$ and $\tan \alpha' = \frac{\omega(L_1 + L_2)}{R_1 + R_2}$.

Thus $$i_1 = \frac{\{R_2^2 + \omega^2 L_2^2\}^{\frac{1}{2}}I}{\{(R_1 + R_2)^2 + \omega^2(L_1 + L_2)^2\}^{\frac{1}{2}}}\cos(\omega t + \alpha_2 - \alpha'),$$

and $$i_2 = \frac{\{R_1^2 + \omega^2 L_1^2\}^{\frac{1}{2}}I}{\{(R_1 + R_2)^2 + \omega^2(L_1 + L_2)^2\}^{\frac{1}{2}}}\cos(\omega t + \alpha_1 - \alpha').$$

If $L_1/R_1 = L_2/R_2$, then $\alpha_1 = \alpha_2 = \alpha'$, and i_1 and i_2 are in phase with i. If L_1/R_1 is greater than L_2/R_2, then α_1 is greater than α_2, and α' is greater than α_2, but less than α_1. Hence the current i_1 lags behind i, but the phase of the current i_2 is in advance of i.

Again, since $[e] = [\rho_1][i_1] = [\rho_2][i_2]$, we have

$$[\rho] = \frac{[\rho_1][\rho_2]}{[\rho_1] + [\rho_2]},$$

and
$$R + L\omega\iota = \frac{Z_1 Z_2}{Z'} \epsilon^{(\alpha_1 + \alpha_2 - \alpha')\iota}$$

$$= \frac{Z_1 Z_2}{Z'}\{\cos(\alpha_1 + \alpha_2 - \alpha') + \iota\sin(\alpha_1 + \alpha_2 - \alpha')\}.$$

Thus
$$R = \frac{Z_1 Z_2}{Z'}\cos(\alpha_1 + \alpha_2 - \alpha'),$$

and
$$L\omega = \frac{Z_1 Z_2}{Z'}\sin(\alpha_1 + \alpha_2 - \alpha'),$$

and
$$Z = \frac{Z_1 Z_2}{Z'},$$

where R is the equivalent resistance, L the equivalent inductance, and Z the equivalent impedance of the two branches in parallel. We see that i lags behind e by an angle $\alpha_1 + \alpha_2 - \alpha'$.

These formulae for the effective resistance and inductance of the branched circuit may be written in the form

$$R = \frac{R_1 R_2}{R_1 + R_2} + \frac{\omega^2}{R_1 + R_2} \cdot \frac{(L_1 R_2 - L_2 R_1)^2}{Z'^2},$$

$$L = \frac{L_1 R_2^2 + L_2 R_1^2}{(R_1 + R_2)^2} - \frac{(L_1 R_2 - L_2 R_1)^2}{L_1 + L_2}\left\{\frac{1}{(R_1 + R_2)^2} - \frac{1}{Z'^2}\right\},$$

where $Z' = \{(R_1 + R_2)^2 + \omega^2(L_1 + L_2)^2\}^{\frac{1}{2}}$.

Hence, as the frequency increases, the apparent resistance increases from $R_1 R_2/(R_1 + R_2)$, its minimum value, to

$$(L_2^2 R_1 + L_1^2 R_2)/(L_1 + L_2)^2,$$

its maximum value, and the apparent self inductance diminishes from $(L_1 R_2^2 + L_2 R_1^2)/(R_1 + R_2)^2$, its maximum value, to

$$L_1 L_2/(L_1 + L_2),$$

its minimum value.

It is also worth noticing that when the time constants of the circuits are equal, R and L are independent of the frequency.

Suppose (Fig. 70) that

Inductive coil in series with a choking coil (o, L) shunted by a non-inductive resistance x. we maintain the effective value of the P.D. between A and B constant. Suppose also that the resistance of the inductive coil is r, and its self inductance l, then by the preceding formulae,

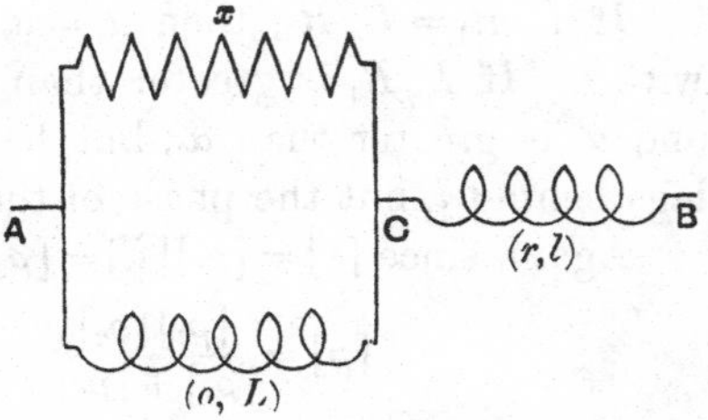

Fig. 70. Current in CB is a minimum for a particular value of x when the effective value of the applied P.D. is constant.

$$Z^2 = \left\{ \frac{x\omega^2 L^2}{x^2 + \omega^2 L^2} + r \right\}^2 + \omega^2 \left\{ \frac{Lx^2}{x^2 + \omega^2 L^2} + l \right\}^2,$$

where Z is the impedance of the circuit AB.

Now by the differential calculus this is a maximum or a minimum when

$$x^2 - \frac{\omega^2 L (L + 2l)}{r} x - \omega^2 L^2 = 0,$$

i.e. when $x = \frac{\omega^2 L (L + 2l)}{2r} + \frac{\omega L}{2r} \{\omega^2 (L + 2l)^2 + 4r^2\}^{\frac{1}{2}}.$

Since x must be positive, we have prefixed the positive sign to the radical. It is easy to see that this value of x makes Z a maximum, and therefore the current through the inductive coil a minimum. Hence shunting a choking coil with a non-inductive resistance sometimes increases the apparent resistance of a circuit, a result which has been noticed in practical work.

Let $(R_1, L_1), (R_2, L_2)...$ be the resistances and self inductances

The apparent resistance and inductance of branched circuits. of the n branches, then, if we neglect mutual inductance,

$$[i_1] = \frac{1}{[\rho_1]} [e],$$

$$[i_2] = \frac{1}{[\rho_2]} [e],$$

. . . .

$$[i_n] = \frac{1}{[\rho_n]} [e].$$

If i be the current in the main, then

$$[i] = [i_1] + [i_2] + \ldots\ldots + [i_n]$$

$$= \left\{ \frac{1}{[\rho_1]} + \frac{1}{[\rho_2]} + \ldots\ldots \right\}[e]$$

$$= \left\{ \frac{1}{R_1 + L_1\omega\iota} + \frac{1}{R_2 + L_2\omega\iota} + \ldots\ldots \right\}[e]$$

$$= \left\{ \Sigma \frac{R_1}{R_1{}^2 + L_1{}^2\omega^2} - \Sigma \frac{L_1\omega}{R_1{}^2 + L_1{}^2\omega^2}\iota \right\}[e]$$

$$= \frac{[e]}{R + L\omega\iota},$$

where R is the equivalent resistance and L the equivalent inductance of the n branches in parallel.

Therefore

$$\frac{R}{R^2 + L^2\omega^2} = \Sigma \frac{R_1}{R_1{}^2 + L_1{}^2\omega^2} = \frac{1}{a}.$$

And

$$\frac{L\omega}{R^2 + L^2\omega^2} = \Sigma \frac{L_1\omega}{R_1{}^2 + L_1{}^2\omega^2} = \frac{1}{b}.$$

Thus

$$\frac{1}{R^2 + L^2\omega^2} = \frac{1}{a^2} + \frac{1}{b^2}.$$

Hence

$$R = \frac{1}{a} \Big/ \left(\frac{1}{a^2} + \frac{1}{b^2} \right),$$

and

$$L\omega = \frac{1}{b} \Big/ \left(\frac{1}{a^2} + \frac{1}{b^2} \right).$$

We also see that the current in the main lags behind the applied P.D. by an angle θ where $\tan\theta = a/b$.

Let (R_1, L_1), (R_2, L_2) be the resistances and inductances of the two branches, and let M be their mutual inductance, then with the usual notation

The currents in a branched circuit when mutual inductance is taken into account.

$$e = (R_1 + L_1 D)i_1 + MDi_2 \atop e = (R_2 + L_2 D)i_2 + MDi_1 \Big\} \quad \ldots\ldots\ldots(a).$$

Assuming that the functions are simple harmonic, we may write $\omega\iota$ for D, and solving the equations we find

$$i_1\{R_1 R_2 + (M^2 - L_1 L_2)\omega^2 + (L_2 R_1 + L_1 R_2)\omega\iota\}$$
$$= \{R_2 + (L_2 - M)\omega\iota\}\, e,$$

$$i_2\{R_1 R_2 + (M^2 - L_1 L_2)\omega^2 + (L_2 R_1 + L_1 R_2)\omega\iota\}$$
$$= \{R_1 + (L_1 - M)\omega\iota\}\, e.$$

Hence since i the current in the main equals $i_1 + i_2$, we get

$$e = \frac{c + d\iota}{a + b\iota} \cdot i$$

$$= \left\{ \frac{ac + bd}{a^2 + b^2} + \frac{ad - bc}{a^2 + b^2} \iota \right\} i$$

$$= \{R + L\omega\iota\} i,$$

where $a = R_1 + R_2,$ $c = R_1 R_2 + (M^2 - L_1 L_2)\, \omega^2,$

$\qquad\qquad b = (L_1 + L_2 - 2M)\, \omega,$ $d = (L_1 R_2 + L_2 R_1)\, \omega.$

Hence

$$R = \frac{ac + bd}{a^2 + b^2}$$

$$= \frac{R_1 R_2 (R_1 + R_2) + \omega^2 \{(L_2 - M)^2 R_1 + (L_1 - M)^2 R_2\}}{(R_1 + R_2)^2 + \omega^2 (L_1 + L_2 - 2M)^2},$$

$$L = \frac{L_1 R_2^2 + L_2 R_1^2 + 2M R_1 R_2 + (L_1 L_2 - M^2)(L_1 + L_2 - 2M)\,\omega^2}{(R_1 + R_2)^2 + \omega^2 (L_1 + L_2 - 2M)^2}.$$

Hence, as the frequency increases, the effective resistance increases from

$$\frac{R_1 R_2}{R_1 + R_2} \quad \text{to} \quad \frac{(L_2 - M)^2 R_1 + (L_1 - M)^2 R_2}{(L_1 + L_2 - 2M)^2}.$$

It is to be noticed that in the equations above, M may be either positive or negative, depending on how the coils are connected with the mains. When the frequency is zero

$$L = L_1 \left(\frac{R_2}{R_1 + R_2} \right)^2 + L_2 \left(\frac{R_1}{R_1 + R_2} \right)^2 + 2M \frac{R_1 R_2}{(R_1 + R_2)^2}.$$

If, for example, we measured the self inductance of the two coils in parallel by Maxwell's method, the self inductance L found would be the L given by this formula. A strict proof of this particular formula can be given by other methods. As the frequency increases the effective inductance L diminishes, and it has its least possible value

$$\frac{L_1 L_2 - M^2}{L_1 + L_2 - 2M}$$

when the frequency is infinite. We have shown on page 38 that $L_1 L_2 - M^2$ is zero or positive. The case when it is zero need not be considered, as it is impossible in practice to arrange two circuits so as to satisfy this condition. In practice, $L_1L_2 - M^2$ is positive, and therefore, since M is less than $\sqrt{L_1L_2}$, $L_1 + L_2 - 2M$ is positive.

The effective values A_1 and A_2 of the currents in a branched
Graphical   circuit may be written down from the equations given
solution.   above. They can be found more simply, however,
from the following graphical constructions.

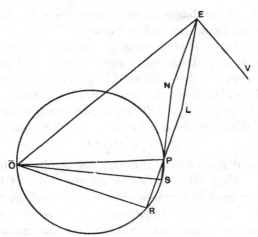

Fig. 71. Currents in a divided circuit.

$$OE = V; \quad OS = R_2A_2; \quad OR = R_1A_1.$$
$$M \text{ is less than } L_1 \text{ or } L_2.$$

The equations (a) given above show that the vector of the effective value of the applied potential difference V is the resultant of the component vectors R_1A_1, ωL_1A_1 and ωMA_2 and also, that it is the resultant of R_2A_2, ωL_2A_2 and ωMA_1. The vectors ωL_1A_1 and ωMA_2 are each perpendicular to R_1A_1 and the vectors ωL_2A_2 and ωMA_1 are each perpendicular to R_2A_2.
In Fig. 71 M is less than either L_1 or L_2.

$$OE = \text{the applied P.D.} = V,$$
$$OR = R_1A_1, \qquad OS = R_2A_2,$$
$$RL = \omega L_1A_1, \qquad SN = \omega L_2A_2,$$
$$LE = \omega MA_2, \qquad NE = \omega MA_1.$$

Now PN and LE are parallel, and so also are NE and PL. Therefore $NPLE$ is a parallelogram and thus PL equals ωMA_1 and PN equals ωMA_2. Hence PR is $\omega(L_1 - M)A_1$ and PS is $\omega(L_2 - M)A_2$. Since the angles at R and S are right angles, the circle described on OP as diameter passes through R and S. Hence

$$(R_1 A_1)^2 + \{\omega(L_1 - M)A_1\}^2 = (R_2 A_2)^2 + \{\omega(L_2 - M)A_2\}^2$$
$$= OP^2.$$

To construct the diagram when the values of R_1, R_2, L_1, L_2, M, f, and V are given, we proceed as follows.

Draw any line OR and, choosing a convenient scale, make its length equal to R_1. Draw RL at right angles to OR and equal to ωL_1 where ω equals $2\pi f$. Make LP equal to ωM so that RP equals $\omega(L_1 - M)$. Join OP and describe a circle on OP as diameter. Make the angle POS equal to the angle whose tangent is $\omega(L_2 - M)/R_2$. Produce SP to N so that SN equals

$$\{L_2/(L_2 - M)\}\ SP.$$

Through N and L draw lines NE and LE parallel to RL and SN respectively. Then OE will be the effective value of the applied potential difference which produces unit current in the branch (R_1, L_1). Now the phase differences and the relative magnitudes of the various vectors are independent of the absolute value of the applied potential difference, provided that the frequency does not alter. Hence if we choose the scale of the diagram so that OE represents V, then, on this scale, OR will represent $R_1 A_1$ and OS will represent $R_2 A_2$ in magnitude and phase, and thus, by dividing these lengths by R_1 and R_2 respectively, we get A_1 and A_2.

If ϕ be the phase difference between A_1 and A_2 and $\tan \alpha$ and $\tan \beta$ be equal to

$$\omega \frac{L_1 - M}{R_1} \quad \text{and} \quad \omega \frac{L_2 - M}{R_2}$$

respectively, we have

$$\tan \phi = \tan(\alpha - \beta)$$
$$= \frac{\omega \dfrac{L_1 - M}{R_1} - \omega \dfrac{L_2 - M}{R_2}}{1 + \omega^2 \dfrac{L_1 - M}{R_1} \cdot \dfrac{L_2 - M}{R_2}} \dots\dots\dots\dots\dots(b).$$

We can see at once from equation (b) that when the frequency is
very low the angle of phase difference between A_1 and A_2 is very
small. When the frequency is very high, the phase difference is
again very small. It is easy to show that it is a maximum when

$$\omega^2 = \frac{R_1 R_2}{(L_1 - M)(L_2 - M)} \quad\quad\dots\dots\dots\dots\dots(c).$$

If we divide the lengths of the lines NE, LE and PE by ωM
we get the magnitudes of the currents A_1, A_2 and A. If we draw
$\dot{E}V$ (Fig. 71) perpendicular to OE, then the angles which NE, LE
and PE make with EV are the respective phase differences between
the currents and the applied P.D.

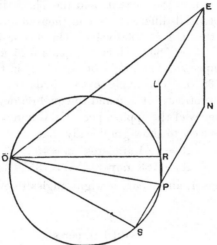

Fig. 72. Currents in a divided circuit.
$OE =$ applied P.D.; $\quad OR = R_1A_1$; $\quad OS = R_2A_2$.
M is greater than L_1 and less than L_2.

Fig. 72 gives the graphical construction when M is greater than L_1
and less than L_2. In this case A_1 and A_2 are in quadrature when

$$\omega^2 = \frac{- R_1 R_2}{(L_1 - M)(L_2 - M)}$$

and the phase difference increases as the frequency increases.
When the frequency is high, the currents are nearly in opposition
in phase, and hence the current in the main is nearly equal to their
difference. If L_1 be approximately equal to L_2 then the current in

the main may be very much smaller than the branch currents. The inductive effects produce as it were a whirlpool of current in the two branches. Large circulating currents produced in this way are often useful in alternating current work.

The diagram in Fig. 72 is drawn to scale for the case of two coils (100, 0·5), (100, 1), the mutual inductance being 0·6 henry and the frequency being 16, so that ω is nearly 100. On the given scale, OR is equal to 100, RL equals 50 and RP equals 10. A circle is described on OP as diameter. The angle POS equals the angle whose tangent is 0·4 and SN equals $\frac{5}{2}SP$. The parallelogram $PNEL$ is then completed and OE is joined.

If we wish to find the currents and the phase differences when the applied potential difference is one thousand volts, we choose a new scale so that OE is 1000 volts. On this scale OR is found to be 625 and OS is 583. Thus A_1 equals 6·25 amperes and A_2 equals 5·83 amperes. The angle of lag of A_1 is the angle EOR and it equals 38°·5. The angle EOS equals the lag of A_2 and measured by a protractor it is found to equal 66 degrees.

If the frequency of the applied potential difference be increased to 80, then it is easy to show graphically that

$$A_1 = 1·41 \text{ amperes}, \quad \alpha_1 = 45°,$$
$$A_2 = 2·83 \text{ amperes}, \quad \alpha_2 = 135°.$$

Thus the vectors A_1 and A_2 are at right angles to one another and so

$$A = \sqrt{A_1{}^2 + A_2{}^2}$$
$$= 3·16 \text{ amperes}.$$

In Fig. 73 M is negative.

It follows from (b) that ϕ is zero when ω is zero and again when ω is infinite.

The angle ϕ has its maximum value when the frequency is determined by (c). The diagram in Fig. 73 is worked out for the case of two coils (53, 0·29), (117, 0·18) and M equal to $-0·14$. The frequency has been taken equal to 80 so that ω is nearly equal to 500. We find by measuring the lines and angles that

$$A_1 = 9·3 \text{ amperes}, \quad \alpha_1 = 40°,$$
$$A_2 = 10·2 \text{ amperes}, \quad \alpha_2 = 18°,$$
$$A = 19·2 \text{ amperes}, \quad \text{Impedance} = 52 \text{ ohms}.$$

It is to be noticed that all the theorems and formulae in this chapter, deduced by the method of the complex variable, are proved on the assumption of a sine wave of E.M.F. Moreover they only give the particular integrals of the differential equations involved, and hence it is best to regard them merely as giving a first approximate solution of the general problem. As a rule, it is better for engineers to solve problems graphically, as the graphical method is quite accurate enough and far more instructive than the mere algebraical manipulation of complex quantities in which the physical laws are not so apparent. In the next chapter we will consider the graphical method and its limitations.

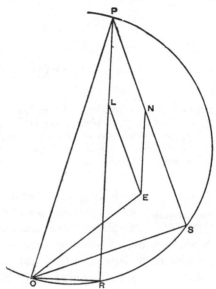

Fig. 73.   Currents in a divided circuit.
OE = applied P.D. ;
$OR = R_1 A_1$;
$OS = R_2 d_2$.
M is negative.

REFERENCES

G. CHRYSTAL, 'Complex Numbers.' *Text Book on Algebra*, Chapter XII.

S. L. LONEY, 'Geometrical Representation of Complex Quantities.' *Plane Trigonometry*, Chapter XII.

LORD RAYLEIGH, 'On Forced Harmonic Oscillations of Various Periods.' *Phil. Mag.* p. 379, May, 1886.

A. RUSSELL, 'Alternating Currents in a Divided Circuit.' *The Electrician*, Vol. 33, p. 595, 1894.

CHAPTER XII

WHEN we are dealing with harmonically varying quantities, we **Parallelogram of vectors.** can represent them graphically by lines drawn in a plane. For example, if we have two E.M.F.s $E_1 \sin 2\pi ft$ and $E_2 \sin (2\pi ft + \alpha)$ we see that their values are the projections of two lines OP_1 and OP_2, which are inclined to one another at an angle α and rotate round O 'f' times per second, upon a fixed line OA drawn at right angles to the initial position of OP_1. The resultant got by adding the two E.M.F.s together, $E_1 \sin 2\pi ft + E_2 \sin (2\pi ft + \alpha)$, is easily shown to be the projection upon OA of the diagonal OR of the parallelogram constructed with OP_1 and OP_2 for adjacent sides. Since in practice we are generally only concerned with the R.M.S. values of the E.M.F.s and their phase differences, and since the R.M.S. value of a simple harmonic quantity bears a constant ratio to its maximum value, it is usual to represent the E.M.F.s by lines OP_1, OP_2 and OR equal to their R.M.S. values and inclined to one another at angles equal to their phase differences. These lines are called vectors and the above theorem is the parallelogram of vectors. It can obviously be generalised into the polygon of vectors.

In alternating current work we seldom have to do with **Extension of theorem.** harmonic waves of P.D. and current, but still graphical methods are found to be convenient in many cases. We have therefore to investigate the limitations of these methods.

Let us first consider the case of two periodic quantities, for example alternating electromotive forces. Let e_1 and e_2 be the instantaneous values of the two E.M.F.s and let e be their resultant, then

$$e = e_1 + e_2.$$

Now the curves of e_1 and e_2 may be dissimilar and that of e may be dissimilar to both. Hence the R.M.S. values of the E.M.F.s do not bear a constant ratio to their maximum values.

In all cases however we have

$$e^2 = e_1{}^2 + e_2{}^2 + 2e_1e_2,$$

and thus

$$V^2 = V_1{}^2 + V_2{}^2 + 2V_1 V_2 \cos \alpha_{1.2},$$

where

$$\cos \alpha_{1.2} = \frac{\int_0^T e_1 e_2 \partial t}{\left\{ \int_0^T e_1{}^2 \partial t \cdot \int_0^T e_2{}^2 \partial t \right\}^{\frac{1}{2}}} \quad \dots\dots\dots\dots(1).$$

Hence the parallelogram construction will still hold good for compounding lines representing the effective values of two periodic quantities if we define phase difference by equation (1). This definition we have already given in Chapter X. It is customary to call the lines representing V_1, V_2 and V vectors. The two vectors and their resultant are lines in one plane and the angles between the various vectors will give their phase differences in accordance with definition (1).

In general if a linear relation hold between the instantaneous values of three periodic alternating functions e_1, e_2 and e_3, each of which has the same period, then their R.M.S. values can be represented by lines drawn in one plane, the angles between the lines being the phase differences. For example, suppose that at every instant

Condition that three vectors lie in a plane.

$$le_1 + me_2 + ne_3 = 0,$$

where l, m and n are constants and e_1, e_2 and e_3 are periodic functions of the time, then

$$l^2 e_1{}^2 = m^2 e_2{}^2 + n^2 e_3{}^2 + 2mn e_2 e_3,$$

and thus

$$l^2 V_1{}^2 = m^2 V_2{}^2 + n^2 V_3{}^2 + 2mn V_2 V_3 \cos \alpha_{2.3}.$$

Hence $l V_1$ is equal in magnitude to the resultant got by compounding two lines inclined to one another at an angle $\alpha_{2.3}$ and the lengths of which are $m V_2$ and $n V_3$ respectively, by the parallelogram construction. The line representing $l V_1$ must however be drawn in the opposite direction to the resultant, since

$$le_1 = -(me_2 + ne_3).$$

We also have $$\alpha_{2.3} + \alpha_{3.1} + \alpha_{1.2} = 2\pi$$

and, just as in statics,

$$\frac{l V_1}{\sin \alpha_{2.3}} = \frac{m V_2}{\sin \alpha_{3.1}} = \frac{n V_3}{\sin \alpha_{1.2}}.$$

These lines $l V_1$, $m V_2$ and $n V_3$ are called vectors in engineering practice, but the instantaneous values of e_1, e_2 and e_3 can no longer be represented by the projections of lines rotating with constant angular velocity.

It follows from the definition of phase difference given in (1) that the cosine of the phase difference between a **Vector of a constant quantity.** constant and a periodic quantity is always zero, and hence the phase difference is ninety degrees. Hence the R.M.S. value of the sum of an alternating and a direct current, for example, is represented in magnitude and phase by the diagonal of the rectangle constructed with the R.M.S. values of the direct and alternating components as adjacent sides. If however we have two alternating current components which are not in the same phase, then the line representing the direct current component must be drawn at right angles to the plane containing the two lines representing the alternating current components, so as to be at right angles to the three lines representing the two alternating components and their resultant. We must therefore, if we are going to use graphical methods in this case, have recourse to solid geometry.

In general, when we have three periodic functions and there is **Vectors in space.** no linear relation connecting them, we can represent their R.M.S. values graphically by three lines drawn in space, the angles between the lines being the phase differences as determined by equation (1). In order to prove this we have to show that α, β and γ, the angles of phase difference, can always form a solid angle. We have to prove therefore that $\alpha + \beta + \gamma$ can never be greater than 2π, and also that the sum of any two of the angles cannot be less than the third.

Let x, y and z be the three periodic functions and let α be the phase difference between y and z, β between z and x, and γ between x and y respectively.

From the definition of phase difference we have

$$\cos \alpha = \frac{\int_0^T yz\, \partial t}{\left\{ \int_0^T y^2 \partial t . \int_0^T z^2 \partial t \right\}^{\frac{1}{2}}}.$$

Divide the period T into a large number (n) of equal intervals, and let $x_1, x_2, \ldots x_n$; $y_1, y_2, \ldots y_n$; $z_1, z_2, \ldots z_n$ be the values of the functions at the end of successive intervals.

Then from the meaning of integration we have, when n is infinitely large,

$$\cos^2 \alpha = \frac{(y_1 z_1 + y_2 z_2 + \ldots + y_n z_n)^2}{(y_1^2 + y_2^2 + \ldots + y_n^2)(z_1^2 + z_2^2 + \ldots + z_n^2)},$$

with corresponding values for $\cos^2 \beta$ and $\cos^2 \gamma$.

Let $X = x_1^2 + x_2^2 + \ldots + x_n^2$, $A = y_1 z_1 + y_2 z_2 + \ldots + y_n z_n$,

$\quad\quad Y = y_1^2 + y_2^2 + \ldots + y_n^2$, $B = z_1 x_1 + z_2 x_2 + \ldots + z_n x_n$,

and $Z = z_1^2 + z_2^2 + \ldots + z_n^2$, $C = x_1 y_1 + x_2 y_2 + \ldots + x_n y_n$.

Then $1 - \cos^2 \alpha - \cos^2 \beta - \cos^2 \gamma + 2 \cos \alpha \cos \beta \cos \gamma$

$$= \frac{XYZ - XA^2 - YB^2 - ZC^2 + 2ABC}{XYZ} \quad \ldots(a).$$

Now $X(XYZ - XA^2 - YB^2 - ZC^2 + 2ABC)$

$$= (XY - C^2)(XZ - B^2) - (XA - CB)^2.$$

Also $XY - C^2 = (x_1 y_2 - x_2 y_1)^2 + (x_1 y_3 - x_3 y_1)^2 + \ldots \quad \ldots(b),$

$\quad\quad XZ - B^2 = (x_1 z_2 - x_2 z_1)^2 + (x_1 z_3 - x_3 z_1)^2 + \ldots \quad \ldots\ldots(c),$

and $XA - CB = (x_1 y_2 - x_2 y_1)(x_1 z_2 - x_2 z_1)$

$$+ (x_1 y_3 - x_3 y_1)(x_1 z_3 - x_3 z_1) + \ldots \quad \ldots(d).$$

Now it is easy to show that

$$(P^2 + Q^2 + \ldots)(p^2 + q^2 + \ldots)$$

is not less than $(Pp + Qq + ...)^2$. It therefore follows from (b), (c) and (d) that

$$(XY - C^2)(XZ - B^2) - (XA - CB)^2$$

is never negative.   Hence from (a)

$$1 - \cos^2\alpha - \cos^2\beta - \cos^2\gamma + 2\cos\alpha\cos\beta\cos\gamma$$

is never negative.   Now the trigonometrical expression can be written in the form

$$\{\cos\gamma - \cos(\alpha + \beta)\}\{\cos(\alpha - \beta) - \cos\gamma\}.$$

Hence we see that if it vanishes, the sum of two of the angles equals the third or the sum of the three equals four right angles. The three vectors in this case are therefore in one plane.

It may also be written in the form

$$4\sin\frac{\alpha + \beta + \gamma}{2}\sin\frac{\beta + \gamma - \alpha}{2}\sin\frac{\gamma + \alpha - \beta}{2}\sin\frac{\alpha + \beta - \gamma}{2} \quad...(e).$$

By definition, the values of α, β and γ each lie between $0°$ and $180°$. As the expression (e) is positive, all the terms may be positive, or two of them may be negative, or all four of them may be negative.

If they are all positive, then $\alpha + \beta + \gamma$ is less than four right angles, and also any two of the angles are together greater than the third. Suppose, now, that the first two terms of (e) are negative. Then $\alpha + \beta + \gamma$ is greater than 2π, and $\alpha - \beta - \gamma$ is greater than zero. Therefore α is greater than π, which is contrary to the definition. Similarly, no other two terms can be negative and à fortiori the whole four cannot be negative. In all cases, therefore, we see that the sum of the three angles is not greater than four right angles, and that the sum of two of them is never less than the third. Hence, the three angles can always form a solid angle.

We shall apply the theorem given above to find graphically the **Resultant of three vectors.** effective value of the sum of three alternating periodic functions. Let e_2, e_3 and e_4 be the three functions, and let e_1 be their resultant, then

$$e_1 = e_2 + e_3 + e_4,$$

and $\quad e_1^2 = e_1 e_2 + e_1 e_3 + e_1 e_4,$

and thus $\quad V_1^2 = V_1 V_2 \cos\alpha_{1.2} + V_1 V_3 \cos\alpha_{1.3} + V_1 V_4 \cos\alpha_{1.4}.$

Hence $\qquad V_1 = V_2 \cos \alpha_{1.2} + V_3 \cos \alpha_{1.3} + V_4 \cos \alpha_{1.4}$

similarly $V_1 \cos \alpha_{1.2} = V_2 \qquad\quad + V_3 \cos \alpha_{2.3} + V_4 \cos \alpha_{2.4}$

$\qquad\quad V_1 \cos \alpha_{1.3} = V_2 \cos \alpha_{2.3} + V_3 \qquad\quad + V_4 \cos \alpha_{3.4}$

$\qquad\quad V_1 \cos \alpha_{1.4} = V_2 \cos \alpha_{2.4} + V_3 \cos \alpha_{3.4} + V_4$

$$\left. \right\} \ldots (f),$$

and $\qquad V_1^2 = V_2^2 + V_3^2 + V_4^2 + 2V_3V_4 \cos \alpha_{3.4} + 2V_4V_2 \cos \alpha_{2.4}$

$$+ 2V_2V_3 \cos \alpha_{2.3} \ldots\ldots\ldots (g).$$

In these equations V_n is the effective value of e_n, and $\alpha_{n.m}$ is the phase difference between e_n and e_m.

Construct the solid angle at O (see Fig. 74), so that the angles POQ, QOR and ROP equal

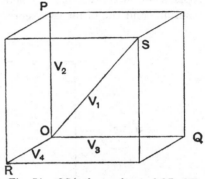

$\alpha_{2.3}$, $\alpha_{3.4}$ and $\alpha_{4.2}$ respectively. Complete the parallelepiped $OPQRS$ and let OS be the diagonal. Then from (g) we see that OS equals V_1. Hence from the equations (f) we see that the angles POS, QOS and ROS equal $\alpha_{1.2}$, $\alpha_{1.3}$ and $\alpha_{1.4}$ respectively. We can thus extend graphical methods to solid geometry for the case of three periodic quantities.

Fig. 74. OS is the resultant of OR, OP and OQ.

In this case we can show from equations (f) that

$$\frac{V_1}{\sin(\alpha_{2.3}, \alpha_{2.4}, \alpha_{3.4})} = \frac{V_2}{\sin(\alpha_{1.3}, \alpha_{1.4}, \alpha_{3.4})}$$

$$= \frac{V_3}{\sin(\alpha_{1.2}, \alpha_{1.4}, \alpha_{2.4})}$$

$$= \frac{V_4}{\sin(\alpha_{1.2}, \alpha_{1.3}, \alpha_{2.3})},$$

where $\sin(\alpha, \beta, \gamma) = \{1 - \cos^2\alpha - \cos^2\beta - \cos^2\gamma + 2\cos\alpha\cos\beta\cos\gamma\}^{\frac{1}{2}}$.

In general when four periodic quantities of the same frequency

Condition that four vectors can be represented graphically. are connected by a linear relation,

$$le_1 + me_2 + ne_3 + pe_4 = 0 ;$$

then their R.M.S. values can be represented in magnitude and phase by lines drawn from a point in space.

In statics these lines would represent a system of four forces in equilibrium, and we shall see later on that many statical theorems have their counterpart in electrical theory.

If no linear relation connects the four periodic quantities then, as a rule, we cannot represent them graphically by lines drawn in space. For example, suppose that one of them is constant, then it would have to be represented by a line drawn at right angles to the other three, which is impossible when they form a solid angle. If however we replace two of the vectors by their resultant, we can represent this resultant and the other two vectors graphically.

Failure of graphical methods.

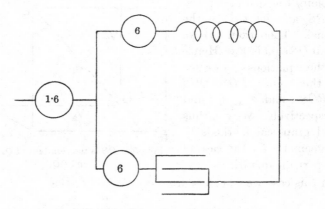

Fig. 75. Resonance of currents.

A variable condenser (Fig. 75) was used to shunt a choking coil on a 2000 volt circuit; the power factor of the choking coil was 0·041 and of the condenser 0·124. The current in the choking coil was 6 amperes, and the current in the main was 1·6 amperes when the condenser was adjusted so that the current in its circuit was 6 amperes. From these data let us find the minimum possible power factor of the shunted choking coil.

Numerical example.

Let OV (Fig. 76) represent the P.D. across the terminals of the choking coil and the condenser. Now if α and β be the phase differences between the choking coil current and the P.D. and between the condenser current and the P.D., then (Chapter X)

$$\cos \alpha = 0\cdot041, \quad \cos \beta = 0\cdot124.$$

Thus α is $87° 39'$ and β is $82° 53'$.

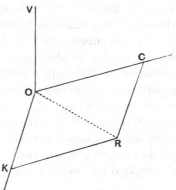

In Fig. 76 OV represents the applied voltage, OC the choking coil current and OK the condenser current. The angle VOC is α and the angle VOK is β. Now at every instant

$$i = i_1 + i_2,$$

where i is the current in the main and i_1, i_2 are the currents in

Fig. 76.

the branches. Hence a linear relation connects the three currents and the three current vectors will be in one plane. Let γ be the phase difference between i_1 and i_2, then, since their effective values are each equal to 6,

$$(1\cdot6)^2 = 6^2 + 6^2 + 2 \cdot 6^2 \cdot \cos \gamma;$$

and therefore $\gamma = 164° 26'$.

Since $\alpha + \beta$ equals $170° 32'$ it follows that OV, OC and OK are not in one plane. But OR, which represents the current in the main, is always in the plane COK.

The diagram (Fig. 76) shows us exactly what happens when we increase or diminish OK, since R always lies on CR which is a line drawn parallel to OK. We suppose that the phase differences between the applied potential difference, the condenser current and the choking coil current remain constant for all values of the condenser current. Hence the angles forming the solid angle at O remain constant. The power factor of the combined circuit is $\cos \phi$ where ϕ is the angle VOR. Now the minimum value of ϕ is got when the planes VOR and COK are at right angles to one another. In this case, by the formulae of spherical trigonometry, we get

$$\cos \phi = \frac{\sin \omega}{\sin \gamma},$$

where $\sin \omega = 2 \sqrt{\sin \sigma \sin (\sigma - \alpha) \sin (\sigma - \beta) \sin (\sigma - \gamma)},$

and σ is half the sum of the angles α, β and γ. Substituting the values of α, β and γ in this equation, we find that ϕ is $52° 24'$ and therefore the maximum possible value of the power factor is $0·61$. We have assumed that the shape of the wave of the applied P.D. does not alter as the capacity of the condenser is altered. An alteration in its shape would alter the power factors and phase differences, and hence the angles α, β and γ would vary with the shape of the wave. The diagram (Fig. 76) shows why it is in general impossible to make the power factor of a choking coil unity by shunting it with a condenser. If it were unity the angle VOR would be zero, and the vectors representing the two currents and the applied P.D. would lie in one plane.

REFERENCES

W. E. SUMPNER, 'The Vector Properties of Alternating Currents and other Periodic Quantities.' *Proceedings of the Royal Society*, Vol. 61, p. 465, 1897.

A. RUSSELL, *The Electrician*, Vol. 44, p. 49, 1897.

—— 'The Limitations of Graphical Methods in Electrical Theory.' *Journal of the Inst. of El. Eng.* Vol. 31, p. 440, 1901.

W. M. MORDEY, 'Capacity in Alternate Current Working.' *Journal of the Inst. of El. Eng.* Vol. 30, p. 864, 1901.

CHAPTER XIII

The measurement of power. The quadrant electrometer. The electrostatic wattmeter. Electrostatic wattmeter shunted. The electromagnetic wattmeter. Electromagnetic wattmeter, with mutual inductance. Watt-hour meters. The effects of friction and inertia. Reisz's method of power measurement. The three voltmeter method. The three ammeter method. Transformer methods. Resonance method. References.

WE have seen in Chapter X that if W be the mean power in watts expended in an alternating current circuit, V and A the effective volts and amperes, and $\cos \alpha$ the power factor, then

The measure-ment of power.

$$W = VA \cos \alpha.$$

The maximum possible value of the power is VA, and it has this value only when the phase difference α is zero, and this can only occur when the ratio of the instantaneous volts to the instantaneous amperes (e/i) is always constant (Chapter X). If, as is generally the case, α is not zero, then in order to find W we should need to know the shapes of the volt and ampere curves and their time lag relatively to one another. This could be done by means of an oscillograph, as it is not difficult to find the mean value of ei, that is W, from the curves. In practice, however, W is best found by means of some form of wattmeter, or by some of the methods described below. Before describing the electrostatic wattmeter we will give the theory of the quadrant electrometer, as the principle of the two instruments is the same.

In the quadrant electrometer, invented by Lord Kelvin, use is made of electrostatic attractions and repulsions to measure potential differences. In the ordinary form of this instrument there is a flat plate of aluminium shaped like

The quadrant electrometer.

the figure **8** inside an insulated cylindrical metal box which is completely divided into four quadrants (Fig. 77). The flat plate 3 is suspended by a torsion fibre perpendicular to its plane, and its position of equilibrium when the quadrants are at the same potential is shown in Fig. 77. The opposite quadrants 1 and 1′ are permanently connected by wires, so that they are always at the same potential, and so also are the quadrants 2 and 2′.

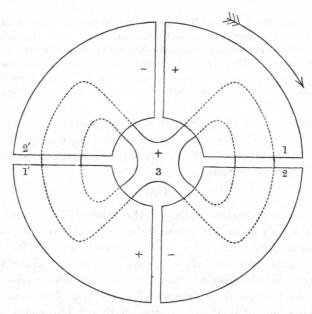

Fig. 77. The Kelvin Quadrant Electrometer.

Let the potential of the quadrants 1 and 1′ be V_1, of the quadrants 2 and 2′ be V_2 and of the plate be V_3. If V_1 lies in value between V_2 and V_3, there will evidently be repulsion between 3 and 1 and attraction between 3 and 2; and hence there will be a torque in the direction of the arrowhead. Now if Q_1, Q_2, Q_3 and Q_0 be respectively the quantities of electricity on the two pairs of quadrants, on the plate and on the inside of the earthed screen, which in the case of an electrostatic wattmeter is a metal cover practically enclosing the quadrants, then

$$Q_1 + Q_2 + Q_3 + Q_0 = 0 \ldots\ldots\ldots\ldots\ldots(1).$$

By Chapter v, we have, since V_0 is zero

$$Q_0 = K_{0.1}V_1 + K_{0.2}V_2 + K_{0.3}V_3,$$

$$Q_1 = K_{1.1}V_1 + K_{1.2}V_2 + K_{1.3}V_3,$$

$$Q_2 = K_{2.1}V_1 + K_{2.2}V_2 + K_{2.3}V_3,$$

$$Q_3 = K_{3.1}V_1 + K_{3.2}V_2 + K_{3.3}V_3.$$

Now (1) must be satisfied for all values of V_1, V_2 and V_3.

Therefore $\quad K_{0.1} + K_{1.1} + K_{2.1} + K_{3.1} = 0$ ....(2)

and $\qquad K_{0.2} + K_{1.2} + K_{2.2} + K_{3.2} = 0$ ....(3)

Now let the plate turn through an angle θ, and let $K'_{1.1}$, $K'_{1.2}$,... be the new values of the coefficients. These coefficients must still satisfy equations (2) and (3). Now, owing to the shape of the plate, we see, since the gaps between the quadrants are very narrow, that so long as the edges of 1 and 3 or 1' and 3' are not brought too close together

$$K'_{3.3} = K_{3.3} = \text{constant},$$

for the motion merely brings a different part of 3 opposite the gaps between 1 and 2 and between 1' and 2'. Also since the motion of the plate does not appreciably alter the coefficients of mutual induction for electrostatic charges between 1 and 2, 0 and 1, and 0 and 2, we have

$$K'_{1.2} = K_{1.2}, \quad K'_{0.1} = K_{0.1}, \quad K'_{0.2} = K_{0.2}.$$

Hence from (2)

$$K'_{1.1} + K'_{1.3} = -K'_{0.1} - K'_{1.2}$$

$$= \text{constant}$$

$$= K_{1.1} + K_{1.3} \dots\dots\dots\dots\dots(4).$$

Similarly $\quad K'_{2.2} + K'_{2.3} = K_{2.2} + K_{2.3} \dots\dots\dots\dots\dots(5).$

As θ increases, the parts of the quadrants 1 and 1' opposed to the plate 3 diminish uniformly, and hence we may write

$$K'_{1.1} = K_{1.1} - \lambda\theta \dots\dots\dots\dots\dots(6),$$

where λ is a constant. For the same reason

$$K'_{2.2} = K_{2.2} + \lambda\theta \dots\dots\dots\dots\dots(7).$$

From (4) and (6) $K'_{1.3} = K_{1.3} + \lambda\theta,$

and from (5) and (7) $K'_{2.3} = K_{2.3} - \lambda\theta.$

The electrical energy W of the system in the new position is given by (see p. 147)

$$W = \tfrac{1}{2} K'_{1.1} V_1^2 + \tfrac{1}{2} K'_{2.2} V_2^2 + \tfrac{1}{2} K'_{3.3} V_3^2$$
$$+ K'_{2.3} V_2 V_3 + K'_{3.1} V_3 V_1 + K'_{1.2} V_1 V_2.$$

Now the work done by the electrical forces on the plate when it turns through an angle $\partial\theta$

= The moment of the forces about its axis $\times \partial\theta$,

but, when the potentials are kept constant, the gain ∂W in the electrical energy (see p. 150) is equal to the work done by the electric forces on the moving plate and is therefore given by

$$\partial W = \text{Torque} \times \partial\theta.$$

Hence $\text{Torque} = \dfrac{\partial W}{\partial\theta}.$

If therefore V_1, V_2 and V_3 are kept constant, we see that the torque equals

$$\tfrac{1}{2} V_1^2 \frac{\partial K'_{1.1}}{\partial\theta} + \tfrac{1}{2} V_2^2 \frac{\partial K'_{2.2}}{\partial\theta} + V_2 V_3 \frac{\partial K'_{2.3}}{\partial\theta} + V_3 V_1 \frac{\partial K'_{3.1}}{\partial\theta}$$
$$= -\tfrac{1}{2}\lambda V_1^2 + \tfrac{1}{2}\lambda V_2^2 - \lambda V_2 V_3 + \lambda V_3 V_1$$
$$= \lambda (V_1 - V_2) \{ V_3 - \tfrac{1}{2} (V_1 + V_2)\}.$$

If the needle be put in metallic connection with the quadrants, 1 and 1', then V_3 equals V_1 and the formula becomes

$$\text{Torque} = \frac{\lambda}{2} (V_1 - V_2)^2.$$

Used in this manner the electrometer becomes a voltmeter. Equilibrium is attained when the torsional couple is equal to the electrical couple. With direct currents the reading is proportional to the square of the P.D. between its terminals, and with alternating currents it is proportional to the mean value of the square of the P.D. The most satisfactory way of reading the deflection is by means of a ray of light reflected on to a scale from a light mirror fixed to the axis of the plate. If the scale is direct reading, then the divisions on the upper part of the scale will be much larger than those lower down, because the deflections increase as the square of the applied voltage.

This instrument, shown diagrammatically in Fig. 78, is practically a slight modification of the quadrant electrometer described above.

The electrostatic wattmeter.

In the figure, Q_1 represents one pair of the quadrants, Q_2 another pair, and N the needle or plate. The plate

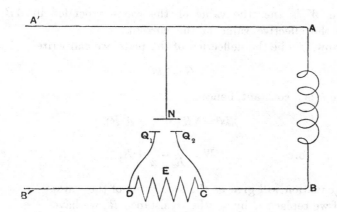

Fig. 78.   Connections of the Electrostatic Wattmeter.

is generally connected to an insulated terminal on the case of the instrument by means of a phosphor bronze suspension strip which replaces the torsion fibre in the electrometer. Suppose that we wish to measure the electric power being expended in the coil AB. Place a small resistance R (DC) in the main BB' and connect the plate N of the instrument to A. Let v_1 be the instantaneous value of the potential of A, v_2 the potential of B, and v_3 the potential of D. Then, by what we have shown above, the torque g acting on the plate will be given by

$$g = \lambda\,(v_2 - v_3)\,\{v_1 - \tfrac{1}{2}\,(v_2 + v_3)\},$$

where λ is a constant depending on the instrument.

If DC be non-inductive, then $v_2 - v_3 = Ri$, where i is the instantaneous value of the current. Hence

$$g = \lambda Ri\,\{v_1 - \tfrac{1}{2}\,(v_2 + v_3)\}.$$

Now, if E be the middle point of DC, its potential is $\tfrac{1}{2}\,(v_2 + v_3)$

and $v_1 - \frac{1}{2}(v_2 + v_3)$ is the P.D. between A and E. Therefore, if G be the mean value of the torque g,

$$G = \lambda R \times \text{watts expended in } AE$$

$$= \lambda R (W + \frac{1}{2} A^2 R),$$

where W is the true value of the watts expended in AB and A is the effective value of the current.

Now, if θ be the deflection of the plate, we can write

$$G = \lambda k\theta,$$

where k is a constant, hence

$$\lambda k\theta = \lambda R (W + \frac{1}{2} A^2 R),$$

and therefore $$W = \frac{k\theta}{R} - \frac{1}{2} A^2 R \dots\dots\dots\dots\dots\dots(8).$$

This equation will give us the true value of the watts.

If we replace R by another resistance R', we have

$$W = \frac{k\theta}{R'} - \frac{1}{2} A^2 R'.$$

Thus we can increase the range of the loads that the wattmeter can measure by making a suitable set of resistances to put in series with the mains. In practice it is not desirable that the potential drop at full load across the series resistance should be more than one per cent. of the potential difference between the mains.

The constant k can be found by using the instrument to measure the power expended in a non-inductive circuit. For example, if V be the effective voltage across AB, and θ_1 be the deflection in this case, then

$$k = \frac{R}{\theta_1} (VA + \frac{1}{2} A^2 R).$$

In practice it is not convenient to have a P.D. of more than about 200 volts between the plate and the quadrants. Hence for measuring power in high tension circuits the instrument has to be used with a shunt.

The connections for this case are shown in Fig. 79. $A'B'$

Electrostatic wattmeter shunted.

represents a non-inductive shunt which is put across the mains and the plate is connected to a point L in the shunt. Let the resistance of $A'L$ be R_1 and

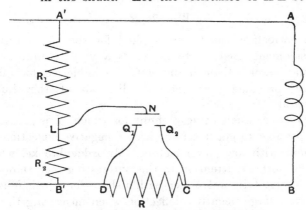

Fig. 79. Electrostatic Wattmeter with shunt.

of LB' be R_2. Let v_1 and v_1' be the potentials of A' and L respectively, then, since the shunt is non-inductive,

$$\frac{v_1 - v_3}{R_1 + R_2} = \frac{v_1' - v_3}{R_2};$$

and thus

$$v_1' - v_3 = \frac{1}{N}(v_1 - v_3),$$

where

$$N = \frac{R_1 + R_2}{R_2}$$

= the multiplying power of the shunt.

As formerly

$$g = \lambda(v_2 - v_3)\{v_1' - \tfrac{1}{2}(v_2 + v_3)\}$$
$$= \lambda Ri\{v_1' - v_3 - \tfrac{1}{2}(v_2 - v_3)\}$$
$$= \lambda Ri\left\{\frac{1}{N}(v_1 - v_3) - \frac{1}{2}(v_2 - v_3)\right\}$$
$$= \frac{\lambda R}{N}(w + Ri^2) - \frac{1}{2}\lambda R^2 i^2.$$

Thus

$$G = \frac{\lambda R}{N}(W + RA^2) - \frac{1}{2}\lambda R^2 A^2,$$

and

$$W + RA^2 = \frac{N}{\lambda R}\left(\lambda k\theta + \frac{1}{2}\lambda R^2 A^2\right);$$

and therefore

$$W = Nk\frac{\theta}{R} + \frac{N-2}{2}RA^2 \quad\ldots\ldots\ldots\ldots\ldots(9).$$

In this formula for W, k has the same value as when the wattmeter is unshunted.

An interesting case arises when we use the wattmeter with a shunt multiplier of 2. The formula then becomes

$$W = (2k/R)\,\theta \quad\dots\dots\dots\dots\dots\dots(10)$$

and no correction has to be applied for the series resistance. The instrument used in this way is a very accurate one for measuring power. Even if the shunt be inductive, formula (10) is correct provided that the two divisions of the shunt are symmetrical.

In practice, when using a shunt for which N is greater than 2, it is possible to get a zero or even a negative deflection of the instrument with low power factors. Mr Addenbrooke, who first called the author's attention to this interesting fact, showed him that an electrostatic wattmeter with a shunt for which N is equal to 10 gave a large negative deflection when measuring the power absorbed by a condenser load. It is easy to see that when W is less than $\{(N-2)/2\}\,RA^2$ we must get a negative reading on the instrument.

When the deflection is zero, then

$$W = \{(N-2)/2\}\,RA^2 \dots\dots\dots\dots\dots(11).$$

Hence the instrument could be used to measure power by means of a null method. R_2 could easily be made variable so that it could be adjusted until the deflection were zero. Then since $N = (R_1 + R_2)/R_2$, W could be found from (11) and the reading of an ammeter in the circuit.

In the general form of electromagnetic wattmeter we have two coils, of which one is fixed and carries the main current while the other, which is movable and in series with a high nearly non-inductive resistance, is placed as a shunt across the circuit. The coils are placed with their axes at right angles, so that when currents are passing through both there is a couple tending to turn them so as to make their axes coincide. The torque on the movable coil is proportional to the product of the two currents so long as that coil is kept in the same position. The movable coil is brought back

The electromagnetic wattmeter.

to its initial position by means of a torsion head connected to it by a spiral spring, so that the angle turned round by the head to which the pointer is attached is proportional to the torque acting on the coil. If g be the torque, we may write

$$g = \lambda i i_1,$$

where i and i_1 are the currents in the series and shunt coils respectively. With direct currents, if θ be the angle turned through by the torsion head, then g equals $k'\lambda\theta$, where k' is a constant, and hence

$$k'\lambda\theta = \lambda i i_1.$$

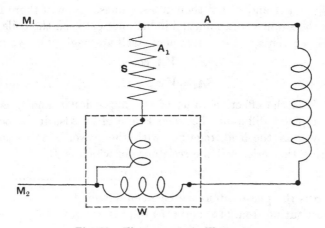

Fig. 80. Electromagnetic Wattmeter.

Now if S be the total resistance of the shunt circuit (Fig. 80) and e the P.D. applied at its terminals,

$$k'\lambda\theta = \lambda i \, (e/S).$$

Therefore $$\theta = (1/Sk') \, W,$$

where W is the true power expended in the load, together with the power Ri^2 expended in heating the series coil the resistance of which is R, and thus

$$W = k\theta \quad \dots\dots\dots\dots\dots\dots\dots\dots(12),$$

where k equals Sk' and is constant. The watts then will be proportional to the reading of the instrument, and, since in this case they are equal to the product of the volts and the amperes, k can easily be determined.

With alternating currents we still have

$$g = \lambda i i_1,$$

where g, i and i_1 are the instantaneous values of the torque and amperes respectively. Therefore the mean value of g (G) is given by

$$G = \lambda \times \text{the mean value of } i i_1$$

$$= \lambda A A_1 \cos \alpha \quad \dots\dots\dots\dots\dots\dots\dots(13),$$

where α is the phase difference between the periodic functions i and i_1, and A and A_1 are their R.M.S. values. Now if there is no iron in the shunt coil, no mutual inductance between the coils, and the eddy currents in the instrument itself are negligible, we have

$$S A_1{}^2 = V A_1 \cos \gamma,$$

and thus

$$S A_1 = V \cos \gamma \quad \dots\dots\dots\dots\dots\dots(14),$$

where V is the effective value of the applied P.D. and γ is the angle of phase difference between V and A_1. Also if W be the true power in the load, together with the power $R A^2$ expended in heating the series coil the resistance of which is R,

$$W = V A \cos \beta \quad \dots\dots\dots\dots\dots\dots(15),$$

where β is the phase difference between V and A.

Substituting from (14) and (15) in (13), we get

$$G = \frac{\lambda}{S} W \frac{\cos \alpha \cos \gamma}{\cos \beta}.$$

Now G equals $\lambda k'\theta$, where θ is the mean value of the deflection. With the frequencies used in practice, θ is constant, and hence

$$W = k\theta \times \frac{\cos \beta}{\cos \alpha \cos \gamma} \quad \dots\dots\dots\dots(16).$$

Now the wattmeter is calibrated with direct currents, and therefore indicates $k\theta$ watts. We see that, when the wattmeter is used on an alternating current circuit, the readings of the instrument must be multiplied by

$$\frac{\cos \beta}{\cos \alpha \cos \gamma}.$$

In Fig. 81, where V, A and A_1 are represented graphically, it must be remembered that we cannot assume that their vectors are in one plane. In general therefore β will be less than $\alpha + \gamma$. If there were no phase difference between the applied P.D. and the current in the shunt coil, then γ would be zero, and β would be equal to α. In this case the instrument would indicate alternating current power correctly. In practice however, γ, although small, cannot be neglected, as it may introduce a large error when measuring the electric power in circuits with a low power factor.

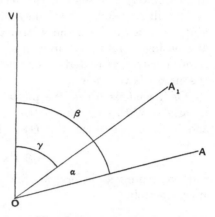

Fig. 81.   Currents in shunt and series coils of wattmeter. Vectors are in different planes.

Let $$\alpha = \beta - \gamma + x,$$

then from the geometry of Fig. 81 we see that $\beta + \gamma$ cannot be less than α, and therefore the maximum possible value of x is 2γ. Hence if γ is small x also is small. Therefore, if these angles be expressed in circular measure, we may write

$$\cos \gamma = \cos (\gamma - x) = 1, \quad \sin \gamma = \gamma \quad \text{and} \quad \sin (\gamma - x) = \gamma - x$$

very approximately. The correcting factor becomes

$$\frac{\cos \beta}{\cos \alpha \cos \gamma} = \frac{\cos \beta}{\cos \{\beta - (\gamma - x)\}}$$

$$= \frac{\cos \beta}{\cos \beta + (\gamma - x) \sin \beta} \quad \ldots\ldots\ldots(17).$$

For very low power factors, when β is nearly ninety degrees, (17) may have almost any value, as it depends chiefly on the value of $\gamma - x$. If the potential difference and current waves are sine-shaped, OV, OA and OA_1 will be in the same plane, and if the inductance of the shunt coil be greater than its capacity, as it generally is in practice, OA and OA_1 will lie on the same side of

21—2

OV for inductive loads and x will be zero. If the waves are approximately sine-shaped, the vectors will be nearly in one plane and x will be small. It will be seen from (17) that if γ is appreciable, then in this case the wattmeter generally reads too high in alternating current measurements. The electromagnetic wattmeter as ordinarily constructed cannot be used when the power factor of the circuit is very low.

On a condenser load the phase difference α between A and A_1 is large. With sine waves, OA and OA_1 would lie in one plane and be on opposite sides of OV. In this case we may therefore write

$$\alpha = \beta + \gamma - x,$$

and proceeding as before we find that the correcting factor is approximately

$$\frac{\cos \beta}{\cos \beta - (\gamma - x)\sin \beta}\dots\dots\dots\dots\dots(18).$$

When $\gamma - x = \cot \beta$ we see that the wattmeter reads zero, and when $\gamma - x$ is greater than $\cot \beta$ we get a negative deflection. Now when $\cos \beta$ is very small so also is $\cot \beta$. The wattmeter therefore does not give trustworthy readings on a condenser load when the power factor is low. The errors of this instrument are as a rule greater the higher the frequency, as γ is practically proportional to the frequency.

In the theory of the electromagnetic wattmeter discussed

Electromagnetic wattmeter, with mutual inductance. above we have assumed that the shunt and series coils are so situated that the mutual inductance between them is zero. We shall now consider the case when the mutual inductance cannot be neglected. Let the constants of the shunt coil be (S, L_1), of the series coil (R, L) and let M be the mutual inductance between the coils. Then if e and v are the instantaneous values of the potential differences across the shunt coil and across the load respectively, our equations are

$$e = Ri + v + L\frac{\partial i}{\partial t} + M\frac{\partial i_1}{\partial t}\dots\dots\dots\dots(19),$$

$$e = Si_1 + L_1\frac{\partial i_1}{\partial t} + M\frac{\partial i}{\partial t}\dots\dots\dots\dots(20).$$

Multiplying both sides of equations (19) and (20) by i and taking mean values, we get

$$VA \cos \beta = W + \frac{M}{T} \int_0^T i \frac{\partial i_1}{\partial t} \partial t$$

and

$$VA \cos \beta = SAA_1 \cos \alpha + \frac{L_1}{T} \int_0^T i \frac{\partial i_1}{\partial t} \partial t.$$

Thus eliminating the integral we have

$$W = \frac{MS}{L_1} AA_1 \cos \alpha + \frac{L_1 - M}{L_1} VA \cos \beta$$

$$= \frac{M}{L_1} k\theta + \frac{L_1 - M}{L_1} VA \cos \beta \quad \dots\dots\dots\dots(21).$$

When M is zero this reduces to the formula (15) which we have discussed above. In the particular case when M equals L_1 the formula (21) becomes

$$W = k\theta,$$

and therefore, when the self inductance of the shunt coil equals the mutual inductance between the coils, the instrument reads correctly.

The above theorem suggests the following method of constructing an electromagnetic wattmeter. Suspend the shunt coil so that its self inductance equals the mutual inductance between the coils, and adjust the controlling spring till the pointer reads zero in this position. When it is used to measure power, and currents are passing through both coils, the shunt coil is brought back to its initial position by means of a torsion head and the pointer reads the power. The scale is evenly divided, and one reading with direct current is sufficient to find the constant of the instrument. In order that M may be equal to L_1 it is necessary that the self inductance of the series coil should be greater than the self inductance of the shunt coil.

Various kinds of watt-hour meters are used in practice to
Watt-hour   measure the total energy expended in a given time
meters.   in an installation. If the meter be theoretically
correct, the number of revolutions of the spindle in a given time must be proportional to the watt-seconds expended. If r be the rate of the meter, that is, the number of revolutions per second per

watt, we have $r = n/(Wt)$, where n is the number of revolutions in the time t seconds, when W watts are being taken by the consumer.

Hence $$W = n/(rt).$$

Therefore if r be known W can be found.

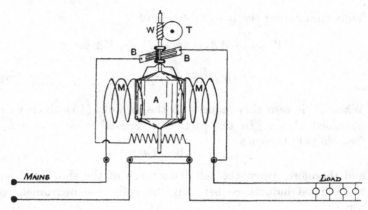

Fig. 82.

Most watt-hour meters cannot, however, be trusted to read accurately when the power factor of the circuit is low unless very special precautions are taken in their manufacture. Consider for example the well-known Elihu Thomson watt-hour meter (Fig. 82). In this meter the shunt circuit consists of a drum-wound armature A, the axis of which is the spindle. The worm wheel W on the spindle drives the toothed wheel T which actuates the registering mechanism. The armature is in series with a high nearly non-inductive resistance. It is placed in the field due to the series coils, the current entering and leaving it by the brushes B, which press on a commutator. It rotates for the same reason that the armature of a motor rotates. Now, if A_1 be the current in the shunt circuit, then, since the field due to the series coils is proportional to the current in them, that is to the main current A, we see that the mean value of the driving torque is proportional to $A A_1 \cos \alpha$. The retarding torque when the spindle rotates is due to the eddy currents induced by permanent magnets in a disc of copper fixed to the spindle. Owing to the inductance of the circuits in which the eddy currents flow, they do not come instantaneously into existence, neither do they die away instantaneously. By Lenz's law they retard the motion of the disc and, since the field due to the permanent magnets is constant, their values are proportional to the angular velocity ($2\pi n$) of the spindle. Therefore, neglecting friction, the retarding torque is proportional to n, and the driving torque to $AA_1 \cos \alpha$. Proceeding as in the case of the electromagnetic wattmeter, and assuming that the mutual inductance between the shunt and series circuits is negligible, we find that

$$W = \frac{n}{rt} \frac{\cos \beta}{\cos \alpha \cos \gamma},$$

where r is the rate found by direct currents. Hence we see that unless γ is zero, *i.e.* unless the shunt circuit is absolutely non-inductive, the reading on an inductive load will in general be too high. Also on a condenser load if $\gamma - x$ is greater than cot β, the meter will run backwards.

In the preceding section we have neglected the effects of both the friction of the bearings and the wind friction on the rotating part. We have also neglected the inertia of the moving parts of the meter. It is obvious that this inertia prevents it from instantaneously acquiring its proper speed when the current is switched on, and that it also prevents it from stopping instantaneously when the current is switched off. It will thus under-register at the start and over-register at the finish. It is important, therefore, in those cases where the current through the meter is continually fluctuating as, for example, when it is connected with a main supplying a traction load, to consider how far these errors counteract one another.

It is found by experiment that the retarding torque due to 'solid friction,' that is, to the friction of the bearings, is not quite constant. We shall denote it by $\mu (\partial\theta/\partial t) + \tau$, where μ and τ are constants and θ is the angle turned through by the rotor, that is, the rotating part of the meter. At the low speeds used in practice the torque due to the wind friction is approximately proportional to $\partial\theta/\partial t$; we shall denote it by $\lambda (\partial\theta/\partial t)$. By the construction of the meter the driving torque is proportional to W, where W is the power expended on the load. We can denote it therefore by aW, where a is a constant. In addition there is usually a compensating coil in the shunt circuit through which a small current is always flowing. This produces a driving torque τ' insufficient to start the rotor but which is adjusted so as nearly to annul that part of the solid friction which is independent of the velocity. The equation of motion is therefore

$$K\,\frac{\partial^2\theta}{\partial t^2} + (\lambda + \mu)\,\frac{\partial\theta}{\partial t} + \tau = a\dot{W} + \tau',$$

where K is the moment of inertia of the rotor.

Hence $K\,\dfrac{\partial\theta}{\partial t} + (\lambda + \mu)\,\theta = (aW + \tau' - \tau)\,t + A,$

where A is a constant. Since both θ and $\partial\theta/\partial t$ are zero initially, A is zero. Hence if the driving torque due to the load be aW_1 for t_1 seconds, we have

$$K\omega_1 + (\lambda + \mu)\,\theta_1 = (aW_1 + \tau' - \tau)\,t_1,$$

where θ_1 and ω_1 are the values of θ and $\partial\theta/\partial t$ after t_1 seconds. Similarly if aW_2 be the driving torque for the next t_2 seconds during which the rotor turns through an angle θ_2, we have

$$K\omega_2 + (\lambda + \mu)\,\theta_2 = (aW_2 + \tau' - \tau)\,t_2 + K\omega_1.$$

Hence also
$$K\omega_3 + (\lambda + \mu)\,\theta_3 = (aW_3 + \tau' - \tau)\,t_3 + K\omega_2,$$
....
and finally
$$K\omega_n + (\lambda + \mu)\,\theta_n = (aW_n + \tau' - \tau)\,t_n + K\omega_{n-1}.$$

If ω_n is zero, we see by adding these equations that

$$(\lambda + \mu)\,\Sigma\theta = a\Sigma Wt + (\tau' - \tau)\,\Sigma t.$$

Since ΣWt is the energy taken by the consumer and $\Sigma\theta$ is proportional to the energy recorded by the meter, we see that the meter can be made to register correctly provided that

$$(\tau' - \tau)\,\Sigma t$$

is negligible compared with $a\Sigma Wt$. We also see that the reading of the meter is quite independent of the value of the moment of inertia of its rotor.

If however the current fluctuate with such rapidity that it is comparable to an ordinary alternating current, then the preceding section shows that errors will be introduced, as the driving torque is no longer exactly proportional to the load. Meters sometimes read incorrectly owing to their being placed on switchboards where they are subjected to vibration and stray magnetic fields arising from currents in neighbouring cables. The vibration alters the bearing friction and the stray magnetic fields the driving torque, and thus both affect the rate of the meter. As a rule also the rate of the meter varies slightly with temperature owing to the alteration in the resistance of the shunt windings.

In Reisz's method of power measurement we require an **Reisz's method of power measurement.** electrostatic voltmeter, an ammeter and two high non-inductive resistances of known values R_1 and R_2. Let BAC (Fig. 83) be the circuit in which we wish to measure the power.

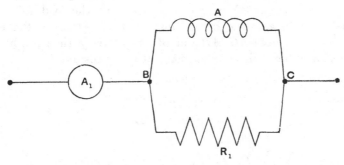

Fig. 83. Reisz's method.

Place an ammeter A_1 in the main circuit, and place R_1 as a shunt across the circuit BAC. Then if i_1 be the main current, e the P.D. across BC and i the current in BAC, we have

$$i_1 = i + e/R_1.$$

Hence
$$i_1^2 = i^2 + e^2/R_1^2 + 2w/R_1,$$

where $w = ei =$ the instantaneous watts expended in BAC.

By summation $A_1^2 = A^2 + V^2/R_1^2 + 2W/R_1.$

Similarly $\qquad A_2^2 = A^2 + V^2/R_2^2 + 2W/R_2.$

Therefore $\qquad W = b\,(A_1{}^2 - A_2{}^2) - aV^2$ ....(22),

where $\qquad a = (1/2)\,(1/R_1 + 1/R_2)$,

and $\qquad b = R_1 R_2/\{2\,(R_2 - R_1)\}$.

If R_2 is infinite, then (22) becomes

$$W = (R_1/2)\,(A_1{}^2 - A_2{}^2) - V^2/(2R_1) \quad \ldots\ldots(23).$$

If the power factor were zero, W would be zero, and we see from (22) that $b\,(A_1{}^2 - A_2{}^2)$ would then have its minimum value aV^2. For low power factors, therefore, W is the difference of two large quantities and a very small error in the ammeter or voltmeter readings may introduce a large error in the value of W calculated from (22). It is also to be noticed that we make the assumption that the shape of the wave of the applied potential difference is not altered by shunting the circuit by the high non-inductive resistances.

In the three voltmeter method a non-inductive resistance R (Fig. 84) is put in series with the load BC. Let

The three voltmeter method. e, e_1 and e_2 be the instantaneous values of the P.D.s across AC, AB and BC respectively, then if i be the current in the circuit, we shall have at every instant

$$e = e_1 + e_2.$$

Therefore $\qquad e^2 = e_1{}^2 + e_2{}^2 + 2e_1 e_2$

$$= e_1{}^2 + e_2{}^2 + 2Rie_2.$$

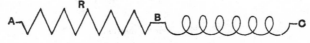

Fig. 84. Three Voltmeter method.

If, therefore, W be the average value of the power ie_2 expended in BC, we have by summation

$$V^2 = V_1{}^2 + V_2{}^2 + 2RW,$$

and thus $\qquad W = \{1/(2R)\}\,(V^2 - V_1{}^2 - V_2{}^2)$

$$= \{A/(2V_1)\}\,(V^2 - V_1{}^2 - V_2{}^2) \quad \ldots\ldots\ldots\ldots(24),$$

where A is the effective value of the current measured by an ammeter and V, V_1 and V_2 are the readings of voltmeters placed across

AC, AB and BC respectively. If the supply voltage is sufficiently steady, one voltmeter can be used to read V, V_1 and V_2.

Suppose that three voltmeters are used and that $\pm a$, $\pm b$, $\pm c$ are the percentage errors of the readings V, V_1 and V_2 and that the ammeter reads correctly, then it can be shown that the maximum possible error in the value of W given by (24) is a minimum when

$$V_2 = V_1 \{(a + c)/(a + b)\}^{\frac{1}{2}},$$

and in this case the maximum percentage error in W is

$$2a + b + (2/\cos \phi) \sqrt{(a + b)(a + c)},$$

where $\cos \phi$ is the power factor of the circuit. Suppose that the three voltmeters are all equally trustworthy, then the maximum percentage error is

$$3a + 4a/\cos \phi.$$

If the voltmeters could be trusted to read correctly to within ± 0.2 per cent., then for a power factor of unity the maximum error would be 1.4 per cent. and for a power factor of 0.01 it would be 80.6 per cent. Owing to the unavoidable errors of observation, with ordinary commercial instruments and on circuits where the voltage is unsteady, formula (24) often gives negative values for W when $\cos \phi$ is small.

The practical objection to this method is that the testing pressure applied between A and C has to be nearly double the working pressure applied between B and C, and this increased pressure is not always available. Another difficulty is that the shape of the wave of the applied potential difference is generally altered when we put a non-inductive resistance in series with the load. The first of these difficulties can be met by using the three ammeter method.

In the three ammeter method a non-inductive resistance R (Fig. 85) is placed as a shunt across the load, and ammeters are placed in the branch circuits and in the main. If i be the instantaneous value of the current in the main, then

The three ammeter method.

$$i = i_1 + i_2.$$

Therefore
$$i^2 = i_1{}^2 + i_2{}^2 + 2i_1 i_2$$
$$= i_1{}^2 + i_2{}^2 + (2/R)\, e i_2,$$

where e is the applied P.D. Hence by summation

$$A^2 = A_1{}^2 + A_2{}^2 + (2/R)\, W.$$

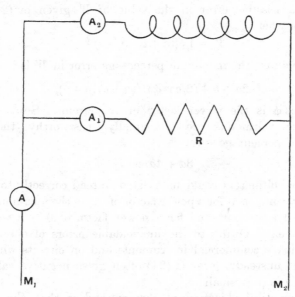

Fig. 85. Three Ammeter method.

Thus
$$W = (R/2)\,(A^2 - A_1{}^2 - A_2{}^2)$$
$$= \{V/(2A_1)\}\,(A^2 - A_1{}^2 - A_2{}^2) \dots\dots\dots\dots(25),$$

where V is the applied P.D. and A_1 the current in the non-inductive resistance. When the power factor is very small, W is small and hence, from (25), A^2 is nearly equal to $A_1{}^2 + A_2{}^2$. Therefore in this case a small error in measuring any of the currents may introduce a large error into the value of W calculated from (25).

There are several methods of measuring power by means of
Transformer   transformers. These are in general based on the fact
methods.   that the phase difference between the primary and the
secondary voltage of a transformer on a light load is almost exactly
180 degrees. Let e_1 and e_2 be the primary and secondary voltages,

i_1 and i_2, r_1 and r_2, n_1 and n_2 be the currents, resistances and turns respectively; then if ϕ be the flux in the core and there is no magnetic leakage,

$$e_1 = r_1 i_1 + n_1 \, (\partial\phi/\partial t),$$

$$e_2 + r_2 i_2 = - n_2 \, (\partial\phi/\partial t) \, ;$$

and therefore $e_1 + (n_1/n_2) \, e_2 = r_1 i_1 - (n_1/n_2) \, r_2 i_2 \ \ldots\ldots\ldots\ldots(26).$

Now, in a well designed closed iron circuit potential transformer on open secondary, the maximum value of e_1 is about ten thousand times greater than the maximum value of $r_1 i_1$, and therefore, in this case, we can write

$$e_2 = - (n_2/n_1) \, e_1,$$

without appreciable error. This equation shows that e_2 and e_1 are similar curves and differ in phase by 180 degrees.

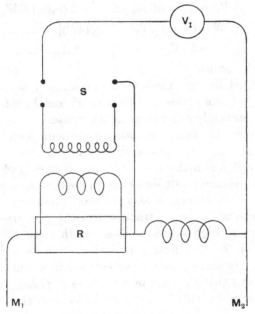

Fig. 86. Campbell's method.

The method shown in Fig. 86 is due to Albert Campbell. A small non-inductive resistance R is placed in series with the load, and the primary of a suitable transformer is placed across it.

By means of a reversing switch S the secondary voltage e_2 of this transformer can be added to, or subtracted from, the P.D. e applied across the load, and the resultant measured by an electrostatic voltmeter. Let v_1 and v_2 be the two resultant voltages, then

$$v_1 = e + e_2$$
$$= e - (n_2/n_1)\, e_1$$
$$= e - (n_2/n_1)\, Ri'',$$

where i' is the current in R.

Now, if the magnetising current of the transformer be very small compared with the current i in the load, then i' is approximately equal to i and

$$v_1 = e - (n_2/n_1)\, Ri.$$

Thus $$V_1^2 = V^2 + (n_2^2/n_1^2)\, R^2 A^2 - 2\,(n_2/n_1)\, R W.$$

Similarly $$V_2^2 = V^2 + (n_2^2/n_1^2)\, R^2 A^2 + 2\,(n_2/n_1)\, R W.$$

Hence $$W = (n_1/n_2)\, \{(V_2^2 - V_1^2)/(4R)\} \dots\dots\dots\dots(27)$$
$$= k\,(V_2 - V_1)\,(V_2 + V_1) \dots\dots\dots\dots(28),$$

where k is a constant.

This method, like the three voltmeter method, fails when W is very small; but as a series of values of V_2 and V_1 can rapidly be taken, it is practically convenient in other cases.

Elihu Thomson, Aron and other watt-hour meters for high tension circuits are often provided with a 'potential' transformer for the volt coil, and make use of the fact that for light loads the primary and secondary volt waves are in opposition in phase.

The series coils of meters also are sometimes connected across the secondaries of 'current' transformers the primaries of which are in series with high tension mains. In this case the magnetising current of the transformer must be small, as it is assumed that the primary and secondary currents are in a constant ratio to one another and that they are in opposition in phase. The theory of these transformers will be discussed in Volume II.

For measuring the dielectric losses in condensers, Rosa and
Resonance method. Smith make use of the principle of resonance. A drum of flexible cable AD (Fig. 87) is put in series with the condenser K, which, for example, may be a concentric or

a polyphase cable on open circuit. *A* and *B* are connected to
low pressure supply mains, and a wattmeter *W* is placed in the
circuit so that it measures the power taken by both the inductive
coil and the condenser. If the core of the cable be made of
many strands of very fine wire and there be no iron or metal near
and the frequency be not too high, then the power consumed by
the coil will be very nearly A^2R, where *A* is the current in the
circuit and *R* is the resistance of the coil. Now if the inductance

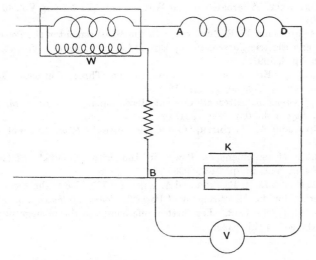

Fig. 87. Resonance method.

of the coil be adjusted by unwinding some of the cable, then, for a
certain inductance, resonance of the first harmonic will ensue and
the voltage *V* across the condenser may be 20 or 30 times greater
than the P.D. applied across *AB*. In this case the power factor
of the resonant circuit may be nearly unity and the wattmeter
will read the load *W* accurately. Therefore, subtracting the power
expended in the inductive coil, we find $W - A^2R$ as the condenser
loss. In this case, although the power factor may be as small as
0·01, the loss could be measured with fair accuracy.

REFERENCES

J. HOPKINSON, 'On the Quadrant Electrometer.' *Proceedings of the Physical Society of London*, Vol. 7, p. 7, March 14, 1885.

J. SWINBURNE, 'The Electrometer as a Wattmeter.' *Phil. Mag.* Vol. 31, p. 504, 1891.

G. L. ADDENBROOKE, 'Alternate Current Measurement by Electrometers.' *Electrician*, Vol. 45, p. 901, 1900.

C. V. DRYSDALE, 'Alternate Current Wattmeter.' *Electrician*, Vol. 46, p. 774, 1901.

W. E. AYRTON and W. E. SUMPNER, 'On the Measurement of the Power given by any Electric Current to any Circuit.' *Proc. Roy. Soc.* Vol. 49, p. 434, 9th April, 1891.

A. RUSSELL, 'Errors of Observation in the Three Voltmeter Method.' *Electrician*, Vol. 30, p. 241, 1892.

ALBERT CAMPBELL, 'Alternate Current Measurement.' *Journ. of the Inst. of El. Eng.* Vol. 30, p. 889, 1901.

E. B. ROSA and A. W. SMITH, 'Condenser Losses.' *Phys. Rev.* Vol. 8, p. 1, 1899.

E. REISZ, 'Measurement of Power in Inductive Circuits.' *Elektrotechn. Zeitschr.* Vol. 21, p. 713, 1900.

C. C. PATERSON, E. H. RAYNER and A. KINNES, 'The Use of the Electrostatic Method for the Measurement of Power.' *Journ. of the Inst. of El. Eng.* Vol. 51, p. 294, 1913. For further references see the bibliography given at the end of this paper.

CHAPTER XIV

I<small>F</small> we have two neighbouring electric circuits and an alternating
The air core   current be flowing in one of them, then, owing to
transformer.   their mutual induction, an alternating E. M. F. will be
induced in the other, and if this be a closed circuit an alternating
current will be set up in it. The magnitude of this induced
current will depend on the relative position of the two circuits and
the permeability of the medium in which they are placed. We
shall for the present consider that there are no magnetic materials
near the circuits, so that we may take the permeability of the
medium to be unity; the two circuits will then have a constant
mutual inductance. This is the problem of the air core transformer
and, as its solution is of fundamental importance in alternating
current theory, we shall first of all attempt to solve it without
making any assumption as to the shape of the wave of the
applied P. D.

Let e be the instantaneous value of the P. D. applied to the
Equivalent   primary circuit, whose resistance and inductance are
net-work. R_1 ohms and L_1 henrys respectively, and let R_2 and
L_2 be the resistance and inductance of the secondary circuit, and
M the mutual inductance of the two circuits. If i_1 and i_2 be the

instantaneous values of the currents in the primary and secondary circuits, we have

$$e = R_1 i_1 + L_1 \frac{\partial i_1}{\partial t} + M \frac{\partial i_2}{\partial t} \quad\text{.................(1),}$$

$$0 = R_2 i_2 + L_2 \frac{\partial i_2}{\partial t} + M \frac{\partial i_1}{\partial t} \quad\text{.................(2).}$$

These equations may be written in the forms

$$e = R_1 i_1 + L_1 \frac{\partial I}{\partial t} \quad\text{.................(3),}$$

$$- M \frac{\partial I}{\partial t} = R_2 i_2 + L_2 \sigma \frac{\partial i_2}{\partial t} \text{.................(4),}$$

where $I = i_1 + (M/L_1)\, i_2,$

and $\sigma = 1 - M^2/L_1 L_2.$

The quantity σ is called the leakage factor of the transformer. We shall see in Volume II that a knowledge of its value is of fundamental importance in the theory of transformers.

Eliminating $\partial I/\partial t$ between (3) and (4), we get

$$- \frac{M}{L_1} (e - R_1 i_1) = R_2 i_2 + L_2 \sigma \frac{\partial i_2}{\partial t} \quad\text{.............(5),}$$

and thus $e - R_1 i_1 = \frac{L_1^{\,2}}{M^2} R_2 i'' + \frac{L_1^{\,2}}{M^2} L_2 \sigma \frac{\partial i''}{\partial t} \text{.........(6),}$

where $i'' = - (M/L_1)\, i_2 \text{.....................(7).}$

Hence finally

$$i_1 = I + i'' \text{................................(8),}$$

$$e - R_1 i_1 = \frac{L_1^{\,2}}{M^2} R_2 i'' + \frac{L_1^{\,2}}{M^2} L_2 \sigma \frac{\partial i''}{\partial t} \quad\text{.............(9),}$$

and $e - R_1 i_1 = L_1 \frac{\partial I}{\partial t} \quad\text{...........................(10).}$

These three equations (8), (9), and (10), show us that the problem of finding i_1 is the same as that of finding the current in the coil AB in Fig. 88, where the resistance of AB is R_1, the resistance and inductance of BCE are $(L_1^{\,2}/M^2)\, R_2$ and $(L_1^{\,2}/M^2)\, L_2 \sigma$ respectively, and the inductance of the choking coil BDE is L_1.

The solution of the transformer problem is therefore identical with the solution of the problem of finding the currents in the comparatively simple (seeing that there is no mutual induction)

net-work shown in Fig. 88, where the primary P.D. (e) is supposed
to be applied between A and E. By (7) we see that the current
in the secondary circuit is in exact opposition in phase to the
current in BCE and its magnitude is L_1/M times this current.

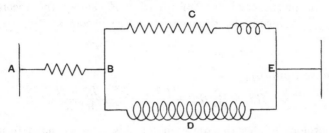

Fig. 88. Equivalent net-work of a transformer. Resistance of AB is R_1; resistance of BCE is $(L_1{}^2/M^2)\,R_2$ and inductance of BCE is $(L_1{}^2/M^2)\,L_2\sigma$ which is
equal to $(L_1{}^2/M^2)\,L_2\{1-M^2/L_1L_2\}$. Inductance of BDE is L_1. Current in the
primary is the same as the current in AB.

Again, if there is a non-inductive resistance across the secondary
terminals, the P.D. between the terminals will be in phase with
the secondary current and therefore in opposition to the phase of
the current in BCE.

This construction still applies when R_2 is a variable resistance,
Variable as, for example, a spark gap. Hence we should expect
secondary load. that when a transformer is supplying an electric arc,
the peculiar shape of the P.D. wave at the terminals of the arc
would be transmitted through the transformer into the primary
circuit. This has been proved experimentally by Duddell and
Marchant.

Again, if we put an inductive load on the secondary, then, from
Secondary equation (2), we alter the value of L_2, and therefore
load also the value of the leakage factor. Suppose that
inductive. the inductance of the secondary load is L_2'. Then
the inductance of the imaginary coil BCE (Fig. 88) is

$$\frac{L_1{}^2}{M^2}\,(L_2 + L_2')\,\sigma' = \frac{L_1{}^2}{M^2}\left(L_2 + L_2' - \frac{M^2}{L_1}\right) = \frac{L_1{}^2}{M^2}\,(L_2\sigma + L_2').$$

Therefore the leakage factor becomes $\sigma + L_2'/L_2$. Thus the effect
of adding inductance to the secondary is exactly the same as

22—2

the effect of increasing the magnetic leakage. If $\omega L_2'$ be great compared with R_2 we see that the primary of the transformer will act like a choking coil, its power factor being very low.

If we put a condenser of capacity K_2 across the secondary
Condenser in secondary load. terminals, then equation (2) becomes

$$0 = R_2 i_2 + L_2 \frac{\partial i_2}{\partial t} + M \frac{\partial i_1}{\partial t} + \frac{\int i_2 \partial t}{K_2}.$$

Hence from (6)

$$e - R_1 i_1 = \frac{L_1^2}{M^2} R_2 i'' + \frac{L_1^2}{M^2} L_2 \sigma \frac{\partial i'}{\partial t} + \frac{L_1^2}{M^2} \frac{\int i'' \partial t}{K_2}.$$

Therefore to get the diagram for this case we have to put a condenser of capacity $(M^2/L_1^2) K_2$ in series with an inductive coil of resistance $(L_1^2/M^2) R_2$ and inductance $(L_1^2/M^2) L_2\sigma$.

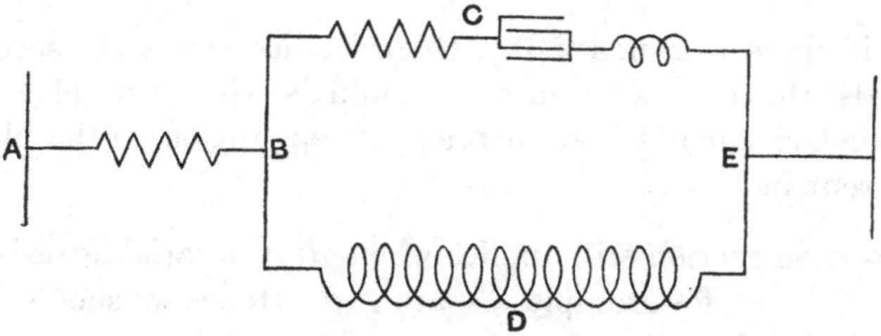

Fig. 89. Equivalent net-work of transformer with a condenser load. The circuit BCE has resistance $(L_1^2/M^2) R_2$, inductance $(L_1^2/M^2) L_2\sigma$ and capacity $(M^2/L_1^2) K_2$. Inductance of BDE equals L_1. Current in the primary equals the current in AB.

When studying formulae connected with transformers it will often be found convenient to employ diagrams similar to Figs. 88 and 89. Then, merely by inspection, we can see what the effect, for example, will be of diminishing the leakage factor or putting inductance in the secondary circuit.

In the particular case when the resistance R_1 of the primary
Circle diagram. circuit is zero, we see that, when the applied potential difference V_1 is maintained constant, the current I in the choking coil BDE in the equivalent net-work is always the

same whatever may be the load on the secondary of the transformer.

If we denote its value by i_0, then, by (8),

$$L_1 i_1 + M i_2 = L_1 i_0.$$

There is therefore a linear relation connecting i_1, i_2 and i_0, and their vectors can be represented by lines which form a triangle.

In Fig. 90 let OA, OB and BA represent the effective values of i_0, i_1 and $(M/L_1) i_2$ in magnitude and phase respectively, and let OV represent V the P.D. across the primary terminals. OV will be at right angles to OA but it will only be in the same plane as

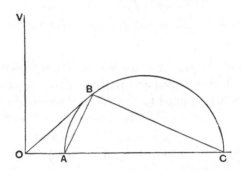

Fig. 90. The circle ABC is the locus of the extremity of the primary current vector of a leaky transformer on certain assumptions. $AC = A_0 \{(1-\sigma)/\sigma\}.$

the triangle OAB in particular cases. We will however make this assumption in order to simplify the diagram. Draw BC perpendicular to AB. Now if θ be the angle between OV and AB, then

$$\sin \theta = \frac{\alpha\omega\,(L_1^2/M^2)\,L_2\sigma A'}{V}$$

$$= \frac{\alpha\omega\,(L_1/M)\,L_2\sigma A_2}{V},$$

where $\omega/2\pi$ is the frequency and α is a quantity that depends on the shape of the current wave in the secondary, but cannot be less than unity.

Similarly $V = \alpha'\omega L_1 A_0.$

Now if we describe a circle on AC as diameter, it will pass through B. If AC equals D, then by trigonometry

$$\frac{D}{A_0} = \frac{(M/L_1) A_2}{A_0 \sin BCA}$$

$$= \frac{\alpha' M^2 \sin \theta}{\alpha L_1 L_2 \sigma \sin BCA}$$

$$= \frac{\alpha' M^2}{\alpha L_1 L_2 \sigma}.$$

If e follows the harmonic law, $\alpha = \alpha' = 1$, and thus

$$\frac{D}{A_0} = \frac{M^2}{L_1 L_2 \sigma} = \frac{M^2}{L_1 L_2 - M^2} = \frac{1 - \sigma}{\sigma}.$$

In the elementary theory of the induction motor this ratio is important.

We shall now assume that the waves of E.M.F. and current are all sine-shaped and that the effective value V_1 of the E.M.F. applied to the primary terminals is constant.

The circle diagram for a constant potential transformer.

Writing D for $\partial/\partial t$, equations (1) and (2) become

$$e_1 = (R_1 + L_1 D) i_1 + M D i_2,$$

and

$$0 = (R_2 + L_2 D) i_2 + M D i_1.$$

Hence eliminating i_2, we get

$$e_1 = (R_1 + L_1 D) i_1 - \frac{M^2 D^2}{R_2 + L_2 D} i_1.$$

Since the waves are sine-shaped we can write $D^2 = - \omega^2$, and thus

$$e_1 = (R_1 + L_1 D) i_1 + \frac{M^2 \omega^2}{R_2 + L_2 D} i_1$$

$$= (R_1 + m^2 R_2) i_1 + (L_1 - m^2 L_2) D i_1 \quad \ldots\ldots(11),$$

where

$$m^2 = M^2 \omega^2 / (R_2^2 + L_2^2 \omega^2).$$

We see at once that in this case the effective resistance and inductance of the primary coil are $R_1 + m^2 R_2$ and $L_1 - m^2 L_2$ respectively. Hence as we diminish the secondary resistance R_2 of the transformer from infinity to zero, the effective primary resistance at first increases from R_1 to $R_1 + (M^2/2L_2) \omega$, a maximum value which it attains when $R_2 = L_2 \omega$, and then diminishes to the value R_1, when R_2 is zero.

We also see that as we diminish the secondary resistance the effective primary inductance diminishes from L_1 to $L_1\sigma$. If we increase the value L_2 of the inductance in the secondary circuit, M remaining constant, the effective primary resistance diminishes from $R_1 + M^2R_2\omega^2/(R_2^2 + L_2^2\omega^2)$ to R_1.

If L_2 be less than R_2/ω initially, then as L_2 increases the effective inductance diminishes from

$$L_1 - M^2L_2\omega^2/(R_2^2 + L_2^2\omega^2) \quad \text{to} \quad L_1 - M^2/2L_2,$$

a minimum value which it attains when $L_2 = R_2/\omega$, it then increases to L_1.

We shall now find the locus of the extremity of the primary current vector. Equation (11) may be written in the form

$$i_1 = e_1/\{(R_1 + m^2R_2) + (L_1 - m^2L_2) D\}$$

$$= \frac{R_1 + m^2R_2}{Z_1^2} e_1 - \frac{L_1 - m^2L_2}{Z_1^2} De_1,$$

where $\qquad Z_1^2 = (R_1 + m^2R_2)^2 + (L_1 - m^2L_2)^2\omega^2.$

Hence if α be the phase difference between e_1 and i_1,

$$A_1 \cos\alpha = \frac{R_1 + m^2R_2}{Z_1^2} V_1,$$

and $\qquad A_1 \cos\left(\frac{\pi}{2} + \alpha\right) = -\frac{L_1 - m^2L_2}{Z_1^2} \omega V_1.$

If OY, therefore (Fig. 91 a), give the phase of V_1, and (x, y) be the coordinates of the extremity P of the vector A_1, we have

$$y = A_1 \cos\alpha = \frac{R_1 + m^2R_2}{Z_1^2} V_1, \text{ and } x = A_1 \sin\alpha = \frac{L_1 - m^2L_2}{Z_1^2} \omega V_1.$$

If we eliminate R_2 between these two equations we shall find the curve on which P must lie.

Squaring and adding, we get $x^2 + y^2 = V_1^2/Z_1^2$, and thus

$$\frac{V_1 x}{x^2 + y^2} - L_1\omega = -\frac{M^2\omega^2}{R_2^2 + L_2^2\omega^2} \cdot L_2\omega \quad\dots\dots\dots(12),$$

and $\qquad \dfrac{V_1 y}{x^2 + y^2} - R_1 = \dfrac{M^2\omega^2}{R_2^2 + L_2^2\omega^2} \cdot R_2 \quad\dots\dots\dots(13).$

Squaring and adding, we get

$$\left(\frac{V_1 x}{x^2 + y^2} - L_1\omega\right)^2 + \left(\frac{V_1 y}{x^2 + y^2} - R_1\right)^2 = \frac{M^4\omega^4}{R_2^2 + L_2^2\omega^2}$$
$$= -\frac{M^2\omega}{L_2}\left\{\frac{V_1 x}{x^2 + y^2} - L_1\omega\right\},$$

and simplifying, we have finally

$$(x - \bar{x})^2 + (y - \bar{y})^2 = r^2 \quad \dots\dots\dots\dots(14),$$

where
$$\bar{x} = V_1 L_1 \omega \,(1 + \sigma)/\{2\,(R_1^2 + L_1^2\omega^2\sigma)\},$$
$$\bar{y} = V_1 R_1/\{R_1^2 + L_1^2\omega^2\sigma\},$$
$$r = V_1 L_1 \omega \,(1 - \sigma)/\{2\,(R_1^2 + L_1^2\omega^2\sigma)\}.$$

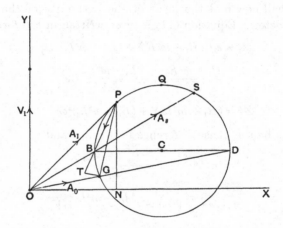

Fig. 91 *a*. The locus of the extremity of the primary current vector in a constant potential transformer is the arc *GS*. $OG = A_0$, $(R_2 = \text{infinity})$. $OP = A_1$. $OS = A_s$, $(R_2 = 0)$. *BCD* is parallel to *OX*. *GT* is at right angles to *PG*; *PBT* gives the phase of A_2 and $PT = (M/L_1)\,A_2$. $V_1 \cdot PN$ is the power taken by the transformer.

Thus the extremity of the current vector always lies on the circumference of the circle the coordinates of the centre of which are $(\bar{x}, \bar{y})$ and the radius of which is r.

If (x_0, y_0) and (x_s, y_s) be the coordinates of the extremities G and S (Fig. 91 *a*) of the primary current vector when R_2 is infinite and zero respectively, then

$$x_0 = V_1 L_1 \omega/(R_1^2 + L_1^2\omega^2), \quad \text{and} \quad y_0 = V_1 R_1/(R_1^2 + L_1^2\omega^2),$$
$$x_s = V_1 L_1 \sigma\omega/(R_1^2 + L_1^2\sigma^2\omega^2), \quad \text{and} \quad y_s = V_1 R_1/(R_1^2 + L_1^2\sigma^2\omega^2).$$

If β be the angle which a line drawn through C and G makes with OX, then

$$\tan \beta = \frac{\overline{y} - y_0}{\overline{x} - x_0} = \frac{2R_1/L_1\omega}{1 - R_1^2/L_1^2\omega^2} = \tan 2\,(\pi/2 - \psi_0) = \tan 2G\hat{O}X,$$

where ψ_0 is the angle YOG.

Hence, if BCD be the diameter of the circle parallel to OX, D lies in OG produced. Similarly we can show that B lies on OS.

We can see at once from the diagram (Fig. 91 a) that the phase difference ψ_1 between V_1 and A_1 is a minimum and therefore the power factor $\cos \psi_1$ of the primary circuit is a maximum, when OP is a tangent to the circle GPQ.

In this case $OP = V_1/(R_1^2 + L_1^2\omega^2\sigma)^{1/2}$, and

$$(\cos \psi_1)_{\text{max.}} = \frac{4R_1\,(R_1^2 + L_1^2\omega^2\sigma)^{1/2} + L_1^2\omega^2\,(1 - \sigma^2)}{4R_1^2 + L_1^2\omega^2\,(1 + \sigma)^2}.$$

Since $PN = A_1 \cos \psi_1$, it follows that $V_1 . PN$ is the power expended on the transformer. Hence this power is a maximum when P coincides with Q the highest point of the circle. The maximum power expended is therefore

$$V_1\,(r + \overline{y}) = V_1\,\{2V_1R_1 + V_1L_1\omega\,(1 - \sigma)\}/\{2\,(R_1^2 + L_1^2\omega^2\sigma)\}.$$

From Fig. 91 a it will be seen that OP has a minimum value OM when OMC is a straight line. Hence

$$(A_1)_{\text{min.}} = OC - r = (\overline{x}^2 + \overline{y}^2)^{1/2} - r$$

$$= \frac{2V_1}{\{4R_1^2 + L_1^2\omega^2\,(1 + \sigma)^2\}^{1/2} + L_1\omega\,(1 - \sigma)}.$$

Let us now consider the secondary circuit. Equation (1) may be written

$$R_1\,(i_1 - i_0) + L_1 D\,(i_1 - i_0) + MDi_2 = 0,$$

where i_0 is the value of the primary current when the secondary resistance is infinite. Hence

$$i_0 - i_1 = \frac{MD}{R_1 + L_1 D}\,i_2$$

$$= \frac{M\omega}{(R_1^2 + L_1^2\omega^2)^{\frac{1}{2}}}\,\epsilon^{(\pi/2 - \psi_0)\iota}\,i_2.$$

Now by the triangle of vectors PG is the vector of $i_0 - i_1$ and this equation shows that A_2 lags behind PG by an angle $\pi/2 - \psi_0$ which equals GOX in Fig. 91 a. Now

$$G\hat{O}X = G\hat{D}B = G\hat{P}B.$$

Hence PB gives the phase of the secondary current A_2 when P is above BD and BP gives the phase when it is below BD. When R_2 is very large the primary and secondary currents are practically in quadrature, that is, are at right angles to one another, and when R_2 is zero they are in opposition in phase.

If we draw GT at right angles to PG,

$$PT = PG \text{ cosec } \psi_0 = PG\,(R_1{}^2 + L_1{}^2\omega^2)^{1/2}/L_1\omega\,;$$

but, by the preceding paragraph, $PG = M\omega A_2/(R_1{}^2 + L_1{}^2\omega^2)^{1/2}$. Hence $PT = (M/L_1)\,A_2$. Thus PT represents the secondary current in phase and is proportional to it in magnitude. Since GP obviously always increases as R_2 diminishes, so also does A_2.

From equation (11) we see that the power expended on the transformer is $(R_1 + m^2 R_2)\,A_1{}^2$ and since this must equal

$$R_1 A_1{}^2 + R_2 A_2{}^2,$$

we see that

$$R_2 A_2{}^2 = \frac{m^2 R_2 V_1{}^2}{(R_1 + m^2 R_2)^2 + (L_1 - m^2 L_2)^2\,\omega^2}$$

and this has its maximum value when

$$R_2{}^2 = L_2{}^2\omega^2 - L_1{}^2 L_2{}^2\omega^4\,(1 - \sigma^2)/(R_1{}^2 + L_1{}^2\omega^2).$$

If we put condensers K_1 and K_2 in both primary and secondary circuits, the locus of the extremity of the current vector is still a circle. To find the coordinates of its centre and its radius we have merely to substitute $L_1 - 1/K_1\omega^2$ for L_1 and $L_2 - 1/K_2\omega^2$ for L_2 in the formulae given above.

To find the locus of the extremity of the current vector when L_2 varies, R_2 remaining constant, we eliminate L_2 **Varying the secondary inductance.** from the equations (12) and (13) given above. It is easy to see that

$$\{V_1 x/(x^2 + y^2) - L_1\omega\}^2 + \{V_1 y/(x^2 + y^2) - R_1\}^2 = M^4\omega^4/(R_2{}^2 + L_2{}^2\omega^2)$$
$$= (M^2\omega^2/R_2)\,\{V_1 y/(x^2 + y^2) - R_1\},$$

and as this equation can be written in the form

$$(x - \bar{x})^2 + (y - \bar{y})^2 = r^2,$$

where $\quad \bar{x} = L_1 \omega V_1/\Omega^2, \quad \bar{y} = (R_1 + M^2\omega^2/2R_2) V_1/\Omega^2,$

$r = (M^2\omega^2 V_1/2R_2)/\Omega^2,$ and $\Omega^2 = R_1^2 + L_1^2\omega^2 + M^2\omega^2 R_1/R_2,$

we see that the locus is a circle (Fig. 91 b) of radius r, the coordinates of the centre of which are $\bar{x}$ and $\bar{y}$.

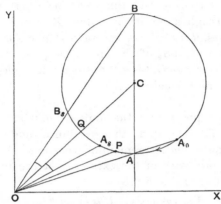

Fig. 91 b.   As L_2 varies from infinity to M^2/L_1, P moves from A_0 towards A_s.

If a line drawn from O cut the circle in points P and P', we have

$$OP \cdot OP' = OC^2 - r^2 = \bar{x}^2 + \bar{y}^2 - r^2 = V_1^2/\Omega^2.$$

Let A and B be the extremities of the vertical diameter of the circle APB (Fig. 91 b), then we have

$$OA \cdot OA_0 = V_1^2/\Omega^2,$$

and since $\quad OA^2 = (\bar{y} - r)^2 + \bar{x}^2 = V_1^2(R_1^2 + L_1^2\omega^2)/\Omega^4,$

it follows that $\quad OA_0 = V_1/(R_1^2 + L_1^2\omega^2)^{1/2}.$

Hence OA_0 is the current vector when L_2 is infinity.

If Z_1 denote the primary impedance, we have

$$Z_1^2 = (R_1 + m^2R_2)^2 + (L_1 - m^2L_2)^2\omega^2$$
$$= R_1^2 + L_1^2\omega^2 + M^2\omega^2(M^2\omega^2 + 2R_1R_2 - 2L_1L_2\omega^2)/(R_2^2 + L_2^2\omega^2).$$

Hence as L_2 diminishes Z_1 continually increases and therefore the current vector OP continually diminishes. If M^2/L_1, the least possible value of L_2, were practically zero, Z_1^2 would equal

$$(R_1 + M^2\omega^2/R_2)^2 + L_1^2\omega^2.$$

Now from the figure we see that

$$OB^2 = (\bar{y} + r)^2 + \bar{x}^2 = (V_1^2/\Omega^4)\{(R_1 + M^2\omega^2/R_2)^2 + L_1^2\omega^2\},$$

and thus $OB_s = V_1/\{(R_1 + M^2\omega^2/R_2)^2 + L_1^2\omega^2\}^{1/2}$ which equals the numerical value of the primary current when L_2 is zero. Hence, if we draw (Fig. 91 b) OA_s, so that the angle COA_s = the angle COB_s, OA_s is a limiting position of the primary current vector. It is clear that OB_s cannot be this vector as otherwise the current would pass through a minimum value OQ which is impossible. As L_2 diminishes, R_2 remaining constant, P moves from A_0 towards A_s along the arc $A_0 A A_s$ of the circle, the power factor attaining a minimum value, when OP is a tangent to the circle, in which case the primary current equals V_1/Ω.

Let us now suppose that the mutual inductance between the coils is varied, the resistance and inductance of the secondary remaining constant. If (x, y) be the coordinates of the extremity of the primary current vector, we have by (12) and (13)

Varying the mutual inductance.

$$V_1 x/(x^2 + y^2) - L_1\omega = - m^2 L_2\omega,$$

and

$$V_1 y/(x^2 + y^2) - R_1 = m^2 R_2.$$

Hence

$$\{V_1 x/(x^2 + y^2) - L_1\omega\}/\{V_1 y/(x^2 + y^2) - R_1\} = - L_2\omega/R_2,$$

and simplifying we get

$$(x - \bar{x})^2 + (y - \bar{y})^2 = r^2,$$

where $\bar{x} = R_2 V_1/2\Omega'$, $\bar{y} = L_2\omega V_1/2\Omega'$, $r = V_1(R_2^2 + L_2^2\omega^2)^{1/2}/2\Omega'$, and $\Omega' = (R_1 L_2 + R_2 L_1)\omega$.

The locus of the extremity of the primary current vector in this case is therefore a circle which passes through the origin (Fig. 91 c).

The equation to determine Z_1^2 may be written in the form

$$Z_1^2 = \frac{\{M^2\omega^2 - (L_1 L_2\omega^2 - R_1 R_2)\}^2 + (R_1 L_2 + R_2 L_1)^2\,\omega^2}{R_2^2 + L_2^2\omega^2}.$$

Hence if $L_1 L_2\omega^2$ be greater than $R_1 R_2$, Z_1 attains a minimum value when $M^2 = L_1 L_2 - R_1 R_2/\omega^2$. In this case the current vector is OT (Fig. 91 c) and its initial position when M is zero is OA_0, where A_0 lies between S and T. When M has its maximum value $(L_1 L_2)^{1/2}$, the current vector is OA_m, where A_m lies between T and Q. Hence as M increases in value, P the extremity of the current vector moves from A_0 to A_m, the value of the power factor continually increasing.

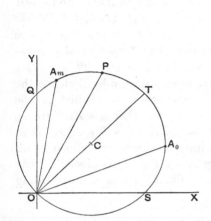

Fig. 91 c. As M increases P moves from A_0 to A_m.

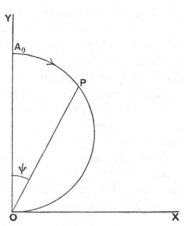

Fig. 92. As f varies from 0 to infinity P moves from A_0 to O, R_2 is supposed to be zero.

Let us now consider the locus of the extremity of the primary current vector when the frequency of the applied potential difference is made to vary. In this case we have to eliminate ω from the equations (12) and (13). Eliminating $M^2\omega^2/(R_2^2 + L_2^2\omega^2)$ between (12) and (13) we obtain a simple equation to find ω. Substituting this value in (13), we get after some reduction

Varying the frequency.

$$\{V_1 y - R_1(x^2 + y^2)\}\,[\{V_1 L_2 - (L_1 R_2 + L_2 R_1)\,y\}^2$$
$$+ (L_2 R_1 + L_1 R_2)^2\, x^2] = M^2 R_2 V_1^2 x^2 \ \ldots(15).$$

The equation to the curve is therefore of the fourth degree. In the particular case when R_2 is zero, this curve becomes the circle (Fig. 92)

$$R_1(x^2 + y^2) - V_1 y = 0.$$

The diameter of this circle equals V_1/R_1, its centre is on OY and it passes through the origin. The current diminishes from

V_1/R_1 to zero as the frequency increases from zero to infinity. The power factor also diminishes from 1 to 0 in this case.

In the general case if $\cos \psi$ be the power factor we have

$$\tan \psi = (L_1 - m^2 L_2)\, \omega/(R_1 + m^2 R_2),$$

and if R_2 be not zero this expression can only vanish when $\omega = 0$. When $x = 0$, therefore, y cannot equal $V_1 L_2/(L_1 R_2 + L_2 R_1)$, as otherwise ψ would vanish. The expression inside the square brackets therefore is always greater than zero and the curve given by (15) lies inside the circle shown in Fig. 92.

The theorems given above have been extended by Eccles to the case when a slightly damped periodic E.M.F. is applied to the primary terminals. In this case we assume that $e_1 = E\epsilon^{-bt} \sin \omega t$, where b is small compared with ω. Writing $-b + \omega\iota$ for D in equations (1) and (2) and proceeding as above we deduce in all except the last case circle diagrams to give the locus of the extremity of A_1 at any instant, the scale of the diagram depending on the instant chosen.

We shall now suppose that the primary coil of the air core transformer is placed in a circuit in which the current is maintained constant. In this case, we have as before

The circle diagram for a constant current transformer.

$$e_1 = (R_1 + m^2 R_2)\, i_1 + (L_1 - m^2 L_2)\, D i_1.$$

Let us assume (Fig. 93) that OX gives the phase of the current A_1, and that x and y are the coordinates of the extremity of OP, the vector of e_1. We have

$$x = (R_1 + m^2 R_2)\, A_1,$$

and $$y = (L_1 - m^2 L_2)\, \omega A_1.$$

Hence
$$(x - R_1 A_1)^2 + (y - L_1 \omega A_1)^2 = \frac{M^4 \omega^4}{R_2^2 + L_2^2 \omega^2}\, A_1^2$$

$$= -\frac{M^2}{L_2}\, \omega\, (y - L_1 \omega A_1)\, A_1,$$

and thus
$$(x - R_1 A_1)^2 + [y - L_1 \omega A_1 \{(1 + \sigma)/2\}]^2 = [L_1 \omega A_1 \{(1 - \sigma)/2\}]^2.$$

Hence the locus of P is a circle the coordinates $(\bar{x}, \bar{y})$ of the centre of which and the radius r are given by

$$\bar{x} = R_1 A_1, \quad \bar{y} = L_1 \omega A_1 \{(1 + \sigma)/2\}, \quad r = L_1 \omega A_1 \{(1 - \sigma)/2\}.$$

When R_2 is infinite we have
$$e_0 = R_1 i_1 + L_1 D i_1.$$
In general $$e_1 = R_1 i_1 + L_1 D i_1 + M D i_2.$$
As i_1 is the same in each of these equations, we get
$$e_1 - e_0 = M D i_2 = M\omega \epsilon^{(\pi/2)\iota}\, i_2.$$
Now by the triangle of vectors, BP (Fig. 93) is the vector of $e_1 - e_0$, and the form of the equation shows that the vector of i_2 lags $90°$ behind the vector of BP. Hence the direction PD gives the phase of A_2. The equation also shows us that $BP = M\omega A_2$. It is therefore proportional to A_2. It is also easy

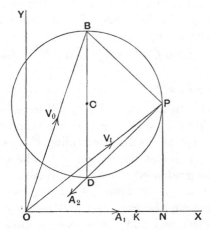

Fig. 93. Diagram for the constant current transformer. $OK = A_1 = $ a constant.
$OB = V_0$, $(R_2 = \text{infinity})$. $OD = V_s$, $(R_2 = 0)$. $OP = V_1$. $BP = M\omega A_2$. PD gives the phase of A_2.

to see from equation (2) that BD gives the phase of the induced E.M.F. in the secondary circuit, the magnitude of which is $M\omega A_1$. Hence if we choose our scale so that $BD = M\omega A_1$, DP will equal $R_2 A_2$, and BP will equal $L_2 \omega A_2$.

We see at once that V_1 has its greatest value when OP passes through C, and hence
$$(V_1)_{\text{max.}} = (A_1/2)\,[\{4R_1^2 + L_1^2\omega^2\,(1+\sigma)^2\}^{1/2} + L_1\omega\,(1-\sigma)].$$
The power expended on the transformer equals $A_1 . ON$, where PN is perpendicular to OX and this is a maximum when CP is

parallel to OX. The power factor of the primary is a maximum when OP is a tangent to the circle and in this case

$$V_1 = A_1 (R_1{}^2 + L_1{}^2 \omega^2 \sigma)^{1/2}.$$

In order to measure the potential difference between high pressure mains it is customary to use a 'potential' transformer the high pressure terminals of which are connected across the mains and the low pressure terminals with a voltmeter which takes a negligible current. The transformer is so constructed that the ratio of the P.D. between the high voltage terminals to the P.D. between the low voltage terminals is practically constant. In practice the transformers have iron cores but the following simple theory for air core transformers is useful. From the theory given above

Instrument transformers.

$$V_1 = (R_1{}^2 + L_1{}^2 \omega^2)^{1/2} A_0 \quad \text{and} \quad V_2 = M \omega A_0.$$

Hence

$$V_1/V_2 = (R_1{}^2 + L_1{}^2 \omega^2)^{1/2}/M\omega = (L_1/M) \{1 + R_1{}^2/(2L_1{}^2 \omega^2) + \ldots\}.$$

Thus the ratio of transformation V_1/V_2 will be independent of the frequency when $R_1{}^2/(2L_1{}^2 \omega^2)$ can be neglected compared with unity. Multiplying therefore the voltmeter reading V_2 by L_1/M_1 a constant which is found experimentally, we get V_1. The scale of the voltmeter can also be graduated to read V_1 directly.

Similarly, for measuring large alternating currents, we use a 'current' transformer, the current to be measured being passed through its primary coil and the secondary circuit being closed through a low resistance ammeter. In this case

$$A_1/A_2 = (R_2{}^2 + L_2{}^2 \omega^2)^{1/2}/M\omega = (L_2/M) \{1 + R_2{}^2/(2L_2{}^2 \omega^2) + \ldots\}.$$

This ratio is therefore independent of the frequency when $R_2{}^2/(2L_2{}^2 \omega^2)$ can be neglected compared with unity. Hence A_1 can be found by multiplying A_2 by L_2/M the constant of the current transformer.

Current and potential transformers are also used in conjunction with wattmeters and watt-hour meters. In this case the phases as well as the ratios of the voltages and the currents have to be taken into account and so the theory is more complex. We shall return to this subject in Volume II when we discuss transformers with iron cores.

A most interesting and important case arises when we put
a condenser across the secondary terminals of the
transformer. We can see from the diagram (Fig. 89)
that, when the condenser is small, it will supply some
of the current required to magnetise the choking
coil, and hence the primary current will be diminished. As we
increase the capacity of the condenser the primary current attains
a minimum value and then begins to increase again. When
the primary current is a minimum, the phase difference between it
and the applied P.D. is small, but it is only zero when the P.D.
wave is sine-shaped. As we increase the capacity the angle of
lead of the current attains a maximum value and then begins to
diminish again. Finally, since a very large condenser acts simply
like a short circuit, the current in the primary ends by lagging
nearly ninety degrees in phase behind the P.D. If there were no
magnetic leakage, it would be possible to obtain leading primary
currents of any magnitude.

Leading primary current with condenser load.

If the resistance of the primary circuit is negligible, and A
is the effective value of the primary current, it
easily follows, by applying the method of the
complex variable to Fig. 89, that

Constant potential transformer with constant primary current.

$$\frac{A^2}{V^2} = \frac{1}{L_1^2 \omega^2} - \frac{M^2}{L_1^2} \frac{2 - K_2 L_2 (1 + \sigma) \omega^2}{K_2 L_2 \omega^2 \left\{ R_2^2 + \left(L_2 \sigma - \frac{1}{K_2 \omega^2} \right)^2 \omega^2 \right\}}.$$

If we make

$$K_2 = \frac{2}{L_2 (1 + \sigma) \omega^2},$$

then

$$A = \frac{V}{L_1 \omega}.$$

whatever be the value of R_2.

If ϕ be the angle of phase difference between the primary
current and the applied P.D., then we have in this case

$$\tan \phi = \frac{4 (L_1^2 / M^2) R_2^2 - \omega^2 M^2}{4 R_2 L_1 \omega}.$$

Hence as R_2 diminishes from infinity to zero, the lag of the
primary current changes from $+90$ degrees to -90 degrees,
although the current itself remains constant all the time.

Such an air core transformer would be useful as a phase regulator for testing whether a wattmeter reads correctly on loads of various power factors. On open circuit it would read nearly zero; as we diminished the resistance in the secondary circuit, the readings would increase to a maximum value AV and then gradually diminish, attaining a minimum value when the resistance in the secondary was zero. We should have to make sure however that the harmonics in the applied P.D. wave were negligible.

REFERENCES

CLERK MAXWELL, 'A Dynamical Theory of the Electromagnetic Field.' *Phil. Trans.* 1865, p. 475.

J. A. FLEMING, *The Alternate Current Transformer.*

E. ARNOLD and J. L. LA COUR, 'Die Wechselstromtechnik,' Vol. 2, 'Die Transformatoren.'

W. H. ECCLES, 'Application of Heaviside's Resistance Operators to the theory of the Air Core Transformer and Coupled Circuits in General.' *Proc. Phys. Soc.* Vol. 24, p. 260, 1912.

CHAPTER XV

THE fundamental principle of the commonest type of three phase
Three phase alternators. alternator is illustrated in Fig. 94. Insulated wire is wound evenly round a laminated iron ring and connections are made at three points on the wire at equal angular distances apart, with the three terminals of the machine. A powerful magnet NS rotates inside this ring, and as the magnetic circuit is completed through both halves of the ring it will be seen that the flux inside the coils will change in direction every half revolution, and hence an E.M.F.

23—2

will be set up in each coil. If everything is symmetrical, the effective P.D.s between any two of the three terminals must be equal. The ring represents the armature of the three phase alternator and the magnet the revolving field magnets. In this case the armature is said to be mesh wound and the three windings are in series with one another. Any want of balance amongst the E.M.F.s generated in the three windings of the

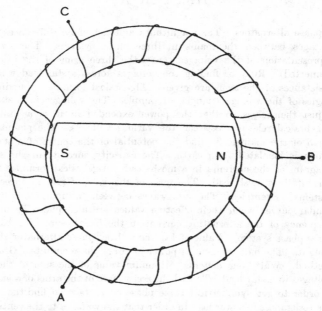

Fig. 94.   Three phase machine with mesh-connected armature.

armature will cause alternating currents to flow in the windings even at no load, and will thus lower the efficiency of the machine. It is therefore important to make the armature as symmetrical as possible.

The mains are connected with the three terminals of the machine and the load can be placed across two of the mains or divided symmetrically between the three of them either in mesh fashion like the windings of the armature or in star fashion like the armature shown in Fig. 95.

The principle of three phase motors can be understood from Fig. 94. If we were to connect A, B and C to three phase supply mains, then currents would flow round the three windings of the armature, and we shall see in Chapter xvIII that a rotating magnetic field will be produced which will cause the magnet to make f revolutions per second, where f is the frequency of the alternating current.

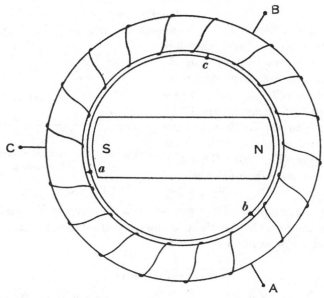

Fig. 95. Three phase machine with star-connected armature.

A diagrammatic sketch of a three phase alternator with its armature star wound is shown in Fig. 95. A, B and C are the terminals of the machine, and a, b and c, the other ends of the three windings, are joined together. In this case there is obviously no current flowing in the coils when the machine is on open circuit.

When the three P.D.s across the terminals of a three phase alternator are all equal, then the load is said to be balanced. In practice it is not always possible to obtain a balance, and so we shall first consider the relations that hold between the effective P.D.s and their phase differences in the general case.

Let v_1 be the potential of No. 1 main, v_2 the potential of No. 2 main and v_3 the potential of No. 3 main.

The magnitudes and the phase differences of the voltages between the mains in three phase systems.

Let also

$v_{1.2} = v_1 - v_2 =$ the P.D. between mains 1 and 2.

$v_{2.3} = v_2 - v_3 =$ the P.D. between mains 2 and 3.

$v_{3.1} = v_3 - v_1 =$ the P.D. between mains 3 and 1.

Then obviously

$$v_{1.2} + v_{2.3} + v_{3.1} = 0 \quad \dots\dots\dots(a).$$

Hence a linear relation connects the instantaneous values of the three P.D.s, and therefore their effective values can be represented by lines drawn in one plane (Chapter XII). Since any one of them is equal and opposite to the resultant of the other two, we see that they form a triangle (Fig. 96). This also follows at once from (a), since

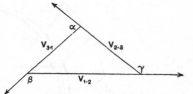

Fig. 96. The voltage triangle for three phase mains.

$$V^2_{1.2} = V^2_{2.3} + V^2_{3.1} + 2V_{2.3}V_{3.1}\cos\alpha,$$

where α is the phase difference between the vectors $V_{2.3}$ and $V_{3.1}$.

All the formulae for the trigonometry of a triangle can now be applied. For example,

$$\frac{V_{1.2}}{\sin\alpha} = \frac{V_{2.3}}{\sin\beta} = \frac{V_{3.1}}{\sin\gamma}.$$

Let 1, 2 and 3 be the three mains, and let O be the centre of the star load (Fig. 97). If p, q and r be the

The graphical representation of the voltages across the three arms of the load (star connected).

resistances of the three arms (non-inductive) and e_1, e_2 and e_3 be the P.D.s across the arms, then, if O is earthed and the capacity currents in the sheath can be neglected, since by Kirchhoff's law the algebraical sum of the instantaneous values of the currents at O must be zero, we have

$$\frac{e_1}{p} + \frac{e_2}{q} + \frac{e_3}{r} = 0.$$

Hence, as before,

$$\frac{E_1{}^2}{p^2} = \frac{E_2{}^2}{q^2} + \frac{E_3{}^2}{r^2} + 2\frac{E_2 E_3}{qr}\cos\theta_{2.3} \dots\dots\dots\dots(1)$$

and two similar equations, where E_1 is the voltage across the arm
$1O$, and $\theta_{2.3}$ is the angle of phase difference between e_2 and e_3.

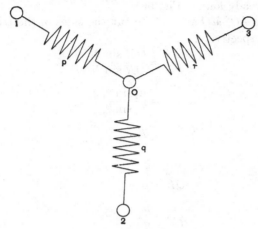

Fig. 97.   Star load.

We see, as in the last theorem, that E_1/p, E_2/q and E_3/r form
a triangle whose exterior angles are $\theta_{2.3}$, $\theta_{3.1}$ and $\theta_{1.2}$ respectively.
These angles can therefore be easily found when E_1, E_2, E_3, p, q
and r are known.

It follows by geometry that
$$\theta_{1.2} + \theta_{2.3} + \theta_{3.1} = 360^\circ \dots\dots\dots\dots(2).$$

It also follows by the 'rule of sines' that
$$\frac{E_1}{p \sin \theta_{2.3}} = \frac{E_2}{q \sin \theta_{3.1}} = \frac{E_3}{r \sin \theta_{1.2}} \dots\dots\dots(3).$$

From any point O (Fig. 98) draw OL equal to E_1.   Make the

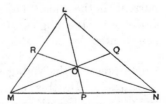

angle LON equal to $\theta_{3.1}$ and make
ON equal to E_3.   Make the angle
NOM equal to $\theta_{2.3}$ and OM equal
to E_2.   Then by (2) the angle LOM
equals $\theta_{1.2}$.

Now, $v_{1.2} = v_1 - v_2 = e_1 - e_2$,
and therefore
$$V^2_{1.2} = E_1^2 + E_2^2 - 2E_1E_2 \cos \theta_{1.2} \quad (4)$$
$$= LM^2 \text{ in Fig. 98.}$$

Fig. 98. O is the c. of G. of
masses $1/p$, $1/q$ and $1/r$ at L, M
and N respectively.

Therefore LM equals $V_{1.2}$. Similarly, MN and NL equal $V_{2.3}$ and $V_{3.1}$ respectively. Hence the triangle LMN is the voltage triangle already found (Fig. 96).

Produce LO, MO and NO to cut the sides of the triangle in P, Q and R respectively.

Then
$$\frac{MP}{PN} = \frac{OM \sin MOP}{ON \sin NOP}$$

$$= \frac{E_2 \sin \theta_{1.2}}{E_3 \sin \theta_{3.1}}$$

$$= \frac{q}{r} \text{ from (3).}$$

Similarly
$$\frac{NQ}{QL} = \frac{r}{p},$$

and
$$\frac{LR}{RM} = \frac{p}{q}.$$

It follows from these relations that O is the centre of gravity of three particles of masses $1/p$, $1/q$ and $1/r$ placed at L, M and N respectively.

Rule for finding the voltages across a star load when the resistances of the arms are given. Construct a triangle LMN the lengths of whose sides represent the voltages between the mains. Find the centre of gravity O of masses $1/p$, $1/q$ and $1/r$ placed at L, M and N. Then OL equals E_1, the voltage across the terminals of the resistance p (Fig. 97), OM equals E_2, and ON equals E_3. The angles LOM, MON and NOL give the angles of phase difference between the various voltages and also, since the arms are non-inductive, between the currents in the arms. The supplements of the angles of the triangle LMN give the angles of phase difference between the voltages represented by the sides of the triangle.

Algebraical formula for finding the angles of the voltage triangle. In practice the voltages between the mains are nearly equal. In this case, if the voltages be V, $V + xV$ and $V + yV$, we can find the angles of the triangle in degrees by their approximate values

$$60 - 33(x + y); \quad 60 + 33(2y - x); \quad 60 + 33(2x - y).$$

Suppose that the P.D.s.between the mains are 2000, 2060 and
2140 volts respectively. The algebraical expressions
make the angles of the voltage triangle 56·7, 59·7 and
63·6 degrees. Their true values are 56·6, 59·6 and 63·8 degrees.
The supplements of these angles will give the phase differences
between the three voltages.

Example.

Since O (Fig. 98) is the centre of gravity of masses $1/p$, $1/q$
and $1/r$ placed at L, M and N, it follows that the
moment of inertia of these masses about an axis
through O perpendicular to the plane of the paper is
less than their moment of inertia about any other
parallel axis, and hence

The voltages in a star load adjust themselves so that the power expended in it is a minimum.

$$\frac{OL^2}{p} + \frac{OM^2}{q} + \frac{ON^2}{r}$$

is a minimum.

Since p, q and r are non-inductive, this expression represents
the power being expended in the star load.

Since

$$\frac{e_1}{p} + \frac{e_2}{q} + \frac{e_3}{r} = 0,$$

we have

$$e_1\left(\frac{1}{p} + \frac{1}{q} + \frac{1}{r}\right) = \frac{v_{1.2}}{q} - \frac{v_{3.1}}{r}.$$

Algebraical relations between the various voltages.

Hence, squaring and taking mean values,

$$E_1^2\left(\frac{1}{p} + \frac{1}{q} + \frac{1}{r}\right)^2 = \frac{V^2_{1.2}}{q^2} + \frac{V^2_{3.1}}{r^2} - \frac{2V_{1.2}V_{3.1}\cos\beta}{qr}.$$

But $\quad 2V_{1.2}V_{3.1}\cos\beta = V^2_{2.3} - V^2_{3.1} - V^2_{1.2}.$

Hence $E_1^2\left(\frac{1}{p} + \frac{1}{q} + \frac{1}{r}\right)^2 = \left(\frac{V^2_{1.2}}{q} + \frac{V^2_{3.1}}{r}\right)\left(\frac{1}{q} + \frac{1}{r}\right) - \frac{V^2_{2.3}}{qr}$ ...(5).

We can write down two similar equations by symmetry.
Similarly

$$\frac{V^2_{1.2}}{pq} = \left(\frac{E_1^2}{p} + \frac{E_2^2}{q}\right)\left(\frac{1}{p} + \frac{1}{q}\right) - \frac{E_3^2}{r^2} \quad\ldots\ldots\ldots(6)$$

and two other similar equations. Hence, also,

$$\left(\frac{E_1^2}{p} + \frac{E_2^2}{q} + \frac{E_3^2}{r}\right)\left(\frac{1}{p} + \frac{1}{q} + \frac{1}{r}\right) = \frac{V^2_{1.2}}{pq} + \frac{V^2_{2.3}}{qr} + \frac{V^2_{3.1}}{rp} \quad\ldots(7).$$

It follows that if p, q and r are all equal, then
$$V^2_{1.2} + V^2_{2.3} + V^2_{3.1} = 3\,(E_1^2 + E_2^2 + E_3^2) \ \ldots\ldots\ldots(8).$$
If in addition $\qquad V_{1.2} = V_{2.3} = V_{3.1},$

then $\qquad\qquad V_{1.2} = \sqrt{3}E_1 \ \ldots\ldots\ldots\ldots\ldots\ldots\ldots(9).$

This can also easily be proved from Fig. 98, since in this case LMN is an equilateral triangle. It is to be noticed that no assumption is made as to the shape of the P.D. waves.

Let v_1, v_2 and v_3 be the potentials of the three mains, and let **The potentials to earth of the mains.** f_1, f_2 and f_3 be their fault resistances to earth. By the fault resistance of a main we mean the resistance of all those leakage paths from it to earth which do not pass through the other mains. Then if the capacity currents in the sheath can be neglected, the sum of the leakage currents to earth must be zero, and therefore
$$\frac{v_1}{f_1} + \frac{v_2}{f_2} + \frac{v_3}{f_3} = 0.$$
Proceeding as before, we see that if O is the centre of gravity of masses $1/f_1$, $1/f_2$ and $1/f_3$ placed at the angles L, M, N of the voltage triangle, then OL, OM and ON represent V_1, V_2 and V_3 respectively in magnitude and phase.

It follows also that, if the fault resistances of the mains vary, then their potentials adjust themselves so that the power expended in leakage currents is a minimum. An analogous theorem holds true for the case of three wire direct current systems.

Let p, q and r be the resistances of the arms of the star load, **To find the potential of the centre of a star load which is insulated from earth.** and let x be the potential of the centre. Then
$$\frac{v_1 - x}{p} + \frac{v_2 - x}{q} + \frac{v_3 - x}{r} = 0,$$
and therefore
$$x\left(\frac{1}{p} + \frac{1}{q} + \frac{1}{r}\right) = \frac{v_1}{p} + \frac{v_2}{q} + \frac{v_3}{r}.$$

Now as the P.D. waves have different shapes, a little consideration will show that the frequency of the alternating potential x is an unknown quantity.

If V denote the effective value of x, then

$$V^2\left(\frac{1}{p}+\frac{1}{q}+\frac{1}{r}\right)^2 = \frac{V_1^2}{p^2} + \dots + 2\,\frac{V_2 V_3}{qr}\cos\phi_{2.3} + \dots \quad\dots\dots(10).$$

Also, if f_1, f_2 and f_3 be the fault resistances of the mains to earth, and if the capacity currents in the sheath can be neglected, then

$$\frac{V_1^2}{f_1^2} = \frac{V_2^2}{f_2^2} + \frac{V_3^2}{f_3^2} + 2\,\frac{V_2 V_3}{f_2 f_3}\cos\phi_{2.3}$$

and two similar equations.

By means of these equations we can eliminate the cosines from (10) and we get finally

$$V^2\left(\frac{1}{p}+\frac{1}{q}+\frac{1}{r}\right)^2 = \frac{V_1^2}{f_1^2}\left(\frac{f_1}{p}-\frac{f_2}{q}\right)\left(\frac{f_1}{p}-\frac{f_3}{r}\right)$$
$$+ \frac{V_2^2}{f_2^2}\left(\frac{f_2}{q}-\frac{f_3}{r}\right)\left(\frac{f_2}{q}-\frac{f_1}{p}\right)$$
$$+ \frac{V_3^2}{f_3^2}\left(\frac{f_3}{r}-\frac{f_1}{p}\right)\left(\frac{f_3}{r}-\frac{f_2}{q}\right).$$

Hence if
$$\frac{f_1}{p}=\frac{f_2}{q}=\frac{f_3}{r},$$

we see that V is zero and x also is zero.

Let us now consider the case of a three core cable with a lead **The capacity currents in the sheath.** sheath. In practice the sheath is earthed and so, approximately at least, it is at zero potential. We saw in Chapter V that, so far as capacity effects are concerned, we can, in theoretical diagrams, replace a three core cable by three small conductors joined by six condensers (see Fig. 57). The effects of leakage can be shown on this diagram by joining $S1$, $S2$ and $S3$ with non-inductive resistances f_1, f_2 and f_3. Since by Kirchhoff's law the sum of the currents flowing to and from the sheath must be zero, we have

$$\frac{v_1}{f_1}+\frac{v_2}{f_2}+\frac{v_3}{f_3}+ K_{1.0}\frac{dv_1}{dt}+K_{2.0}\frac{dv_2}{dt}+K_{3.0}\frac{dv_3}{dt}=0,$$

and thus
$$\frac{v_1}{f_1}+\frac{v_2}{f_2}+\frac{v_3}{f_3}+\frac{dq_0}{dt}=0,$$

where q_0 is the total charge on the sheath at the time t. We may call $\partial q_0/\partial t$ the capacity current in the sheath. It is to be noted that it is the sum of a number of small leakage currents taking place along paths which may be at considerable distances from each other and therefore the value of the capacity current in the sheath may vary largely at different points along its length. A formula for the effective value C_0 of $\partial q_0/\partial t$ can be found as follows. We have

$$-\frac{dq_0}{dt} = \frac{v_1}{f_1} + \frac{v_2}{f_2} + \frac{v_3}{f_3}.$$

Thus
$$C_0{}^2 = \frac{V_1{}^2}{f_1{}^2} + \ldots + 2\,\frac{V_2 V_3}{f_2 f_3}\cos\phi_{2.3} + \ldots,$$

and therefore
$$C_0{}^2 = \frac{V_1{}^2}{f_1{}^2} + \ldots + \frac{V_2{}^2 + V_3{}^2 - V_{2.3}^2}{f_2 f_3} + \ldots.$$

Therefore, if we know the star and mesh voltages and the fault resistances of the mains, we can calculate C_0.

Let a_1, a_2 and a_3 be the instantaneous values of the currents in the mains (Fig. 99),

Diagram of the currents in a mesh load.

and let i_1, i_2 and i_3 be the currents in the mesh windings. The arrowheads are drawn pointing round the same way so as to get algebraical symmetry in our equations. At any instant however either one or two of the quantities i_1, i_2 and i_3 must be negative.

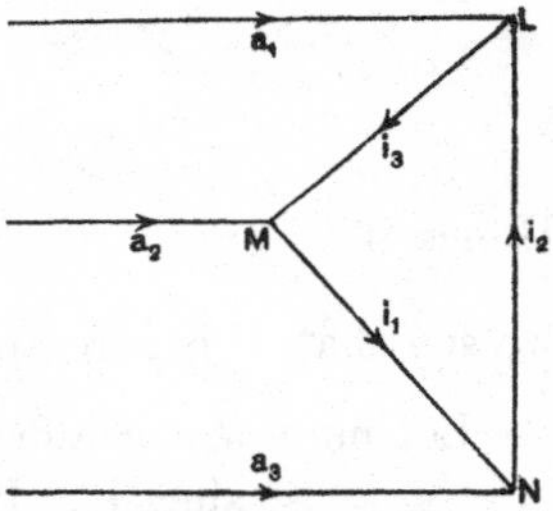

Fig. 99.　Currents in a mesh-connected load.

From the figure we see that
$$a_1 = i_3 - i_2, \qquad a_2 = i_1 - i_3$$
and
$$a_3 = i_2 - i_1 \quad\ldots\ldots\ldots(11).$$
Hence
$$a_1 + a_2 + a_3 = 0.$$

Thus, proceeding as before, we see that A_1, A_2 and A_3 form a triangle (Fig. 100), the exterior angles of which give the phase differences between the three vectors.

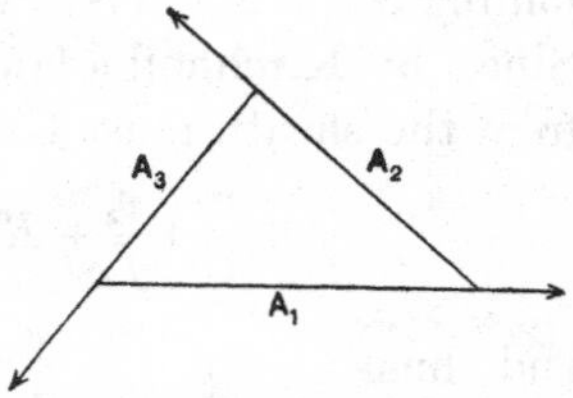

Again, if r_1, r_2 and r_3 be the resistances of the arms MN, NL and LM in Fig. 99, and if they are non-inductive,

Fig. 100.　The phase differences of the currents in the mains.

$$r_1 i_1 = v_2 - v_3,$$
$$r_2 i_2 = v_3 - v_1,$$
$$r_3 i_3 = v_1 - v_2.$$

Therefore $r_1 i_1 + r_2 i_2 + r_3 i_3 = 0 \ \ldots\ldots\ldots\ldots\ldots\ldots\ldots\ldots\ldots\ldots(12).$

We have supposed the three arms of the load non-inductive, but equation (12) is also true in other cases. Suppose, for example, that the three arms of the load are the primary coils of a three phase transformer (see Volume II), each coil being wound with n turns of wire, then

$$r_1 i_1 + n \frac{\partial \phi_1}{\partial t} = v_2 - v_3, \text{ etc.},$$

where ϕ_1 is the magnetic flux in the first limb of the transformer. Thus

$$r_1 i_1 + r_2 i_2 + r_3 i_3 + n \frac{\partial}{\partial t} (\phi_1 + \phi_2 + \phi_3) = 0.$$

If there is no magnetic leakage, $\phi_1 + \phi_2 + \phi_3$ must equal zero, since the lines of induction are closed curves, and therefore, as before,

$$r_1 i_1 + r_2 i_2 + r_3 i_3 = 0.$$

Hence, as formerly,

$$\frac{r_1 I_1}{\sin \theta_{2.3}} = \frac{r_2 I_2}{\sin \theta_{3.1}} = \frac{r_3 I_3}{\sin \theta_{1.2}} \quad \ldots\ldots\ldots\ldots\ldots(13),$$

where $\theta_{2.3}$ is the phase difference between I_2 and I_3. If $r_1 I_1$, $r_2 I_2$ and $r_3 I_3$ represent forces acting at a point O, equations (13) show that they will be in equilibrium.

Now it is known from statics that O is the centre of gravity of three equal masses placed at the extremities of $r_1 I_1$, $r_2 I_2$ and $r_3 I_3$ respectively. Hence O will also be the centre of gravity of masses proportional to r_1, r_2 and r_3 placed at the extremities of lines equal in length to I_1, I_2 and I_3 respectively.

Now from (11)

$$A_1{}^2 = I_3{}^2 + I_2{}^2 - 2 I_3 I_2 \cos \theta_{2.3}.$$

If we draw therefore from O (Fig. 101) lines proportional to I_1, I_2 and I_3 and inclined to one another at angles $\theta_{2.3}$, $\theta_{3.1}$ and $\theta_{1.2}$, we see that the lines joining the extremities of these lines give us A_1, A_2 and A_3 respectively. Hence the triangle is the same triangle as in Fig. 100.

Knowing the currents in the mains and the resistances of the arms of the mesh, we can find the currents in the arms by the following construction.

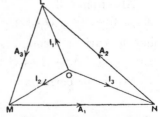

Fig. 101. The currents in the mains and in the arms of the mesh load.

Make a triangle LMN (see Fig. 101) whose sides are equal to A_1, A_2 and A_3 respectively. Find the centre of gravity O of masses r_1, r_2 and r_3 placed at L, M and N respectively. Then OL will give I_1, OM will give I_2, and ON will give I_3.

Again, if in Fig. 101 we produce OL to L', etc. so that $OL' = r_1 . OL$, $OM' = r_2 . OM$, and $ON' = r_3 . ON$, then if the resistances are non-inductive, $L'M'N'$ is the triangle giving the voltages between the mains and OL', OM' and ON' give the voltages between the mains and the centre of the star load.

We can prove the following algebraical relations between the currents in the same way as we proved the relations between the P.D.s:

Algebraical formulae for the currents in a mesh load.

$$r_2 r_3 A_1^2 = (r_2 I_2^2 + r_3 I_3^2)(r_2 + r_3) - r_1^2 I_1^2 \quad \ldots \ldots (14)$$

and two similar equations. Similarly,

$$(r_1 + r_2 + r_3)^2 I_1^2 = (r_2 A_3^2 + r_3 A_2^2)(r_2 + r_3) - r_2 r_3 A_1^2 \quad \ldots (15)$$

and two similar equations. Hence, also,

$$(I_1^2 r_1 + I_2^2 r_2 + I_3^2 r_3)(r_1 + r_2 + r_3)$$
$$= A_1^2 r_2 r_3 + A_2^2 r_3 r_1 + A_3^2 r_1 r_2 \ldots \ldots (16).$$

The case when r_1, r_2 and r_3 are all equal is important. O (Fig. 102) is now the centre of gravity of the triangle LMN.

Hence, from geometry or from equations (14) and (15) above,

$$A_1^2 = 2(I_2^2 + I_3^2) - I_1^2 \quad \ldots \ldots (17),$$
$$9I_1^2 = 2(A_2^2 + A_3^2) - A_1^2 \quad \ldots (18),$$

and four similar equations.

Also from (16)

$$A_1^2 + A_2^2 + A_3^2 = 3I_1^2 + 3I_2^2 + 3I_3^2 \ldots (19),$$

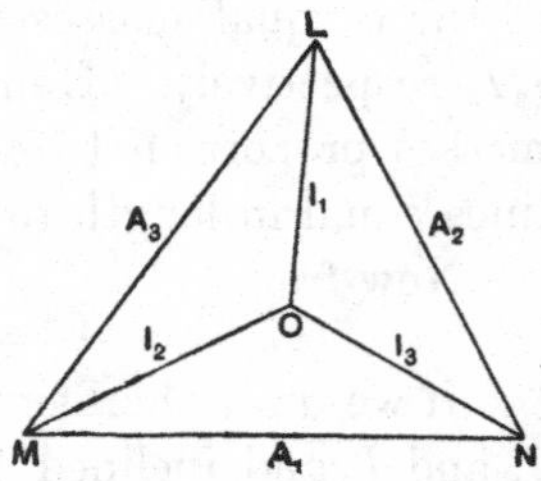

Fig. 102. The relations between the currents.

and the curious theorem is also true that

$$A_1^4 + A_2^4 + A_3^4 = (3I_1^2)^2 + (3I_2^2)^2 + (3I_3^2)^2 \ldots \ldots \ldots (20).$$

When the currents in the mains are all equal,

$$A = \sqrt{3} I \ldots \ldots \ldots \ldots \ldots \ldots \ldots \ldots (21).$$

It is to be noted that these formulae are perfectly general, no assumptions having been made as to the shape of the wave.

When we have a star winding, the currents in the branches are of course equal to the currents in the mains. Their phase differences can be got by constructing a triangle whose sides are proportional to the currents. The exterior angles of this triangle are the phase differences required.

If we have both a star and a mesh winding, the graphical representation of all the currents by a figure drawn in one plane is rarely possible.

We will first suppose that the effective P.D.s between the mains are all equal to one another. Let $Vf(t)$ represent the

The form of the wave of P.D. in three phase systems.

instantaneous value of one of these P.D.s, where V is the effective voltage and t is the time in seconds from the era of reckoning. Then the other two will be represented by $Vf(t + T/3)$, and $Vf(t + 2T/3)$, where T is the period of the alternating current. Since the sum of the instantaneous values of the P.D.s must always be zero, we must have

$$f(t) + f(t + T/3) + f(t + 2T/3) = 0 \quad \ldots\ldots\ldots\ldots(22).$$

This functional equation limits the possible forms of $f(t)$. Solving it by Laplace's method, we get

$$f(t) = X \sin (2\pi t/T + Y),$$

where X and Y are constants or functions of t whose values do not alter when $t + T/3$ or $t + 2T/3$ is substituted for t.

Since, however, $f(t)$ is an alternating periodic function, we have another equation to satisfy, namely

$$f(t) = -f\left(t + \frac{T}{2}\right).$$

This adds the condition that X and Y do not change when $t + T/2$ is written for t in them. Hence X and Y are functions of t that do not alter when $t + T/3$, $t + T/2$, or $t + 2T/3$ is written for t.

In Fig. 103, X has been taken equal to 1, and Y equal to

Examples.

$(1/2) \sin \{6 (2\pi t)/T\}$, so that the equation to the curve is

$$y = \sin \left(\frac{2\pi t}{T} + \frac{1}{2} \sin 6 \frac{2\pi t}{T}\right).$$

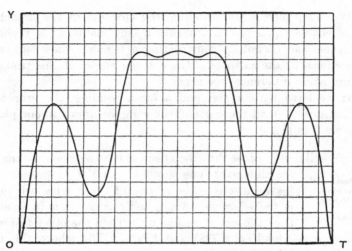

Fig. 103. $y = \sin\left(\dfrac{2\pi t}{T} + \dfrac{1}{2}\sin 6\,\dfrac{2\pi t}{T}\right).$

Possible form of P.D. wave in balanced three phase system.
Positive half only shown.

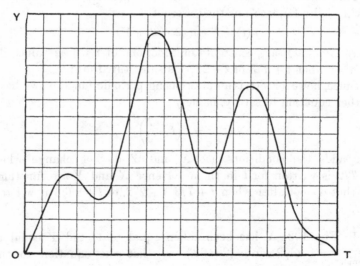

Fig. 104. $y = \left(1 + \dfrac{1}{2}\sin 6\,\dfrac{2\pi t}{T}\right)\sin\dfrac{2\pi t}{T}.$

Another possible form.

In Fig. 104 the equation to the curve is

$$y = \left(1 + \frac{1}{2}\sin 6\,\frac{2\pi t}{T}\right)\sin \frac{2\pi t}{T}.$$

The equation (22) can also be solved by Fourier's method. The Fourier series in this case may be written

$$f(t) = \Sigma A_{6n\pm1}\sin\left\{(6n\pm1)\frac{2\pi t}{T} + \alpha_{6n\pm1}\right\} \quad\ldots\ldots\ldots(23).$$

As a rule it will be found that the expression $X\sin(2\pi t/T + Y)$ is more convenient to work with than the series given by (23).

The P.D. waves between the mains can only be similar curves when their effective values are all equal.

If it is possible for the potential difference waves between the mains to be similar curves when their effective values are different, let the P.D.s between the mains be represented by

$$V_1 f(t), \quad V_2 f(t + T/3) \quad \text{and} \quad V_3 f(t + 2T/3)$$

respectively, where V_1, V_2 and V_3 are positive. Then,

at time t,
$$V_1 f(t) + V_2 f(t + T/3) + V_3 f(t + 2T/3) = 0,$$

at time $t + T/3$,
$$V_3 f(t) + V_1 f(t + T/3) + V_2 f(t + 2T/3) = 0,$$

and at time $t + 2T/3$,
$$V_2 f(t) + V_3 f(t + T/3) + V_1 f(t + 2T/3) = 0.$$

Eliminating the functions by determinants or otherwise, we get

$$(V_1 + V_2 + V_3)\{(V_2 - V_3)^2 + (V_3 - V_1)^2 + (V_1 - V_2)^2\} = 0,$$

and therefore
$$V_1 = V_2 = V_3,$$

since $V_1 + V_2 + V_3$ must be positive.

Hence, if we make the assumption that the P.D. waves between the terminals of a three phase alternator are all similar, we also make the assumption that their values are all equal.

By an almost identical proof we can show that the P.D. waves between the mains and earth can only be similar curves when the leakage currents to earth are all equal to one another.

R. I.   24

In Fig. 105 let A, B and C be the three terminals of a three

To find the
frequency of
the alternating
current in the
fourth wire of
a balanced
three phase
system.

phase alternator, the armature of which is star wound. Let also AD, BE and CF be the mains. The points O and L are sometimes put to earth or connected by a wire. We shall investigate the frequency of the current in this wire when the loads on the three arms are equal. Let

$$Cf(t), \quad Cf(t + T/3) \quad \text{and} \quad Cf(t + 2T/3)$$

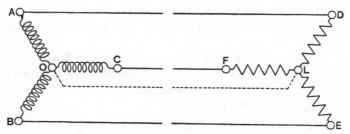

Fig. 105. Frequency of the current in the fourth wire equals $3(2n+1)f$.

be the instantaneous values of the currents in the mains and let $-C'\phi(t)$ be the current in the fourth wire. Then

$$C'\phi(t) = C\{f(t) + f(t + T/3) + f(t + 2T/3)\},$$

and

$$\begin{aligned}
C'\phi(t + T/6) &= C\{f(t + T/6) + f(t + T/2) + f(t + 5T/6)\} \\
&= -C\{f(t + 2T/3) + f(t) + f(t + T/3)\} \\
&= -C'\phi(t),
\end{aligned}$$

and thus

$$\phi(t + T/6) = -\phi(t).$$

Hence

$$\phi(t + T/3) = -\phi(t + T/6) = \phi(t).$$

It follows that, if the frequency of the currents in the mains be f, the frequency of the current in the fourth wire is of the form $3(2n + 1)f$, where n is a positive integer. Its lowest value is therefore $3f$.

It is easy to show that, if the current waves in the mains be triangular or parabolic in shape, then the current wave in the fourth wire is also triangular or parabolic, and, since its frequency is three times as great as that of the wave in any main, its effective value is one-third that of the current in each main. If the currents in the mains were sine curves, its value would be zero. Since a sine curve and a parabola differ very little in shape from one another (see Fig. 64, p. 281), this shows how important a small modification in the shape of the wave may be. In the general case the shape of the resultant wave in the fourth wire is quite different from the shape of the component waves in the mains.

These results prove that in practical three phase working slight causes may considerably alter the shape of the E.M.F. and current waves. For example, a slight variation in the resistance or, *a fortiori*, in the inductance of the fourth wire, will alter the shapes of the current waves in the other three wires.

We have shown that, when everything is symmetrical in a three phase system, the potentials of the mains can be expressed by functions of the form

Value of load on a three phase alternator at every instant.

$$X \sin (2\pi t/T + Y),$$

where X and Y are periodic functions of t whose frequency is $6/T$ or $6f$. If the mesh load between the mains be three non-inductive resistances each equal to R, and the star load arms be each equal to r, then the power at any instant is

$$\frac{1}{R} \{X \sin (2\pi t/T + Y) - X \sin (2\pi t/T + Y + 2\pi/3)\}^2 + \ldots + \ldots$$

$$+ \frac{1}{r} X^2 \sin^2 (2\pi t/T + Y) + \ldots + \ldots$$

$$= \frac{3}{2} \left(\frac{1}{r} + \frac{3}{R}\right) X^2,$$

where X is a function of t whose frequency is $6f$. Hence the instantaneous value of the power is only constant in special cases.

Suppose that there is both a mesh and a star load as in Fig. 106. Let a_1, a_2 and a_3; i_1, i_2 and i_3; i_x, i_y and i_z be the instantaneous values of the currents in the mains, in the arms of the mesh load and in the arms of the star load respectively.

The measurement of power in three phase circuits.

Then

$$a_1 = i_3 - i_2 + i_x,$$
$$a_2 = i_1 - i_3 + i_y,$$

and

$$a_3 = i_2 - i_1 + i_z.$$

Thus $\qquad a_1 + a_2 + a_3 = i_x + i_y + i_z = 0,$

if the three phase machine is insulated from the earth and the leakage currents from the mains are inappreciable.

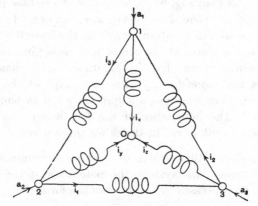

Fig. 106. The measurement of power.

Let w be the instantaneous value of the watts, then

$$w = v_{1.2} i_3 + v_{2.3} i_1 + v_{3.1} i_2$$
$$+ v_1 i_x + v_2 i_y + v_3 i_z.$$

Now $\qquad v_{3.1} = - v_{1.2} - v_{2.3},$

and $\qquad i_y = - i_x - i_z.$

Thus $\qquad w = v_{1.2} (i_3 - i_2 + i_x) + v_{3.2} (i_2 - i_1 + i_z)$

$$= v_{1.2} a_1 + v_{3.2} a_3$$

by symmetry $\qquad = v_{2.3} a_2 + v_{1.3} a_1 \qquad \dots\dots\dots\dots\dots\dots(24).$

and $\qquad = v_{3.1} a_3 + v_{2.1} a_2$

Similarly $\qquad w = v_1 a_1 + v_2 a_2 + v_3 a_3 \dots\dots\dots\dots\dots\dots\dots(25).$

Let us next consider the case when the point O is not maintained at zero potential and let v_x be its potential. Then the power being expended in all the paths of the current a_1 from the point 1 where the potential is v_1 to the point or points where the potential is v_x is $(v_1 - v_x) a_1$. Similarly $(v_2 - v_x) a_2$ and $(v_3 - v_x) a_3$ will be the values of the power in the paths of the two other main currents respectively. Thus the total power w in the load is given by

$$w = (v_1 - v_x) a_1 + (v_2 - v_x) a_2 + (v_3 - v_x) a_3 \dots\dots\dots(26).$$

If we make the supposition that O is insulated from the earth, then $a_1 + a_2 + a_3$ must be zero and we get

$$w = v_1 a_1 + v_2 a_2 + v_3 a_3.$$

Similarly $\qquad w = v_{1.3} a_1 + v_{2.3} a_2$

and two similar equations. Therefore equations (24) and (25) still hold in this case.

When O is connected with the earth by a wire and the leakage currents from the mains are appreciable, or when the centre O' of the star winding of the armature is also connected with the earth, or when O and O' are joined by a fourth wire, then $a_1 + a_2 + a_3$ is not necessarily zero. Let i_0 be the current in the wire joining O to the earth, then we must always have

$$a_1 + a_2 + a_3 + i_0 = 0,$$

and thus by (3) $\qquad w = v_1 a_1 + v_2 a_2 + v_3 a_3 + v_x i_0 \dots\dots\dots\dots\dots(27).$

Therefore $v_1 a_1 + v_2 a_2 + v_3 a_3$ is not equal to the power in the load in this case. We also have

$$w = v_{1.3} a_1 + v_{2.3} a_2 + v_{x.3} i_0 \dots\dots\dots\dots\dots(28).$$

When i_0 is zero these formulae simplify to (25) and (24) given above.

The formulae (24) and (26) give two methods of measuring power in three phase circuits. The first method is to use two wattmeters. The ampere coil of one of them is put in No. 1 main and the volt coil is connected across 1 and 2. The ampere coil of the other is put in No. 3 main and the volt coil is connected across 3 and 2. Suppose that w_1 is the reading on one meter and that w_2 is the reading on the other, and suppose that w_1 is greater than w_2. Then the power given to the circuit is $w_1 \pm w_2$. If the phase difference between a_3 and $v_{3.2}$ is less than 90 degrees, w_2 is positive, but if greater, w_2 is negative. We must be careful, when measuring, to note whether we have to reverse the shunt connections or not in order that the needle may deflect the right way. For example, if the two wattmeters are exactly similar and if it is found necessary to reverse the shunt leads of the meter reading w_1, but not those of the meter reading w_2, then the true power is $w_1 - w_2$. If neither or both have to be reversed, the true power is $w_1 + w_2$.

The second method is to use three wattmeters, their ampere

coils being put in the main circuits and their volt coils across O and 1, O and 2, and O and 3 respectively.

It will be seen from (26) that the three wattmeter method is applicable to three phase systems which use a fourth wire.

When there is a fourth wire, equation (28) shows that the two wattmeter method cannot be applied. Attempts have been made in practice to get over this difficulty by winding supplementary coils round the ordinary ampere coils and connecting these coils in series with the fourth wire. The meters now register w, where

$$w = v_{1.3}\,(a_1 + \lambda i_0) + v_{2.3}\,(a_2 + \mu i_0)\,\ldots\ldots\ldots\ldots(29),$$

and λ and μ are constants which can be adjusted.

In order that (28) and (29) may be identical,

$$\lambda v_{1.3} + \mu v_{2.3} = v_{x.3}.$$

This equation obviously cannot be always true since the value of x varies with the load. The method, therefore, is not an accurate one.

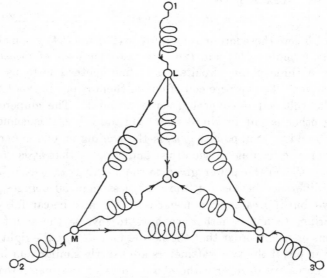

Fig. 107.   Possible connections.

It is to be noticed that the equation (27) is always true no matter how the coils forming the load are interconnected. For (Fig. 107) if v_x be the potential of O, the power given to the load

by main 1 is $(v_1 - v_x)\, a_1$ and similar expressions hold for the power
given to the load by the mains 2 and 3.   Thus

$$w = (v_1 - v_x)\, a_1 + (v_2 - v_x)\, a_2 + (v_3 - v_x)\, a_3$$
$$= v_1 a_1 + v_2 a_2 + v_3 a_3 + v_x i_0.$$

If all the mains are equally loaded and the loads are non-
inductive, one meter is sufficient.   If the volt coil be connected
across two of the mains, we multiply the reading by 2.   If it be
connected from one main to the centre of the system, then the
multiplying factor is 3.

An important case arises when the volt coil of the meter forms
one of the arms of a star load, made up of two equal high resistances connected with the volt coil.   An arrangement of this kind
is generally called a star-box.   If the three arms are of equal
resistance, the multiplying factor is 3.   If, however, as is often
the case in practice, the resistance of the volt coil be different
from that of the other two arms, then the multiplying factor for
balanced loads is $2 + r/R$, where R is the resistance of the volt coil
and r that of either of the other arms of the box.

Let O (Fig. 108) be the centre of gravity of masses $1/R$, $1/r$
and $1/r$ placed at A, B and C, the
corners of the voltage triangle, respectively.   Then OA, OB and OC will be
the three P.D.s to the centre of the
star-box.

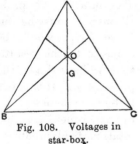

Also by (5), p. 361,

$$OA^2 \left(\frac{1}{R} + \frac{1}{r} + \frac{1}{r} \right)^2$$
$$= \left(\frac{CA^2}{r} + \frac{AB^2}{r} \right) \left(\frac{1}{r} + \frac{1}{r} \right) - \frac{BC^2}{r^2},$$

Fig. 108.   Voltages in
star-box.

and thus, since the triangle is equilateral,

$$OA^2\, (2 + r/R)^2 = 3\,.\,AB^2 = 9\,.\,GA^2,$$

where G is the centre of gravity of the triangle ABC;
and therefore, $3\,.\,GA/OA = 2 + r/R.$

Now if the arms had been equal, GA would have been the P.D.
across the volt coil.   Hence the required multiplying factor is
$3\,.\,GA/OA$, and this we have shown equals $2 + r/R$.

In the case of a three wire system, we have

Graphical
method.
$$a_1 + a_2 + a_3 = 0,$$

at every instant, where a_1, a_2 and a_3 are the currents in the three mains. The vectors OA_1, OA_2 and OA_3 (Fig. 109) drawn to represent these currents lie, therefore, in one plane.

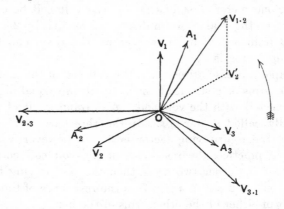

Fig. 109.   Diagram of the voltages and the currents in a three phase system.
Case when the phase of V_2 is in advance of the phase of V_1.

The sum of the projections of their lengths also on any line through O must be zero. If they represented forces acting on O these forces would be in equilibrium, each being equal and opposite to the resultant of the other two. It is to be noticed that we have drawn the vector of a_2 as if it were in advance of the phase of a_1. If this is not the case, then OA_3 and OA_2 would have to be interchanged. In practice this question can be determined by experiment as, for example, by noticing the direction of a rotating field produced by the currents (Chap. XVIII).

Similarly we have
$$v_{1.2} + v_{2.3} + v_{3.1} = 0,$$

and so the three vectors $OV_{1.2}$, $OV_{2.3}$ and $OV_{3.1}$ lie in one plane. But this plane need not necessarily coincide with the plane containing the three current vectors OA_1, OA_2 and OA_3. In the general case it is not possible to represent these six vectors by lines drawn in space, the three current vectors being in one plane and the three voltage vectors in another. We are therefore forced to make the assumption that all the currents and electromotive forces follow the harmonic law, and this limits the value of the graphical method. In this case all the vectors shown in Fig. 109 are in the same plane. Since $v_1 + v_2 + v_3$ is only zero in a very special case, OV_1, OV_2 and OV_3 would not necessarily represent a system of forces in equilibrium. If we produce V_2O to V_2' and make $OV_2' = OV_2$, $OV_{1.2}$ will be the diagonal of the parallelogram constructed on OV_1 and OV_2' as adjacent sides.

If w denote the instantaneous value of the power given to the load, we have

$$w = v_1 a_1 + v_2 a_2 + v_3 a_3,$$

and so

$$W = V_1 A_1 \cos(v_1 a_1) + V_2 A_2 \cos(v_2 a_2) + V_3 A_3 \cos(v_3 a_3),$$

where W is the mean power and $v_1 a_1$ denotes the angle between OV_1 and OA_1. Projecting OA_1, OA_2, and OA_3, on OV_2, we get

$$A_1 \cos(a_1 v_2) + A_2 \cos(a_2 v_2) + A_3 \cos(a_3 v_2) = 0,$$

and thus

$$W = A_1 \{ V_1 \cos(v_1 a_1) - V_2 \cos(v_2 a_1) \}$$
$$+ A_3 \{ V_3 \cos(v_3 a_3) - V_2 \cos(v_2 a_3) \}$$
$$= A_1 V_{1.2} \cos(v_{1.2} a_1) + A_3 V_{3.2} \cos(v_{3.2} a_3).$$

This proves the two wattmeter method when the currents and electromotive forces follow the sine law.

Let us now assume that the load on the system is balanced. In this case $V_1 = V_2 = V_3$, and $A_1 = A_2 = A_3$, and $V_1 \hat{O} V_{1.2} = 30°$. Let ψ also be the angle by which A_1, A_2 and A_3 lag behind V_1, V_2 and V_3 respectively. We see at once from the diagram Fig. 109 that

$$W_1 = V_{1.2} A_1 \cos(30° - \psi) \text{ and } W_2 = V_{3.2} A_3 \cos(30° + \psi).$$

Hence, since $V_{1.2} = V_{3.2}$, we get

$$W = V_{1.2} A_1 \{ \cos(30° - \psi) + \cos(30° + \psi) \}$$
$$= \sqrt{3} V_{1.2} A_1 \cos\psi = 3 V_1 A_1 \cos\psi.$$

In practice it is customary to call $\cos\psi$ the power factor of the balanced load. The following table shows how the wattmeter readings vary with the power factor.

ψ	$\cos\psi$		$W_1/V_{1.2}A_1$	$W_2/V_{1.2}A_1$	$W/V_{1.2}A_1$
$-90°$	0	lag	-0.5	0.5	0
$-60°$	0.5	,,	0	0.866	0.866
$-30°$	0.866	,,	0.5	1	1.5
$0°$	1	,,	0.866	0.866	1.732
$30°$	0.866	lead	1	0.5	1.5
$60°$	0.5	,,	0.866	0	0.866
$90°$	0	,,	0.5	-0.5	0

It is useful to remember that, on a balanced load, if both W_1 and W_2 are less than $0.866\,V_{1.2}A_1$, we subtract the smaller of the readings, W_1 or W_2, from the larger. If either of the readings is greater than $0.866\,V_{1.2}A_1$, the true power is $W_1 + W_2$.

In Fig. 110, G is the generator and M is the motor. If we **Watt-hour meters.** assume that the three arms are equally loaded, then the energy expended can be measured by means of an ordinary single phase watt-hour meter W and a star-box (S in

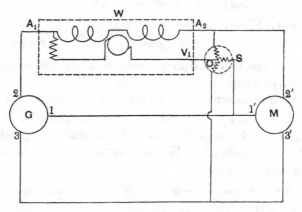

Fig. 110.   Connections of watt-hour meter to measure balanced loads.

the figure). The ampere coil $A_1 A_2$ (see Chapter XIII) is put in one of the mains $22'$ and the volt coil $A_1 V_1$ is connected to $22'$ and the centre of the star-box. The units registered by the meter multiplied by three will give the total energy expended on the motor. If the arms are unequally loaded, a watt-hour

meter has to be put in each arm and the volt coils connected
to O. The sum of the three readings will then give the energy
expended in the given time. It will be seen that the three
wattmeters can easily be made into a single one containing a
star-box and having six ampere terminals A_1, A_2, etc., three being
connected with the supply mains and three being connected with
the mains for the load, the volt connections being permanently
made inside the meter.

If we connect two single phase watt-hour meters, as in Fig. 111,
then by the formulae given above the sum of their readings will

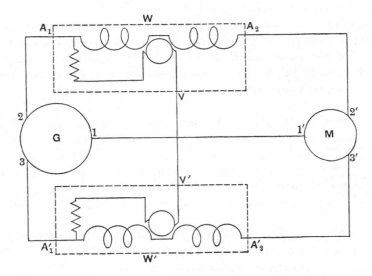

Fig. 111.   Connections of Watt-hour Meters to measure all loads.

The ampere coils of the two meters are put in two of the mains and the
volt coils are connected with the third main.

give the energy expended. If one of them runs backwards, then
its reading has to be subtracted from the reading of the other,
which will always be the greater. A true three phase meter can
be made by combining the two meters (see Chapter XVII). It will
have four ampere terminals and one volt terminal. If the load be
balanced and non-inductive, the reading on either instrument shown
in Fig. 111 multiplied by two will give the true units expended.

It can be shown mathematically that if we take a point O within a triangle LMN (Fig. 112), then

$$OL + OM + ON$$

Minimum value of the sum of the three voltages in a star load.

is a minimum when the angles LOM, MON and NOL are each equal to 120 degrees, provided, of course, that no angle of the triangle is equal to or greater than 120 degrees. If one angle of the triangle is equal to 120 degrees, then, to make $OL + OM + ON$ a minimum, O must coincide with this angle.

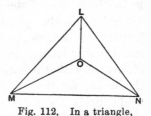

Fig. 112. In a triangle,
$OL + OM + ON$
is a minimum when the angles
at O are all equal.

It follows therefore from Fig. 98, p. 359, that the sum of the three voltages across the arms of a star-box is a minimum when their phase differences are each equal to 120 degrees. In this case the currents in the arms are all equal to one another.

To find the resistances of the arms of a star load in order to get symmetrical three phase currents.

Describe the voltage triangle LMN. On each side of it describe a segment of a circle which will contain an angle of 120 degrees. These will intersect in a single point O. Produce LO to cut MN in P, then the ratio of MP to PN will be the ratio of the resistances q and r in the desired load.

To find the ratio of the resistances in a star load in order that the voltages to the centre may be equal.

In this case $E_1 = E_2 = E_3$ and O is therefore the centre of the circumscribing circle. Let A, B and C be the angles of the voltage triangle, then

$$\theta_{2.3} = 2A, \text{ etc.}$$

Also

$$\frac{E_1}{p \sin \theta_{2.3}} = \frac{E_2}{q \sin \theta_{3.1}} = \frac{E_3}{r \sin \theta_{1.2}}.$$

Hence

$$p \sin 2A = q \sin 2B = r \sin 2C.$$

It will be seen that three phase problems, considered graphically, generally resolve themselves into problems connected with the trigonometry of a triangle. When the arms of the load are inductive, then, as a rule, only approximate solutions can be got graphically, as the vectors can no longer be accurately represented geometrically (see Chapter XII).

REFERENCES

A. E. KENNELLY, 'On the Equivalence of Triangles and Stars in Conducting Networks.' *Electrical World and Engineer*, Vol. 34, p. 413, 1899.

A. RUSSELL, 'The Elements of Three Phase Theory.' *The Electrician*, Vol. 47, p. 639, 1901.

—— 'P.D. Wave Forms in Three Phase Systems.' *The Electrician*, Vol. 48, p. 487, 1902.

For Laplace's Method of Solving Functional Equations, see GEORGE BOOLE, *The Calculus of Finite Differences*.

CHAPTER XVI

Two phase systems. The magnitudes and the phase differences of the potential differences between the mains. The voltage tetrahedron. The graphical representation of the voltages in the four arms of a star load. Rule for finding the voltages across the arms of a star load when the resistances of the arms are given. The voltages to the centre of the load so adjust themselves that the power expended on it is a minimum. The potentials to earth of the mains. The currents in a mesh load. The conditions under which it is possible for the P.D. waves between the four mains and earth to be similar curves. The P.D. waves between the mains are only similar in a special case. To find the phase differences between the opposite voltages and between the diagonal voltages in a two phase system. The measurement of power in two phase circuits. Two phase meters. When the currents in a star load are equal the sum of the voltages to the centre is a minimum. Polyphase Systems.

DIAGRAMS 94 and 95 in the last chapter illustrate the funda-

Two phase systems.

mental principle of three phase machines with mesh and star wound armatures respectively. The principle of two phase machines is the same, but instead of the armature being wound in three sections it is wound in four. There is, however, in this case a third method of winding, which is illustrated in Fig. 113. *A, B, C,* and *D* are the four terminals of the machine. *A* and *C* are the terminals of one winding of the armature and *B* and *D* are the terminals of the other. From symmetry the P.D.s generated between *A* and *C* and between *B* and *D* when the field magnet revolves will differ in phase by ninety degrees. The currents flowing in circuits connected across the mains to *A* and *C* and in similar circuits connected across the mains to *B* and *D* will also differ in phase by a right angle, and we shall see later on that it is easy to produce a rotary magnetic field by means of these currents, and hence two phase motors are simple to construct.

In practice instead of having four mains connected to A, B, C and D respectively it is customary to have only three, one of which is connected to two adjacent terminals, as for example A and B. In this case the P.D.s between A and C and between B

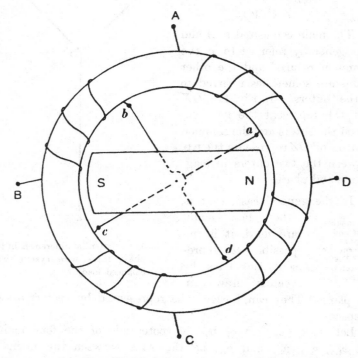

Fig. 113. Two Phase Machine with two separate windings on armature.

or A and D will still differ in phase by a right angle, and they will from symmetry be equal to one another. Now with our usual notation

$$v_{CD} = v_C - v_D$$
$$= -(v_A - v_C) - (v_D - v_A)$$
$$= -v_{AC} - v_{DA}.$$

Thus $$v_{CD} + v_{DA} + v_{AC} = 0,$$

and $$v_{CD} + v_{DB} + v_{AC} = 0.$$

Therefore $\qquad V^2{}_{CD} = V^2{}_{DB} + V^2{}_{AC},$

since V_{DB} and V_{AC} are in quadrature. They are also equal, and thus

$$V_{CD} = \sqrt{2}\, V_{DB}$$
$$= \sqrt{2}\, V_{AC}.$$

The main connected to A and B is generally referred to as the 'common return' and the other mains are sometimes referred to as the 'outers.' In Fig. 114, OX and OY represent the P.D.s between the outers and the common return and OZ represents the P.D. between the two outers in magnitude and phase.

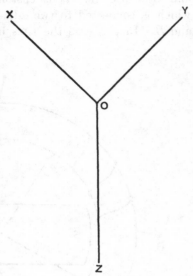

Fig. 114. Potential differences in two phase three wire system with balanced load.

In the general case, that is, when four mains **The magnitudes and the phase differences of the P.D.s between the mains.** are used, it is impossible to represent the P.D.s by vectors drawn in one plane. They can, however, be represented by vectors drawn in space.

Let v_1, v_2, v_3 and v_4 be the potentials of the four mains. Let $v_{1.2}$, $v_{2.3}$, $v_{3.4}$ and $v_{4.1}$ be the P.D.s between the mains 1 and 2, 2 and 3, 3 and 4, and 4 and 1 respectively.

Then
$$v_{1.2} = v_1 - v_2,$$
$$v_{2.3} = v_2 - v_3,$$
$$v_{3.4} = v_3 - v_4,$$
and
$$v_{4.1} = v_4 - v_1.$$
Hence
$$v_{1.2} + v_{2.3} + v_{3.4} + v_{4.1} = 0.$$

A linear relation therefore connects the four instantaneous values, and hence by Chapter XII their effective values can be represented by lines drawn in space. In Fig. 74, p. 309, if we produce SO backwards to T making OT equal to OS, then OT will represent V_1 in this case, and the relations between the voltages and phase differences can be written down by the formulae given in that chapter.

If N be the fourth corner of the parallelogram ORQ (Fig. 74, page 309) and we join ON and SQ, we get the voltage tetrahedron 1, 2, 3, 4 (Fig. 115). The angles between the lines drawn in this figure are the supplements of the phase differences between the P.D.s these lines represent. In Fig. 115 12, 23, etc. represent the P.D.s between the mains 1 and 2, between the mains 2 and 3, etc.

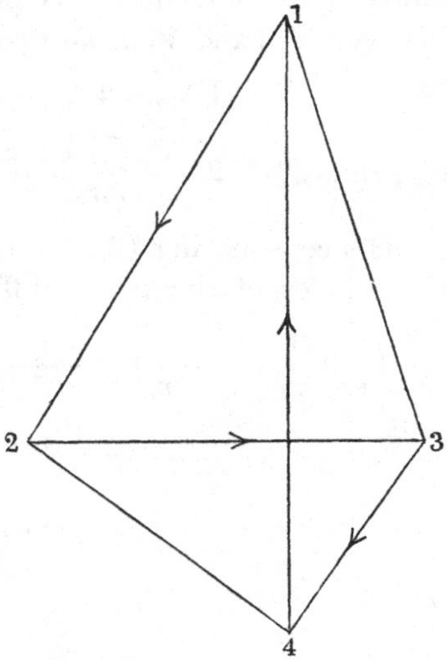

The voltage tetrahedron.

Fig. 115. The Voltage Tetrahedron.

Let 1, 2, 3 and 4 (Fig. 116) be the four mains and let O be the centre of the star load. Let r_1, r_2, r_3 and r_4 be the resistances (non-inductive) of the four arms, and let e_1, e_2, e_3 and e_4 be the P.D.s between the mains and the centre of the load. Then, since the algebraical sum of the currents at O must be zero, we have

The graphical representation of the voltages in the four arms of a star load.

$$\frac{e_1}{r_1} + \frac{e_2}{r_2} + \frac{e_3}{r_3} + \frac{e_4}{r_4} = 0 \ldots (1).$$

Hence $\dfrac{E_1}{r_1}$, $\dfrac{E_2}{r_2}$, $\dfrac{E_3}{r_3}$ and $\dfrac{E_4}{r_4}$ can be represented by vectors in the ways shown in Fig. 115 and in Fig. 74, page 309.

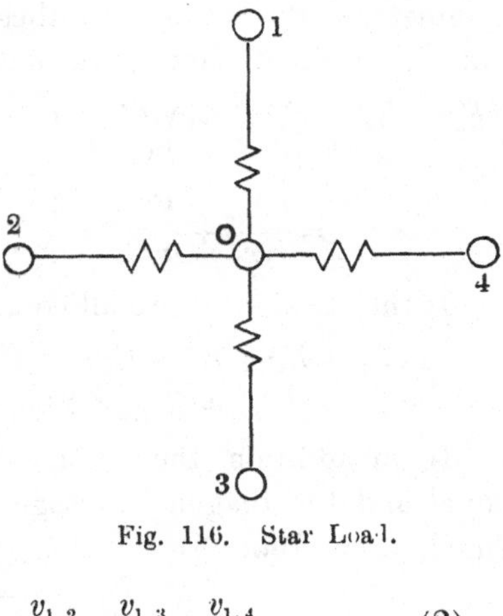

Fig. 116. Star Load.

We may write (1) in the form

$$e_1 \left(\frac{1}{r_1} + \frac{1}{r_2} + \frac{1}{r_3} + \frac{1}{r_4} \right) = \frac{v_{1.2}}{r_2} + \frac{v_{1.3}}{r_3} + \frac{v_{1.4}}{r_4} \ldots \ldots \ldots (2),$$

where $v_{1.2} =$ the P.D. between the mains 1 and 2

$\qquad\qquad = e_1 - e_2.$

Now by the voltage tetrahedron (Fig. 115) we see that $V_{1.2}$, $V_{2.3}$ and $V_{3.1}$ form a triangle. Hence if α be the angle of phase difference between $V_{1.2}$ and $V_{1.3}$, we find by trigonometry that

$$V^2_{2.3} = V^2_{1.2} + V^2_{1.3} - 2 V_{1.2} V_{1.3} \cos \alpha,$$

and thus $2 \dfrac{V_{1.2} V_{1.3} \cos \alpha}{r_2 r_3} = \dfrac{V^2_{1.2} + V^2_{1.3} - V^2_{2.3}}{r_2 r_3}.$

Hence, squaring (2), taking mean values and substituting for the cosines of the phase differences, we get

$$E_1^2 \left(\frac{1}{r_1} + \frac{1}{r_2} + \frac{1}{r_3} + \frac{1}{r_4} \right)^2 = \frac{V^2_{1.2}}{r_2^2} + \frac{V^2_{1.3}}{r_3^2} + \frac{V^2_{1.4}}{r_4^2} + \frac{V^2_{1.2} + V^2_{1.3} - V^2_{2.3}}{r_2 r_3}$$

$$+ \frac{V^2_{1.3} + V^2_{1.4} - V^2_{3.4}}{r_3 r_4} + \frac{V^2_{1.4} + V^2_{1.2} - V^2_{4.2}}{r_4 r_2}$$

$$= \left\{ \frac{V^2_{1.2}}{r_2} + \frac{V^2_{1.3}}{r_3} + \frac{V^2_{1.4}}{r_4} \right\} \left\{ \frac{1}{r_2} + \frac{1}{r_3} + \frac{1}{r_4} \right\}$$

$$- \frac{V^2_{2.3}}{r_2 r_3} - \frac{V^2_{3.4}}{r_3 r_4} - \frac{V^2_{4.2}}{r_4 r_2} \quad \dots\dots\dots\dots(3).$$

The other three equations giving E_2, E_3 and E_4 in terms of the six P.D.s between the mains can easily be written down by symmetry. By adding up these four equations and cancelling out the common factor, we deduce

$$\left(\frac{E_1^2}{r_1} + \frac{E_2^2}{r_2} + \frac{E_3^2}{r_3} + \frac{E_4^2}{r_4} \right) \left(\frac{1}{r_1} + \frac{1}{r_2} + \frac{1}{r_3} + \frac{1}{r_4} \right)$$

$$= \frac{V^2_{1.2}}{r_1 r_2} + \frac{V^2_{1.3}}{r_1 r_3} + \frac{V^2_{1.4}}{r_1 r_4} + \frac{V^2_{2.3}}{r_2 r_3} + \frac{V^2_{2.4}}{r_2 r_4} + \frac{V^2_{3.4}}{r_3 r_4} \quad \dots\dots(4).$$

If the resistances are all equal,

$$4 (E_1^2 + E_2^2 + E_3^2 + E_4^2)$$
$$= V^2_{1.2} + V^2_{1.3} + V^2_{1.4} + V^2_{2.3} + V^2_{2.4} + V^2_{3.4} \dots\dots\dots(5).$$

If, in addition, the voltages between adjacent mains are all equal and the diagonal voltages, namely $V_{1.3}$ and $V_{2.4}$, are also equal, then from (3)

$$E_1 = E_2 = E_3 = E_4.$$

And from (5) $8 E_1^2 = 2 V^2_{1.2} + V^2_{1.3} \quad \dots\dots\dots\dots(6).$

If, finally, $V_{1.2}$ and $V_{2.3}$ are in quadrature,

$$V^2_{1.3} = V^2_{1.2} + V^2_{2.3}$$

$$= 2V^2_{1.2}.$$

Hence $\qquad\qquad \sqrt{2}E_1 = V_{1.2}$ ....(7).

In getting the equation (7) we have made no assumption as to the shapes of the waves of P.D., but the shapes are restricted by equation (1).

Looking back at Fig. 74, page 309, we see that if OP, OQ, OR and OT (drawn equal and opposite to OS) represent forces acting at O, they will be in equilibrium. Now, if G be the centre of gravity of equal particles placed at P, Q, R and T, the resultant of the forces OP, OQ, OR and OT will be $4 . OG$; but since they are in equilibrium, this resultant must be zero, and therefore G must coincide with O. Hence O is the centre of gravity of equal masses placed at P, Q, R and T.

Let OP represent $\dfrac{E_1}{r_1}$, and let OQ, OR and OT represent $\dfrac{E_2}{r_2}$, $\dfrac{E_3}{r_3}$ and $\dfrac{E_4}{r_4}$. Let also $OP' = r_1 . OP$, $OQ' = r_2 . OQ$, etc., then $P'Q'R'T'$ will be the tetrahedron that gives the P.D.s between the mains. To prove this it is sufficient to notice that

$$v_{1.2} = e_1 - e_2.$$

Therefore $\qquad V^2_{1.2} = E_1{}^2 + E_2{}^2 - 2E_1E_2\cos\phi_{1.2},$

and hence the rest follows by trigonometry.

Rule for finding the voltages across the arms of a star load when the resistances of the arms are given. Construct the tetrahedron that gives the voltages between the mains. In order to do this we need to take six voltage readings, namely, the four readings between adjacent mains and the two diagonal readings. Find the centre of gravity G of masses $1/r_1$, $1/r_2$, $1/r_3$ and $1/r_4$ placed at the four angular points of the tetrahedron. Then the lines joining G to the four angular points will give the phase differences and the magnitudes of the required voltages.

Let L, M, N and R be the angular points of the tetrahedron representing the voltages between the mains, and we suppose that the dynamo maintains these voltages constant. Then, by a well-known theorem in statics, if O be any point in space,

The voltages to the centre of the load so adjust themselves that the power expended on it is a minimum.

$$\frac{OL^2}{r_1} + \frac{OM^2}{r_2} + \frac{ON^2}{r_3} + \frac{OR^2}{r_4}$$

is a minimum when O is the centre of gravity of masses $1/r_1$, $1/r_2$, $1/r_3$ and $1/r_4$ placed at L, M, N and R respectively. But since the resistances are non-inductive, this expression represents the power expended on the load, and hence the theorem follows.

Let f_1, f_2, f_3 and f_4 be the fault resistances of the four mains; then, by Kirchhoff's law, when the condenser currents in the sheath are negligible, we have

The potentials to earth of the mains.

$$\frac{v_1}{f_1} + \frac{v_2}{f_2} + \frac{v_3}{f_3} + \frac{v_4}{f_4} = 0.$$

Proceeding as above, we see that, if G be the centre of gravity of masses $1/f_1$, $1/f_2$, $1/f_3$ and $1/f_4$ placed at the vertices of the voltage tetrahedron $LMNR$, then GL, GM, GN and GR represent V_1, V_2, V_3 and V_4 respectively in magnitude and phase.

It follows that if the fault resistances of the mains vary, then their potentials adjust themselves so that the power lost in leakage currents is a minimum.

Let a_1, a_2, a_3 and a_4 (Fig. 117) be the currents in the mains, and let i_1, i_2, i_3 and i_4 be the currents in the mesh load.

The currents in a mesh load.

Then
$$a_1 = i_1 - i_4, \qquad a_2 = i_2 - i_1 \atop a_3 = i_3 - i_2, \qquad a_4 = i_4 - i_3 \right\} \quad \dots\dots\dots\dots(8).$$

Therefore $a_1 + a_2 + a_3 + a_4 = 0.$

This also follows at once since there is no accumulation of electric charge in the load.

Hence, like the voltages, A_1, A_2, A_3 and A_4 can be represented by lines drawn from a point in space, and this point is the centre of gravity of four equal particles placed at their extremities. The currents can also be represented by a skew quadrilateral, the angles of which are the supplements of the angles of phase difference between the four currents.

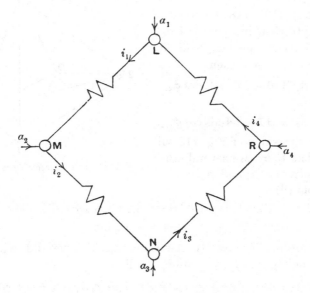

Fig. 117.   The Currents in a Mesh Load.

Again, if r_1, r_2, r_3 and r_4 be the resistances of the arms LM, MN, NR and RL in Fig. 117, then

$$r_1 i_1 = v_1 - v_2, \qquad r_2 i_2 = v_2 - v_3 \Big\}$$
$$r_3 i_3 = v_3 - v_4, \qquad r_4 i_4 = v_4 - v_1 \Big\} \cdot$$

Therefore $\qquad r_1 i_1 + r_2 i_2 + r_3 i_3 + r_4 i_4 = 0 \ \dots\dots\dots\dots\dots (9).$

Hence, if $r_1 I_1$, $r_2 I_2$, $r_3 I_3$ and $r_4 I_4$ be represented by lines drawn from a point O, this point will be the centre of gravity of equal masses placed at the extremities of these lines. The lines joining these extremities give the voltage tetrahedron of the P.D.s between the mains.

Again, if we draw I_1, I_2, I_3 and I_4 (Fig. 118) from the point O, it can easily be proved that O is the centre of gravity of particles, whose masses are proportional to r_1, r_2, r_3 and r_4, placed at the extremities of I_1, I_2, I_3 and I_4. Equations (8) show us that the lines joining the extremities of I_1 and I_4, I_2 and I_1, I_3 and I_2, and I_4 and I_3 represent A_1, A_2, A_3 and A_4.

Let A_5 (Fig. 118) be equal to the length of the line joining 1 to 3 and let A_6 be the length of the line joining 2 to 4, then

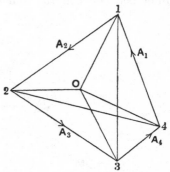

$$A_5^2 = A_2^2 + A_3^2 + 2A_2 A_3 \cos \phi_{2.3},$$

and

$$A_6^2 = A_1^2 + A_2^2 + 2A_1 A_2 \cos \phi_{1.2}.$$

An inspection of Fig. 118 will show that there are several other expressions for A_5 and A_6.

Fig. 118.   The Current Tetrahedron.

From (9)

$$(r_1 + r_2 + r_3 + r_4)\, i_1 = r_2 i_{1.2} + r_3 i_{1.3} + r_4 i_{1.4},$$

where $\qquad i_{1.2} = i_1 - i_2 = -a_2$, etc.

Proceeding in exactly the same way as we did with the corresponding voltage equation, we find

$$I_1^2 (r_1 + r_2 + r_3 + r_4)^2 = (r_2 A_2^2 + r_3 A_5^2 + r_4 A_1^2)(r_2 + r_3 + r_4)$$
$$- r_2 r_3 A_3^2 - r_3 r_4 A_4^2 - r_4 r_2 A_6^2 \ldots (10),$$

and three similar equations.

Hence also

$$(I_1^2 r_1 + I_2^2 r_2 + I_3^2 r_3 + I_4^2 r_4)(r_1 + r_2 + r_3 + r_4)$$
$$= r_1 r_2 A_2^2 + r_1 r_3 A_5^2 + r_1 r_4 A_1^2 + r_2 r_3 A_3^2 + r_3 r_4 A_4^2 + r_4 r_2 A_6^2 \ldots (11).$$

If the resistances are all equal, then

$$4(I_1^2 + I_2^2 + I_3^2 + I_4^2) = A_1^2 + A_2^2 + A_3^2 + A_4^2 + A_5^2 + A_6^2 \ldots (12).$$

If, in addition, the currents in the mains are all equal and the phases are such that A_5 equals A_6, then from (10)

$$I_1 = I_2 = I_3 = I_4.$$

Hence $\qquad 8I_1^2 = 2A_1^2 + A_5^2 \quad \ldots\ldots\ldots\ldots\ldots (13).$

If the currents in the mains are in quadrature, then (Fig. 118)

$$A_5^2 = A_2^2 + A_3^2 = 2A_1^2.$$

Hence $\qquad A_1 = \sqrt{2}\, I_1 \ldots\ldots\ldots\ldots\ldots\ldots\ldots\ldots (14).$

In order to get the factor $\sqrt{2}$, which is used in practical work, we have therefore to make important assumptions. The skew quadrilateral reduces simply to a square in this case (Fig. 119) and if O be the intersection of the diagonals the magnitudes and phases of the currents are as shown in this figure.

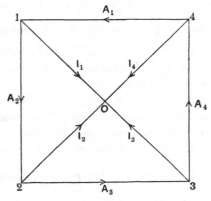

Fig. 119.   The currents in a balanced load.

Let the equation to the wave of P.D. between a main and earth be of the form

The conditions under which it is possible for the P.D. waves between the four mains and earth to be similar curves.

$$e = Vf(t).$$

Let r_1, r_2, r_3 and r_4 be the fault resistances of the mains and let their potentials be

$$V_1 f(t), \quad V_2 f(t + T/4), \quad V_3 f(t + T/2) \quad \text{and} \quad V_4 f(t + 3T/4)$$

respectively, then

$$\frac{V_1}{r_1}f(t) + \frac{V_2}{r_2}f\left(t + \frac{T}{4}\right) + \frac{V_3}{r_3}f\left(t + \frac{T}{2}\right) + \frac{V_4}{r_4}f\left(t + \frac{3T}{4}\right) = 0 \dots(15).$$

Now this is true for all values of t, hence writing $t + T/4$, $t + T/2$, and $t + 3T/4$ for t in (15) successively, we get four equations from which the functions can easily be eliminated by determinants and we get that

$$\left(\frac{V_1}{r_1} + \frac{V_2}{r_2} + \frac{V_3}{r_3} + \frac{V_4}{r_4}\right)\left(\frac{V_1}{r_1} - \frac{V_2}{r_2} + \frac{V_3}{r_3} - \frac{V_4}{r_4}\right)$$
$$\left\{\left(\frac{V_1}{r_1} - \frac{V_3}{r_3}\right)^2 + \left(\frac{V_2}{r_2} - \frac{V_4}{r_4}\right)^2\right\} = 0.$$

Hence either $V_1/r_1 + V_3/r_3 = V_2/r_2 + V_4/r_4$ ....(16),

or $V_1/r_1 = V_3/r_3$ and $V_2/r_2 = V_4/r_4$ ....(17).

In what precedes the only assumption we have made is that $f(t)$ is a periodic function. If we make the additional assumption that it is an alternating function, *i.e.* that

$$f(t) = -f(t + T/2)$$

and $f(t + T/4) = -f(t + 3T/4),$

we get from (15) that

$$\left(\frac{V_1}{r_1} - \frac{V_3}{r_3}\right) f(t) + \left(\frac{V_2}{r_2} - \frac{V_4}{r_4}\right) f\left(t + \frac{T}{4}\right) = 0.$$

As this has to be true for all values of t, we must have

$$V_1/r_1 = V_3/r_3 \quad \text{and} \quad V_2/r_2 = V_4/r_4 \quad \dots\dots\dots(17).$$

Now V_1, V_2, V_3 and V_4 are determined by finding the centre of gravity of masses $1/r_1$, $1/r_2$, $1/r_3$ and $1/r_4$ placed at the angular points of the voltage tetrahedron and joining this point to the angular points. Hence (17) can only be true in very special cases. We are therefore not justified in assuming that the potential waves between the mains and earth are similar curves.

Suppose that the P.D. waves between the mains are

The P.D. waves between the mains are only similar in a special case.

$$V_1 f(t), \quad V_2 f(t + T/4), \quad V_3 f(t + T/2)$$

and

$$V_4 f(t + 3T/4).$$

Then since the sum of them must be zero and

$$f(t) = -f(t + T/2),$$

we get

$$(V_1 - V_3) f(t) + (V_2 - V_4) f(t + T/4) = 0.$$

Hence V_1 must equal V_3 and V_2 must equal V_4.

Let $LMNR$ (Fig. 120) be the voltage tetrahedron. Let A, B, C, D, E and F be the middle points of its edges.

To find the phase differences between the opposite voltages and between the diagonal voltages in a two phase system.

Then from geometry we see that $ABCD$, etc. are parallelograms, and that AC, BD and EF intersect in a point which is the centre of gravity of the tetrahedron. We shall first find ϕ the angle of phase difference between the diagonal voltages LN and MR.

We have

$$2LN \cdot MR \cos\phi = 4 \cdot 2AB \cdot BC \cos ABC$$
$$= 2(AC^2 - BD^2)$$
$$= 2\{AC^2 + EF^2 - (BD^2 + EF^2)\}$$
$$= 4\{AE^2 + EC^2 - (EB^2 + ED^2)\}$$
$$= MN^2 + LR^2 - (LM^2 + NR^2),$$

and therefore $$\cos\phi = \frac{V_{2.3}^2 + V_{4.1}^2 - (V_{1.2}^2 + V_{3.4}^2)}{2V_{1.3}V_{2.4}} \quad \dots\dots(18).$$

Similarly if ϕ' be the angle of phase difference between the opposite voltages $V_{1.2}$ and $V_{3.4}$, then

$$\cos \phi' = \frac{V^2_{1.3} + V^2_{2.4} - (V^2_{2.3} + V^2_{4.1})}{2V_{1.2}V_{3.4}} \qquad \ldots\ldots\ldots(19).$$

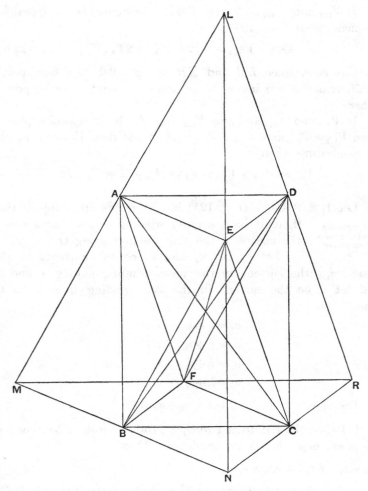

Fig. 120.   Finding the Phase Differences in a Two Phase System.

The phase differences between the other voltages can easily be got from the triangular faces of the voltage tetrahedron (Fig. 120).

If the diagonal voltages be in quadrature,

$$V^2_{2.3} + V^2_{4.1} = V^2_{1.2} + V^2_{3.4} \quad \ldots\ldots\ldots\ldots(20).$$

Conversely, if (20) be true, then from (18), the phase difference of the diagonal voltages is 90 degrees.

If $V_{1.2}$ and $V_{3.4}$ (LM and NR in the figure) be in opposition in phase, then from (19)

$$V^2_{1.3} + V^2_{2.4} = V^2_{2.3} + V^2_{4.1} + 2V_{1.2}V_{3.4} \quad \ldots\ldots\ldots(21).$$

In this case, since LM and NR are parallel, the four points L, M, N and R are in one plane, and (21) could easily be proved otherwise.

If $V_{1.2}$ and $V_{3.4}$ and also $V_{2.3}$ and $V_{4.1}$ be in opposite phases, then $V_{1.2} = V_{3.4}$ and $V_{2.3} = V_{4.1}$. If, in addition, $V_{1.3}$ and $V_{2.4}$ be in quadrature, then

$$V_{1.2} = V_{2.3} = V_{3.4} = V_{4.1} = V_{1.3}/\sqrt{2} = V_{2.4}/\sqrt{2}.$$

Let 1, 2, 3 and 4 (Fig. 121) be the mains and suppose that **The measurement of power in two phase circuits.** there is both a mesh winding and a star winding, the centre of the star winding being O.

Let a_1, a_2, a_3 and a_4 be the currents in the mains, $i_{1.2}$ the current in the mesh winding joining 1 and 2, and let i_1 be the current in the star winding from 1 to O. Then

$$\left.\begin{aligned} a_1 &= i_1 + i_{1.2} - i_{4.1} \\ a_2 &= i_2 + i_{2.3} - i_{1.2} \\ a_3 &= i_3 + i_{3.4} - i_{2.3} \\ a_4 &= i_4 + i_{4.1} - i_{3.4} \end{aligned}\right\}.$$

Therefore $a_1 + a_2 + a_3 + a_4 = 0.$

Let v_1 be the P.D. from 1 to O, $v_{1.2}$ the P.D. from 1 to 2, and w the power expended in the windings.   Then

$$\begin{aligned} w &= v_1 i_1 + v_2 i_2 + v_3 i_3 + v_4 i_4 \\ &\quad + (v_1 - v_2) i_{1.2} + (v_2 - v_3) i_{2.3} + (v_3 - v_4) i_{3.4} + (v_4 - v_1) i_{4.1} \\ &= v_1 a_1 + v_2 a_2 + v_3 a_3 + v_4 a_4 \quad \ldots\ldots\ldots\ldots\ldots\ldots\ldots\ldots\ldots\ldots\ldots(22), \\ &= v_{1.4} a_1 + v_{2.4} a_2 + v_{3.4} a_3 \quad \ldots\ldots\ldots\ldots\ldots\ldots\ldots\ldots\ldots\ldots\ldots(23). \end{aligned}$$

Formulae (22) and (23) give the two methods of measuring power in two phase circuits. For the first method we require four wattmeters, the ampere coils being put in series with the mains, and the volt coils being connected from the mains to the centre of the star system. This method gives the power correctly whether a fifth wire be used or not. For the second method we require

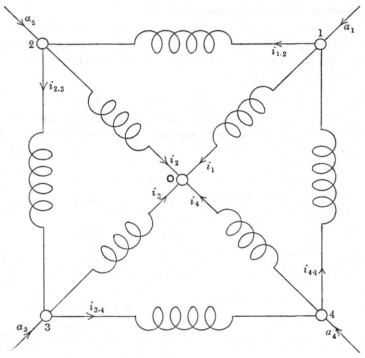

Fig. 121. The Measurement of Power.

three wattmeters. The ampere coils are put in any three of the mains and the volt coils of the three wattmeters are connected from these mains with the other main. Care must be taken to find out whether any of the wattmeters are reading negatively.

If the system is symmetrical, then

$$w = 4v_1 a_1 \quad \dots\dots\dots\dots\dots\dots\dots\dots\dots(24),$$

and

$$w = 2v_{1.4} a_1 + v_{2.4} a_2 \quad \dots\dots\dots\dots\dots\dots(25).$$

Hence one wattmeter is required for the first method, the multiplying factor being 4, and for the second two wattmeters are required. Their ampere coils are put in adjacent mains, No. 1 and No. 2 for example, and their volt coils are connected between No. 1 and No. 4, and No. 2 and No. 4 respectively. Twice the reading of the first wattmeter added to the reading of the second will give us the watts expended in the circuit. The first of these methods is obviously the preferable one.

When part of the load is in series with the mains as in Fig. 107, page 374, the formulae still apply. It follows from first principles that formula (22) is true however complicated may be the connections of the load.

Just as in the case of three phase circuits, if we have an Two phase ordinary watt-hour meter and the arms are equally meters. loaded, we can measure the energy expended by connecting up as in Fig. 122, where S is a star-box connected

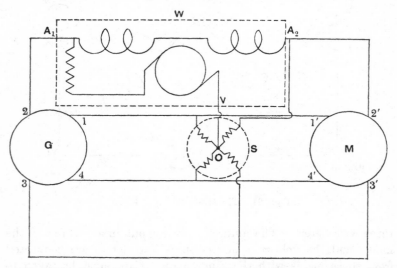

Fig. 122. Connections of watt-hour meter for balanced working. A_1 and A_2 are the ampere terminals of the meter and V is the volt terminal. G is the generator, M the motor and O the centre of the star-box S.

with the four mains. The energy recorded multiplied by four will give the total energy expended in the given time. If the arms are unequally loaded, we should require a watt-hour meter in each arm and the sum of the meter readings would give the required energy.

If we connect up as in Fig. 123, then the sum of the three readings will measure the units expended. Thus we can make a two phase meter with six ampere terminals and one volt terminal.

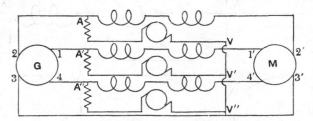

Fig. 123.　Connections of three watt-hour meters for measuring the energy expended in any two phase circuit.

If the arms are equally loaded and the load be non-inductive, then the reading of either of the meters AV and $A''V''$ (Fig. 123) multiplied by four or the reading of the meter $A'V'$ multiplied by two will give the required energy.

Fig. 124 shows the connections for the case of a common return. The sum of the two watt-hour meter readings will obviously give the units expended. The two meters could be combined into a single instrument having four ampere terminals and one volt terminal which has to be connected to the common return. By comparison with Fig. 111, page 379, it will be seen that this form of meter measures the units consumed either on three phase circuits or on two phase circuits with a common return. It would also measure the units consumed in a three wire direct current system.

Let $LMNR$ be the voltage tetrahedron, the magnitude of
which we suppose fixed, and O any point, then
$OL + OM + ON + OR$ is a minimum when any two
of the opposite edges subtend equal angles at O.
In this case the solid angles (see Chapter XII)
formed by any three of the vectors OL, OM, ON and
OR are equal, and hence

When the currents in a star load are equal the sum of the voltages to the centre is a minimum.

$$OL/r_1 = OM/r_2 = ON/r_3 = OR/r_4,$$

i.e. the currents in the four arms of the star load are equal.

It is interesting to notice that the equations for the electrical methods of measuring power in two and three phase circuits, when interpreted geometrically, give rise to numerous theorems connected with the trigonometry of triangles and tetrahedrons.

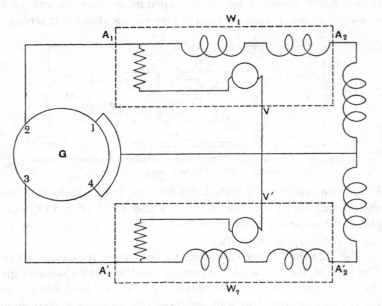

Fig. 124.   Two meters sufficient when there is a common return.

For example, if ABC be the voltage triangle for a three phase system and D, E and F are the middle points of the sides, then if R be the resistance (non-inductive) of each arm of a mesh load, the watts expended in it are $(V^2_{1.2} + V^2_{2.3} + V^2_{3.1})/R$. Hence by the two wattmeter formula

$$V_{1.2}A_1 \cos DAB + V_{3.2}A_3 \cos BCF = (V^2_{1.2} + V^2_{2.3} + V^2_{3.1})/R,$$

or interpreting trigonometrically

$$2(c \cdot AD \cos DAB + a \cdot CF \cos BCF) = a^2 + b^2 + c^2,$$

a known theorem. Similarly, by taking the three wattmeter formula and noting that the star load can be negligible, we get

$$OA \cos(\phi \pm OAG) + OB \cos(\phi \pm OBG)$$
$$+ OC \cos(\phi \pm OCG) = 3 \cdot AG \cos \phi,$$

where ABC is an equilateral triangle, G its centre, O any point within it and ϕ any angle. The proper signs to be prefixed to the angles OAG, etc. depend on the position of O. For example, if O be inside the triangle AGE where E is the middle point of AC, then the positive sign must be taken in the first two terms and the negative in the third.

It will be noticed that in the general 'two phase' system discussed in this chapter we have *four* supply mains, whilst in the three phase system discussed in the preceding chapter we have only three mains. Strictly speaking therefore the 'two phase' system ought to be called a four phase system, and this nomenclature is adopted by engineers when they refer to an n phase system. In such a system the phases of the voltages between the n consecutive mains equal definite fractions of the periodic time. In balanced systems these fractions are all equal.

Polyphase systems.

If $\dot{v}_1, v_2, \ldots v_n$ be the potentials of the n mains in a polyphase system and $a_1, a_2, \ldots a_n$ be the currents in the mains, then the output w is obviously given by

$$w = v_1 a_1 + v_2 a_2 + \ldots + v_n a_n \quad \ldots\ldots\ldots\ldots(26).$$

If the mains be insulated from the earth, we have

$$a_1 + a_2 + \ldots + a_n = 0,$$

and thus, substituting for a_n its value got from this equation in (26), we get

$$w = v_{1.n} a_1 + v_{2.n} a_2 + \ldots + v_{n-1.n} a_{n-1} \quad \ldots\ldots\ldots(27).$$

Equations (26) and (27) prove the n wattmeter and the $(n-1)$ wattmeter methods of measuring power in a polyphase system of n phases.

CHAPTER XVII

Conversion of a two phase system to a three phase system. Conversion of a three phase system to a two phase system. The power factor of a three phase system. Wattmeter method of finding cos ϕ. Phase indicator. Meter for the wattless current. Induction type watt-hour meters. References.

THE following method, due to C. F. Scott, of converting a two phase system of supply into a three phase system is sometimes useful. In the diagram (Fig. 125) 1 and 3, 2 and 4 are the terminals of the primaries of two transformers which are connected to the two phase mains.

Conversion of a two phase system to a three phase system.

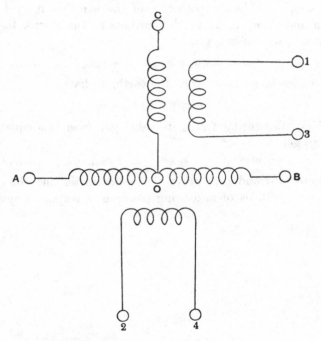

Fig. 125. Two phase to three phase transformer. Scott's method.

AOB and OC are the secondaries of the two transformers, O being the middle point of AB. If the number of windings be so arranged that the P.D. between A and B is to the P.D. between C and O in the ratio of 2 to $\sqrt{3}$, then the P.D.s between A and B, B and C, and C and A will be equal and the phase differences between these P.D.s will also be equal. The following investigation shows how the new P.D. wave shape may differ from the old P.D. wave shape and also gives a rigorous proof of the method.

We will suppose that there is no magnetic leakage and that the resistances of the primary and secondary windings are negligible. In this case the ratio of the potential difference applied to the primary to the potential difference between the secondary terminals will be equal to the ratio of the number of turns in the primary to the number of turns in the secondary. We will also suppose that the P.D. between 1 and 3 is a quarter of a period in advance of the period of the P.D. between 2 and 4. We can represent therefore the P.D. between A and B by $Vf(t)$ and that between O and C by

$$(\sqrt{3}/2)\ Vf(t + T/4).$$

Hence

the P.D. between A and $B = Vf(t)$

$$\left.\begin{array}{l} \text{,,}\qquad\text{,,}\qquad B \text{ and } C = -\dfrac{V}{2}f(t) + \dfrac{\sqrt{3}}{2}Vf\left(t + \dfrac{T}{4}\right) \\[2mm] \text{,,}\qquad\text{,,}\qquad C \text{ and } A = -\dfrac{V}{2}f(t) - \dfrac{\sqrt{3}}{2}Vf\left(t + \dfrac{T}{4}\right) \end{array}\right\}...(1).$$

If v_{AB} represent the P.D. between A and B, then

$$v_{AB} + v_{BC} + v_{CA} = 0,$$

and therefore V_{AB}, V_{BC} and V_{CA} form a triangle.

Also, if $e = f(t)$ be a symmetrical alternating curve (see page 285),

$$\int_0^T f(t)f(t + T/4)\,dt = 0.$$

Hence, from (1), we find for the effective values

$$V_{AB} = V_{BC} = V_{CA} = V \ \dots\dots\dots\dots\dots(2).$$

The voltage triangle is therefore equilateral and the phase differences are 120 degrees.

It is impossible to determine the values of v_A, v_B and v_C from (1), but we can write

$$v_A = \frac{V}{2}f(t) + \frac{V}{2\sqrt{3}}f\left(t + \frac{T}{4}\right) + K$$

$$v_B = -\frac{V}{2}f(t) + \frac{V}{2\sqrt{3}}f\left(t + \frac{T}{4}\right) + K \quad\quad\quad\text{......}(3),$$

$$v_C = -\frac{V}{\sqrt{3}}f\left(t + \frac{T}{4}\right) + K$$

where K is a constant or a function of the time.

Let the fault resistances of the three mains connected with A, B and C be f_1, f_2 and f_3 respectively.

Then, when we can neglect the capacity current in the sheath,

$$\frac{v_A}{f_1} + \frac{v_B}{f_2} + \frac{v_C}{f_3} = 0.$$

Substituting from (3), we find that

$$K\left(\frac{1}{f_1} + \frac{1}{f_2} + \frac{1}{f_3}\right) = \frac{Vf(t)}{2}\left(\frac{1}{f_2} - \frac{1}{f_1}\right) + \frac{V}{2\sqrt{3}}f\left(t + \frac{T}{4}\right)\left\{\frac{2}{f_3} - \frac{1}{f_1} - \frac{1}{f_2}\right\}.$$

This determines K, and substituting in (3) and squaring we find the values of V_A, V_B and V_C.

If $f_1 = f_2 = f_3$, then K is zero and

$$V_A = V_B = V_C = V/\sqrt{3}\quad\quad\quad\text{......................}(4).$$

If the three phase system, obtained by Scott's method from a two phase supply, were exactly the same as the system got directly from a three phase alternator, we ought to be able to get v_{AB} from v_{BC} in (1) by writing $t - T/3$ for t in the latter. Similarly we ought to be able to deduce v_{BC} from v_{CA}.

Thus, we must have

$$f(t) = -\tfrac{1}{2}f\left(t - \frac{T}{3}\right) + \frac{\sqrt{3}}{2}f\left(t - \frac{T}{12}\right),$$

and
$$-\tfrac{1}{2}f(t) + \frac{\sqrt{3}}{2}f\left(t + \frac{T}{4}\right) = -\tfrac{1}{2}f\left(t - \frac{T}{3}\right)\quad\quad\text{........}(5).$$

$$-\frac{\sqrt{3}}{2}f\left(t - \frac{T}{12}\right)$$

These equations are only true for special values of $f(t)$.

We saw, on page 369, that, with a symmetrically made three phase machine, only the $(6n \pm 1)$th harmonics can be present. In an ordinary two phase machine all the odd harmonics may be present, and hence $f(t)$ may contain the third and other harmonics which are not present in the waves from a three phase machine. If, therefore, the Scott system is to yield currents similar to those obtainable from a three phase machine, a first essential condition is that the two phase alternator must be such that the third, ninth, fifteenth,... harmonics are absent from its electromotive force wave. Equations (5) give the other essential condition.

Again, when the system is balanced, K is zero and we ought to be able to get v_A from v_B in (3) by writing $t - T/3$ for t in the latter. Similarly we ought to be able to deduce v_B from v_C.

Hence

$$-\frac{1}{2}f\left(t - \frac{T}{3}\right) + \frac{1}{2\sqrt{3}}f\left(t - \frac{T}{12}\right) = \frac{1}{2}f(t) + \frac{1}{2\sqrt{3}}f\left(t + \frac{T}{4}\right),$$

or
$$\left.\begin{array}{l}\sqrt{3}\left\{f(t) + f\left(t - \frac{T}{3}\right)\right\} = f\left(t - \frac{T}{12}\right) - f\left(t + \frac{T}{4}\right). \\[3mm] \text{Similarly} \qquad 2f\left(t - \frac{T}{12}\right) = \sqrt{3}f(t) - f\left(t - \frac{T}{4}\right).\end{array}\right\} \quad \ldots(6).$$

We can verify that

$$f(t) = X' \sin\left(2\pi t/T + Y'\right)$$

is a solution of the equations (5) and (6), where X' and Y' are constants or functions of t that do not alter when $t + \dfrac{T}{12}$ is substituted for t in them. Comparing this solution with the solution arrived at for a symmetrical three phase system (page 367) we see that the only difference is that it is possible for X and Y to have double the period of X' and Y'.

As a general rule, therefore, the shape of the waves in a three phase system obtained by Scott's method will not be similar to the shape of those obtained from a three phase machine. In practice, however, this is a matter of little importance. The only objection to the method is that there can be a greater number of harmonics in the electromotive force waves, and thus the chance of resonance occurring at particular frequencies may be greater.

The dangers from resonance arise from the high potential differences set up in various parts of the distributing net-work. The insulation of cables, transformers and armature windings, etc. is sometimes broken down from this cause.

If we connect A, B and C (Fig. 125) with the three phase

Conversion of a three phase system to a two phase system.

mains, we can obviously get two phase currents from 1, 3 and 2, 4. There are other ways of doing this which are suggested by elementary geometrical considerations.

Scott's arrangement can also be generalised without difficulty. For example, if the primaries of four transformers be magnetised by the currents in the arms of a two phase mesh load and each transformer have two secondary windings, the middle points of one set being attached to four terminals, then we could either have four sets of three phase terminals, two sets of six phase terminals, or two sets of three phase terminals and one set of two phase terminals.

Let the terminals of the two secondary windings of the transformer, the primary coil of which forms the first arm of the two phase mesh load, be A_1, $A_1{}'$ and B_1, $B_1{}'$ respectively, and let the ratio of the turns of wire in the coil $A_1A_1{}'$ to the turns in the coil $B_1B_1{}'$ be as 2 is to $\sqrt{3}$. Let also a terminal O_1 on this transformer be connected with the middle of the coil $A_1A_1{}'$. If the primary coils of three transformers similar to this form the other arms of the two phase load, then we shall have twenty secondary terminals A_1, A_2, A_3, A_4, B_1,.... If we join O_1 and B_2, then by the theorem given above A_1, $A_1{}'$ and $B_2{}'$ are three phase terminals; and if we join O_3 and B_4, then A_3, $A_3{}'$ and $B_4{}'$ are also three phase terminals. It is easy to see that a six phase star load or a six phase mesh load can be connected across A_1, $B_4{}'$, $A_1{}'$, A_3, $B_2{}'$ and $A_3{}'$. Similarly many other combinations may be obtained from the twenty terminals.

Consider first the case of a mesh connected load. Let W_1, W_2

The power factor of a three phase system.

and W_3 be the power expended in the three arms, and let I_1, I_2 and I_3 be the effective currents in them. Then, if $\cos\phi_1$, $\cos\phi_2$, and $\cos\phi_3$ be the power factors of the three arms, we have

$$\cos \phi_1 = \frac{W_1}{V_{2.3}I_1}, \quad \cos \phi_2 = \frac{W_2}{V_{3.1}I_2}, \quad \cos \phi_3 = \frac{W_3}{V_{1.2}I_3}.$$

The power factor ($\cos \phi$) of the load may be defined as the ratio of the true watts to the sum of the apparent watts, and hence

$$\cos \phi = \frac{W_1 + W_2 + W_3}{V_{2.3}I_1 + V_{3.1}I_2 + V_{1.2}I_3} \dots\dots\dots\dots(7).$$

It therefore lies in value between the greatest and the least values of the power factors of the arms.

When the load is balanced,

$$\cos \phi = \frac{W_1 + W_2 + W_3}{\sqrt{3}\,V_{2.3}A_1} = \frac{W}{\sqrt{3}\,VA} \dots\dots\dots\dots(8),$$

where W represents the power expended in the balanced load, V the mesh voltage and A the current in a main.

In order to find $\cos \phi$ we must therefore measure the power expended in the load by the two wattmeter method; if a star-box is available, one wattmeter will be sufficient. We must also measure the pressure between two mains and the current in either of them.

For a star connected load we find in a similar manner

$$\cos \phi = \frac{W_1 + W_2 + W_3}{E_1A_1 + E_2A_2 + E_3A_3}.$$

When the load is balanced, we have

$$\cos \phi = \frac{W_1 + W_2 + W_3}{\sqrt{3}\,V_{2.3}A_1} = \frac{W}{\sqrt{3}\,VA}$$

When only one wattmeter is available, the following approximate method of finding the power factor is sometimes used.

Wattmeter method of finding cos φ. Let the P.D.s and currents of the balanced load be represented by the lines drawn in Fig. 126. We have made the assumption that the current and P.D. vectors can be represented by lines drawn in one plane. This assumption is true when the currents and potential differences follow the harmonic law.

Let ϕ be the angle (Fig. 126) between V_1 and A_1, and let W be the power expended on the load. Let W_1 be the wattmeter reading when the ampere coil is placed in series with the main 1 and the volt coil is connected between the mains 1 and 2. Let also W_2 be the reading of the wattmeter when we disconnect the

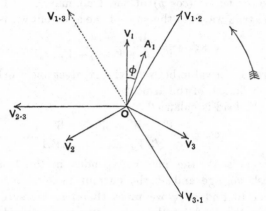

Fig. 126. Voltages and currents in a balanced three phase load. The phase of V_2 is assumed to be in advance of the phase of V_1.

volt coil terminal which is joined with 2 and connect it on to 3, leaving the other connections as before. We see at once from the diagram that

$$W_1 + W_2 = V_{1.2}A_1 \cos(30° - \phi) + V_{1.3}A_1 \cos(30° + \phi)$$
$$= V_{1.2}A_1 \{\cos(30° - \phi) + \cos(30° + \phi)\}$$
$$= \sqrt{3}\, V_{1.2}A_1 \cos \phi = 3\, V_1 A_1 \cos \phi = W.$$

Hence on the given assumptions the sum of the two wattmeter readings will give us the power expended on a balanced load.

This method of measuring the power expended on a balanced load is sometimes used in practice and it will be seen from formula (b) given below that it is applicable when the current and the electromotive force waves do not follow the harmonic law.

Dividing W_1 by W_2, we get

$$\frac{W_1}{W_2} = \frac{\cos(30° - \phi)}{\cos(30° + \phi)}$$
$$= \frac{\sqrt{3} + \tan \phi}{\sqrt{3} - \tan \phi},$$

and therefore $\qquad\qquad \tan\phi = \sqrt{3}\,\dfrac{W_1 - W_2}{W_1 + W_2},$

and $\qquad\qquad\qquad \cos\phi = \dfrac{1 + x}{2\,(1 - x + a^2)^{\frac{1}{2}}},$

where x equals W_2/W_1, the ratio of the wattmeter readings.

This method gives the value of $\cos\phi$, when the waves follow the harmonic law, by means of two wattmeter readings only. In other cases the formula is only approximate.

We shall now consider the theory of the phase indicator, which

Phase indicator.
is an instrument to measure the power factor of a balanced load on a polyphase system. The scale of this instrument may be graduated by making its readings equal the phase differences calculated from the readings of a wattmeter, an ammeter and a voltmeter when they are all connected in the proper manner with a balanced load the power factor of which can be varied; and this is the method adopted in practice. It is instructive, however, to consider the theoretical basis for the design of instruments of this class, as this enables us to find out their limitations and shows how some of their defects may be remedied.

Let us take the case of a three phase instrument. We saw on page 372 that the instantaneous value w of the total power taken by a three phase load is given by

$$w = v_1 a_1 + v_2 a_2 + v_3 a_3.$$

If W be the mean value of w, then when the load is balanced we have, from symmetry,

$$W = 3V_1 A_1 \cos\phi$$
$$= \sqrt{3}\,V_{1.2} A_1 \cos\phi \quad\ldots\ldots\ldots\ldots\ldots(a),$$

where $\cos\phi$ is the power factor of the balanced load and ϕ is the phase difference between v_1 and a_1.

We get in a similar manner, from formula (24), page 372, that

$$W = V_{1.2} A_1 \cos\phi_1 + V_{3.2} A_3 \cos\phi_3',$$

where ϕ_1 is the phase difference between $v_{1.2}$ and a_1 and ϕ_3' is the phase difference between $v_{3.2}$ and a_3. Similarly we have

$$W = V_{2.3} A_2 \cos\phi_2 + V_{1.3} A_1 \cos\phi_1'.$$

If the load be balanced, we have, by symmetry,

$$V_{1.3}A_1 \cos \phi_1' = V_{3.2}A_3 \cos \phi_3',$$

and thus we have

$$W = V_{1.2}A_1 \cos \phi_1 + V_{1.3}A_1 \cos \phi_1' \quad \dots\dots\dots\dots(b).$$

By equating (a) and (b) we get, since $V_{1.2}$ and $V_{1.3}$ are equal on a balanced load,

$$\cos \phi_1 + \cos \phi_1' = \sqrt{3} \cos \phi \quad \dots\dots\dots\dots\dots(c).$$

We shall now make the assumption that $v_{1.2}$, $v_{3.1}$ and a_1 follow the harmonic law so that the three vectors $V_{1.2}$, $V_{3.1}$ and A_1 are in one plane (Fig. 126). We see at once from Fig. 126 that

$$\phi_1 = 30° - \phi, \text{ and } \phi_1' = 30° + \phi \dots\dots\dots\dots(d).$$

Hence if we make the assumption that the currents and the potential differences follow the harmonic law, the angle between $V_{1.2}$ and A_1 is 30° – ϕ. Numerous attempts have been made to construct a phase indicator which will utilise this geometrical property of the vectors.

In the phase indicator described below we have two fixed coils and one movable coil. The two fixed coils have the same number of turns, and each is put in series with a large non-inductive resistance R. In Fig. 127 let OA and OB be the axes of fixed coils which are connected between the mains 1 and 2 and between the mains 1 and 3 respectively. In practice, it is customary to arrange the coils so that if a current flowing from 1 to 2 produces a magnetic field in the direction

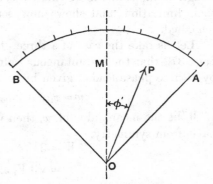

Fig. 127.   Phase Indicator.

OA, a current flowing from 3 to 1 will produce a magnetic field in the direction BO. Since R is large, the current in the fixed coil which has OA for its axis is in phase with $v_{1.2}$ and the current in the other fixed coil is in phase with $v_{3.1}$. Hence, if the time be reckoned from the instant when the potential difference between the mains 1 and 2 is a maximum, the current in the fixed coil between 1 and 2, and therefore the strength of the magnetic field in the direction OA due to this current, is proportional to $\cos \omega t$ and we shall denote it by $h_1 \cos \omega t$. Similarly the strength of the field in the direction OB due to the current in the other fixed coil may be denoted by $- h_1 \cos (\omega t - \psi)$, where ψ is the phase difference between $v_{1.2}$ and $v_{3.1}$.

Let the angle AOB in Fig. 127 be denoted by α, and let OM bisect the angle AOB. Then, if h_x and h_y be the components of the resultant magnetic field due to the currents in the two fixed coils perpendicular to and along OM respectively, we have

$$h_x = h_1 \cos \omega t \sin (\alpha/2) + h_1 \cos (\omega t - \psi) \sin (\alpha/2)$$
$$= 2h_1 \sin (\alpha/2) \cos (\psi/2) \cos (\omega t - \psi/2),$$

and
$$h_y = h_1 \cos \omega t \cos (\alpha/2) - h_1 \cos (\omega t - \psi) \cos (\alpha/2)$$
$$= - 2h_1 \cos (\alpha/2) \sin (\psi/2) \sin (\omega t - \psi/2).$$

We see that when α equals ψ, h_x and h_y are the projections of a line of length $2h_1 \sin (\alpha/2) \cos (\alpha/2)$ or $h_1 \sin \alpha$, which revolves with a constant angular velocity ω. If, therefore, the angle AOB is made equal to ψ, the resultant magnetic field due to the currents in the fixed coils is of constant strength and rotates with uniform angular velocity. Since the indicator is only used to find the power factor on balanced loads, ψ is constant and equals 120 degrees. Let us suppose that the angle AOB is made equal to this, so that we always have a pure rotating field.

The strength h of the magnetic field resolved along OP is given by

$$h = h_x \sin \phi' + h_y \cos \phi'$$
$$= h_1 \sin \alpha \sin \{\phi' + (\alpha/2) - \omega t\},$$

and since α equals $120°$ we get

$$h = (\sqrt{3}/2) h_1 \cos (\omega t + 30° - \phi') \dots\dots\dots\dots(e).$$

Now the movable coil carries an alternating current which is in phase with a_1. We can therefore suppose that it is replaced by a magnet the polarity of which alternates with the current. If the coil be small, the action of this magnet may be imitated exactly by two small permanent magnets which rotate in opposite directions with equal angular velocities ω and the moments of which are half the maximum moment of the alternating current magnet.

To prove this, consider two small magnets each of moment $H/2$ rotating side by side in opposite directions with equal angular velocities ω, and let the time be reckoned from the instant when the magnetic fields due to each are pointing in the same direction. Then at the time t, the component field in this direction at unit distance is $H\cos\omega t + H\cos(-\omega t)$, that is, $2H\cos\omega t$, and the component field perpendicular to this direction is

$$\tfrac{1}{2}H\sin\omega t + \tfrac{1}{2}H\sin(-\omega t)$$

or zero (see page 19). We see therefore that they are equivalent to the single alternating current magnet which, at unit distance, produces the field $2H\cos\omega t$ in the direction of its axis.

Now the action of the rotating field on the equivalent permanent magnet which is rotating in the same direction as itself is to produce a couple which acts so as to diminish the angle between this magnet and the direction of the rotating field. The mean couple produced by the rotating field on the equivalent magnet which rotates in the other direction is zero. Thus the movable coil is in a position of equilibrium when the direction of the rotating field due to the fixed coils coincides with the direction of the equivalent magnet which rotates in the same direction.

Let OP (Fig. 127) be the position of the axis of the movable coil when in equilibrium. At the instant when OP is in the direction of the rotating field, h, which acts along this direction, has its maximum value and both the equivalent permanent magnets are also pointing in this direction. Thus h is in phase with a_1, and therefore from (e) the phase difference between a_1 and $v_{1.2}$ is $30° - \phi'$. But, by the equation (d), this angle is equal to $30° - \phi$ when the load is balanced, and thus ϕ' must be equal to ϕ. If therefore the scale of the instrument be divided into degrees, the cosine of the reading will give the power factor, when the potential differences and the currents follow the harmonic law.

If the axes of the fixed coils be not placed so as to include an angle of 120 degrees, then, in general, both the angular velocity and the magnitude of the rotating field due to the fixed coils vary at different instants even when the applied waves are sine-shaped. It is found however that the pointer gives a definite reading for a load of given power factor, and so, with the aid of a wattmeter, an ammeter and a voltmeter, the scale can be marked.

In Fig. 128 are shown the connections of the Heap phase indicator which is constructed on the above principle. R and R are resistances in series with the fixed coils, and the movable coil carries the whole current. In the Heap phase indicators for high pressure working the fixed coils are the secondaries of transformers

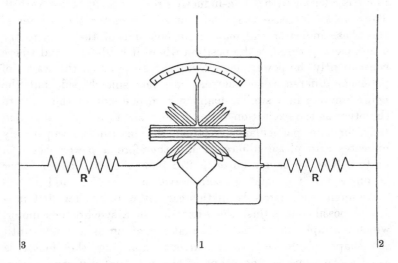

Fig. 128. Connections of Heap Phase Indicator for three phase working.

the primaries of which are connected between 2 and 1 and between 1 and 3. The movable coil sometimes carries only a fraction of the current in the main 1, and sometimes it is the secondary of a transformer whose primary consists of one or two turns placed in series with the main 1. It is thus possible to arrange the indicator in the circuit so that only low pressure wires come to the instrument.

Let us suppose that the phase indicator is connected in the usual manner with the three mains supplying power to the windings of the armature of a three phase synchronous motor (see Vol. II, Chap. v). Then, it will be proved in Vol. II that, when the potential differences and the currents follow the harmonic law, we can alter the power factor by varying the direct current excitation required by the machine, although the mechanical power given out by the motor is maintained at a constant value. For a particular excitation we can show that the power factor is unity. For excitations less than this the current in a winding lags behind the potential difference applied at the terminals of the winding, and for excitations greater than this it leads the potential difference, the winding of one phase acting like a condenser with a non-inductive resistance in series with it. Thus, as we increase the excitation of the motor, the pointer of the phase indicator will move from one side of the scale to the other, passing through the position where it reads zero, and where consequently the power factor is unity. In practice the waves of potential difference and current are not sine-shaped, but the pointer moves in a similar manner from one side of the scale to the other as the excitation varies from a low to a high value, and thus for one particular excitation the instrument apparently indicates zero phase difference and therefore a power factor of unity. Now we have seen on p. 275 that when the power factor is unity the volt and ampere waves are similar, and if the instrument reads correctly in this case, it would follow that it is always possible to adjust the excitation of a synchronous motor, which is simply an ordinary alternator running as a motor, until the shape of the waves of current supplying the motor is exactly the same as the shape of the potential difference waves between the mains. As a matter of fact, however, a zero reading of a phase indicator only corresponds to a power factor of unity when the waves are sine-shaped. In other cases there is no simple relation between the reading of the instrument and the power factor, although when the instrument is always used in the same supply circuit its readings are of value to the engineer.

If we make the shunt circuit of an electromagnetic watt-meter, which is constructed on the dynamometer principle and has no mutual inductance between its coils in the zero position, very inductive, then the current A_1 in it will lag in phase by nearly ninety degrees behind the applied potential difference V. Such an instrument may be used to measure the wattless current (see page 287). The reading

Meter for the
wattless
current.

of the instrument will be proportional to $AA_1 \cos \alpha$, where A is the main current and α is the phase difference between A and A_1. If the power factor of the load be $\cos \phi$, A will lag behind V by ϕ degrees and A_1 will lag behind V by ninety degrees. When the applied potential difference and the currents follow the harmonic law, the three vectors representing V, A and A_1 lie in one plane (see Chapter XII), and therefore α is $90° - \phi$. Since A_1 is proportional to V, the reading of the instrument is proportional to $VA \cos(90° - \phi)$, that is, to $VA \sin \phi$. Thus when V is known we can find the wattless current $A \sin \phi$.

If one or more of the waves do not follow the harmonic law, the three vectors form a solid angle, and $\phi + \alpha$ is therefore greater than ninety degrees. Let us suppose that

$$\phi + \alpha = 90° + x$$

and that x is small. Then the reading of the instrument divided by V will be proportional to

$$A \cos \{90° - (\phi - x)\},$$

that is, to

$$A \sin \phi - x \,.\, A \cos \phi \text{ approximately.}$$

Thus, in practice, the reading of the meter will always be low, and so the power calculated from the reading and a knowledge of the value of A will be greater than its true value.

A watt-hour meter in which the action depends on the induction of eddy currents in a metal disc or drum which is free to rotate, and on the forces produced by moving magnetic fields on the disc carrying these currents, is generally called an induction watt-hour meter. The principle of action of this meter will best be understood by considering the simplest form. We shall suppose that there is an aluminium disc fixed on the spindle and that this disc is free to rotate in the horizontal air-gaps of two **C**-shaped electromagnets placed side by side at its circumference. The winding of one of these magnets is in series with a choking coil and the combination is connected between the mains. The current in the winding of this magnet, which we shall call the shunt magnet, lags by nearly ninety degrees behind the potential difference applied to the load.

Induction type watt-hour meters.

The other magnet—the series magnet—is excited by the main current itself or by a current in phase with it.

Let us suppose that the disc is fixed. When currents flow in the windings of both magnets, the alternating magnetic fluxes in the air-gaps will induce eddy currents in the disc. The phases of the tubes or stream lines of current generated in the disc will depend on the resistance and inductance of their paths. The eddy currents produced by the field due to the shunt magnet will be acted on by the magnetic field due to the series magnet and *vice versâ*. The mean values of the electromagnetic attractions and repulsions do not in general balance one another and thus a torque is produced. If the disc rotate, the shape and position of the stream lines relatively to the poles will be altered, and electromotive forces will be set up in the metal of the disc by its motion. As a rule, however, we may neglect the electromotive forces due to the motion of the metal in comparison with the electromotive forces set up by the alternating flux. In practice, even at full load, the disc makes less than one revolution per second. The alternating flux, on the other hand, will, if the frequency of the supply current is fifty, alter its direction one hundred times per second. We may therefore in finding an approximate formula neglect the effects of rotation.

Let i_1 be the strength at a particular instant of one of the tubes of current induced in the disc by the field of the shunt magnet. The torque produced by the action of this tube of current on the series magnet, if we assume that the permeability of the iron is constant, must be of the form $k_1 n_2 I_2' i_1$, where n_2 is the number of turns in the series winding, k_1 is a constant that depends on the relative positions of the tube of eddy current and the series magnet, and I_2' is the instantaneous value of the current exciting the series magnet. Thus, taking the sum of the torques produced by the series magnet on all the tubes of current generated in the disc by the shunt magnet, we get $n_2 I_2' \Sigma k_1 i_1$ for the total torque. Similarly $- n_1 I_1' \Sigma k_1' i_2$ will represent the total torque produced by the shunt magnet on the eddy currents in the disc generated by the series magnet. Thus if g represent the instantaneous value of the resultant torque on the disc, we have

$$g = n_2 I_2' \Sigma k_1 i_1 - n_1 I_1' \Sigma k_1' i_2 \quad \dots\dots\dots\dots(9).$$

Let us now suppose that the polar faces of the magnets are equal, and that they are similarly situated with respect to the disc so that for every tube of current i_1 generated by one of them there is a corresponding tube of current i_2 generated by the other, and the constant k_1 for a circuit of one set of stream lines equals the constant k_1' for a circuit of the other set. We shall also suppose that the currents follow the harmonic law. We may write, therefore,

$$I_1' = I_1 \cos(\omega t - \alpha), \qquad I_2' = I_2 \cos(\omega t - \beta),$$
$$i_1 = n_1 l_1 I_1 \cos(\omega t - \alpha_1), \quad i_2 = n_2 l_1 I_2 \cos(\omega t - \beta_1),$$
$$i_1' = n_1 l_2 I_1 \cos(\omega t - \alpha_2), \quad i_2' = n_2 l_2 I_2 \cos(\omega t - \beta_2),$$
$$\dots\dots\dots\dots\dots\dots\dots\dots\dots\dots\dots\dots .$$

The values of l_1, l_2, ..., α_1, β_1, ... depend on the values of the inductance and resistance of the paths of the eddy currents. Their values also depend on the frequency.

Let the phase difference between the currents in the magnet windings be γ, then we shall have

$$\gamma = \beta - \alpha = \beta_1 - \alpha_1 = \dots .$$

Let also δ_1 be the phase difference between i_1 and I_1'; then δ_1 will also be equal to the phase difference between i_2 and I_2'. Thus we have

$$\delta_1 = \alpha_1 - \alpha = \beta_1 - \beta.$$

Let also
$$\delta_2 = \alpha_2 - \alpha = \beta_2 - \beta,$$
$$\dots\dots\dots\dots\dots\dots$$

Substituting in (9), we get

$$g = n_2 n_1 I_2 I_1 \cos(\omega t - \beta)\{k_1 l_1 \cos(\omega t - \alpha_1) + k_2 l_2 \cos(\omega t - \alpha_2) + \dots\}$$
$$- n_1 n_2 I_1 I_2 \cos(\omega t - \alpha)\{k_1 l_1 \cos(\omega t - \beta_1) + k_2 l_2 \cos(\omega t - \beta_2) + \dots\}.$$

If G denote the mean torque, we have

$$G = \frac{n_2 n_1 I_2 I_1}{2}\{k_1 l_1 \cos(\alpha_1 - \beta) + k_2 l_2 \cos(\alpha_2 - \beta) + \dots$$
$$- k_1 l_1 \cos(\beta_1 - \alpha) - k_2 l_2 \cos(\beta_2 - \alpha) + \dots\}$$
$$= n_1 n_2 I_1 I_2 \left\{k_1 l_1 \sin\frac{\alpha_1 - \alpha + \beta_1 - \beta}{2}\sin\frac{\beta_1 - \alpha_1 + \beta - \alpha}{2} + \dots\right\}$$
$$= n_1 n_2 I_1 I_2 \sin\gamma\{k_1 l_1 \sin\delta_1 + k_2 l_2 \sin\delta_2 + \dots\}.$$

The values of δ_1, δ_2, ... depend on the conductivity of the metal disc. If the temperature and the frequency remain constant, the expression $k_1 l_1 \sin \delta_1 + k_2 l_2 \sin \delta_2 + ...$ will be a constant. We thus see that, when we make the sine curve assumption, G will be proportional to $I_1 I_2 \sin \gamma$.

Now, since we can neglect the actions of the small eddy currents, I_1 is proportional to the voltage V applied to the load and I_2 is proportional to the effective value A of the load current. The accelerating torque G is therefore proportional to $VA \sin \gamma$. The angle γ is the phase difference between the currents in the windings of the shunt and series magnets. On a non-inductive load it is nearly ninety degrees; and since we are assuming that all the currents follow the harmonic law, γ will equal $90° - \phi$ when $\cos \phi$ is the power factor of the load. It therefore follows that the resultant accelerating torque is proportional to $VA \cos \phi$, that is, to the power being expended on the load.

The retarding torque is produced by the eddy currents generated in the rotating disc by the field due to permanent magnets. These magnets are generally C-shaped, and the circumference of the disc rotates in the air-gap. The retarding torque will therefore be proportional to the angular velocity of the spindle. It is to be noted that, since the eddy currents due to the series and shunt magnets are alternating, the mean effect of the permanent magnets on them is zero. The shunt and series magnets also have on the average no effect on the eddy currents due to the permanent magnets. When the motion is steady, the accelerating torque is equal to the retarding torque, and thus the power being expended on the load is proportional to the angular velocity of the spindle. The spindle is connected with the counting mechanism by worm gearing and so a record is made of the energy expended.

Instead of a disc, a light hollow cylinder of aluminium is sometimes used, the shunt and series magnets being placed near to one another and facing the circumference of the cylinder. As before the retarding torque is produced by the eddy currents due to permanent magnets. The theory of the action of this instrument is practically identical with that of the disc meter.

In meters which have an aluminium disc, a compound magnet made up of ⋒-shaped iron stampings is often used instead of two

separate magnets for the shunt and series windings.   The shunt coil
is wound on the middle limb and the series coils on one or both
of the outer limbs.   The compound magnet is placed near the
circumference of the aluminium disc, and strips of iron are fixed
underneath the disc so as to reduce the reluctance of the magnetic
circuits of the compound magnet.   When there is a current in the
series coils, the mean values of the magnetic fluxes in the two
outer limbs are unequal, as the magnetising force of the shunt
coil increases the resultant magnetising force in one limb and
diminishes it in the other.   This effect alters the ratio of the
angular velocity of the disc to the effective current in the series
coils, and tends to make the meter read inaccurately.   It can
be neutralised by placing a few turns of series winding on the
middle limb.

This type of meter can easily be adapted to register the energy
expended in a two or three phase load.   All that we need do is to
apply the two wattmeter method (see Chapters XV and XVI) and
combine the two watt-hour meters into one instrument.   Two ⋒-
shaped magnets are arranged in this case to act on the same
aluminium disc.   They are placed facing the top surface on opposite
sides of the centre of the disc.   If they are suitably wound and
connected with the mains as in the two wattmeter method, the sum
of the accelerating torques, that is, the resultant torque, will be
proportional to the total power being expended on the load.

At first sight it would appear that the temperature error in
this type of meter ought to be large owing to the large temperature
coefficient of the conductivity of the metal disc in which the eddy
currents are developed.   As the mean accelerating and retarding
torques are both affected by the change of the conductivity of the
metal, it might happen that they are both affected in the same ratio,
in which case the rate would not be altered.   It has to be remembered,
however, that the time lags of the eddy currents are altered and
so the average torques may not be altered in the same ratio.

The meters are adjusted for alternating current of a given
frequency.   Any alteration of the frequency, and hence also any
alteration of the wave shape of the alternating current, will affect
their rate.

In other meters of the induction type, we have a metal disc

placed in a rotating magnetic field. The principle of these meters will easily be understood from the theory of the induction motor developed in Volume II.

REFERENCES

C. F. Scott, 'Polyphase Transmission.' *The Electrician*, Vol. 32, p. 640, 1894.

W. H. Browne, 'Power-Factor Indicators.' *Trans. of the Amer. Inst. El. Eng.*, Vol. 18, p. 475, 1901.

'A Direct-reading Power-factor Indicator.' *The Electrician*, Vol. 51, p. 168, 1903.

K. Edgcumbe, *Industrial Electrical Measuring Instruments.*

H. G. Solomon, *Electricity Meters.*

CHAPTER XVIII

Rotating magnetic fields. Resultant field due to n equal and symmetrically placed poles produced by n phase currents. Equal poles unevenly spaced ; phase differences of magnetic vectors equal to their angular distances apart. General case. Properties of rotating and alternating magnetic fields. Magnetic field in the air-gap of polyphase machines. Rotating field in the air-gap of a polyphase induction motor. Gliding magnetic fields. Rotating magnetic fields when the currents are not sine-shaped. Extension to three phase theory. Rotating magnetic field producing a constant effective E.M.F. in a search coil placed with its plane perpendicular to the field. Arnò's phase indicator. References.

MAGNETIC forces are compounded by the parallelogram law, and hence we may apply statical constructions in order to find their resultant. For example, suppose that the magnetic forces at the point O (Fig. 129) are represented in magnitude and direction by the lines OA_1, OA_2, OA_3, and OA_4. Then, if G be the centre

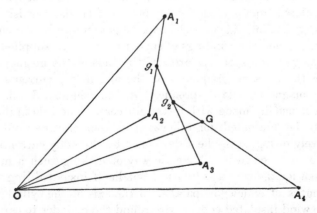

Fig. 129. Resultant of OA_1, OA_2, OA_3 and $OA_4 = 4 . OG$.

of gravity of equal masses placed at A_1, A_2, A_3 and A_4, the resultant of the forces will be represented in magnitude by $4 \cdot OG$, and in direction by OG. If there had been n forces, then the resultant would have been equal to $n \cdot OG$.

To find G, we bisect $A_1 A_2$ in g_1, and then make

$$g_1 g_2 = \tfrac{1}{3} g_1 A_3, \quad g_2 g_3 = \tfrac{1}{4} g_2 A_4, \text{ etc.}$$

This construction is very simple in practice. It is easy to see that it is true whether the lines are in one plane or not. The necessary and sufficient condition that the forces are in equilibrium is that the centre of gravity of equal masses placed at the extremities of the lines representing the forces coincides with O. We have seen that this theorem also holds for alternating current vectors, and we have already made use of it in the chapters on two and three phase theory.

In a rotating magnetic field the direction of the magnetic force is continually revolving. If we move a strong permanent magnet round a small compass needle, keeping the same end of the magnet always pointing to the needle, we produce a rotating magnetic field at the centre of the small compass. The needle at any instant points out the direction of the magnetic field, and its angular velocity measures the velocity of rotation of the field.

Rotating magnetic fields.

If the core of an electromagnet be made up of thin strips of soft iron insulated from one another by means of shellac varnish, or by paper pasted on one side of each strip, we get an alternating current magnet. When the windings of such a magnet are supplied with alternating currents, the polarity of the ends of the magnet alternates with the same frequency as the alternating currents and a varying magnetic field is produced in the neighbourhood. The magnet is usually made with a straight core. The iron in the core needs to be laminated, for otherwise its temperature would rise excessively owing to the heat developed by the eddy currents that would be induced in it. A simple way of making such a magnet is to take for the core a cylindrical bundle of insulated iron wires, the lengths of which are parallel to the axis of the cylinder, and then to wind insulated copper wire round the cylinder to carry the alternating current required to magnetise the core.

Now place two equal cylindrical coils carrying alternating currents with their axes perpendicular and pointing in the directions AOA' and BOB' respectively, and let their centres be at equal distances from O (Fig. 130). If the magnetic force at O due to the first be $H\cos\omega t$ and that due to the second be

$$H\cos(\omega t - \pi/2),$$

and if (Fig. 130)

$$OQ = H\cos\omega t,$$

and

$$OP = H\cos(\omega t - \pi/2)$$
$$= H\sin\omega t,$$

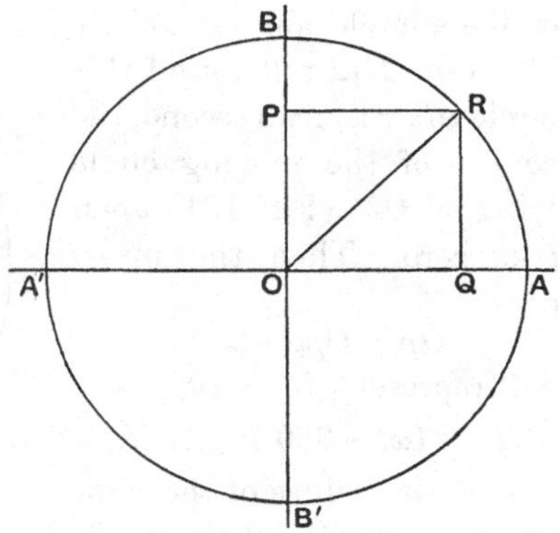

Fig. 130. Pure Rotating Field produced by two alternating fields in quadrature.

then, if OR be the resultant of OP and OQ, we have

$$OR^2 = OP^2 + OQ^2 = H^2;$$

and therefore $\qquad OR = H = \text{a constant},$

and the angle which OR makes with OA will be ωt, so that OR will rotate with constant angular velocity ω. In this case, then, we obtain a pure rotating magnetic field, that is, one which is not only constant in magnitude but rotates with constant angular velocity. If the cylindrical coils were supplied with sine-shaped current waves from a two phase machine, we could produce this field. The rotating fields produced in practice, however, although often constant in magnitude, rarely rotate with constant angular velocity. For the present we will consider the case of pure rotating fields, and consequently we make the supposition that the alternating currents inducing the fields are sine-shaped, and that the permeability of the cores of the electromagnets is constant.

Resultant field due to n equal and symmetrically placed poles produced by n phase currents.

Consider the case of the field produced by n equal poles arranged evenly round a circle, the currents producing the poles differing in phase by $360°/n$ degrees.

Let the strength of the field at O, due to the pole P_1, be $H\cos\omega t$ in the direction Op_1. Similarly let the field at O, due to P_2, be $H\cos(\omega t - 360°/n)$ in the direction Op_2, and so on for the remaining poles.   With centre

O and radius equal to H describe a circle, and let a circle of radius $H/2$ roll round this circle $\omega/2\pi$ times a second, the centre of the moving circle lying on OP_1 (Fig. 131) when t is zero. Then the intercepts

$$Op_1, \quad Op_2, \quad \ldots\ldots$$

will represent $H\cos\omega t$,

$$H\cos(\omega t - 360^\circ/n), \quad \ldots\ldots,$$

that is the values of the component strengths of the fields. It is easily shown that $p_1, p_2, \ldots$ are fixed points on the moving

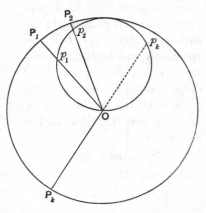

Fig. 131.   Resultant Field of n phase currents equals $(n/2)\,H$.

circle.   Since the angles p_1Op_2, $p_2Op_3, \ldots$ are all equal, $p_1p_2\ldots p_n$ is a regular polygon.   If C be the centre of the rolling circle, C is always the centre of gravity of equal masses placed at $p_1, p_2, \ldots$. Therefore the resultant magnetic force is represented in magnitude and direction by $n.OC$. Hence the resulting field is a pure rotating one of magnitude $(n/2)\,H$.

As in the last case, suppose

Equal poles unevenly spaced ; phase differences of magnetic vectors equal to their angular distances apart. that a circle rolls inside another of double its diameter.   Then

$$p_1p_2, \quad p_2p_3, \quad \ldots$$

(Fig. 132) sub-

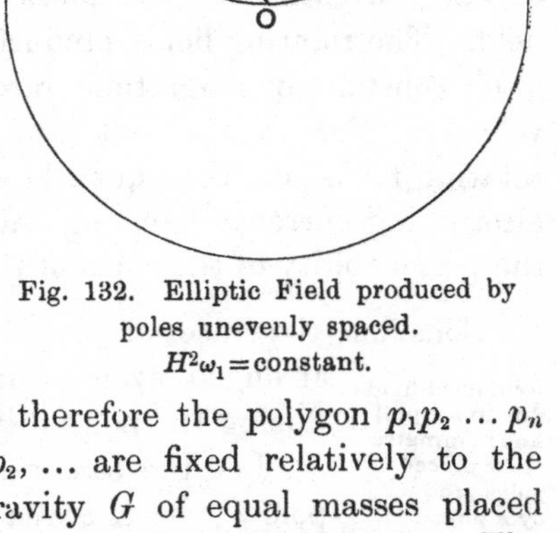

Fig. 132.   Elliptic Field produced by poles unevenly spaced. $H^2\omega_1 = \text{constant}$.

tend constant angles at the circumference of the circle, and therefore the polygon $p_1p_2\ldots p_n$ is of constant size.   Since $p_1, p_2, \ldots$ are fixed relatively to the moving circle, the centre of gravity G of equal masses placed at $p_1, p_2, \ldots$ is fixed with respect to the rolling circle.   The angular velocity of G about C is double the angular velocity of C about O and is in the reverse direction. Let CG equal c, and let x and y be the coordinates of G with respect to two axes through O at right angles to one another. Then, the axis of x passing through G when G is at its greatest distance from O, it is easy to see that

$$x = (H/2 + c)\cos \omega t, \quad \text{and} \quad y = (H/2 - c)\sin \omega t.$$

Thus G lies on an ellipse whose centre is O and axes $H + 2c$ and $H - 2c$ respectively. Since the field is represented in direction and magnitude by $n.OG$, we see that we do not get a pure rotating field in this case. If ω_1 be the angular velocity of OG about O, then

$$\omega_1.OG = \frac{\partial y}{\partial t}\cdot\frac{x}{OG} - \frac{\partial x}{\partial t}\cdot\frac{y}{OG},$$

and thus $\qquad\qquad OG^2.\omega_1 = (H^2/4 - c^2)\,\omega.$

It is proved in Thomson and Tait's *Natural Philosophy*, Part I, § 66, that a point, whose motion is the **General case.** resultant of any number of simple harmonic motions of the same period in any directions and with any phases, moves on the circumference of an ellipse, and that its radius vector sweeps out equal areas in equal times. Hence, if H be the strength of the field at a point due to any number of magnetic poles of any strengths with any directions, and with any phase differences between them, provided they follow the sine law, then

$$H^2\omega_1 = \text{constant,}$$

where ω_1 is the angular velocity with which the resultant field revolves.

The important case in practice is when the vectors are all in one plane. Let us suppose that the direction of the alternating field $H_1 \cos(\omega t - \alpha_1)$ makes an angle ψ_1 with the axis of x. Then, if x and y be the coordinates of the extremity of the vector representing the resultant field, we have

$$x = \Sigma\, H \cos(\omega t - \alpha)\cos\psi$$
$$= a \cos \omega t + b \sin \omega t,$$

and $\qquad\qquad y = \Sigma\, H \cos(\omega t - \alpha)\sin\psi$
$$= c \cos \omega t + d \sin \omega t,$$

where a, b, c and d are quantities which do not vary with the time. Solving these equations for cos ωt and sin ωt we get

$$\cos \omega t = \frac{dx - by}{da - bc},$$

and

$$\sin \omega t = \frac{cx - ay}{cb - ad}.$$

Hence, we get

$$(dx - by)^2 + (cx - ay)^2 = (ad - bc)^2,$$

which shows that the locus of the extremity of the vector representing the resultant field is an ellipse.

In the special case when

$$a = b \quad \text{and} \quad c = d$$

the ellipse becomes a straight line, and thus the resultant field is purely alternating. Again, when

$$a = \pm d \quad \text{and} \quad b = \mp c$$

the ellipse becomes a circle and the strength of the resultant field is constant at every instant. Also, since

$$r^2 \frac{\partial \theta}{\partial t} = x \frac{\partial y}{\partial t} - y \frac{\partial x}{\partial t},$$

we have

$$r^2 . \omega_1 = \omega (ad - bc),$$

where ω_1 is the angular velocity of the vector of the resultant field; and thus, if r is constant, ω_1 is also constant.

If the rotating field be produced by currents of different frequencies, we get all manner of varying fields. The curves that would be described by the extremity of the vector representing the instantaneous value of the resultant magnetic force in some of these cases are given in treatises on the theory of sound under the head of Lissajous's figures, and various mechanical devices are described for drawing them.

We shall now consider a few of the properties of rotating and alternating magnetic fields. By a pure rotating field we mean one whose strength and angular velocity are both constant, and by an alternating field we mean one whose direction is constant but whose strength is a periodic function of the time obeying the

Properties of rotating and alternating magnetic fields.

sine law. When we consider more than one field, all the fields
will be supposed to be parallel to one plane and the frequency
both of the rotations of the rotating fields and of the alternations
of the alternating fields will be supposed to have the same value.
We will suppose the fields represented by rotating and alternating
vectors, which may be drawn in one plane. The following theorems
will be found useful in practice.

(1) *Two vectors rotating in the same direction in one plane
are equivalent to a single rotating vector.*

Let Op and Oq be the two vectors. Since they rotate with the
same angular velocity, the angle pOq remains constant. Construct
a parallelogram on Op and Oq as adjacent sides, and let OR
be the diagonal. This represents the resultant field, which will
obviously rotate with the same angular velocity as Op and Oq.
Hence two vectors rotating in the same direction compound into
a single rotating vector.

(2) *Two equal vectors rotating in opposite directions in one
plane are equivalent to a single alternating vector, and conversely
a single alternating vector is equivalent to two equal vectors
rotating in opposite directions.*

Let Op and Oq be the two equal vectors. At any instant their
resultant is $2 \cdot Or$, where r is the middle point of pq. Also since
Or is perpendicular to pq, and Op and Oq are rotating with equal
angular velocities in opposite directions, Or is fixed in direction.
Thus two equal vectors rotating in opposite directions compound
into a pure alternating vector the amplitude of which equals
$2 \cdot Op$. Similarly an alternating vector whose amplitude is Or
may be replaced by two equal rotating vectors whose magnitudes
are each equal to $\frac{1}{2} \cdot Or$.

(3) *Two unequal vectors rotating in opposite directions are
equivalent to an alternating vector and a rotating vector, and, conversely, an alternating vector and a rotating vector are equivalent
to two unequal vectors rotating in opposite directions.*

Let Op and Oq be the magnitudes of the two vectors. Make
Oq' equal to Op. Then, since Oq is equivalent to the sum of two
rotating fields, in phase with one another, whose magnitudes are
Oq' and $q'q$ respectively, and since by (2) Op and Oq' compound

into a vector alternating along the line Oa (Fig. 133), which bisects the angle pOq, and having the amplitude $2 . Op$, we see that the given vectors compound into an alternating vector of amplitude $2 . Op$ and a rotating vector whose magnitude is $Oq - Op$.

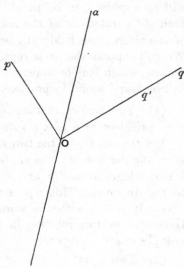

(4) *Two unequal alternating vectors can, in general, be replaced by a simple alternating vector and a rotating vector.*

By (2) we can replace the alternating vectors whose directions are Or and Or' (Fig. 134) by four rotating vectors Op, Oq and Op', Oq' of which Op and Op' rotate in one direction, and Oq and Oq' in the other, and where Op is half the maximum magnitude of Or and Op' is half the maximum magnitude of Or'.

Fig. 133. Two unequal vectors rotating in opposite directions are equivalent to an alternating and a rotating vector.

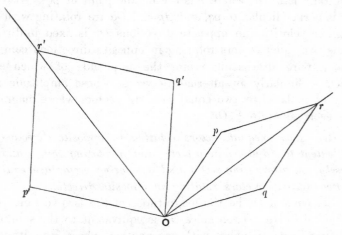

Fig. 134. Two alternating vectors can be replaced by an alternating and a rotating vector.

Hence, by (1), we can replace Op and Op' by OP, and Oq and Oq' by OQ, where OP and OQ are vectors rotating in opposite directions. If OP and OQ are equal to one another, the resultant field is, by (2), a purely alternating one. If either OP or OQ be zero, the field is a purely rotating one. In all other cases we see, by (3), that it can be represented by an alternating vector and a rotating vector.

When OP equals OQ the resultant field is purely alternating, and it is easy to show that in this case the two component fields must be in phase with one another. To get a purely rotating field we must have either OP or OQ equal to zero. Hence either $Op = Op'$ or $Oq = Oq'$; in either case the amplitudes of the two alternating fields must be equal in magnitude. In the first case (Fig. 135) Op and Op' must be in the same straight line but

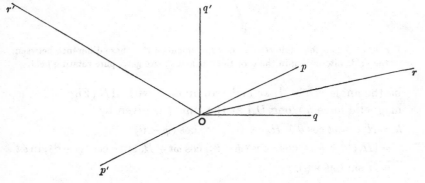

Fig. 135. Op and Op' are in the same line, when the alternating vectors produce a pure rotating field.

pointing in opposite ways, so that the angle between Op and Op' is π.

Hence $$\phi + \alpha = \pi,$$

where ϕ is the angle between the directions of the two alternating vectors and α is their phase difference. For if (Fig. 135) Op be in the same line as Op', then

$$\phi + \alpha = r\hat{O}r' - r\hat{O}p + r'\hat{O}p' = r'\hat{O}p + r'\hat{O}p' = \pi.$$

We can prove this important theorem analytically as follows.

Let the strengths of the fields in the directions sr and $s'r'$ (Fig. 136) be given by $H_1 \cos \omega t$ and $H_2 \cos (\omega t - \alpha)$ respectively, and let ϕ

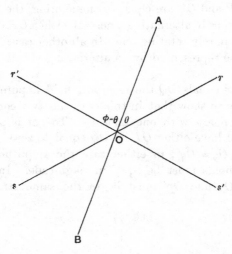

A
B

Fig. 136.   When the angle rOr' is the supplement of the phase difference between the fields alternating in the directions sr and $s'r'$ we get a pure rotating field.

be the angle rOr'.   If sr make an angle θ with AB (Fig. 136) the magnetic force h along OA at the time t is given by

$$h = H_1 \cos \omega t \cos \theta + H_2 \cos (\omega t - \alpha) \cos (\phi - \theta)$$
$$= \{H_1 \cos \theta + H_2 \cos \alpha \cos (\phi - \theta)\} \cos \omega t + H_2 \sin \alpha \cos (\phi - \theta) \sin \omega t$$
$$= H \sin (\omega t + \gamma),$$

where

$$H^2 = H_1^2 \cos^2 \theta + H_2^2 \cos^2 (\phi - \theta) + 2 H_1 H_2 \cos \alpha \cos \theta \cos (\phi - \theta),$$

and
$$\tan \gamma = \frac{H_1 \cos \theta + H_2 \cos \alpha \cos (\phi - \theta)}{H_2 \sin \alpha \cos (\phi - \theta)}.$$

Now if h be independent of θ, that is, if the amplitude of the alternating field in every direction at O (Fig. 136) be the same, the value of H must be the same when θ is $\pi/2$ and when it is $\phi - \pi/2$.   Hence we must have H_1 equal to H_2 and

$$H^2 = H_1^2 \{\cos^2 \theta + \cos^2 (\phi - \theta) + 2 \cos \alpha \cos \theta \cos (\phi - \theta)\}.$$

Now $\cos^2 \theta = \sin^2 \phi + \cos^2 \theta - \sin^2 \phi$
$$= \sin^2 \phi + \cos (\phi + \theta) \cos (\phi - \theta).$$

XVIII]   PROPERTIES OF VECTORS   429

Therefore $H^2 = H_1^2 \left[\sin^2 \phi + \cos (\phi - \theta) \{ \cos (\phi + \theta) + \cos (\phi - \theta) \right.$
$$\left. + 2 \cos \alpha \cos \theta \} \right]$$
$$= H_1^2 \{ \sin^2 \phi + 2 \cos (\phi - \theta) \cos \theta (\cos \phi + \cos \alpha) \}.$$

Thus, when $\cos \phi + \cos \alpha$ is zero, the value of H is $H_1 \sin \phi$, and is therefore the same for all values of θ. In this case we must have

$$2 \cos \frac{\phi + \alpha}{2} \cos \frac{\phi - \alpha}{2} = 0,$$

and therefore, since α is not greater than π, $\phi + \alpha$ must be equal to π.

In this case $h = H_1 \sin \phi \sin (\omega t + \phi - \theta)$
and $\gamma = \phi - \theta.$

(5) *Any number of alternating vectors are in general equivalent to a simple alternating vector and a rotating vector.*

We can replace every alternating vector of amplitude Or by two equal vectors Op and Oq rotating in opposite directions. All the component vectors like Op rotating in the same direction can be replaced by a single rotating vector OP. Similarly the Oq components compound into OQ. Hence the theorem follows from (3).

When OP and OQ are equal, the resultant field is a purely alternating one. When either OP or OQ is zero, the resultant field is a purely rotating one. If OP is zero, the resultant of all the Op components is zero, and thus a closed polygon can be constructed whose sides are equal and parallel to all the Op components, and a similar theorem holds when OQ is zero.

(6) *If p_1, p_2, ... and q_1, q_2, ... be any vectors and $\phi_{r.s}$ be the angle between p_r and q_s, then*

$$\Sigma\Sigma p_r q_s \cos \phi_{r.s} = PQ \cos \Phi,$$

where P is the resultant of all the p vectors, Q is the resultant of all the q vectors, and Φ is the angle between P and Q.

Resolve all the p vectors along q_1, then by projections

$$p_1 \cos \phi_{1.1} + p_2 \cos \phi_{2.1} + \dots + p_n \cos \phi_{n.1} = P \cos \Psi_1.$$

Hence $q_1 \Sigma p_r \cos \phi_{r.1} = P q_1 \cos \Psi_1.$

Similarly $q_2 \Sigma p_r \cos \phi_{r.2} = P q_2 \cos \Psi_2,$
....
Therefore $\Sigma\Sigma p_r q_s \cos \phi_{r.s} = P \Sigma q \cos \Psi.$

But $\Sigma q \cos \Psi$ is the sum of the projections of all the q components upon P, and it therefore equals $Q \cos \Phi$.

Therefore $\qquad \Sigma\Sigma p_r q_s \cos \phi_{r.s} = PQ \cos \Phi.$

(7) *If p be the amplitude of an alternating vector and q be a rotating one, the mean value of $pq \cos \phi$ is $(pq/2) \cos \delta$, where δ is the angle between the directions of p and q when p has its maximum value.*

We can replace the alternating vector p by two vectors, rotating in opposite directions, the magnitudes of which are each equal to $p/2$. The mean value of the product of q and the component vector $p/2$ rotating in the same direction and the cosine of the angle between them is $(pq/2) \cos \delta$, because the angle between them is always equal to δ. The mean value of the corresponding product for the other vector is zero, since, for a complete revolution, the mean value of $\cos(\omega t + \delta)$ is zero.

(8) *If p and q be two alternating vectors whose directions are inclined to one another at an angle ϕ, then the mean value of $pq \cos \phi$ is*

$$(pq/4)\{\cos(\phi + \beta - \alpha) + \cos(\phi + \alpha - \beta)\},$$

where $\alpha - \beta$ is the phase difference between the two vectors.

The angle between p and q is always ϕ, so that we have only to find the mean value of the instantaneous value of the product of the two vectors. This can be done by (7), for we can replace p by two vectors each equal to $p/2$ rotating in opposite directions. The mean value of one of these components multiplied by q and the cosine of the angle between them is by the preceding theorem

$$(pq/4) \cos \delta,$$

and δ equals $\phi + \beta - \alpha$ or $\phi + \alpha - \beta$ depending on which component we take. The sum of the two will obviously give the mean value of $pq \cos \phi$. This mean value, by adding the cosines, may be written in the shape

$$(pq/2) \cos \phi \cos(\alpha - \beta).$$

(9) *If p and q be vectors which rotate with different angular velocities, the mean value of $pq \cos \phi$ is zero.*

Suppose that they are rotating in the same direction with angular velocities ω_1 and ω_2; then in the time t, where $(\omega_1 - \omega_2)t$

equals 2π, the angle ϕ will have increased from 0 to 2π, and therefore its mean value taken over the time t will be zero. Hence, if the mean value be taken over a time t_1 which is large compared with t, it will be zero if t_1 be a multiple of t and will be very small in other cases.

If the vectors rotate in opposite directions, then the time t which the angle ϕ takes to increase from 0 to 2π is given by

$$(\omega_1 + \omega_2)\,t = 2\pi$$

and the mean value over a time t_1, large compared with t, will be zero or very nearly zero.

Magnetic field in the air-gap of polyphase machines. When we are dealing with the magnetic fields in the air-gaps between the rotating and the stationary parts (the rotor and the stator) of several kinds of polyphase machines, a modification of the above method becomes necessary. The following way of treating the problem is due to A. Potier.

In the case considered, we have a hollow laminated cylinder of soft iron with slots along the interior, parallel to the axis, which carry evenly distributed windings for the polyphase currents. When a current flows in one of these windings, it divides the interior into p segments of North and p segments of South polarity. If l be the breadth of one of these segments, then $2pl$ is the inner circumference of the cylinder. We shall suppose that the next winding is displaced from the first by a distance l/q on the circumference of the cylinder, where q is the number of phases. The rotor consists of a laminated iron cylinder axial with the stator and suitably wound with copper conductors. Like the stator it will have $2p$ poles, but the phases of the currents in it may have any values. The air-gap between the two cylinders may be as small as half a millimetre.

Rotating field in the air-gap of a polyphase induction motor. Let us consider the case of a polyphase induction motor (Vol. II, Chap. XII). We shall suppose that the windings of the stator are connected with the polyphase supply mains, and that the windings of the rotor consist of closed coils so that the currents in them are due to induction only. When this kind of motor is running, the frequency of the alternating currents in the rotor is small compared with the frequency of the currents in the stator. We shall for the present neglect the magnetising effects of the rotor currents.

Take some point O (Fig. 137) on the inner surface of the stator and make

$$OA_1 = A_1A_2 = A_2A_3 = \ldots = l,$$

where the points A_1, A_2, A_3, ... are on the circumference and the distances between them are measured along the circumference. If the radial magnetic force at O be zero at the instant when t is zero, the magnitude of the first harmonic of the radial component of the field may be indicated by the wavy line in the figure, the force being positive from O to A_1, negative from A_1 to A_2, etc.

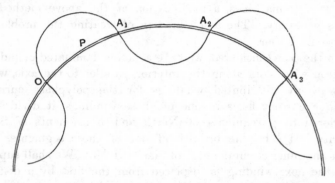

Fig. 137.   Field in the air-gap of a polyphase induction motor.

Consider the magnetic force perpendicular to the surface of the stator at a point P, where OP equals x, due to a single alternating current in one of the windings. We may evidently write this in the form $f(t)\,\phi(x)$ where $f(t)$ is some function of the time, and $\phi(x)$ is a function depending on the shape of the rotor and the stator, the method of winding, the permeability of the iron, the air-gap, etc. These functions are alternating periodic functions of the time and space respectively; we have, for example,

$$f(t) = -f(t + T/2) = f(t + T) = \ldots$$

and $$\phi(x) = -\phi(x + l) = \phi(x + 2l) = \text{etc.}$$

Suppose that the second winding is similar to the first and that it is displaced to the right to a distance l/q, and that the current in it lags by an interval $T/(2q)$ behind the current in the first winding. Then the radial component at P due to this winding will be

$$f\{t - T/(2q)\}\ \phi\ (x - l/q).$$

Hence, when all the windings are excited by polyphase currents, we find that the radial force H at P is given by

$$H = f(t)\ \phi\ (x) + f\{t - T/(2q)\}\ \phi\ (x - l/q) + \cdots$$
$$+ f\{t - (q-1)\ T/(2q)\}\ \phi\ \{x - (q-1)l/q\}\ldots\ldots(1).$$

It is easily seen that, when t is increased by $T/(2q)$ and x by l/q the value of H is not altered. If, therefore, we only examine the field at intervals of time $T/(2q)$, it will appear to glide round the air-gap, without altering in shape, with a linear velocity $2l/T$. If $2p$ be the number of poles and r the radius of the rotor, $2\pi r = 2pl$, and hence the angular velocity of the field is $2l/(Tr)$ or ω/p, where ω is 2π times the frequency of the alternating current.

If $\qquad f(t) = B \sin \omega t$ and $\phi(x) = C \sin(\pi x/l)$,

then, substituting in (1) and summing by the ordinary trigonometrical formula, we find

$$H = \frac{q}{2} BC \cos \{\omega t - \pi x/l\}\ \ldots\ldots\ldots\ldots\ldots(2).$$

Hence, when the magnetic force due to the current follows the harmonic law and when the distribution of the flux round the air-gap is a sine function of the space, the resultant radial field is sine-shaped and glides round the air-gap with constant speed. In this case also, if A be the effective current in each phase, the maximum value of the resultant magnetic field would equal that produced by a current $(q/2)\ A$ in one phase, provided that the permeability of the iron were constant.

When $f(t)$ and $\phi(x)$ are not sine functions, they may be developed in two series of sine functions by Fourier's theorem. The field will thus be decomposed into a series of magnetic fields rotating with angular velocities $m\omega/(np)$, where m and n are integers. Some of these fields, also, may rotate in the reverse direction. We shall consider this case more fully in Vol. II, Chapters XII and xiv, where also the shape of the magnetic fields produced by various simple windings is considered. It is shown that the induced currents in the rotor windings produce a magnetic field which rotates in space with the same angular velocity as the magnetic field due to the polyphase currents in the stator windings. The magnetic field which is the resultant of the fields due to the currents in the stator and rotor windings respectively must rotate in space with the same angular velocity as its two components. Thus when we take into account the rotor currents, we still have in the air-gap a magnetic field in which the distribution of the magnetic flux is always exactly the same at intervals of time which differ by multiples of $T/(2q)$. At any instant also we have p segments of North and p segments of South polarity. This kind of magnetic field we shall call a gliding magnetic field.

If one phase only of the stator windings be excited and the rotor be stationary, then we have as before p segments of North and p segments of South polarity. The polarity of these segments alternates with the frequency of the supply current, but the field is a stationary one, the lines dividing the segments being fixed in position and the magnetic force along these lines being always zero. When we are discussing the magnetic field in the air-gap of an alternating current machine, we shall refer to a field of this kind as an alternating field. Unlike the alternating fields considered earlier in this chapter, the amplitude of the magnetic force is different at different points of the field.

We will now consider what happens when gliding fields which
Gliding mag- follow the harmonic law are superposed on one
netic fields. another.

(1) *Two fields gliding in the same direction. Time lag but no space lag.*

In this case

$$H = H_1 \sin(\omega t - \pi x/l) + H_2 \sin(\overline{\omega t - \alpha} - \pi x/l)$$

$$= R \sin(\overline{\omega t - \beta} - \pi x/l),$$

where $R^2 = H_1^2 + H_2^2 + 2H_1 H_2 \cos \omega\alpha$

and $\tan \omega\beta = \dfrac{H_2 \sin \omega\alpha}{H_1 + H_2 \cos \omega\alpha}.$

Hence the resultant field glides in the same direction with the same velocity, and its magnitude and phase are given by the parallelogram construction.

(2) *Two fields gliding in the same direction. Space lag but no time lag.*

We have

$$H = H_1 \sin (\omega t - \pi x/l) + H_2 \sin \{\omega t - \pi (x - a)/l\}$$
$$= R \sin \{\omega t - \pi (x - b)/l\},$$

where $R^2 = H_1^2 + H_2^2 + 2H_1 H_2 \cos (\pi a/l)$

and $\tan \dfrac{\pi b}{l} = \dfrac{H_2 \sin (\pi a/l)}{H_1 + H_2 \cos (\pi a/l)}.$

The resultant field therefore glides in the same direction with the same velocity, and its magnitude and-phase can be got graphically by the parallelogram construction.

(3) *Two equal fields gliding in opposite directions are equivalent to an alternating field.*

In this case

$$H = H_1 \sin (\omega t - \pi x/l + \alpha_1) + H_1 \sin (\omega t + \pi x/l + \alpha_2)$$
$$= 2H_1 \sin \{\omega t + (\alpha_1 + \alpha_2)/2\} \cos \{\pi x/l + (\alpha_2 - \alpha_1)/2\}.$$

The resultant field is therefore a purely alternating one, H always being zero at the points given by

$$x = (2n + 1) l/2 + (\alpha_1 - \alpha_2) l/(2\pi),$$

where n is an integer. The greatest value of H is obviously $2H_1$.

(4) *An alternating magnetic field whose maximum value is H is equivalent to two equal fields (whose maximum values are each $H/2$) gliding in opposite directions.*

This follows from the theorem that

$$H \sin \omega t \cos (\pi x/l) = (H/2) \sin (\omega t + \pi x/l) + (H/2) \sin (\omega t - \pi x/l).$$

The angular velocities of the fields are ω/p and $- \omega/p$, where $2p$ is the number of poles.

(5) *The resultant of any number of alternating and gliding fields of the same period is in general two fields gliding in opposite directions.*

This follows from (1), (2) and (4). It is to be noticed however that sine curve assumptions are made.

If the two systems of currents represented by

$$i_1 = I_1 \sin(\omega t - \pi x/l + \alpha_1)$$

and $$i_2 = I_2 \sin(\omega t + \pi x/l + \alpha_2)$$

are superposed in the distributed conductors, then the heat generated in the conductors will be that due to each system of currents separately, for the mean value of $(i_1 + i_2)^2$ is $I_1^2/2 + I_2^2/2$. We shall call i_1 a system of currents turning to the right.

(6) *The mean torque produced by a magnetic field turning to the left on a system of currents turning to the right is zero.*

The currents may be flowing, for example, in the windings of the rotor of an induction motor.

The torque on any conductor will be proportional to the product

$$\sin(\omega t - \pi x/l + \alpha_1) \sin(\omega t + \pi x/l + \alpha_2).$$

The mean value of this expression from t equal to 0 to t equal to T is

$$(1/2)\cos(2\pi x/l + \alpha_2 - \alpha_1)\dots\dots\dots\dots\dots(3).$$

And the mean value of (3) from x equal to 0 to x equal to $2pl$ is obviously zero, and thus the mean torque between the field and the system of currents is zero. This theorem is analogous to the theorems in the undulatory theory of optics regarding the non-interference of circular vibrations in opposite directions. Several other theorems might be adopted from optics, and it will be found that these theorems are useful when we come to the theory of asynchronous motors. The artifice of replacing a fixed alternating field by two fields rotating in opposite directions, used first by Fresnel in optics and adapted by Ferraris to alternating current theory, is invaluable in this connection.

Consider the case of two coils the axes of which intersect at some point O. We will consider the field at O in the plane determined by the two axes. Let h_1, h_2 be the values at O of the magnetic forces along the axes Op and Oq respectively, and let the angle pOq equal α. Now if Op and Oq be equal to h_1 and h_2 and r be the middle point of pq, $2Or$ will be the resultant force, and it will be seen that, as h_1 and h_2 alter, the locus of r may be a

Rotating magnetic fields when the currents are not sine-shaped.

complicated curve and the angular velocity of Or may vary in an erratic manner. Draw any line Oa in the plane pOq, and let the angle pOa equal θ. Then, if h be the resultant magnetic force resolved along this line,

$$h = h_1 \cos \theta + h_2 \cos (\alpha - \theta).$$

Therefore
$$h^2 = h_1^2 \cos^2 \theta + h_2^2 \cos^2 (\alpha - \theta)$$
$$+ 2h_1 h_2 \cos \theta \cos (\alpha - \theta).$$

Now if the currents in the coils be adjusted until the effective or root mean square value of h_1 equals the R.M.S. value of h_2, and if capital letters denote the R.M.S. values, then

$$H^2 = H_1^2 \{\cos^2 \theta + \cos^2 (\alpha - \theta) + 2 \cos \theta \cos (\alpha - \theta) \cos \phi\},$$

where ϕ is the phase difference between h_1 and h_2 (see page 278).
Noticing that

$$\cos^2 \theta = \sin^2 \alpha + \cos (\alpha + \theta) \cos (\alpha - \theta),$$

we can easily prove that the above equation may be written in the form

$$H^2 = H_1^2 \{(\cos \alpha + \overline{\cos \alpha - 2\theta})(\cos \alpha + \cos \phi) + \sin^2 \alpha\} \quad \ldots(4).$$

Now if
$$\alpha = \pi - \phi,$$
so that
$$\cos \alpha + \cos \phi = 0,$$
then
$$H = H_1 \sin \alpha.$$

Thus, if the phase difference between the magnetic forces h_1 and h_2, or between the currents in the coils when no iron is used, be supplementary to the angle between their axes, then the effective value of the resultant magnetic force resolved along any line in the plane determined by the axes of the coils is constant. If this relation does not hold, then it is easy to show from (4) that H is a maximum when θ is $\alpha/2$ or $\pi + \alpha/2$, and a minimum when θ is $\pi/2 + \alpha/2$ or $-\pi/2 + \alpha/2$. If we plot out a curve showing the values of H for various values of θ, we get a curve which is very similar to an ellipse.

This theorem can also be extended to three phase theory. **Extension to three phase theory.** Suppose that we have three equal cylindrical coils arranged round a circle at 120 degrees apart, their axes all pointing to the centre of the circle. Let h_1, h_2 and h_3 be the strengths of the fields produced by them at the centre, when they are connected to the mains of a three phase system in star fashion, then by Kirchhoff's law

$$i_1 + i_2 + i_3 = 0,$$

and hence if there be no iron in the coils and if they be equal and similar

$$h_1 + h_2 + h_3 = 0.$$

Therefore $\qquad H_1^2 = H_2^2 + H_3^2 + 2H_2 H_3 \cos \phi_{2.3}.$

But $\qquad\qquad H_1 = H_2 = H_3.$

Thus $\qquad\qquad \cos \phi_{2.3} = -1/2,$

and hence $\qquad\qquad \phi_{2.3} = 120°.$

If h be the magnetic force resolved along a line drawn through the centre making an angle θ with h_1, then

$$h = h_1 \cos \theta + h_2 \cos (\theta - 2\pi/3) + h_3 \cos (\theta - 4\pi/3).$$

Thus $\quad H^2 = H_1^2 \{\cos^2 \theta + \cos^2 (\theta - 2\pi/3) + \cos^2 (\theta - 4\pi/3)$

$$- \cos \theta \cos (\theta - 2\pi/3) - \cos (\theta - 2\pi/3) \cos (\theta - 4\pi/3)$$

$$- \cos (\theta - 4\pi/3) \cos \theta\},$$

and hence $\qquad\qquad H = (3/2) H_1.$

Therefore H is independent of θ and is the same for every line in the plane of the circle.

In practical work, when the coils have iron cores, $h_1 + h_2 + h_3$ is not zero at every instant. We can, however, represent H_1, H_2 and H_3 by lines forming a solid angle (Chap. XII) and if the coils be equal and symmetrical and the currents magnetising them be equal, then, from symmetry, the phase differences between any two will be equal and will be less than 120 degrees. Hence we may put

$$\cos \phi_{2.3} = -m/2,$$

where m is less than 1, and thus

$$H^2 = H_1^2 [\cos^2 \theta + \dots - m \{\cos \theta \cos (\theta - 2\pi/3) + \dots\}]$$

$$= H_1^2 \{3/2 + 3m/4\},$$

and hence $\quad H = (H_1/2) \sqrt{6 + 3m}.$

Therefore, in this case also, H is independent of θ.

If h be the strength of the field perpendicular to the plane of the coil, the E.M.F. induced in it will obviously be proportional to $\partial h/\partial t$. When we have two coils, we find in the same manner as before that

$$E^2 = E_1^2 \{(\cos \alpha + \cos \overline{\alpha - 2\theta})$$
$$(\cos \alpha + \cos \psi) + \sin^2 \alpha\} \cdots (5),$$

Rotating magnetic field producing a constant effective E.M.F. in a search coil placed with its plane perpendicular to the field.

where the effective values of $\partial h_1/\partial t$ and $\partial h_2/\partial t$ are each equal to E_1, and ψ is the angle of phase difference between them. Hence when $\psi + \alpha$ equals π, the search coil indicates the same effective E.M.F. $E_1 \sin \alpha$, no matter what the angle θ may be, and since α can easily be measured ψ can be found. Now in a choking coil the applied potential difference e is proportional to $\partial h/\partial t$. If, therefore, we have two choking coils and we adjust them until the search coil connected to the terminals of an electrostatic voltmeter indicates the same effective E.M.F. for all positions of the search coil such that its plane is perpendicular to the plane of the axes of the choking coils, then the phase difference between the applied P.D.s is the supplement of the angle between the axes.

Similarly, if three choking coils with their axes at angles of 120 degrees apart be magnetised by a system of three phase currents, we can show that the effective E.M.F. E induced in the search coil, no matter what the value of θ may be, is equal to $(E_1/2)\sqrt{6 + 3m}$, where m is less than unity and E_1 is the E.M.F. that would be induced in the search coil by one of the coils acting alone when the plane of the search coil is perpendicular to the axis of the other coil. Thus it is possible, since E and E_1 can be found, to find m, and thus the phase difference between the rates at which the magnetisations in the coils are altering. This is equal to the phase difference between the electromotive forces.

In the form of phase indicator invented by Riccardo Arnò, two circular coils have a common diameter, about which they can rotate, and the currents, whose phase differences are to be measured, are passed through them. A search coil at the centre of their common axis and entirely enclosed by the two coils can also rotate about this axis which is

Arnò's phase indicator.

in its plane. The search coil is connected to an electro-dynamo-meter or a hot wire ammeter, of negligible resistance, which measures the current in it. Now, in general, for different positions of the search coil, we get different readings in the ammeter. If, however, the current in one of the given coils be varied until the effective values of the magnetic forces produced at the common centre of the two coils be equal, it is possible, by adjusting the angle α between the planes of the two coils carrying the currents, to find a position in which the reading on the ammeter is always the same whatever the position of the search coil. If f be the resolved magnetic force perpendicular to the plane of the search coil, then

$$k\frac{\partial f}{\partial t} = Ri + L\frac{\partial i}{\partial t},$$

where R is the resistance and L the inductance of the search coil circuit and k is a constant. If R be negligible, then

$$k\frac{\partial f}{\partial t} = L\frac{\partial i}{\partial t}.$$

Therefore, integrating, $kf = Li + \text{const.}$,

and, since f and i are alternating functions, the constant must be zero; thus taking effective values

$$kF = LA.$$

Now, since A is constant in all positions of the search coil, F must also be constant, and therefore, by the converse of the theorem proved on p. 437,

$$\phi = 180^\circ - \alpha,$$

where ϕ is the angle of phase difference between f_1 and f_2, i.e. the angle of phase difference between the currents, since there is no iron in the circuit, and α is the angle between the planes of the two coils.

REFERENCES

G. Ferraris, 'Rotazioni Elettro-dynamiche.' *Turin Acad.*, March 1888.

A. Russell, 'Rotary Magnetic Fields.' *The Electrical Review*, June 1893.

André Blondel, 'Quelques Propriétés Générales des Champs Magnétiques Tournants.' *L'Éclairage Électrique*, Vol. 4, p. 241, 1895.

G. Ferraris, 'A Method for the Treatment of Rotating or Alternating Vectors,' translated in the *Electrician*, Vol. 33, p. 110, 1894.

A. Potier, 'Sur les Courants Polyphasés.' *International Physical Congress at Paris*, Report 3, p. 197, 1900.

A. Campbell, 'Test Room Methods of Alternate Current Measurement.' *Journ. Inst. of El. Engin.*, Vol. 30, p. 889, 1901.

S. P. Thompson, 'Hysteresis Loops and Lissajous' Figures.' *Phil. Mag.*, Vol. 20, p. 417, 1910.

CHAPTER XIX

The magnetic fields round polyphase cables. The magnetic field round two long parallel wires carrying equal currents flowing in opposite directions. Bipolar circles. Currents equal and flowing in the same direction. Cassinian ovals. Lines of force when the currents are unequal. How the magnetic field alters round wires carrying single phase currents. Magnetic field round a three phase cable. Two phase cable. Twin concentric cable. Field of force round n parallel wires symmetrically arranged with their axes on a circle. Concentric cable. The strength of the magnetic field round n parallel wires. The strength of the field round a concentric main. The losses in cables. Three core cables. Dielectric losses. Interference with telephone circuits. Duality. References.

THE magnetic fields round polyphase cables when carrying currents are very complex, but the equations to the

The magnetic fields round polyphase cables. lines of force can easily be found and it will be instructive to study them. In practice, we have to investigate whether the field will affect neighbouring telephone or telegraph wires, and whether it will produce appreciable eddy current losses in the lead sheath or in the copper shield, which is sometimes placed immediately inside the lead sheath to insure that in the event of one of the copper conductors getting accidentally into contact with it, the main fuse may act promptly. If there were no earth shield and one of the conductors made contact with the sheath, then, if the resistance of the earth in the neighbourhood of the place of contact were high, the sheath might be maintained at a high potential and be dangerous. In armoured cables we have also to investigate the hysteresis and eddy current losses in the steel strip or galvanised iron wires used to protect the cables.

We will first of all consider the magnetic field round two
parallel wires. This would be the case of the mains
of a two wire direct current system or of a single
phase alternating current system. Let the wires be
perpendicular to the plane of the paper, and let A and
B (Fig. 138) be the points where their axes cut this
plane, the current in A flowing towards the observer
and in B away from him. We suppose that the wires are circular
in section. So far therefore as the magnetic force at points
external to them is concerned we may suppose that the currents
are concentrated along the axes of the wires (p. 55). We shall
only consider the magnetic forces at points external to the conductors. The magnetic force at any point P will be the resultant
of two forces $2i/AP$ and $2i/BP$ which are perpendicular to AP and

*The magnetic
field round two
long parallel
wires carrying
equal currents
flowing in
opposite
directions.*

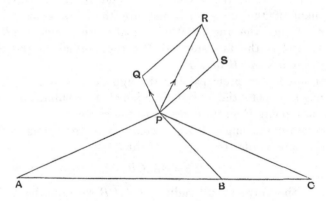

Fig. 138. Currents in opposite directions.

$$F = PR = \frac{2 \cdot AB \cdot i}{r_1 r_2}.$$

BP respectively and in the plane of the paper. Let PQ and PS
represent these forces.

Let $AP = r_1$, $BP = r_2$, $AB = 2a$, $i =$ the value of the current
in each wire in C.G.S. measure, and let $F = PR =$ the resultant
magnetic force at P. Now the angles APB and PSR are each
the supplement of the angle QPS and are therefore equal to one
another.

Also
$$\frac{PS}{RS} = \frac{2i/r_2}{2i/r_1} = \frac{r_1}{r_2} = \frac{AP}{BP}.$$

Hence by Euclid (VI. 6) the triangles RPS and BAP are similar. Therefore the angle SPR equals the angle PAB and the angle QPR equals the angle PBA.

Again,
$$\frac{PR}{PS} = \frac{AB}{AP} = \frac{2a}{r_1},$$

and therefore $F = 4ai/(r_1 r_2).$

If the angle ABP were a right angle, then PS would be parallel to AB and hence APR would be a straight line. Thus at any point P' on the line through B perpendicular to AB, the direction of the resultant magnetic force will be AP'.

In Fig. 138 if we draw PC perpendicular to PR to meet AB produced in C, then the angles CPB and SPR are each the complement of the angle SPC and are therefore equal to one another. Hence the angle CPB is equal to the angle PAB and therefore PC is the tangent and PR the normal to the circle passing through A, B and P.

Therefore every circle passing through A and B is a line of equal magnetic potential as the direction of the resultant magnetic force at any point P on it is normal to the curve.

Also, since the angle PCA is common to the triangles CPB, CAP, they are similar triangles, and thus

$$CP^2 = CA.CB.$$

If, then, with centre C and radius $\sqrt{CA.CB}$ we describe a circle, any radius CP of this circle will be a tangent to the circle through A, B and P; the tangent therefore at every point on the circle, whose centre is C, will be in the direction of the resultant magnetic force at that point. Hence this circle will be a line of force. The points A and B are inverse points with respect to the circle.

It follows that all circles which have A and B for inverse points
Bipolar circles.   (Fig. 139) will be lines of force round the wires. The polar of A with regard to all the circles surrounding B will pass through B and *vice versâ*. In other words,

the chord of contact of the two tangents from A, to any circle surrounding B, passes through B. These circles and the circles passing through A and B are called bipolar circles. The bipolar equation to any circle round B (Fig. 139) is

$$\frac{r_1}{r_2} = \frac{AP}{BP} = \frac{CA}{CP} = \text{constant} = m,$$

or $r_1 = mr_2.$

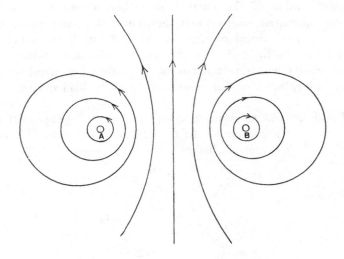

Fig. 139.   The lines of force round wires carrying equal currents flowing in opposite directions are circles.

The equation to any line of equal magnetic potential will obviously be

$$\theta_2 - \theta_1 = \alpha,$$

where α is a constant and θ_2, θ_1 are the angles PBC, PAC respectively.

We have shown above that the magnitude of the resultant magnetic force is given by $4ai/(r_1 r_2)$.

Thus the bipolar equation of any line of equal magnetic force is

$$r_1 r_2 = \text{constant}.$$

These curves are Cassinian ovals and are shown in Fig. 141.

Since with single phase alternating currents the currents in the mains are at every instant equal in magnitude and opposite in direction, the lines of force due to the currents will always be the circles which have A and B for inverse points. Hence at any point P the magnetic field will be fixed in direction, and the vector representing it will simply oscillate backwards and forwards through P. It will therefore be a purely oscillatory field, and its maximum value will be $4aI/(r_1r_2)$, where I is the maximum value of the current in either main. This is also true for concentric mains when the axis of the outer conducting cylinder is not coincident with the axis of the inner conductor. In this case the lines of force external to the outer conductor are bipolar circles, and induction effects are produced on neighbouring wires. Eddy currents will also be set up in the lead sheath.

Let F be the resultant magnetic force at P (Fig. 140) and let the angle APB equal α, then, with the same notation as before,

Currents equal and flowing in the same direction.

$$F^2 = \frac{4i^2}{r_1^2} + \frac{4i^2}{r_2^2} + 2\,\frac{4i^2}{r_1r_2}\cos\alpha$$

$$= \frac{4i^2}{r_1^2} + \frac{4i^2}{r_2^2} + \frac{4i^2}{r_1^2 r_2^2}(r_1^2 + r_2^2 - 4a^2)$$

$$= \frac{8i^2}{r_1^2 r_2^2}(r_1^2 + r_2^2 - 2a^2)$$

$$= \frac{8i^2}{r_1^2 r_2^2} \cdot 2 \cdot OP^2.$$

Therefore $F = \dfrac{4i \cdot OP}{r_1 r_2}$.

Also, since in Fig. 140

$$\frac{AP}{PB} = \frac{PS}{PQ},$$

and the angle APB equals the angle QPS, it follows that the parallelogram described on AP and PB as adjacent sides will be similar to the parallelogram $PQRS$. Thus the diagonals of the two parallelograms will be inclined equally to their respective sides. But the diagonal of the parallelogram described on AP

and PB as adjacent sides is represented by $2 . PO$ in magnitude and direction. Therefore

$$S\hat{P}R = O\hat{P}A \quad \text{and} \quad R\hat{P}Q = O\hat{P}B.$$

Also $\qquad O\hat{P}A = \pi/2 - B\hat{P}R \quad \text{and} \quad O\hat{P}B = A\hat{P}R - \pi/2.$

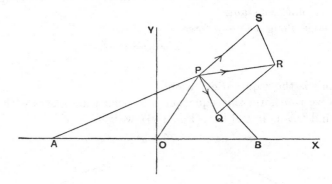

Fig. 140.   Currents flowing in the same direction.

$$F = PR = \frac{4 . OP . i}{r_1 r_2}.$$

Hence, for example, in the particular case when APR is a straight line, OPB is a right angle, and therefore at every point P' on the circle described on OB as diameter, AP' gives the direction of the resultant force.

The bipolar equation to the lines of force can be found as **Cassinian ovals.** follows.

If N_1 be the number of lines of force, per unit length of the wires, crossing OP (Fig. 140) due to the current i in the wire whose axis cuts the plane of the paper perpendicularly at A, we have

$$N_1 = \int \frac{2i}{r} \, dr = [2i \log r],$$

where A is the origin and the limits of r are AO and AP. Thus we get

$$N_1 = 2i \log \frac{AP}{AO}.$$

If N be the total number of lines of force, we obviously have

$$N = 2i \log \frac{AP}{AO} + 2i \log \frac{BP}{BO}$$

$$= 2i \log r_1 r_2 - 2i \log a^2.$$

If P trace out a line of force, the number of lines crossing OP must remain constant.

Hence along a line of force

$$r_1 r_2 = \text{constant}$$
$$= m^2,$$

and this is the required equation.

If we transform this equation into polar coordinates with O as pole and OB as initial line (Fig. 140), we get

$$r^4 - 2a^2 r^2 \cos 2\theta + a^4 - m^4 = 0.$$

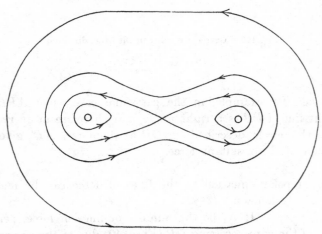

Fig. 141.   The lines of force when equal currents are flowing in the same direction are Cassinian ovals.

If m be less than a, we get two ovals (Fig. 141); when m equals a we get the lemniscate, the lines of force apparently crossing at the origin where the magnetic force is zero; and when m is greater than a we get only one oval. If m be greater than $\sqrt{2}a$ there is no point of inflexion on the oval, and if m be very large the lines of force are approximately circular.

Again, since the sum of the magnetic potentials due to the two currents is a constant at every point on a line of equal magnetic potential, we have by p. 49

$$2i\theta_2 + 2i\theta_1 = \text{a constant}$$
$$= 2i\alpha,$$

where α is a constant.

Thus the lines of equal magnetic potential are given by the equation

$$\theta_2 + \theta_1 = \alpha,$$

which represents a series of equiaxial hyperbolas passing through A and B and cutting the Cassinian ovals at right angles.

Lines of force when the currents are unequal. When the currents in the long parallel wires are unequal in magnitude, the equations to the lines of force can be written down at once in bipolar coordinates by the method above.

Since there is no magnetic force at right angles to a line of force, we have

$$\frac{2i_1}{r_1}\cos\psi_1 + \frac{2i_2}{r_2}\cos\psi_2 = 0,$$

where ψ_1 and ψ_2 are the angles between the radii AP and BP and the resultant magnetic force PR (Fig. 140) which is a tangent to the required curve.

Now $$\cos\psi_1 = \frac{\partial r_1}{\partial s}, \text{ and } \cos\psi_2 = \frac{\partial r_2}{\partial s}.$$

Hence substituting we get

$$\frac{i_1}{r_1}\frac{\partial r_1}{\partial s} + \frac{i_2}{r_2}\frac{\partial r_2}{\partial s} = 0,$$

and therefore $$i_1 \log r_1 + i_2 \log r_2 = \text{constant}.$$

Thus $$r_1^{i_1} r_2^{i_2} = \text{constant}.$$

If ψ_1', ψ_2' be the angles which the normal to the line of force at P makes with r_1 and r_2, then $\psi_1' = \psi_1 - \pi/2$, and $\psi_2' = \psi_2 - \pi/2$, hence substituting we get

$$\frac{2i_1}{r_1}\sin\psi_1' + \frac{2i_2}{r_2}\sin\psi_2' = 0.$$

Now if $\partial s'$ be an element of the equipotential curve through P,

$$\sin \psi_1' = r_1 \frac{\partial \theta_1}{\partial s'}, \quad \text{and} \quad \sin \psi_2' = r_2 \frac{\partial \theta_2}{\partial s'},$$

and therefore $i_1 \partial \theta_1 + i_2 \partial \theta_2 = 0.$

Hence $i_1 \theta_1 + i_2 \theta_2 = \text{constant}.$

In Fig. 142 the lines of force are shown when the currents are flowing in opposite directions and i_1 is four times i_2.

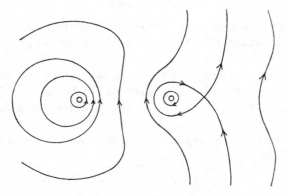

Fig. 142.   Currents flowing in opposite directions, one being
four times greater than the other.

The neutral point, that is, the point where the magnetic force is zero, ought to be specially noticed.   The bipolar equation to the looped line of force through this point is

$$27 r_1^4 = 2048 r_2 a^3.$$

In general, if N be the neutral point and A and B the points where the axes of the wires cut the plane of the paper, then,

$$\frac{AN}{BN} = \frac{i_1}{i_2}.$$

If the currents are flowing in the same direction, N is between the wires, if in opposite directions N is on the side next the weaker current.   When N is at an infinite distance, $i_1 = -i_2$ and the lines of force are the circles shown in Fig. 139.   When $i_1 = -4i_2$ the lines of force are as in Fig. 142.   When i_2 equals zero the lines of force are circles round A.   When $i_1 = 4i_2$, N divides AB

in the ratio of 4 to 1, and a crossed line of force through this point loops A with B. Finally, when $i_1 = i_2$ the lines of force are the Cassinian ovals shown in Fig. 141.

If the ratio of i_1 to i_2 is always constant, that is, if the phase difference between i_1 and i_2 is either zero or 180 degrees, then the lines of force are fixed. In this case the field at any point is a purely oscillatory one.

How the magnetic field alters when i_1 and i_2 are alternating currents.

If however the phase difference between the currents is neither zero nor 180 degrees, the magnetic field is continually changing in direction. It is not difficult to form a mental picture of what happens by considering how the neutral point N oscillates in any given case and by studying the Figures 139, 141 and 142. In general the force at any point is continually changing in magnitude and direction and hence is partly oscillatory and partly rotary.

Magnetic field round a three phase cable.

Suppose that the three copper conductors are parallel cylinders and that their axes cut the paper at the angular points A, B and C of an equilateral triangle whose sides are of length d. Then if i_1, i_2 and i_3 be the instantaneous values of the currents, the magnetic force (f) at any instant will be the resultant of the three magnetic forces $2i_1/r_1$, $2i_2/r_2$ and $2i_3/r_3$ acting at P and perpendicular to AP, BP and CP respectively. By a well-known theorem in Statics, we have

$$f^2 = \frac{4i_1^2}{r_1^2} + \ldots + 2\frac{4i_2 i_3}{r_2 r_3}\cos\theta_{2.3} + \ldots .$$

But

$$\cos\theta_{2.3} = \frac{r_2^2 + r_3^2 - d^2}{2r_2 r_3},$$

and thus

$$f^2 = \frac{4i_1^2}{r_1^2} + \ldots + 4i_2 i_3\left(\frac{1}{r_2^2} + \frac{1}{r_3^2} - \frac{d^2}{r_2^2 r_3^2}\right) + \ldots .$$

Now, in practice, we generally have

$$i_1 + i_2 + i_3 = 0,$$

and therefore

$$2i_2 i_3 = i_1^2 - i_2^2 - i_3^2.$$

29—2

On substituting this value of $2i_2 i_3$ in the equation for f^2 and simplifying, it is easy to see that

$$f^2 = 2i_1^2 d^2 \left(\frac{1}{r_1^2 r_2^2} + \frac{1}{r_1^2 r_3^2} - \frac{1}{r_2^2 r_3^2} \right) + \ldots + \ldots .$$

Let O be the centre of the circle through A, B and C, a its radius and let the angle POA be θ, then we have

$$r_1^2 = r^2 - 2ar \cos \theta + a^2,$$
$$r_2^2 = r^2 - 2ar \cos (\theta - \tfrac{2}{3}\pi) + a^2,$$

and
$$r_3^2 = r^2 - 2ar \cos (\theta - \tfrac{4}{3}\pi) + a^2.$$

Substituting these values of r_1, r_2 and r_3 in the formula for f^2 and noticing that d^2 is $3a^2$ and that

$$r_1^2 r_2^2 r_3^2 = r^6 - 2a^3 r^3 \cos 3\theta + a^6,$$

we get

$$f^2 = \frac{6a^2 (a^2 + r^2)}{r^6 - 2a^3 r^3 \cos 3\theta + a^6} (i_1^2 + i_2^2 + i_3^2)$$

$$+ \frac{24 a^3 r}{r^6 - 2a^3 r^3 \cos 3\theta + a^6} \{ i_1^2 \cos \theta + i_2^2 \cos (\theta - \tfrac{2}{3}\pi) + i_3^2 \cos (\theta - \tfrac{4}{3}\pi) \}.$$

At the centre of the circle through A, B and C, r is zero and thus

$$f^2 = \frac{6 (i_1^2 + i_2^2 + i_3^2)}{a^2}.$$

In order that the field at the centre of the circle may be of constant strength we must have

$$i_1^2 + i_2^2 + i_3^2 = \text{constant}.$$

This is true, for example, when the currents follow the harmonic law.

Let us now suppose that the currents follow the harmonic law so that we can write

$$i_1 = I \cos \omega t, \quad i_2 = I \cos (\omega t - 2\pi/3), \quad i_3 = I \cos (\omega t - 4\pi/3).$$

Then we have

$$i_1^2 + i_2^2 + i_3^2 = \tfrac{3}{2} I^2,$$

and
$$i_1^2 \cos \theta + \ldots + \ldots = \tfrac{1}{2} I^2 \{ (1 + \cos 2\omega t) \cos \theta + \ldots + \ldots \}$$
$$= \tfrac{3}{4} I^2 \cos (2\omega t + \theta).$$

Substituting these values in the formula for f^2, we get

$$f^2 = \left(\frac{3I}{a} \right)^2 \cdot \frac{a^4 \{ a^2 + r^2 + 2ar \cos (2\omega t + \theta) \}}{r^6 - 2a^3 r^3 \cos 3\theta + a^6}.$$

At points on the circumference of the circle passing through ABC, r equals a, and thus

$$\pm f = \frac{3I}{a} \cdot \frac{\cos(\omega t + \theta/2)}{\sin(3\theta/2)}.$$

Let us now consider the magnetic force at a point P on the circumference of this circle when the currents in the cores are i_1, i_2 and i_3 respectively (Fig. 143).

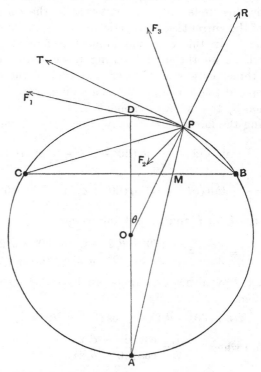

Fig. 143. The currents in the wires are supposed to be flowing in the same direction.

Since ABC is an equilateral triangle the angle CPB is 120° and AP bisects it. Let O be the centre of the circle and let the angle POD be θ and let OA be a. The angle OPA is obviously $\theta/2$ so that the angles CPO and BPO are $60° - \theta/2$ and $60° + \theta/2$ respectively. We have also $r_1 = AP = 2a\cos(\theta/2)$, $r_2 = 2a\cos(60° + \theta/2)$ and $r_3 = 2a\cos(60° - \theta/2)$.

Let PT be the tangent at the point P and let OP be produced to R. The tangential force T along PT will be given by

$$T = \frac{2i_1}{r_1} \cos(\theta/2) + \frac{2i_2}{r_2} \cos(60° + \theta/2) + \frac{2i_3}{r_3} \cos(60° - \theta/2)$$

$$= (i_1 + i_2 + i_3)/a.$$

Hence, at any instant, the tangential force has the same value at all points on this circle which are external to the cores, whatever the value of the currents. In practice we generally have $i_1 + i_2 + i_3$ equal to zero. In this case the tangential force is zero at all points which lie on the circle passing through the axes of the cores of a three phase cable but are external to the cores. The resultant force at these points is therefore represented in magnitude and direction by the radial component R.

Resolving the forces at P radially (Fig. 143) we get

$$- R = \frac{2i_1}{r_1} \sin(\theta/2) + \frac{2i_2}{r_2} \sin(60° + \theta/2) - \frac{2i_3}{r_3} \sin(60° - \theta/2)$$

$$= \frac{1}{a} \{i_1 \tan(\theta/2) + i_2 \tan(60° + \theta/2) - i_3 \tan(60° - \theta/2)\}.$$

Writing $- i_2 - i_3$ for i_1 and simplifying we get

$$- R = \frac{\sqrt{3}}{2a} \cdot \frac{i_2 \cos(60° - \theta/2) - i_3 \cos(60° + \theta/2)}{\cos(\theta/2) \cos(60° - \theta/2) \cos(60° + \theta/2)}.$$

We see that R vanishes and therefore also the resultant force vanishes when

$$i_2 \cos(60° - \theta/2) - i_3 \cos(60° + \theta/2) = 0,$$

and therefore when

$$\frac{i_2}{i_3} = \frac{\cos(60° + \theta/2)}{\cos(60° - \theta/2)}$$

$$= \frac{r_2}{r_3} = \frac{BM}{CM}.$$

This theorem gives us a simple construction for finding the neutral point. Join the axes B and C of the two cores in which the currents are flowing in the same directions. Divide this line at M so that BM is to CM in the ratio of i_2 to i_3. Produce AM to meet the circle at P; then, since PM bisects the angle CPB, P is the neutral point.

If we write $I \cos(\omega t - 2\pi/3)$ and $I \cos(\omega t - 4\pi/3)$ for i_2 and i_3 in the formula above, we get

$$- R = \frac{3I}{a} \cdot \frac{\sin(\omega t - \theta/2)}{\cos(\theta/2)(2\cos\theta - 1)}$$

$$= \frac{3I}{a} \cdot \frac{\sin(\omega t - \theta/2)}{\cos(3\theta/2)}.$$

Writing $180° - \theta$ for θ in this formula, we see, by comparing it with the formula we have found for $\pm f$ (p. 453) that R must be equal to f. This also follows since, in this case, T is zero.

We see that the neutral point is determined by

$$\theta = 2\omega t, \text{ so that } \frac{\partial\theta}{\partial t} = 2\omega,$$

and thus as long as it is outside the cores of the conductors it moves round the circumference of the circle with an angular velocity 2ω.

The amplitude of the magnetic force at a point P on the circumference of the circle is $3I/\{a\cos(3\theta/2)\}$. Thus the magnetic force has the minimum amplitude $3I/a$ where θ is zero. The amplitude of the magnetic force increases as θ increases. When θ is 50° it is nearly four times as large as the minimum value and for higher values of θ it increases very rapidly. The formula is not applicable to points inside the cores.

Let us now consider the magnetic force at any point P, not on the circle ABC, when the currents follow the harmonic law. This force is the resultant of three forces fixed in direction which oscillate according to the same law. The extremity of the line representing the resultant force at P therefore (p. 424) traces out an ellipse, and if f be the value of the resultant force and ω_1 its angular velocity at any instant we have

$$f^2\omega_1 = \text{constant}.$$

This is true even for points inside the metal cores. It follows that if f vanish at any instant it must either be zero always or it must oscillate in a straight line. Since there is no neutral point outside two of the cores when the current in the third vanishes, it follows that there is no point outside the cores where the magnetic force is always zero. Similarly there is no point inside the cores where the magnetic force is always zero. If the force, therefore, ever vanish at a point, the field at that point must be purely oscillatory.

Along the axis of a three phase cable carrying currents which follow the sine law, the field is purely rotary, that is, the extremity of the line representing the resultant force describes a circle. As we approach the circle ABC the fields become elliptical and their eccentricity increases as we approach the circumference. At points on this circle, outside the cores, the magnetic forces are purely oscillatory, the oscillations taking place in the radial direction. Outside this circle the field at any point is elliptical but the eccentricity is practically zero when the distance r of the point from the cable is large compared with the radius a of the circle ABC. We shall see later on that, when a^2/r^2 can be neglected compared with a/r, we get a rotating field the strength of which is $3aI/r^2$ where I is the maximum value of the current in a core.

The locus of the positions of the neutral points *inside* the cores lies outside of the circle ABC. At all points on this curve the field is purely oscillatory.

The equations to the lines of force can easily be found in terms of tripolar coordinates. Since the resultant magnetic force at any point of a line of force is by definition a tangent to the curve, it follows that the sum of the component forces resolved at right angles to a line of force vanishes. At every point on a line of force we therefore have

$$f_1 \cos \psi_1 + f_2 \cos \psi_2 + f_3 \cos \psi_3 = 0,$$

where ψ_1 is the angle between AP and the tangent to the line of force through P, etc., and thus we get

$$\frac{2i_1}{r_1} \cdot \frac{\partial r_1}{\partial s} + \frac{2i_2}{r_2} \cdot \frac{\partial r_2}{\partial s} + \frac{2i_3}{r_3} \cdot \frac{\partial r_3}{\partial s} = 0.$$

Therefore $\quad i_1 \log r_1 + i_2 \log r_2 + i_3 \log r_3 = \text{constant},$

or $\quad r_1^{i_1} r_2^{i_2} r_3^{i_3} = \text{constant}.$

For given values of i_1, i_2 and i_3 we get the equations to all the lines of force by giving different values to the constant.

Since, in practice,

$$i_1 + i_2 + i_3 = 0,$$

the equation to the lines of force may be written

$$r_1{}^{i_1} r_2{}^{i_2} = m r_3{}^{i_1 + i_2},$$

where m is a constant. To illustrate the curves represented by this equation, we shall draw the lines of force in particular cases. Suppose, for example, that

$$i_1 = i_2 = -i_3/2,$$

then, at this instant, all the lines of force are given by

$$r_1 r_2 = m r_3{}^2.$$

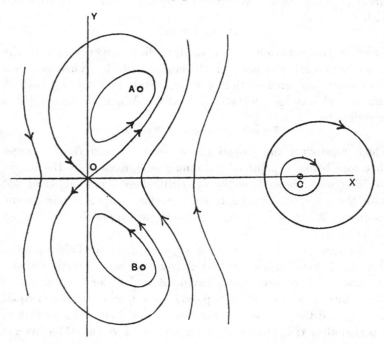

Fig. 144. Lines of force in a three phase cable when
$i_1 = i_2 = -i_3/2$.

If we take as origin (Fig. 144) a point O on the circle circumscribing A, B and C, so that O is equidistant from A and B, and if we take OC as the axis of x, the equation to the lines of force is

$$\{(x - R/2)^2 + (y - \sqrt{3}\,R/2)^2\}\,\{(x - R/2)^2 + (y + \sqrt{3}\,R/2)^2\}$$
$$= m^2\,\{(x - 2R)^2 + y^2\}^2,$$

where R is the radius of the circumscribing circle. Putting this equation into polar coordinates we get

$$(r^2 - rR \cos \theta + R^2)^2 - m^2 (r^2 - 4rR \cos \theta + 4R^2)^2 = 3r^2R^2 \sin^2 \theta.$$

Some of these curves are shown in Fig. 144. When m is greater than unity the curves are ovals round C. When m equals unity we get

$$(2r^2 - 5Rr \cos \theta + 5R^2) (r \cos \theta - R) = Rr^2 \sin^2 \theta$$

as the equation of the line of force through the centre of the circumscribing circle. When r is large this curve almost coincides with

$$2r \cos \theta = 3R,$$

which is the equation to the straight line perpendicular to the axis of x and midway between the centre and C. When m is less than unity but greater than $\frac{1}{4}$ we get ovals surrounding both A and B and having points of inflexion. When m equals $\frac{1}{4}$ the equation is

$$5r^2 = 8Rr \cos \theta - 8R^2 \cos 2\theta,$$

which represents the looped curve round A and B, the origin therefore being a point of zero magnetic force. At the origin $\cos 2\theta$ equals zero so that the tangents there make angles of 45° with the axes; the tangents are therefore perpendicular to one another. When m is less than $\frac{1}{4}$ the curves are ovals surrounding either A or B.

We can now form a picture of the magnetic field round a three core cable. Suppose that i_2 is zero, then i_1 equals $-i_3$ and the lines of force are circles having A and C for inverse points (Fig. 139). The twelfth of a period later i_2 and i_1 are each equal to $-\frac{1}{2}i_3$, and the lines of force are as in Fig. 144. The twelfth of a period after this, they will be circles round B and C as inverse points and so on. The neutral point, outside the cores, makes two complete revolutions round the circle passing through A, B and C during the period of the alternating current.

It must be remembered that in practice the cores, instead of being cylinders, are stranded cables made up of $1 + 6 + 12 + 18 + \dots$ wires, the usual numbers being nineteen and thirty-seven. The outside layers generally have a 'lay' of about twenty times the diameter, that is, the wires make a complete twist round the axis of a core in a length equal to twenty times the core's diameter. In addition, the three stranded conductors or cores are spiralled relatively to one another, each making a complete turn round the axis of the cable in about eight feet. The section of a conductor is sometimes shaped like the sector of a circle, the vertical angle of the sector being 120 degrees (Fig. 51). The main features of the field of force round the cable are, however, similar to the fields we have described above.

In this cable (see Fig. 53) we have four cores and the sections

Magnetic field round a two phase cable. of the axes by the plane of the paper are at the angular points of a square. If r_1, r_2, r_3 and r_4 be the distances of these four points from P and if i_1, i_2, i_3 and i_4 be the values of the currents at any instant, the lines of force will be given by

$$r_1^{i_1} r_2^{i_2} r_3^{i_3} r_4^{i_4} = \text{constant.}$$

During the normal working of the system, we have

$$i_1 + i_3 = 0 \text{ and } i_2 + i_4 = 0,$$

and thus the equation becomes

$$\left(\frac{r_1}{r_3}\right)^{i_1} \left(\frac{r_2}{r_4}\right)^{i_2} = \text{constant.}$$

When i_2 is zero the lines of force are circles having 1 and 3 as inverse points. An eighth of a period later all the currents are equal in magnitude and therefore

$$r_1 r_2 = m r_3 r_4$$

gives the equations to the lines of force. When m equals unity we get a straight line passing through the centre and dividing the field into two symmetrical portions. There are in general two neutral points on the circle circumscribing 1, 2, 3 and 4, and the resultant magnetic force at all other points on this circle, which are external to the cores, is normal to the circle. When i_1 equals i_2, the equation to the looped lines of force passing through the neutral points which are external to the cores is

$$r_1 r_2 = (3 \pm 2\sqrt{2}) r_3 r_4.$$

Hence it is easy to draw a rough diagram of the lines of force in this case. An eighth of a period later, the lines of force are circles having 2 and 4 as inverse points, and so on.

When the currents follow the harmonic law, the neutral points, when outside the cores, travel with uniform speed round the circle which passes through the axes of the four cores. At the instants when i_1 and i_3 or i_2 and i_4 vanish, the lines of force are circles (Fig. 139) and there are neutral points only inside the cores.

In the case of a twin concentric cable (Fig. 55), the phase *Twin con-* difference between the currents in the two inner *centric cable.* conductors is ninety degrees, and the current in the outer cylindrical conductor enveloping them is equal, at every instant, to the sum of the two inner currents. The current in the outer return conductor produces no magnetic field inside its inner radius. The magnetic field inside the outer conductor will therefore be due to the currents in the two inner conductors. The neutral point, due to these two currents, will oscillate on the line joining their axes and it is easy to draw the lines of force inside at any instant.

If O be the centre of the cable and A and B the points where the axes of the inner conductors cut the plane of the paper, O will be the middle point of AB. Let $OA = a$ and $ON = x$, where N is the neutral point outside the cable, then

$$\frac{2i_1}{x+a} + \frac{2i_2}{x-a} = \frac{2(i_1+i_2)}{x}.$$

Therefore $x = a(i_1 + i_2)/(i_1 - i_2)$.

Hence the position of the neutral point is determined and the field outside the cable can be drawn.

When i_2 equals i_1, the equation to the lines of force outside the cable is $r_1 r_2 = m \, OP^2$,

or $(r^2 + a^2)^2 - 4a^2r^2 \cos^2\theta = m^2 r^4$.

When m equals unity this gives an equiaxial hyperbola.

The equation to the lines of force in a plane perpendicular to *Field of force round n parallel wires symmetrically arranged with their axes on a circle.* the axes of the wires, which we may suppose to be the plane of the paper, is

$$r_1^{i_1} r_2^{i_2} \ldots r_n^{i_n} = \text{constant},$$

where $r_1, r_2 \ldots r_n$ are the distances of a point P in this plane from the axes of the wires, and $i_1, i_2 \ldots i_n$

are the currents in the wires. If the currents are all equal and flowing in the same direction this equation becomes

$$r_1 r_2 \ldots r_n = \text{constant.}$$

By De Moivre's property of the circle, we can write

$$\{r^{2n} - 2a^n r^n \cos n\theta + a^{2n}\}^{\frac{1}{2}} = r_1 r_2 \ldots r_n$$
$$= \text{constant,}$$

where a is the radius of the circle, r the distance of P from its centre O, and θ the angle between OP and a line joining O to one of the points of intersection of an axis of the conductor with the plane of the paper.

At the centre of the circle,

$$r_1 = r_2 = \ldots = r_n = a.$$

The equation to the lines of force through the centre is therefore

$$r^n = 2a^n \cos n\theta.$$

This gives us n loops having a multiple point at the centre, where the force is zero.

The equation to the line of force passing through the point

$$(r = a, \ \theta = \pi/n)$$

is $$r^{2n} - 2a^n r^n \cos n\theta + a^{2n} = 4a^{2n}.$$

This is a curve with n ripples, the minimum value of r being a and the maximum value being $3^{1/n}a$. If n were 40, the maximum deviation of this line of force from the circle passing through the axes of the wires would be only 2·8 per cent. of the radius of that circle.

The equation to the line of force passing through the point

$$(r = 1·5a, \ \theta = \pi/n)$$

is $$r^{2n} - 2a^n r^n \cos n\theta + a^{2n} = \{(1·5)^n + 1\}^2 a^{2n}.$$

It is easy to see that 1·5a is the minimum value of r and that $\{(1·5)^n + 2\}^{1/n} a$ is its maximum value. When n is 8 the maximum value of r is 1·514a, and hence this line of force never deviates from the circle whose radius is 1·507a by as much as the half of one per cent. of that radius.

The lines of force when there are eight wires are shown in Fig. 145, the equation to the curves being

$$r^{16} - 2a^8 r^8 \cos 8\theta + a^{16} = m^{16},$$

where m is a constant which cannot be less than a. When m is large the lines of force are practically indistinguishable from circles.

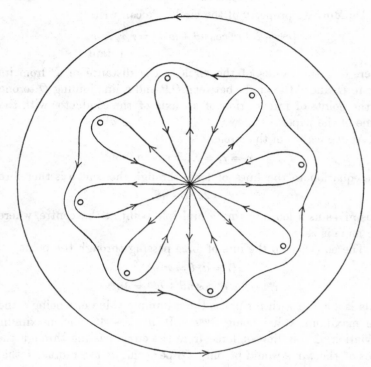

Fig. 145. Lines of force round eight wires carrying
currents flowing in the same direction.

Let the inner conductor be a circular cylinder and let the

Concentric
cable.
outer consist of n wires each of which carries an equal share of the current. Let the distance of a point P from the axis of the inner conductor be r and from the axes of the wires be $r_1, r_2, \ldots$ respectively. The equation to the lines of force will be

$$mr^n = r_1 r_2 \ldots r_n,$$

where m is a constant.

Hence, by De Moivre's property of the circle,

$$m^2 r^{2n} = r^{2n} - 2a^n r^n \cos n\theta + a^{2n}.$$

When m equals unity, we get the curves

$$r^n \cos n\theta = a^n/2.$$

It is easy to see that this represents an open symmetrical curve round each wire, the vertex of which is directed towards the centre, and the equation to the asymptotes is

$$\theta = \frac{2p+1}{2n}\,\pi,$$

where p is an integer. When n is even, r is imaginary when θ lies between $\pi/2n$ and $3\pi/2n$, $5\pi/2n$ and $7\pi/2n$, etc. When n is odd, it is negative between these values.

When m is less than unity, $\cos n\theta$ cannot be less than $\sqrt{1-m^2}$ and hence, in this case, for each value of m we get a loop round each wire, the loops getting narrower and shorter as m diminishes.

When m is greater than unity we get rippled lines of force which only embrace the central wire. The equation to one of those passing through the point

$$\theta = \pi/n, \quad r = b,$$

is $$\left\{\left(\frac{b^n + a^n}{b^n}\right)^2 - 1\right\} r^{2n} + 2a^n r^n \cos n\theta - a^{2n} = 0.$$

It follows from this equation that the minimum value of r for a given value of b is

$$\frac{ab}{\{a^n + 2b^n\}^{1/n}}.$$

Since b is less than a this is approximately equal to

$$b\{1 - (2/n)(b/a)^n\}.$$

Hence we see that when n is large and b is small the amplitude of the ripples is very small.

When b equals a, the minimum value of r is $a/3^{1/n}$. If there were eight wires (Fig. 146) then the maximum and minimum values of r for this line of force would be a and $0\cdot87a$ respectively. Hence it would differ from the circle $r = 0\cdot935a$ by less than 7 per cent. When b is greater than a then the minimum value of b is

approximately $$\frac{a}{2^{1/n}}\left(1 - \frac{a^n}{2nb^n}\right),$$

which is always less than $a/2^{1/n}$. Hence, the lines of force which are close to the point

$$r = a/2^{1/n}, \quad \theta = 0,$$

but pass inside it, have very large ripples, which extend in some cases to great distances beyond the circle $r = a$.

Fig. 146.   Concentric cable in which the outer conductor consists of eight wires.

Some of the lines of force for a concentric cable having eight wires symmetrically arranged for its outer conductor are shown in Fig. 146. If it were carrying direct current, the needle of a little compass would change its direction $2n$ times on being taken round the cable. Since, with alternating currents, the current in the inner conductor is always in exact opposition in phase to the currents in any of the outer wires, the magnetic field at any point in the neighbourhood is in this case a purely oscillatory one.

In general, if we have a series of parallel wires carrying alternating currents of any magnitudes which are always in step with one another, that is, which are either in phase or in exact opposition in phase, then the magnetic field in their neighbourhood is purely oscillatory.

The strength of the magnetic field in the two preceding problems is easily found by the following method, which furnishes an example of the use of Maxwell's Vector Potential.

The strength of the magnetic field round n parallel wires.

Let us first consider the case when the magnetic force due to the current in the return conductor is negligible. Let the current in each of the n wires be i/n and let $A, B, C, \ldots$ (Fig. 147) be the points where the axes of the wires cut the plane of the paper.

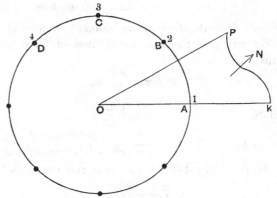

Fig. 147. $A, B, C, D, \ldots$ are the points where the axes of n long parallel wires cut at right angles the plane of the paper.

Consider two lines through P and through a fixed point K, parallel to the wires, and let N be the number of lines of force which pass between these two lines per unit length. Then, if $PA = r_1$, $PB = r_2, \ldots$, $KA = \rho_1$, $KB = \rho_2, \ldots$, $OP = r$ and $OA = a$, we have

$$N = (2i/n)\{\log(\rho_1/r_1) + \log(\rho_2/r_2) + \log(\rho_3/r_3) + \ldots \ldots\}$$
$$= -(2i/n)\log(r_1 r_2 \ldots r_n) + \lambda i,$$

where λ is a constant.

Hence, by De Moivre's property of the circle,

$$N - \lambda i = - (i/n) \log (r^{2n} - 2r^n a^n \cos n\theta + a^{2n}) \ldots\ldots\ldots(\alpha).$$

Now let R (Fig. 148) denote the radial magnetic force at the point P. If the field were uniform it would be measured by the number of lines of force, per unit area, parallel to OP. When it

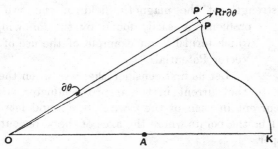

Fig. 148. R is the radial magnetic force.

is not uniform, R multiplied by an element of area at P at right angles to OP will give the number of lines of force through that area.   Thus, we have

$$R.r\partial\theta = \frac{\partial N}{\partial \theta} . \partial\theta,$$

and therefore

$$R = \frac{1}{r} \frac{\partial N}{\partial \theta}.$$

Hence, by (α),

$$R = - \frac{2i}{r} \cdot \frac{r^n a^n \sin n\theta}{r^{2n} - 2r^n a^n \cos n\theta + a^{2n}} \ldots(\beta).$$

Similarly, from Fig. 149, we see that the tangential force T is given by

$$T = - \frac{\partial N}{\partial r}.$$

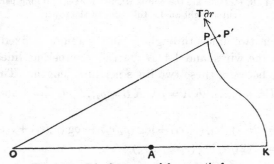

Fig. 149. T is the tangential magnetic force.

Thus, from (α), we get

$$T = \frac{2i}{r} \cdot \frac{r^{2n} - r^n a^n \cos n\theta}{r^{2n} - 2r^n a^n \cos n\theta + a^{2n}} \quad\cdots\cdots\cdots\cdots\cdots(\gamma).$$

If F be the resultant force at any point in the field,

$$F^2 = R^2 + T^2,$$

since R and T are at right angles. Therefore

$$F = \frac{2i}{r} \cdot \frac{r^n}{\{r^{2n} - 2r^n a^n \cos n\theta + a^{2n}\}^{\frac{1}{2}}}$$

$$= \frac{2i}{r} \cdot \frac{r^n}{r_1 r_2 \ldots r_n} \quad\cdots\cdots\cdots\cdots\cdots\cdots\cdots\cdots\cdots\cdots(\delta).$$

Now at all points on a line of force we have

$$r_1 r_2 \ldots r_n = \text{constant},$$

and therefore, along a line of force, F will vary as r^{n-1}. For example, along any of the lines of force shown in Fig. 145 F varies as r^7.

Again, from (γ), we see that the tangential force T, that is, the force at right angles to the radius vector, is zero at all points on the curves

$$r^n = a^n \cos n\theta.$$

These are curves exactly similar to the looped curves shown in Fig. 145, the extremities of the loops passing through the axes of the wires. If P be a point, therefore, on one of these loops, OP is the direction of the magnetic force at that point. For example, when n equals 2, we see that at any point P on the lemniscate $r^2 = a^2 \cos 2\theta$, the direction of the resultant magnetic force is OP and the magnitude of the force is proportional to OP.

In the preceding problem, if we suppose that a current $- i$ flows in a return wire through O, parallel to the n wires, we get the case of a concentric main. The effect of this return current is to add to the tangential force T, due to the currents in the n wires, the amount $- 2i/r$ and to leave the radial component R unaltered. If T' be

The strength of the field round a concentric main.

30—2

the new tangential force, we have

$$T' = T - \frac{2i}{r}$$

$$= \frac{2i}{r} \cdot \frac{r^n a^n \cos n\theta - a^{2n}}{r^{2n} - 2r^n a^n \cos n\theta + a^{2n}}.$$

This vanishes at all points on the curves

$$r^n \cos n\theta = a^n.$$

These curves are similar to the open curves shown in Fig. 146 and they pass through the axes of the wires. At any point P on these curves OP gives the direction of the magnetic force.

If F' be the resultant magnetic force at any point P,

$$F'^2 = R^2 + T'^2,$$

and thus

$$F' = \frac{2i}{r} \cdot \frac{a^n}{r_1 r_2 \ldots r_n}.$$

The magnetic force at any point on a given line of force (Fig. 146) is therefore inversely proportional to r.

Again, from (α), when r is greater than a, we get

$$N - \lambda i = -2i \log r - \frac{i}{n}\left\{\log\left(1 - \frac{a^n}{r^n}\epsilon^{n\theta\iota}\right) + \log\left(1 - \frac{a^n}{r^n}\epsilon^{-n\theta\iota}\right)\right\},$$

where $N - \lambda i$ refers to the n wires only. Thus we have

$$N - \lambda i = -2i \log r + \frac{i}{n}\left(\frac{a^n}{r^n}\epsilon^{n\theta\iota} + \frac{1}{2}\cdot\frac{a^{2n}}{r^{2n}}\epsilon^{2n\theta\iota} + \ldots\right)$$

$$+ \frac{i}{n}\left(\frac{a^n}{r^n}\epsilon^{-n\theta\iota} + \frac{1}{2}\cdot\frac{a^{2n}}{r^{2n}}\epsilon^{-2n\theta\iota} + \ldots\right)$$

$$= -2i \log r + \frac{2i}{n}\left(\frac{a^n}{r^n}\cos n\theta + \frac{1}{2}\cdot\frac{a^{2n}}{r^{2n}}\cos 2n\theta + \ldots\right).$$

Similarly, when r is less than a, we may write

$$N - \lambda i = -2i \log a + \frac{2i}{n}\left(\frac{r^n}{a^n}\cos n\theta + \frac{1}{2}\cdot\frac{r^{2n}}{a^{2n}}\cos 2n\theta + \ldots\right).$$

Therefore, when r is greater than a,

$$R = \frac{1}{r}\frac{\partial N}{\partial \theta} = -\frac{2i}{r}\left(\frac{a^n}{r^n}\sin n\theta + \frac{a^{2n}}{r^{2n}}\sin 2n\theta + \ldots\right).$$

and

$$T' = -\frac{\partial N}{\partial r} - \frac{2i}{r} = \frac{2i}{r}\left(\frac{a^n}{r^n}\cos n\theta + \frac{a^{2n}}{r^{2n}}\cos 2n\theta + \ldots\right).$$

Similarly when r is less than a,

$$R = -\frac{2i}{r}\left(\frac{r^n}{a^n}\sin n\theta + \frac{r^{2n}}{a^{2n}}\sin 2n\theta + \dots\right),$$

and $$T' = -\frac{2i}{r} - \frac{2i}{r}\left(\frac{r^n}{a^n}\cos n\theta + \frac{r^{2n}}{a^{2n}}\cos 2n\theta + \dots\right).$$

The series given above show clearly the degree of approximation of the field to that due to a longitudinal current uniformly distributed over the surface of a hollow cylinder and returning by a thin coaxial solid conductor.

If we have an alternating current, of effective value A, flowing
The losses in in a conductor of resistance R, then RA^2 is the
cables. least possible value of the mean power lost in the
conductor.   Even when the conductor is the inner conductor of a concentric main so that it is shielded from the inductive effects of the return current, yet owing to the skin effect (p. 216) the power expended in it is greater than RA^2.   In polyphase cables the wires that form the cores are not shielded from the inductive effects of the currents in neighbouring wires, or of the currents in neighbouring cores and so, except in the special case when the cores are made up of fine wires not too tightly pressed together, the eddy current losses in them are appreciable.   Let us first consider the simple problem of single phase mains.

When the consumer's terminals are connected with the mains of an alternating current supply company by two conductors run through two separate iron pipes, it is found that, at times of heavy load, there is a great diminution of the voltage between the consumer's terminals.   The presence of the iron magnifies very considerably the back electromotive force due to the inductance of the conductors, and thus the potential difference V_1 between the consumer's terminals is much smaller than the voltage V between the mains.   In this case the potential difference V_2 between the ends of a conductor joining a main to a house terminal, that is, the 'voltage drop' along the conductor, will be large.   In general the phase difference between the voltage drop and the current in a conductor is large.   It will only be small in the exceptional case when the eddy current and hysteresis loss is large.

If the load in the house circuit consist of incandescent lamps so that its power factor is practically unity and if, as is generally the case in practice, the connecting conductors be equal and the losses in them are the same and take place in the same manner, we can easily measure W, the power lost in the conductors and in the pipes surrounding them. Since the voltage drop along one conductor is in phase with that along the other, their resultant value is $2V_2$ and so the power W is given by the three voltmeter formula (p. 330). We have, therefore,

$$W = (A/2V_1)\{V^2 - V_1^2 - (2V_2)^2\},$$

where A is the reading of an alternating current ammeter in the circuit. If we subtract $2A^2R$ from W, where R is the resistance of either of the connecting conductors, we get the power expended in hysteresis and eddy currents in the pipes.

As V, V_1 and V_2 differ in phase, $V - V_1$ is generally much smaller than $2V_2$. For example, we could have

$$V = 260, \quad V_1 = 240, \quad V_2 = 40, \quad A = 100.$$

In this case the value of W, in watts, is given by

$$W = \frac{100}{2 \times 240}\{260^2 - 240^2 - (2 \times 40)^2\} = 750.$$

Since W must be positive we see that when V is 260 and V_1 is 240, V_2 must be less than 50. This follows since

$$260^2 - 240^2 = (2 \times 50)^2.$$

It has been noticed that when the iron pipes are insulated from the earth the voltage drop V_2 for a given current A is less than when the pipes are laid in the earth. This shows that the eddy currents are increased when the pipe is earthed throughout its length.

To avoid this excessive drop in the pressure, the two conductors are put into the same pipe. This is found, in practice, to be a satisfactory solution. Let us suppose that the axes of the conductors and the axis of the pipe are in one plane and that these axes are parallel. Let us also suppose that the pipe is of very thin metal so that the distortion of the field produced by the eddy currents in it can be neglected. The lines of force will then practically coincide with the circles shown in Fig. 139, and an inspection of this figure will show that the plane containing the straight lines of force bisects the pipe. If the pipe be long, then at any instant the induced current will be flowing down one half of the pipe and back along the other half.

The magnetic force at any point P can be easily found. Let N be the number of lines of force per unit length of the system between P and a point on the sheath equidistant from the two wires. Then, as on p. 448, we can easily show that

$$N = -i \log (r^2 - 2ar \cos \theta + a^2) + i \log (r^2 + 2ar \cos \theta + a^2),$$

where i is the current in either conductor, $2a$ the distance between the axes of the conductors and (r, θ) are the polar coordinates of the point P, the origin being on the axis of the pipe and the initial line passing through the axis of a conductor. Thus, if R and T be the radial and tangential components respectively of the magnetic force at P, we get

$$R = \frac{1}{r} \cdot \frac{\partial N}{\partial \theta}$$

$$= -\frac{2i}{r} \cdot \frac{2ar \sin \theta (r^2 + a^2)}{r^4 - 2a^2 r^2 \cos 2\theta + a^4},$$

and

$$T = -\frac{\partial N}{\partial r}$$

$$= \frac{2i}{r} \cdot \frac{2ar \cos \theta (r^2 - a^2)}{r^4 - 2a^2 r^2 \cos 2\theta + a^4}.$$

Hence also the resultant force F is given by

$$F^2 = R^2 + T^2,$$

and thus

$$F = 4ai/(r_1 r_2),$$

a result which agrees with the formula given on p. 444.

The following method of considering the eddy current losses in a pipe, when it is insulated from earth, is instructive. We have seen that the number N of the lines of force between a point $P(r, \theta)$ on the pipe and a point equidistant from the axes of the two conductors is given by

$$N = i \log (r^2 + 2ar \cos \theta + a^2) - i \log (r^2 - 2ar \cos \theta + a^2),$$

when the walls of the pipe are very thin and the metal of which it is made is non-magnetic. Using the method employed on p. 468 and noticing that r is greater than a, we have

$$N = 2i\left\{\frac{a}{r}\cos\theta - \frac{1}{2}\frac{a^2}{r^2}\cos 2\theta + \dots\right\}$$

$$+ 2i\left\{\frac{a}{r}\cos\theta + \frac{1}{2}\frac{a^2}{r^2}\cos 2\theta + \dots\right\}$$

$$= 4i\left\{\frac{a}{r}\cos\theta + \frac{1}{3}\frac{a^3}{r^3}\cos 3\theta + \frac{1}{5}\frac{a^5}{r^5}\cos 5\theta + \dots\right\}.$$

Thus since, by symmetry, the current vanishes at the points on the pipe equidistant from the two wires, it follows that if u be the current density at P and ρ the resistivity of the pipe, we have, making the assumption that every tube of eddy current is in quadrature with the flux it embraces,

$$\rho u = -\frac{\partial N}{\partial t}$$

$$= -4\frac{\partial i}{\partial t}\left\{\frac{a}{r}\cos\theta + \frac{1}{3}\frac{a^3}{r^3}\cos 3\theta + \dots\right\}.$$

If i equal $I\cos\omega t$, the mean value of $(\partial i/\partial t)^2$ is $\frac{1}{2}\omega^2 I^2$ and thus the mean value of $\rho^2 u^2$ is

$$8\omega^2 I^2\frac{a^2}{r^2}\left\{\cos\theta + \frac{1}{3}\frac{a^2}{r^2}\cos 3\theta + \dots\right\}^2.$$

Hence, if the thickness of the sheath be h, we find that the mean power W lost, per unit length of the sheath, is given by

$$W = \frac{8\omega^2 I^2 a^2 h}{r^2\rho}\int_0^{2\pi}\left\{\cos\theta + \frac{1}{3}\frac{a^2}{r^2}\cos 3\theta + \dots\right\}^2 r\partial\theta$$

$$= \frac{16\pi\omega^2 A^2 a^2 h}{r\rho}\left\{1 + \frac{1}{9}\frac{a^4}{r^4} + \frac{1}{25}\frac{a^8}{r^8} + \dots\right\},$$

where A is the effective value of the current. Thus, if $(a^2/3r^2)^2$ be negligible compared with unity, the eddy current loss W in the pipe, in ergs per second, is given by

$$W = (4Aa\omega)^2\frac{\pi h}{r\rho}l,$$

where l is the length of the pipe, and all the quantities are measured in absolute units. If A be in amperes, the eddy current loss in watts is

$$(4Aa\omega)^2(\pi hl/r\rho)10^{-9}.$$

We see that, when the reactance of every eddy current path is negligible compared with its resistance, the eddy current loss, per unit length, varies as the square of the current, the square of the distance between the axes of the wires and the square of the frequency. It also varies directly as the thickness of the pipe and inversely as its radius and its resistivity.

Let us now consider the hysteresis loss in a very thin iron pipe. If r be the radius of the pipe, the magnetic force F at a point P on its circumference is given by

$$F = \frac{4ai}{r_1 r_2}$$

$$= \frac{4ai}{(r^4 - 2a^2 r^2 \cos 2\theta + a^4)^{\frac{1}{2}}},$$

and thus by p. 18

$$F = \frac{4ai}{r^2} \left\{ 1 + \cos 2\theta \, \frac{a^2}{r^2} + \left(\frac{3}{2} \cos^2 2\theta - \frac{1}{2} \right) \frac{a^4}{r^4} + \ldots \right\}.$$

It can be shown that this series is convergent when r is greater than a. F has its maximum value when θ is zero and its minimum value when θ is $\pi/2$. Thus the amplitudes of F in C.G.S. measure at points in the pipe lie between

$$\frac{4I_{\max.} \, a}{r^2 - a^2} \quad \text{and} \quad \frac{4I_{\max.} \, a}{r^2 + a^2},$$

where $I_{\max.}$ denotes the maximum value of the current i. If we know Steinmetz's coefficient for the iron (p. 61) and the permeability curve, we can easily find a superior limit to the hysteresis loss.

The following data concerning three core cables and the
Three core cables. method of their manufacture will show the nature of the problems that arise in practical work. The cores consist of stranded conductors; a 19/14 core, for example, would be one made up of 19 strands of wire of No. 14 Standard Wire Gauge, that is, wire with a diameter of 0·080 of an inch. Strips of paper are wound round the cores, and the cores are spiralled relatively to one another making a complete turn round the axis in from four to eight feet, the space between the three cores being filled either with jute or with a paper core. The

cores are now wrapped together with more strips of paper and, after immersion in a bath of a special compound, are passed to the lead covering press. In the low temperature process, the lead covering is put on under very considerable pressure and at a temperature well below the melting point of lead. In most cases a covering of compounded tape or jute yarn is laid on (served) over the lead by a special machine. The armouring comes over the compounded tape and generally consists of galvanised iron wires laid on longitudinally with sufficient 'spiral' to keep them together. In some cases this armouring is served with compounded jute. The cross section of a 19/14 core is nearly 0·1 of a square inch, so that if we assume that a current density of 800 amperes per square inch is permissible, then each core can carry 80 amperes.

In some cases the armouring consists of steel ribbon or strip. This strip is put on in two layers, the first layer being spiralled in one direction and the next in the opposite direction, thus 'breaking joint' and affording efficient protection for the cable. This is the standard armouring for ordinary direct current cables, but reasons will be given later on why it should not be used for three core cables, unless the hysteresis losses in it when subjected to appreciable magnetising forces are negligible.

The minimum distance of the cores from one another and from the sheath depends on the voltage for which the cable is designed. For voltages up to 5000 it is customary to allow a thickness of 0·05 of an inch of dielectric for every thousand volts. For higher pressures the thicknesses used in practice are less than that given by this rule.

If the cable be for 11,000 volts mesh working, the minimum distance between the cores is generally about 0·4 of an inch. The inner diameter of the lead sheath would be slightly less than 2·5 inches if the cores were 0·2 of a square inch in section and would be about 3 inches if the cores were 0·4 of a square inch in section. The thickness of the lead is generally about 0·16 of an inch (160 mils). The jute between the lead and the armouring is about 50 mils in thickness but in some cables the armouring is placed directly on the lead. When steel strip is used it is generally about 40 mils thick and as there are two strips one over the other this gives a steel cylinder 80 mils thick surrounding the cable and having a diameter somewhat greater than three inches for large cables. If this be covered with jute, the thickness of the jute will be about 100 mils (0·254 cm.).

The section of the cores, in the smallest of the three core cables used by the Underground Electric Railway Company of London, is 0·15 of a square inch and, in the largest, 0·25 of a square inch. The effective voltage between any pair of cores is 11,000 and the minimum distance between the cores is 0·44 of an inch. The coefficient of self induction for electrostatic charges of each core in the smallest size of cable is 0·26 of a microfarad per mile, and in the cable which has cores 0·25 of a square inch in section the coefficient of each core is about 0·3 of a microfarad per mile. The cables are plain lead covered.

On the power station circuit of the Metropolitan Electric Tramways Company of London, the three core cables are sheathed under the lead with copper tape as an earthing shield for protective purposes, and are drawn into earthenware ducts. The cores of most of the cables are 0·1 of a square inch in section, and the minimum distance between the cores is 0·38 of an inch. The coefficient of each core in this class of cable is about 0·27 of a microfarad per mile.

The following are the principal data for a three core cable, for 11,000 volt mesh working, made by the British Insulated Wire Company. The section of each core is roughly similar to the segment of a circle which has a vertical angle of 120 degrees. There are 37 strands of wire in a core, the diameter of each wire being 0·082 of an inch. Each core is wrapped round with paper to a radial depth of 0·18 of an inch and the minimum distance between the cores is about 0·37 of an inch. Over this another layer of paper 0·18 inch thick is wrapped. The radial thickness of the lead sheath is 0·16 of an inch and its outer diameter is 2·5 inches. The lead in this case is served with compounded jute and over this ninety galvanised iron wires are laid, each 0·104 of an inch in diameter. Only sufficient 'spiral' is given to the wires to keep them together. Wire armouring is never put on by coiling the wire round the cable.

The weight of each of the copper cores in this cable is

about 4000 lbs. per mile. The weight of the lead sheath is over thirteen tons per mile and the weight of the wire armouring is about 10,000 lbs. per mile. The coefficient of self induction for electrostatic charges of each core is approximately 0·37 of a microfarad per mile.

In many situations, as for example in mines, mechanical protection is necessary for three core cables. They are therefore either drawn into iron pipes or heavily armoured. When this is done a copper shield under the lead is in general unnecessary.

In addition to the eddy current losses in the cores themselves and in the lead sheath, there may be considerable hysteresis and eddy current losses in the armouring. It is important therefore to know the value of the magnetic forces due to the currents at points near a three phase cable.

Formulae for the magnetic forces due to three phase cables. We shall now find the direction and the magnitude of the magnetic force produced at any point by the currents in the cores of a three phase cable. We shall assume that the load is balanced, and that the cores can be regarded as cylinders whose axes are parallel. Our formulae will be approximately true when the conductors are spiralled. Let i_1, i_2, and i_3 be the currents in the cores. With the notation of p. 465, if we choose the centre of the cable as K, we have

$$N = 2i_1 \log a + 2i_2 \log a + 2i_3 \log a$$
$$- 2i_1 \log r_1 - 2i_2 \log r_2 - 2i_3 \log r_3$$
$$= - 2i_1 \log r_1 - 2i_2 \log r_2 - 2i_3 \log r_3,$$

since $i_1 + i_2 + i_3 = 0$.

Thus, we get

$$N = - 2 (i_1 + i_2 + i_3) \log r$$
$$- i_1 \log \left\{ 1 - 2\frac{a}{r} \cos \theta + \frac{a^2}{r^2} \right\}$$
$$- i_2 \log \left\{ 1 - 2\frac{a}{r} \cos \left(\theta - \frac{2}{3}\pi \right) + \frac{a^2}{r^2} \right\}$$
$$- i_3 \log \left\{ 1 - 2\frac{a}{r} \cos \left(\theta - \frac{4}{3}\pi \right) + \frac{a^2}{r^2} \right\},$$

which by p. 87 $= 2i_1 \left\{ \dfrac{a}{r}\cos\theta + \dfrac{1}{2}\dfrac{a^2}{r^2}\cos 2\theta + \dfrac{1}{3}\dfrac{a^3}{r^3}\cos 3\theta + \ldots \right\}$

$\qquad + 2i_2 \left\{ \dfrac{a}{r}\cos\left(\theta - \dfrac{2}{3}\pi\right) + \dfrac{1}{2}\dfrac{a^2}{r^2}\cos 2\left(\theta - \dfrac{2}{3}\pi\right) + \ldots \right\}$

$\qquad + 2i_3 \left\{ \dfrac{a}{r}\cos\left(\theta - \dfrac{4}{3}\pi\right) + \dfrac{1}{2}\dfrac{a^2}{r^2}\cos 2\left(\theta - \dfrac{4}{3}\pi\right) + \ldots \right\}.$

From this equation R and T can be found at once by the formulae

$$R = \frac{1}{r}\frac{\partial N}{\partial\theta} \quad \text{and} \quad T = -\frac{\partial N}{\partial r}.$$

In order to simplify the formulae for the radial and tangential forces we shall suppose that the currents follow the harmonic law, so that we can write

$i_1 = I\cos\omega t,\quad i_2 = I\cos(\omega t - 2\pi/3)\quad \text{and}\quad i_3 = I\cos(\omega t - 4\pi/3).$

Now, we have

$2\cos\omega t\cos p\theta$

$\qquad = \cos(\omega t + p\theta) + \cos(\omega t - p\theta),$

$2\cos(\omega t - 2\pi/3)\cos p\,(\theta - 2\pi/3)$

$\qquad = \cos\{\omega t + p\theta - 2\,(p+1)\,\pi/3\} + \cos\{\omega t - p\theta + 2\,(p-1)\,\pi/3\},$

and

$2\cos(\omega t - 4\pi/3)\cos p\,(\theta - 4\pi/3)$

$\qquad = \cos\{\omega t + p\theta - 4\,(p+1)\,\pi/3\} + \cos\{\omega t - p\theta + 4\,(p-1)\,\pi/3\}.$

If S be the sum of these quantities, we have

$$S = \frac{\cos\{\omega t + p\theta - 2\,(p+1)\,\pi/3\}\sin(p+1)\,\pi}{\sin(p+1)\,\pi/3}$$

$$+ \frac{\cos\{\omega t - p\theta + 2\,(p-1)\,\pi/3\}\sin(p-1)\,\pi}{\sin(p-1)\,\pi/3}.$$

Now, the value of $\sin x\pi/\sin(x\pi/3)$ is zero, when x is neither zero nor an integral multiple of 3. As x approaches any integral multiple of 3 the quantity approaches the limit 3. Using these results we obtain

$$N = 3I\left\{ \frac{a}{r}\cos(\omega t - \theta) + \frac{1}{2}\frac{a^2}{r^2}\cos(\omega t + 2\theta) \right.$$

$$+ \frac{1}{4}\frac{a^4}{r^4}\cos(\omega t - 4\theta) + \frac{1}{5}\frac{a^5}{r^5}\cos(\omega t + 5\theta)$$

$$\left. + \frac{1}{7}\frac{a^7}{r^7}\cos(\omega t - 7\theta) + \ldots\ldots\ldots\ldots \right\}.$$

Hence we find the following value for R

$$R = \frac{1}{r}\frac{\partial N}{\partial \theta}$$

$$= \frac{3I}{r}\left\{\frac{a}{r}\sin(\omega t - \theta) - \frac{a^2}{r^2}\sin(\omega t + 2\theta)\right.$$

$$+ \frac{a^4}{r^4}\sin(\omega t - 4\theta) - \frac{a^5}{r^5}\sin(\omega t + 5\theta)$$

$$\left. + \ldots\ldots\ldots\ldots - \ldots\ldots\ldots\ldots \right\}.$$

Putting their exponential values for $\sin\theta$, $\cos\theta$, $\sin 2\theta$, etc. we get simple geometrical progressions for the coefficients of $\sin\omega t$ and $\cos\omega t$. Summing the series and replacing the exponential values by sines and cosines, we find that

$$R = \frac{3I}{a} \cdot \frac{a^2\left\{(r^4 + a^4)\sin(\omega t - \theta) - ar(r^2 + a^2)\sin(\omega t + 2\theta)\right\}}{r^6 - 2a^3 r^3 \cos 3\theta + a^6}.$$

It can be shown also, that this formula is true when r is equal to or less than a. Again, we have

$$T = -\frac{\partial N}{\partial r}$$

$$= \frac{3I}{r}\left\{\frac{a}{r}\cos(\omega t - \theta) + \frac{a^2}{r^2}\cos(\omega t + 2\theta)\right.$$

$$+ \frac{a^4}{r^4}\cos(\omega t - 4\theta) + \frac{a^5}{r^5}\cos(\omega t + 5\theta)$$

$$\left. + \ldots\ldots\ldots\ldots + \ldots\ldots\ldots\ldots \right\}$$

$$= \frac{3I}{a} \cdot \frac{a^2\left\{(r^4 - a^4)\cos(\omega t - \theta) + ar(r^2 - a^2)\cos(\omega t + 2\theta)\right\}}{r^6 - 2a^3 r^3 \cos 3\theta + a^6}.$$

This formula is also true when r is less than or equal to a. We see at once that T vanishes when r equals a. The amplitude of T equals

$$\frac{3I}{a} \cdot \frac{a^2(r^2 - a^2)\left\{(r^2 + a^2)^2 + 2ar(r^2 + a^2)\cos 3\theta + a^2 r^2\right\}^{\frac{1}{2}}}{r^6 - 2a^3 r^3 \cos 3\theta + a^6}.$$

For a given value of r this has obviously its greatest value $T_{\max}$.

when $\cos 3\theta$ is 1 and its least value $T_{\min.}$ when $\cos 3\theta$ is -1. Thus we have

$$T_{\max.} = \frac{3I}{a} \cdot \frac{a^2(r+a)}{r^3 - a^3},$$

and

$$T_{\min.} = \frac{3I}{a} \cdot \frac{a^2(r-a)}{r^3 + a^3}.$$

It is easy to show that $T_{\min.}$ has a maximum value when r is $1·679\,a$ and $T_{\min.}$ is then equal to $(3I/a) \times 0·1184$.

An inspection of the following table will show how the coefficients of $\dfrac{3I}{a}$ in the formulae for $T_{\max.}$ and $T_{\min.}$ vary as r increases.

$\dfrac{r}{a}$	1	1·5	2	2·5	3	4	10	100
$\dfrac{a^2(r+a)}{r^3 - a^3}$	∞	1·053	0·429	0·239	0·154	0·079	0·011	0·000
$\dfrac{a^2(r-a)}{r^3 + a^3}$	0	0·114	0·111	0·090	0·071	0·046	0·009	0·000

The formulae for R and T given above show us that at all points which are outside the three cores the field is elliptical. Along the axis of the cable the ellipse is a circle, at points on the circle passing through the axes of the three cores T vanishes and thus the ellipse becomes a straight line pointing to the centre of the circle, and when a^2/r^2 can be neglected in comparison with a/r we again get a pure rotating field the strength of which is $3aI/r^2$.

If O be the centre of the section of the cable, A the point where the axis of the core cuts this section and P a point on OA or OA produced, the values of R and T at P are given by

$$R = \frac{3I}{a} \cdot \frac{a^2(r-a)}{r^3 - a^3} \sin \omega t,$$

and

$$T = \frac{3I}{a} \cdot \frac{a^2(r+a)}{r^3 - a^3} \cos \omega t$$

$$= T_{\max.} \cos \omega t.$$

Thus at all points on OA or OA produced the elliptical field has its major axis perpendicular to OA and equal to $2T_{\max.}$.

Similarly, if OP' bisect the angle BOA where B is a point on the axis of another core, the field at a point P on OP', at a distance r from O, is given by

$$R = \frac{3I}{a} \cdot \frac{a^2(r+a)}{r^3+a^3} \sin(\omega t - \pi/3),$$

and

$$T = \frac{3I}{a} \cdot \frac{a^2(r-a)}{r^3+a^3} \cos(\omega t - \pi/3)$$

$$= T_{\min.} \cos(\omega t - \pi/3).$$

The minor axis of the ellipse representing the field at P equals $2T_{\min.}$ and its direction is perpendicular to OP.

The following table shows how the coefficients of $3I/a$ in the expression for the radial force vary as we move in the direction OA or in the direction OP.

$\dfrac{r}{a}$	1	1·5	2	2·5	3	4	10	100
$\dfrac{a^2(r-a)}{r^3-a^3}$	0·333	0·211	0·143	0·103	0·077	0·048	0·009	0·000
$\dfrac{a^2(r+a)}{r^3+a^3}$	1	0·571	0·333	0·211	0·143	0·077	0·011	0·000

If f be the value of the resultant force at the point (r, θ), we have

$$f^2 = R^2 + T^2$$

$$= \left(\frac{3I}{a}\right)^2 \cdot \frac{a^4\{a^2 + r^2 + 2ar\cos(2\omega t + \theta)\}}{r^6 - 2a^3r^3\cos 3\theta + a^6},$$

which agrees with the result obtained on p. 452. Since the maximum value of $\cos 3\theta$ is 1 and its minimum value is -1, we find that the maximum and minimum values of f at points at a distance r from the axis of the cable are given by

$$f_{\max.} = \frac{3I}{a} \cdot \frac{a^2(r+a)}{r^3-a^3} = T_{\max.}$$

and

$$f_{\min.} = \frac{3I}{a} \cdot \frac{a^2(r-a)}{r^3+a^3} = T_{\min.}.$$

Let us now consider the hysteresis loss in the steel armouring of a three core cable. We shall suppose that the armour consists of very thin steel strip, and we shall neglect the shielding effect of the eddy currents induced in the lead sheath. The integral of the tangential force at any instant taken round the steel cylinder is zero, the lines of induction therefore leave the cylinder and demagnetising effects are produced. If we calculate the hysteresis loss on the assumption that there are no demagnetising effects and that $T_{max.}$ acts on every point of the armouring, we get a superior limit to the hysteresis losses. If A be the effective value of the alternating current in a core, expressed in amperes, $T_{max.}$ will be given in c.g.s. units by the formula

$$T_{max.} = 3\frac{A\sqrt{2}}{10a}\cdot\frac{a^2\,(r+a)}{r^3 - a^3}.$$

As an example we shall take the case of a three core cable designed for 10,000 volt working and having a normal current of about 150 amperes in each core. The area of the cross section of each core is $1\cdot25$ square centimetres, the radius a of the circle circumscribing the axes of the cores $1\cdot8$ cms. and the inner radius of the steel armouring $4\cdot5$ cms. If the effective current in each core be A amperes, the maximum value of the magnetic force in c.g.s. units is $0\cdot056\,A$. If, therefore, A were 150 amperes, the maximum value of the magnetic force would be $8\cdot4$ c.g.s. units. Hence, if we know the permeability curve and Steinmetz's coefficient for the steel (p. 60), we can assign a superior limit to the hysteresis loss.

We have seen that the minimum value of the tangential force at a distance r from the axis is given by the formula

$$T_{min.} = 3\frac{A\sqrt{2}}{10a}\cdot\frac{a^2\,(r-a)}{r^3 + a^3}.$$

For the cable considered above we get

$$T_{min.} = 0\cdot021\,A$$

$$= 3\cdot2 \text{ c.g.s. units,}$$

when A is 150 amperes. This, however, does not enable us to fix a minimum value to the tangential force acting on the steel as we cannot neglect the considerable demagnetising effect produced by the lines of induction leaving the steel.

If we make the assumption that the mean value of the

R. I.   31

maximum flux density is 1000 and take $\eta = 0.002$, we see, by p. 60, that the loss W, in watts per kilogramme of the steel, when the frequency is 50, is given by

$$W = \eta f V B_{\text{max.}}^{1.6} \times 10^{-7}$$
$$= 0.002 \times 50 \times \frac{1000}{7.8} \times (1000)^{1.6} \times 10^{-7}$$
$$= 0.08 \text{ nearly,}$$

assuming that the specific gravity of steel is 7·8. If the steel armouring therefore have a mass of 2000 kilogrammes per mile, this would give a loss of 0·16 of a kilowatt per mile.

When iron wire is used, with its length parallel to the axis of the cable, the magnetic flux produced will be very much smaller than with steel strip owing to the very large demagnetising effects produced by the flux leaving the wires. The hysteresis loss with wire armouring will therefore be quite negligible.

In addition to the hysteresis losses, there are eddy current losses in the cores themselves, in the lead sheath, in the copper earth shield and in the armouring. We shall see in the next chapter that the currents generated near the surface of the sheath screen off the magnetic force from the interior, and thus in the simplest cases even with non-magnetic metals, the calculation of the eddy current losses is difficult. It has to be remembered, in estimating possible eddy current losses in sheaths, that these losses do not vary in any simple manner with the resistivity of the metal of which the sheath is made.

In addition to the losses that take place in the metallic parts Dielectric losses. of the cable, we may have losses in the dielectric itself due to the electrostatic forces. When the cores are very thin it is not difficult to find formulae for the resultant electrostatic force at points in the dielectric by the method of images (p. 191). Let us consider, for example, the electrostatic force R at the axis of the cable. We suppose that all the electrical quantities are given in electrostatic units. The component R_1 of the force in the direction OA, due to the wire whose axis is at A and the image of this wire at A', is given by (p. 161)

$$\lambda R_1 = -\frac{2q_1}{a} + \frac{2q_1}{a'},$$

where $OA = a$, and $OA' = a'$ and λ is the dielectric coefficient. Now when the load is balanced (p. 178), so that

$$v_1 + v_2 + v_3 = 0,$$

then $\qquad q_1 = (K_{1.1} - K_{1.2})\,v_1,$

and thus, since R_1, R_2 and R_3 are inclined to each other at angles of $120°$, we get, by Statics,

$$R^2 = R_1^2 + R_2^2 + R_3^2 - R_2R_3 - R_3R_1 - R_1R_2,$$

and hence, on substituting, we get

$$\lambda^2 R^2 = 4\left(\frac{1}{a} - \frac{1}{a'}\right)^2 (K_{1.1} - K_{1.2})^2 \{v_1^2 + v_2^2 + v_3^2 - v_2v_3 - v_3v_1 - v_1v_2\}$$

$$= 6\left(\frac{1}{a} - \frac{a}{r^2}\right)^2 (K_{1.1} - K_{1.2})^2 (v_1^2 + v_2^2 + v_3^2),$$

since $aa' = r^2$, where r is the inner radius of the sheath.

When the potential differences follow the harmonic law, we get

$$\lambda R = 3\left(\frac{1}{a} - \frac{a}{r^2}\right)(K_{1.1} - K_{1.2})\,E,$$

where E is the maximum value of v_1 and therefore of the star pressure.

Let us take the case of a cable working with an effective pressure of 11,000 volts between the cores. Let $a = 1\cdot5$, $r = 4\cdot5$ and let the capacity between two of the cores be $0\cdot15$ of a microfarad per mile. Then, by p. 176,

$\frac{1}{2}(K_{1.1} - K_{1.2}) = 0\cdot15$ of a microfarad per mile, and thus

$$(K_{1.1} - K_{1.2})\,E = 0\cdot3 \times 10^{-15} \times 11,000 \sqrt{\tfrac{2}{3}} \times 10^8$$

c.g.s. electromagnetic units per mile

$$= 0\cdot3 \times 10^{-7} \times 11,000 \sqrt{\frac{2}{3}} \times \frac{3 \times 10^{10}}{160,900}$$

$$= 50\cdot2 \text{ c.g.s. electrostatic units per centimetre,}$$

and thus $\qquad \lambda R = 89\cdot3.$

If we assume that λ is $2\cdot8$ (see p. 177), we get

$$R = 31\cdot9 \text{ dynes per unit charge nearly.}$$

This corresponds to a rate of variation of the potential, at the axis of the cable, in the direction of R, equal to $31\cdot9 \times 300 \times 2\cdot8$, that is, 26,800 volts per centimetre. If the frequency of the alternating current be 25, then R will make 25 revolutions per second.

31—2

Hitherto we have made the assumptions that the dielectric has zero conductivity and that no power is being expended on it. As the 'conductivity' of dielectrics is exceedingly small, the former assumption is permissible in practically every case. Experiments with alternating pressures, however, prove that with ordinary dielectrics there is a considerable expenditure of power at high frequencies. Fleming and Dyke have shown that the power expended in the dielectric may be expressed in the form

$$a V^2 + b f V^2,$$

where V is the effective value of the applied potential difference, f the frequency, and a and b are functions, independent of V and f, which vary with the temperature. If we suppose that the dielectric has a conductance a, then $a V^2$ would be the loss in accordance with Joule's law. No satisfactory cause has yet been assigned for the loss $b f V^2$. Engineers often refer to this loss as the loss due to 'dielectric hysteresis,' but it has not been proved that it is merely due to a lagging of the polarization behind the electric intensity. The interiors of the dielectrics used in practice are very far from being homogeneous, and the loss is probably due to a variety of causes. Even with direct pressures the phenomena are very difficult to explain. For example the apparent resistance of most of the insulating materials used in practice diminishes as the applied electric pressure is increased. This is probably due to the effects produced by the moisture present in the material.

The assumption is sometimes made that an ordinary condenser is equivalent to an ideally perfect condenser K in series with a resistance r both being shunted by a resistance R. In this case if i_1 be the current in r, i_2 the current in R and i the total current, we get

$$i = i_1 + i_2 = e/(r + 1/KD) + e/R$$

$$= \left\{ \frac{1}{R} + \frac{\omega^2 K^2 r}{1 + \omega^2 K^2 r^2} \right\} e + \frac{KD}{1 + \omega^2 K^2 r^2} e.$$

If this were true, therefore, the power losses in the condenser would be given by

$$\frac{V^2}{R} + \frac{\omega^2 K^2 r}{1 + \omega^2 K^2 r^2} V^2.$$

But this is not of the form $aV^2 + bfV^2$ as $\omega^2 K^2 r/(1 + \omega^2 K^2 r^2)$ can only be proportional to f when r varies as $1/f$ and this is contrary to the hypothesis. The assumption, therefore, leads to results which are not in accordance with experiment and is inadmissible.

Fleming and Dyke have found that, as a rule, the apparent conductivity $a + bf$ of dielectrics increases with rise of temperature. Vulcanised rubber between $-30°$ C. and $20°$ C., gutta-percha above $10°$ C. and celluloid between $-50°$ C. and $0°$ C. are, however, important exceptions, the conductivity decreasing with rise of temperature. They also found that the presence of moisture in the dielectric increases the value of a.

It follows from the formulae we have found for the magnetic force near a three phase cable with straight cores, that there will be an appreciable disturbance set up in telephone circuits if the wires are near the cable and are not twisted together. When the cable is armoured the disturbance will be less. As the armouring is always thin compared with the diameter of the cable and the radial magnetic forces are appreciable, the diminution of the disturbance due to the armouring may, however, only be slight. The intensity of the disturbance at distances greater than a metre from the cable is approximately inversely proportional to the square of the distance, provided that the going and return wires of the telephone circuit be close together.

The formulae we have found for the magnetic force round three core cables also apply to the forces round overhead wires carrying three phase currents, provided that the wires are equidistant from one another. In this case, if the telephone wires be carried on the same posts as the wires for the three phase currents and be not twisted together, a continual hum will be heard in the telephone. This will be slight if the mains are spiralled, and can be made negligible by 'crossing' the telephone wires at regular intervals, that is, making the higher wire the lower wire for the next section and so on. In the event of a short circuit occurring at an insulator and earthing one of the mains, the currents in them will no longer be balanced, and a loud hum will be set up in the telephone.

Interference with telephone circuits.

Duality. The equations and the diagrams for the lines of force and equipotential lines given in this chapter are similar to the equations and the diagrams for the equipotential lines and the lines of force in electrostatics. This is an example of the principle of duality (Chapter XXI). The equipotential line in electrostatics corresponds to the magnetic line of force, and the electrical line of force corresponds to the line of equal magnetic potential. The lines of force in statical electricity are also the same curves as the lines of flow of an electric current through a conducting medium. The diagrams of the lines of force given above can be obtained experimentally by maintaining suitably chosen spots on a sheet of tinfoil at given potentials and mapping out the equipotential lines on the sheets by Kirchhoff's method.

REFERENCES

G. R. KIRCHHOFF, *Poggendorff's Annalen*, Vol. 64, p. 497.

W. ROBERTSON SMITH, *Proc. Roy. Soc. Edin.* 1869—70, p. 79.

G. CAREY FOSTER and OLIVER J. LODGE, *Proc. Phys. Soc. London*, Vol. 1, p. 113.

For Cassinian ovals and dipolar (bipolar) circles see A. G. GREENHILL, *Differential and Integral Calculus.*

For De Moivre's and Cotes' properties of the circle see S. L. LONEY, *Trigonometry.*

For numerous diagrams of equipotential curves got experimentally see JAMIN and BOUTY, *Cours de Physique*, Vol. 4, Part I. p. 172, 4th Edition.

M. B. FIELD, 'A Theoretical Consideration of the Currents Induced in Cable Sheaths, and the Losses occasioned thereby.' *Journ. of the Inst. of El. Eng.* Vol. 33, p. 936, 1904.

J. A. FLEMING and G. B. DYKE, 'On the Power Factor and Conductivity of Dielectrics when tested with Alternating Electric Currents of Telephonic Frequency at Various Temperatures.' *Journ. of the Inst. of El. Eng.* Vol. 49, p. 323, 1912.

S. EVERSHED, 'The Characteristics of Insulation Resistance.' *Journ. of the Inst. of El. Eng.* Vol. 52, p. 51, 1913.

CHAPTER XX

Eddy currents. Short circuited coil in a uniform alternating magnetic field. Currents in the closed secondary of a constant voltage transformer. Eddy currents in the iron plates of the core of a transformer. Eddy currents produced in an infinite iron plate, placed in a uniform harmonically varying magnetic field, with its sides parallel to the lines of force. Analogy with the theory of heat. Approximate formulae. Eddy currents induced in a metallic cylinder forming the core of a long solenoid. References. Tables of $X(x)$, $V(x)$, and $Z(x)$.

THE calculation of the losses due to the eddy currents, which

Eddy currents.

are generated in conductors when placed in variable magnetic fields, is a problem of great practical importance. In most cases only roughly approximate solutions can be obtained. When the conductors are of non-magnetic materials and the strength of the field varies harmonically, the problem is greatly simplified. Solutions can be obtained without much difficulty for the case of an infinite plate when the surface of the plate is parallel to the lines of force of the field and for the case of a solid cylinder of infinite length, whose axis is parallel to the lines of force of a uniform alternating magnetic field. Before discussing these problems we shall consider the problem of the currents induced in a closed coil of insulated wire, when placed in an alternating magnetic field.

Let us suppose that the coil has n turns of wire, that the mean

Short circuited coil in a uniform alternating magnetic field.

area enclosed by the turns is S and that the induction density parallel to the axis of the coil is

$$B \sin \omega t.$$

Let R be the resistance of the coil, L its self inductance and i the induced current, then

$$Ri + L \frac{\partial i}{\partial t} = - n \frac{\partial}{\partial t} (SB \sin \omega t)$$

$$= - n S B \omega \cos \omega t.$$

The effective value A of the current is therefore given by

$$A = \frac{nSB\omega}{\sqrt{2}\sqrt{R^2 + L^2\omega^2}}.$$

Hence for a given maximum induction density B, the effective value of the current will continually increase as ω increases and therefore as the frequency increases, but it can never be greater than $nSB/(L\sqrt{2})$. The power expended in heating the coil is RA^2 and this equals

$$\frac{(nSB\omega)^2 R}{2(R^2 + L^2\omega^2)}.$$

This expression continually increases as the frequency increases and its limiting value is $\{n SB/(L\sqrt{2})\}^2 R$. If we suppose that R is the variable, then the heating for a given frequency will be a maximum when R equals $L\omega$ or $2\pi f L$. Hence, the increase of the resistivity of the metal of the conductor, due to the heating, will increase the power expended in heating the circuit if R is less than $2\pi f L$ but will diminish it if R is greater than $2\pi f L$.

From the equations of the air core transformer it is easy to

Currents in the closed secondary of a constant voltage transformer.

show (p. 338) that

$$-\frac{M}{L_1}(e_1 - R_1 i_1) = R_2 i_2 + L_2\sigma\frac{\partial i_2}{\partial t},$$

where $\sigma = 1 - M^2/L_1 L_2$

= the leakage factor.

Thus, squaring and taking mean values, we find that

$$(M^2/L_1^2)(V_1^2 - 2R_1 W + R_1^2 A_1^2) = R_2^2 A_2^2 + \alpha^2 L_2^2\sigma^2\omega^2 A_2^2,$$

where α is a constant which has its minimum value unity (p. 136) when e_1 follows the harmonic law. Now W is the mean value of the power taken by the primary coil. It must therefore be equal to the mean rate of heat production in both circuits and thus

$$W = R_1 A_1^2 + R_2 A_2^2.$$

Hence, on substituting this value of W in the equation above, we get

$$A_2^2\{R_2^2 + 2(M^2/L_1^2)R_1 R_2 + \alpha^2 L_2^2\sigma^2\omega^2\} = (M^2/L_1^2)(V_1^2 - R_1^2 A_1^2).$$

The power expended on the secondary circuit is therefore

$$\frac{R_2}{R_2{}^2 + 2\,(M^2/L_1{}^2)\,R_1 R_2 + a^2 L_2{}^2 \sigma^2 \omega^2} \cdot \frac{M^2}{L_1{}^2}\,(V_1{}^2 - R_1{}^2 A_1{}^2).$$

Assuming that $R_1 A_1$ is negligible compared with V_1, we see that, when σ is not zero, the power expended in heating the secondary circuit continually diminishes as the frequency increases and vanishes when the frequency is infinite.

If the secondary resistance vary, then, if we neglect $R_1 A_1$ compared with V_1, the power transmitted to the secondary circuit at a given frequency is a maximum when R_2 equals $2\pi f a L_2 \sigma$ and its value is then

$$\frac{1}{2\,(M^2/L_1{}^2)\,R_1 + 2R_2}\left(\frac{M}{L_1}\,V_1\right)^2,$$

provided that the shape of the current wave in the secondary circuit remain constant. When R_2 is greater than $2\pi f a L_2 \sigma$, an increase in the resistance of the circuit diminishes the loss due to the heating in the circuit, but if R_2 is less than $2\pi f a L_2 \sigma$ then an increase of R_2 increases the loss.

When $R_1 A_1$ is small compared with V_1 and the leakage factor is zero, the heating of the secondary circuit is simply equal to

$$\frac{1}{R_2 + 2\,(M^2/L_1{}^2)\,R_1}\left(\frac{M}{L_1}\,V_1\right)^2,$$

and it is therefore independent of the frequency and of the shape of the applied potential difference wave. When the leakage factor is not zero, the heating of the secondary circuit for a given effective voltage V_1 is a maximum when the applied potential difference wave is sine-shaped.

These theorems for insulated fine wire coils, although instructive, do not give us much help in calculating the eddy current losses in masses of metal subjected to an alternating magnetising force. To do this we must find the paths and the magnitudes of the induced currents in the metal.

It is found in practice that, unless the iron plates which form
Eddy currents in the iron plates of the core of a transformer. the core of a transformer are less than about half a millimetre in thickness, the heating due to eddy currents is appreciable. The cores are therefore built up of finely laminated iron. The lamination appears

to have little, if any, effect on the losses due to hysteresis. In the core of a transformer the plates are practically thin strips of iron, the length of the strips being parallel to the direction of the alternating magnetic field. The induced electromotive force will therefore be parallel to the cross section of the strips, and if their thickness be small compared with their breadth, we can suppose that the lines of flow are at right angles to the length of the strips and parallel to the faces of the strips. These currents will alter the value of the magnetising force H, and hence H will have different values at different distances from the surface of the plate. We shall show that there is a diminution in the amplitude of H as we go into the plate. To compensate for this diminution in the value of the magnetising force, we must increase the magnetising current of the transformer, and hence increase the losses in the copper conductors. Again, since by Steinmetz's law the hysteresis loss varies as $B^{1\cdot6}$, the hysteresis loss will be greater than if the magnetic induction were uniform and had its density equal to the mean value of B. The iron at the centre of the plates is screened, to a certain extent, by the eddy currents, and we shall see that this screening effect is greater the thicker the plates and the higher the frequency of the magnetic force.

Let AA' (Fig. 150) be a line perpendicular to the two faces of the plate and let O be the middle point of AA'.

Eddy currents produced in an infinite iron plate, placed in a uniform harmonically varying magnetic field, with its sides parallel to the lines of force. Let OZ be parallel to the direction of the applied magnetic force and let OY be at right angles to OZ and OX. Suppose that the field is produced by harmonic alternating currents flowing in a coil of insulated wire wrapped round the plate, the windings lying in planes parallel to the plane XOY, the coil forming an infinitely long and infinitely broad solenoid with OZ for its axis, so that the lines of force are everywhere parallel to OZ. By making the breadth of the strip great enough in proportion to its thickness, we can consider that the eddy currents, except near the edges, are all flowing in directions parallel to OY. We suppose therefore that every tube of current which cuts the plane XOZ at a distance x from OZ also cuts the same plane at a distance $-x$ from it.

Let i be the current density at a point P (Fig. 150) in the iron, where OP equals x, and let ρ be the resistivity of iron. Consider a small circuit in the plane XOY of breadth ∂x and unit length. The electromotive force round this circuit is

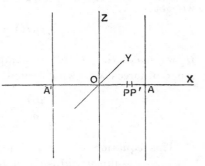

Fig. 150. AA' the thickness of the plate equals $2a$, and YOZ is the median plane.

$$\rho \left(i + \frac{\partial i}{\partial x} \partial x \right) - \rho i,$$

and this, by Faraday's law, must be equal to the rate at which the lines of induction linked with the circuit diminish. Thus, if B denote the induction density at the point P at the time t, we have

$$\rho \left(i + \frac{\partial i}{\partial x} \partial x \right) - \rho i = -\frac{\partial}{\partial t}(B \partial x),$$

and thus

$$\rho \frac{\partial i}{\partial x} = -\frac{\partial B}{\partial t} \quad \ldots\ldots\ldots\ldots\ldots(1).$$

Let us now consider a small circuit in the plane XOZ. Suppose that its breadth is dx and its length parallel to OZ is unity. The current flowing across this section parallel to OY will be $i\partial x$. Hence, equating the two expressions for the work done in taking a unit magnetic pole round the boundary of this circuit, we obtain

$$4\pi . i\partial x = H - \left(H + \frac{\partial H}{\partial x} \partial x \right).$$

Therefore

$$i = -\frac{1}{4\pi} \frac{\partial H}{\partial x} \quad \ldots\ldots\ldots\ldots\ldots\ldots(2).$$

Now the induction density B is not in simple proportion to the magnetising force H. Owing to hysteresis, B is different when H is increasing or diminishing, and the ratio of B to H is also different for different values of H. To take hysteresis into account in our mathematical equations would therefore be very difficult. If we neglect hysteresis we may write μH for B, where

μ is a single-valued function of H. Hence from (1) and (2), we get

$$\frac{\partial}{\partial t}(\mu H) = \frac{\rho}{4\pi}\frac{\partial^2 H}{\partial x^2}.$$

If we make the further assumption that μ is constant for the range of forces with which we are concerned in practice, then

$$\frac{\partial H}{\partial t} = \frac{\rho}{4\pi\mu}\frac{\partial^2 H}{\partial x^2} \quad\dots\dots\dots\dots\dots(3).$$

This equation is identical in form with the equation for the linear diffusion of heat through a conducting plate, namely

Analogy with the theory of heat.

$$\frac{\partial v}{\partial t} = \frac{k}{c}\frac{\partial^2 v}{\partial x^2},$$

where v is the temperature at a distance x from a central plane, k is the thermal conductivity of the substance and c is its thermal capacity per unit volume. Comparing the electrical and thermal equations, we see that resistivity corresponds to thermal conductivity and permeability to thermal capacity. An infinitely good electrical conductor corresponds to a thermal insulator, and a substance having high permeability to a substance having great capacity for heat. The magnetic force will diffuse into the metal according to the same law as heat diffuses into it, provided the applied magnetic forces at the faces of the plate vary according to the same law as the temperatures of the faces. It can be shown that this theorem is true in all cases. Hence we can make use of the known solutions of problems in the theory of conduction of heat to find the values of H and therefore of i in the corresponding electrical problems.

Let the thickness of the plate be $2a$ and let the value of H at the faces of the plate be $H_0\cos\omega t$. Then H must be a function of x and t which satisfies the differential equation (3) and has the value $H_0\cos\omega t$, when x is $+a$ or $-a$.

Let us suppose that

$$H = h\cos\omega t + k\sin\omega t$$

is a solution of (3), where h and k are functions of x.

Substituting this value for H in (3), we get

$$\omega\left(-h\sin\omega t + k\cos\omega t\right) = \frac{\rho}{4\pi\mu}\left(\frac{\partial^2 h}{\partial x^2}\cos\omega t + \frac{\partial^2 k}{\partial x^2}\sin\omega t\right),$$

a relation which must be true for all values of t. Hence we must have

$$h = -\frac{\rho}{4\pi\mu\omega}\frac{\partial^2 k}{\partial x^2}, \quad \text{and} \quad k = \frac{\rho}{4\pi\mu\omega}\frac{\partial^2 h}{\partial x^2},$$

and, therefore,

$$h = -\left(\frac{\rho}{4\pi\mu\omega}\right)^2\frac{\partial^4 h}{\partial x^4} = -\frac{1}{4p^4}\frac{\partial^4 h}{\partial x^4} \quad\ldots\ldots\ldots\ldots(4),$$

where $\qquad p^2 = 2\pi\mu\omega/\rho = (2\pi)^2\,\mu f/\rho,$

where f is the frequency of the magnetising current.

Thus $$p = 2\pi\sqrt{\frac{\mu f}{\rho}} \quad\ldots\ldots\ldots\ldots\ldots\ldots(5).$$

Now assume that h is $A\epsilon^{nx}$ and substitute this value in (4). We find that

$$n^4 = -4p^4;$$
therefore $\qquad n^2 = 2p^2(\pm\sqrt{-1}),$
and $\qquad n = \pm p(1 \pm \sqrt{-1}).$

The complete solution of (4) is therefore

$$h = A\epsilon^{px}\cos px + B\epsilon^{px}\sin px + C\epsilon^{-px}\cos px + D\epsilon^{-px}\sin px,$$

where A, B, C and D are constants.

Again, we have

$$k = \frac{1}{2p^2}\frac{\partial^2 h}{\partial x^2}.$$

On substituting for h and performing the differentiations, we get

$$k = -A\epsilon^{px}\sin px + B\epsilon^{px}\cos px + C\epsilon^{-px}\sin px - D\epsilon^{-px}\cos px.$$

Now $\qquad H = h\cos\omega t + k\sin\omega t,$
and therefore

$$H = A\epsilon^{px}\cos(\omega t + px) + B\epsilon^{px}\sin(\omega t + px)$$
$$+ C\epsilon^{-px}\cos(\omega t - px) - D\epsilon^{-px}\sin(\omega t - px).$$

Again, since, by symmetry, H does not change sign with x for any value of t, we see, by putting ωt equal to zero and equal to

$\pi/2$ successively, that neither h nor k must change sign with x. It will be seen from inspection that

$$A = C \text{ and } B = -D$$

satisfy both these conditions. We get, therefore,

$$H = A \left\{ \epsilon^{px} \cos (\omega t + px) + \epsilon^{-px} \cos (\omega t - px) \right\}$$
$$+ B \left\{ \epsilon^{px} \sin (\omega t + px) + \epsilon^{-px} \sin (\omega t - px) \right\} \dots (6).$$

Now at the faces of the plate H equals $H_0 \cos \omega t$, and thus

$$h = H_0 \text{ and } k = 0, \text{ when } x = \pm a.$$

We get, therefore,

$$H_0 = A \left(\epsilon^{pa} + \epsilon^{-pa} \right) \cos pa + B \left(\epsilon^{pa} - \epsilon^{-pa} \right) \sin pa,$$

and $\qquad 0 = A \left(\epsilon^{pa} - \epsilon^{-pa} \right) \sin pa - B \left(\epsilon^{pa} + \epsilon^{-pa} \right) \cos pa.$

Writing $\cosh pa$ for $\frac{1}{2} \left(\epsilon^{pa} + \epsilon^{-pa} \right)$ and $\sinh pa$ for $\frac{1}{2} \left(\epsilon^{pa} - \epsilon^{-pa} \right)$ and solving the equations, we find that

$$\left. \begin{array}{l} A = \dfrac{H_0 \cosh pa \cos pa}{\cosh 2pa + \cos 2pa} \\[3mm] B = \dfrac{H_0 \sinh pa \sin pa}{\cosh 2pa + \cos 2pa} \end{array} \right\} \dots\dots\dots\dots\dots(7).$$

and

Again, we have

$$h = 2A \cosh px \cos px + 2B \sinh px \sin px,$$

and therefore

$$\frac{h}{H_0} = \frac{\cosh p (a - x) \cos p (a + x) + \cosh p (a + x) \cos p (a - x)}{\cosh 2pa + \cos 2pa}$$

and $k = -2A \sinh px \sin px + 2B \cosh px \cos px,$

and hence

$$\frac{k}{H_0} = \frac{\sinh p (a - x) \sin p (a + x) + \sinh p (a + x) \sin p (a - x)}{\cosh 2pa + \cos 2pa}.$$

Now, $\qquad H = h \cos \omega t + k \sin \omega t$

$$= \sqrt{h^2 + k^2} \cos \left(\omega t - \tan^{-1} \frac{k}{h} \right),$$

and $\quad \dfrac{h^2 + k^2}{H_0^2} = 4 \dfrac{A^2 + B^2}{H_0^2} \left(\cosh^2 px \cos^2 px + \sinh^2 px \sin^2 px \right)$

$$= \frac{4 \left(\cosh^2 pa \cos^2 pa + \sinh^2 pa \sin^2 pa \right) \left(\cosh^2 px \cos^2 px + \sinh^2 px \sin^2 px \right)}{\left(\cosh 2pa + \cos 2pa \right)^2}$$

$$= \frac{\left(\cosh 2pa + \cos 2pa \right) \left(\cosh 2px + \cos 2px \right)}{\left(\cosh 2pa + \cos 2pa \right)^2}$$

$$= \frac{\cosh 2px + \cos 2px}{\cosh 2pa + \cos 2pa}.$$

XX]   MAGNETIC FIELD INSIDE PLATE   495

Thus, on substituting, we get

$$H = H_0 \left(\frac{\cosh 2px + \cos 2px}{\cosh 2pa + \cos 2pa}\right)^{\frac{1}{2}} \cos(\omega t - \gamma) \quad\ldots\ldots\ldots(8),$$

where $\tan \gamma = k/h$

$$= \frac{\sinh p(a-x)\sin p(a+x) + \sinh p(a+x)\sin p(a-x)}{\cosh p(a-x)\cos p(a+x) + \cosh p(a+x)\cos p(a-x)} \quad\ldots(9).$$

Now, since

$$\cosh 2px + \cos 2px = 2\left\{1 + \frac{(2px)^4}{\lfloor 4} + \frac{(2px)^8}{\lfloor 8} + \ldots\ldots\ldots\right\},$$

we see that, as x increases from zero, $\cosh 2px + \cos 2px$ increases continually from its least value 2. Thus, we see from (8) that the amplitude of H diminishes as we approach the centre of the plate, where it has its minimum value, which is given by (11) below. We see also from (9) that the phase of H is different for different values of x. If the thickness of the plate were $2n\pi/p$, that is, if ma equals $n\pi$, then $\sin mx$ would be a factor of both terms in the numerator and γ would be zero when x is 0, π/p, $2\pi/p$, ... $n\pi/p$. Hence at depths $2\pi/p$, $4\pi/p$, ... the magnetising force, in this case, would be in phase with its surface value.

If γ_c be the phase difference between the surface value of H and the value of H at the centre, then by (9)

$$\tan \gamma_c = \tan pa \tanh pa \quad\ldots\ldots\ldots\ldots(10).$$

Also, if H_c be the amplitude of H at the centre, then

$$H_c = \frac{H_0 \sqrt{2}}{(\cosh 2pa + \cos 2pa)^{\frac{1}{2}}} \quad\ldots\ldots\ldots\ldots(11).$$

In the sheet iron used in the cores of transformers ordinary values of ρ and μ are 10,000 c.g.s. units and 2500 respectively. If the frequency were 100, then by (5)

$$p = 2\pi \sqrt{\frac{2500 \times 100}{10,000}}$$
$$= 10\pi.$$

Therefore if pa equal π, that is, if the thickness of the plates were two millimetres, we see from (10) that γ_c would be 180 degrees and from (11) by the aid of mathematical tables that H_c would be $H_0/11\cdot6$. If the plates were four millimetres thick

γ_c would be 360 degrees, that is, the values of H at the surface and the centre of the plate would be in phase with one another but H_c would only be $H_0/268$. If the thickness of the plates were one millimetre, then γ_c would be ninety degrees and H_c would be $H_0/2\cdot4$.

In Fig. 151 the relative values of the maximum amplitudes of H, when the frequency of the alternations is 100, at various depths in iron sheets 2, 1 and $\frac{1}{2}$ a millimetre thick respectively are shown.

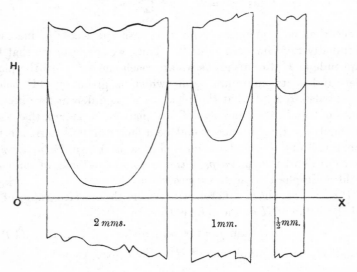

Fig. 151.   The amplitudes of H at various depths in iron sheets which are
2 mms., 1 mm. and $\frac{1}{2}$ mm. thick respectively.

$\mu = 2500.$ $\rho = 10,000.$ $f = 100.$

The diagrams do not show instantaneous values of H. For example, in the two millimetre plate, when H has its maximum positive value at the surface it has its maximum negative value at the centre. The permeability of the iron is supposed to be 2500 and its resistivity 10,000 C.G.S. units. The same result would follow if the permeability were 400 and the resistivity 1600 absolute units, if the frequency be still 100. Increasing the permeability or diminishing the resistivity increases the screening effect of the eddy currents. The diagrams in Fig. 151 show that we should not use iron sheets thicker than half a millimetre for the cores of transformers when the frequency is 100.

From (6), since $i = -\dfrac{1}{4\pi}\dfrac{\partial H}{\partial x}$, we find that

$$i = -\frac{Ap\sqrt{2}}{4\pi}\left\{\epsilon^{px}\cos\left(\omega t + px + \frac{\pi}{4}\right) - \epsilon^{-px}\cos\left(\omega t - px + \frac{\pi}{4}\right)\right\}$$

$$-\frac{Bp\sqrt{2}}{4\pi}\left\{\epsilon^{px}\sin\left(\omega t + px + \frac{\pi}{4}\right) - \epsilon^{-px}\sin\left(\omega t - px + \frac{\pi}{4}\right)\right\}.$$

Hence, by squaring and integrating, we find that the mean value of ρi^2 over a whole period is

$$\rho\,\frac{(A^2 + B^2)\,p^2}{8\pi^2}\,(\cosh 2px - \cos 2px).$$

If therefore the mean value of the eddy current losses per cubic centimetre of the plate is W watts, then W is given by

$$W = \rho\,\frac{(A^2 + B^2)\,p^2 . 10^{-7}}{8\pi^2 \times 2a}\int_{-a}^{+a}(\cosh 2px - \cos 2px)\,dx$$

$$= \rho\,\frac{(A^2 + B^2)\,p . 10^{-7}}{16\pi^2 a}\,(\sinh 2pa - \sin 2pa)$$

$$= \frac{p\rho\,H_0^2 . 10^{-7}}{32\pi^2 a}\left(\frac{\sinh 2pa - \sin 2pa}{\cosh 2pa + \cos 2pa}\right)\dots\dots\dots\dots(12).$$

For values of pa greater than π, that is, when $2a\sqrt{\mu f/\rho}$ is greater than unity, the factor in brackets is practically equal to unity, and the formula becomes

$$W = \frac{\sqrt{\mu f \rho}\,H_0^2 . 10^{-7}}{16\pi a}.$$

We see, therefore, that quadrupling the resistivity ρ, provided that $2a\sqrt{\mu f/\rho}$ still remained greater than unity, would double the eddy current loss.

When pa is small, γ_c is nearly zero and H_c is nearly equal to H_0. Hence the amplitude of H is approximately constant across the plate. Hence, if $B_{\max.}$ be the maximum density of the flux, we may write

$$B_{\max.} = \mu H_0.$$

Since pa is small, we see from (12) that

$$W = \frac{p\rho\,H_0{}^2.10^{-7}}{32\pi^2 a} \cdot \frac{\tfrac{1}{6}(2pa)^3}{1+\tfrac{1}{24}(2pa)^4}$$

$$= \frac{p^4\,\rho a^2\,H_0{}^2.10^{-7}}{24\pi^2} \text{ nearly}$$

$$= \frac{\pi^2}{6}\frac{(2a)^2}{\rho}f^2\,B^2{}_{\max}.\,.10^{-7}$$

$$= 1\cdot64\,\frac{(2a)^2}{\rho}f^2\,B^2{}_{\max}.\,.10^{-7}.$$

When the iron, subjected to the harmonic magnetising forces, consists of thin sheets, this formula gives us the loss of power in watts expended by eddy currents per cubic centimetre of the iron. The thickness $2a$ of each sheet must be expressed in centimetres, the resistivity ρ is in C.G.S. units (about 10,000), f is the frequency and $B_{\max}$ is the maximum value of the flux density. In proving this formula we have supposed that the iron is free from hysteresis and that μ is constant.

We have seen that, when the iron sheets are thick, H varies both in amplitude and phase at different depths in the iron. In finding the flux we must therefore take this into account.

The mean value of the magnetising force for all the points on a line, inside the metal, perpendicular to the faces of the plate will be a harmonic function of the time. Hence, if $H_{\max}$ be the maximum value of the mean magnetic force, we can write

$$H_{\max}.\sin(\omega t + \alpha) = \frac{1}{2a}\int_{-a}^{+a} H\partial x$$

$$= \frac{1}{a}\int_0^a H\partial x.$$

Substituting the value of H got from (6) and performing the integration, we get, after simplifying,

$$H_{\max}. = \frac{H_0}{\sqrt{2}.am}\left(\frac{\cosh 2pa - \cos 2pa}{\cosh 2pa + \cos 2pa}\right)^{\frac{1}{2}} \quad\ldots\ldots\ldots(13).$$

Now, if a depth d on each side of the plate were uniformly magnetised by a force H_0 and if the sum of the fluxes produced in these portions of the plate were equal to the maximum value of the total flux produced in the actual plate by the given magnetising forces, then

$$d . H_0 = a . H_{max.},$$

and hence $\qquad d = \dfrac{1}{p \sqrt{2}} \left(\dfrac{\cosh 2pa - \cos 2pa}{\cosh 2pa + \cos 2pa} \right)^{\frac{1}{2}}$ ....(14).

Ewing calls d the 'equivalent depth of uniform magnetisation.' The maximum value of the total flux produced in a thick iron plate placed in an alternating magnetic field, whose value is given by $H_0 \sin \omega t$, is equal to the total flux which would be produced in two layers of depth d on each side of it uniformly magnetised by the force H_0.

From (14) we see that when $2pa$ is large the equivalent depth d_1 is given by

$$d_1 = 1/(p \sqrt{2}) = \sqrt{\rho}/(2\pi \sqrt{2\mu f}).$$

When $2pa$ is less than $\pi/2$, d is less than d_1. When it is greater than $\pi/2$ and less than $3\pi/2$, d is greater than d_1. Hence we see that d is alternately less and greater than d_1 as the thickness of the plate is increased, the amplitude of the oscillations of d about the value d_1 continually diminishing. Whenever $2pa$ equals $(2n + 1)(\pi/2)$, where n is an integer, d equals d_1. Suppose, for example, that μ is 5000, f is 50 and ρ is 10,000, then

$$p = 2\pi \sqrt{\mu f/\rho} = 10\pi.$$

The following table is calculated using this value of p.

$2a$ in millimetres	0·25	0·50	0·75	1	1·5	2	∞
$2d$ in millimetres	0·248	0·450	0·514	0·491	0·450	0·449	0·450

In order to calculate the hysteresis loss in the iron sheets, we need to know the amplitude of H at various depths in the iron. Ewing has shown that the hysteresis loss in iron sheets, for values of B between 2000 and 8000, may be expressed approximately by a simple empirical formula of the form

$$W' = \alpha H - \beta,$$

where W' gives the loss in ergs per cubic centimetre per cycle due to hysteresis. For example, in a particular sample he found

$$W' = 1340H - 1610.$$

Now if we could find the mean value of the maximum amplitudes of H at all points on a line in the metal perpendicular to the faces of the plate, then this formula would enable us to find the hysteresis loss in the plate. If we make the assumption that the eddy currents are the same when hysteresis is present, we can find the mean value of H by the formula (8) given above. To do this by the integral calculus is difficult, but it can easily be done graphically. We have only to plot a curve, having the amplitude of H for ordinate and the depth (x) for abscissa, similar to those shown in Fig. 151. The required mean value will then be $\frac{1}{a}\int_0^a y\,dx$, that is, the area of the curve divided by its breadth. The area can easily be found by a planimeter.

The curves in Fig. 152 show graphically the relative values of hysteresis and eddy current losses in a transformer core for various thicknesses of the iron plates, obtained by making the given assumptions. The losses for the various thicknesses of the plates were calculated by Ewing by the methods described above. It is interesting to notice that on the given assumptions the hysteresis loss increases with the thickness of the plates owing to the fact that it varies as $B^{1\cdot 6}_{max.}$, and hence it will be greater than if B were constant and equal to its mean value. The diagram also shows that, when the frequency is 100, the eddy current losses become appreciable when the thickness of the plates is greater than a quarter of a millimetre.

As a first rough approximation it is permissible to apply the formulae given above to iron sheets, but as we have neglected the effects of hysteresis in modifying the eddy currents, they may be affected by large errors. An inspection of the curves given on p. 57 will show that the ratio of B to H is far from being constant in practice, and that when H follows the harmonic law B follows some other and more complicated law.

Let us suppose that the amplitude of the magnetic force is constant over the cross section of a thin sheet of iron (Fig. 150), and that its phase is also constant. Let it be represented by $H_0 \cos \omega t$. By Fourier's theorem we can also suppose that B is given by

$$B = B_1 \cos(\omega t - \alpha_1) + B_3 \cos(3\omega t - \alpha_3) + \dots.$$

The intensity i of the current in the metal at a distance x from the central plane is given by

$$\rho i = \frac{\partial}{\partial t}(Bx),$$

and therefore

$$\rho i^2 = \frac{x^2}{\rho}\left(\frac{\partial B}{\partial t}\right)^2.$$

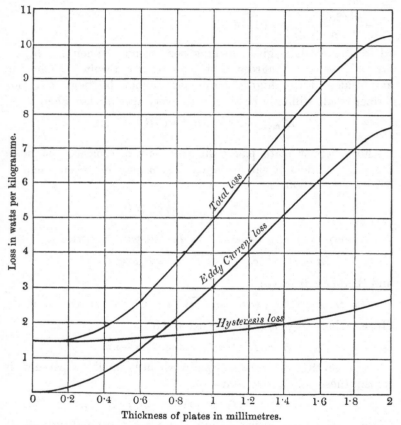

Fig. 152. Losses due to eddy currents and hysteresis in sheet-iron plates when B is 4000 and the frequency is 100.

The mean rate at which heat is being produced at any moment, per unit volume of the sheet,

$$= \frac{1}{2\rho a}\left(\frac{\partial B}{\partial t}\right)^2 \int_{-a}^{a} x^2\, \partial x = \frac{a^2}{3\rho}\left(\frac{\partial B}{\partial t}\right)^2.$$

Thus the average rate of heat production, per unit volume of the plate,

$$= \frac{a^2}{3\rho} \frac{1}{T} \int_0^T \omega^2 \{- B_1 \sin(\omega t - \alpha_1) - 3B_3 \sin(3\omega t - \alpha_3) - \ldots\}^2 \, \partial t$$

$$= \frac{a^2}{6\rho} \omega^2 \{B_1{}^2 + (3B_3)^2 + (5B_5)^2 + \ldots\}$$

$$= \frac{\pi^2}{6} \frac{(2a)^2}{\rho} f^2 \{B_1{}^2 + (3B_3)^2 + \ldots\},$$

where the result is given in ergs per second. When B_3, B_5, ... are all zero, this agrees with our former result. When the amplitudes of the higher harmonics cannot be neglected, our former result will only be true in the very special case when

$$B^2{}_{\text{max.}} = B_1{}^2 + (3B_3)^2 + (5B_5)^2 + \ldots.$$

For a copper plate, placed in a uniform alternating magnetic
Approximate formulae. field, μ will be unity, and ρ may be taken equal to 1600, hence

$$p = \frac{\pi}{20} \sqrt{f} = 0{\cdot}1571 \sqrt{f}.$$

When pa is large, we see from (7) that we may write

$$A = H_0 \epsilon^{-pa} \cos pa, \quad \text{and} \quad B = H_0 \epsilon^{-pa} \sin pa,$$

and therefore, from (6),

$$H = H_0 \{\epsilon^{-p(a-x)} \cos(\omega t - p\overline{a-x}) + \epsilon^{-p(a+x)} \sin(\omega t - p\overline{a+x})\}.$$

Hence, when $a - x$ is small compared with a,

$$H = H_0 \epsilon^{-p(a-x)} \cos(\omega t - p\overline{a-x}).$$

At a depth d in the copper plate, which is small compared with its thickness $2a$, we may write

$$H = H_0 \epsilon^{-pd} \cos(\omega t - pd) \quad \ldots\ldots\ldots\ldots(15).$$

We see, then, that as we go into the plate the amplitude of H diminishes according to the exponential law, and also that at points whose depths differ by a multiple of $2\pi/p$ the values of H are in phase with one another. The maximum values of H at depths 0, $1/p$, $2/p$, $3/p$, ... $10/p$ are

$$1{\cdot}00H_0, \quad 0{\cdot}368H_0, \quad 0{\cdot}135H_0, \quad 0{\cdot}050H_0, \ldots 0{\cdot}000045H_0.$$

We suppose that the windings of the solenoid can be regarded as practically parallel to one plane, so that the magnetic field inside is parallel to the axis and has a value $4\pi nc$, at all points between the solenoid and the core, where c is the instantaneous value of the alternating magnetising current and n is the number of turns per unit length. We shall denote the radius of the core by R. The eddy currents in the core will, from symmetry, be circles whose centres are on the axis of the solenoid. Let r be the radius of a ring whose breadth is ∂r and thickness unity, then, if i be the intensity of the current in it, $4\pi \cdot i\partial r$ will be the amount by which H changes when we pass from r to $r + \partial r$.

Eddy currents induced in a metallic cylinder forming the core of a long solenoid.

Therefore

$$H - \left(H + \frac{\partial H}{\partial r}\,\partial r \right) = 4\pi i\partial r,$$

and thus

$$i = -\frac{1}{4\pi}\frac{\partial H}{\partial r} \quad\dots\dots\dots\dots\dots\dots(a).$$

The total electromotive force round the circle of radius r is $2\pi r i\rho$ and is also equal to the rate at which the total flux of induction through it is diminishing. Hence, assuming a constant permeability, we get

$$2\pi r i\rho = -\mu \int_0^r \frac{\partial H}{\partial t}\,2\pi r\partial r \quad\dots\dots\dots\dots(b).$$

Hence from (b) and (a)

$$-\frac{r\rho}{4\pi}\cdot\frac{\partial H}{\partial r} = -\mu \int_0^r \frac{\partial H}{\partial t}\,r\partial r,$$

and therefore

$$\frac{1}{r}\frac{\partial}{\partial r}\left(r\frac{\partial H}{\partial r} \right) = \frac{4\pi\mu}{\rho}\frac{\partial H}{\partial t} \quad\dots\dots\dots\dots(c).$$

Writing

$$m^2 = 4\pi\mu\omega/\rho = 8\pi^2\mu f/\rho \quad\dots\dots\dots\dots(d),$$

we get

$$\frac{\partial^2 H}{\partial r^2} + \frac{1}{r}\frac{\partial H}{\partial r} = \frac{m^2}{\omega}\frac{\partial H}{\partial t} \quad\dots\dots\dots\dots(e).$$

If we assume that the magnetising current follows the harmonic law, we may write $H_0 \cos \omega t$ for the magnetic force at the surface of the core. We have already seen (p. 213) that

$$H = (A\ \mathrm{ber}\ mr + B\ \mathrm{bei}\ mr) \cos \omega t$$
$$+ (- A\ \mathrm{bei}\ mr + B\ \mathrm{ber}\ mr) \sin \omega t,$$

where A and B are constants, is the solution of (e).

Now, when r equals R, the radius of the cylinder, we must have H equal to $H_0 \cos \omega t$. Hence the boundary conditions give us

$$H_0 = A \text{ ber } (mR) + B \text{ bei } (mR),$$

and $$0 = B \text{ ber } (mR) - A \text{ bei } (mR).$$

Therefore $$A = H_0 \text{ ber } (mR)/X (mR),$$

and $$B = H_0 \text{ bei } (mR)/X (mR),$$

where $X (mR)$ is defined on p. 210.

The complete solution is therefore

$$H = C \{\text{ber } (mR) \text{ ber } (mr) + \text{bei } (mR) \text{ bei } (mr)\} \cos \omega t$$
$$+ C \{\text{bei } (mR) \text{ ber } (mr) - \text{ber } (mR) \text{ bei } (mr)\} \sin \omega t \dots (f),$$

where $$C = H_0/X (mR).$$

The amplitude of H, at a point at a distance r from the axis of the cylinder, is

$$H_0 \{X (mr)/X (mR)\}^{\frac{1}{2}} \dots\dots\dots\dots\dots(g),$$

and this increases as r increases. At the axis of the cylinder, where r equals zero, the amplitude becomes

$$H_0/\{X (mR)\}^{\frac{1}{2}}.$$

If γ be the phase difference between the value of H at a distance r from the axis and its value on the circumference of the cylinder, then

$$\tan \gamma = \frac{\text{bei } (mR) \text{ ber } (mr) - \text{ber } (mR) \text{ bei } (mr)}{\text{ber } (mR) \text{ ber } (mr) + \text{bei } (mR) \text{ bei } (mr)} \dots\dots(h).$$

When r is zero, ber (mr) is 1 and bei (mr) is 0, and then

$$\tan \gamma_c = \frac{\text{bei } (mR)}{\text{ber } (mR)} \dots\dots\dots\dots\dots(j),$$

where γ_c is the value of γ at the axis.

If the frequency of the applied magnetising force be 100 and if the cylinder be made of copper, for which μ equals unity and ρ equals 1600, we have, from (d),

$$m^2 = \frac{8\pi^2 \cdot 100}{1600},$$

and, therefore, $$m = 2 \cdot 22 \text{ nearly.}$$

Now, from the tables, we see that bei (mR) vanishes when mR is zero, it then increases as mR increases and attains a maximum

value for some value of mR between 3·5 and 4. It then diminishes
and vanishes when mR is a little greater than 5. When mR is 5,
R is 2·2 centimetres nearly, $\tan \gamma_c$ is approximately zero and the
maximum value of H along the axis of the cylinder is $H_0/6·2$.
Now, from (j) and the tables (p. 230) for bei (x) and ber (x), we see
that when mR is zero, γ_c is zero. As mR increases $\tan \gamma_c$ increases
and after attaining an infinite positive value it becomes negative
and vanishes when mR is a little greater than 5. We thus see that
when mR is a little greater than 5, γ_c is equal to π and the magnetic
force along the axis is in opposition in phase to the magnetic force
at the surface. In this case also, the amplitude of the magnetic
force along the axis is less than the sixth part of the amplitude of
the magnetic force at the surface.

The rate of generation of heat at any point of the core, per
unit volume, is ρi^2 ergs per second. Hence for a length l of the
core the instantaneous value w of the power expended in eddy
currents is given by

$$w = l \int_0^R \rho i^2 . 2\pi r \partial r.$$

But $$i = -\frac{1}{4\pi}\frac{\partial H}{\partial r},$$

and therefore by (f)

$$-4\pi i = Cm \{\text{ber} (mR) \text{ ber}' (mr) + \text{bei} (mR) \text{bei}' (mr)\} \cos \omega t$$
$$+ Cm \{\text{bei} (mR) \text{ber}' (mr) - \text{ber} (mR) \text{bei}' (mr)\} \sin \omega t,$$

where $$\text{ber}' (x) = \frac{\partial}{\partial x} \{\text{ber} (x)\},$$

and $$\text{bei}' (x) = \frac{\partial}{\partial x} \{\text{bei} (x)\}.$$

The effective value A of the current is therefore given by

$$32\pi^2 A^2 = H_0^2 m^2 \{V (mr)/X (mR)\},$$

and since m^2 equals $8\pi^2 \mu f/\rho$, we get

$$A^2 = (\mu f/4\rho) H_0^2 \{V (mr)/X (mR)\}.$$

Tables of $V (x)$ and $X (x)$ are given on p. 509.

Now, if W watts be the mean power expended in a length l of
the core,

$$W = 10^{-7} l \int_0^R A^2 \rho . 2\pi r \partial r,$$

and thus

$$W = \tfrac{1}{2}\pi\mu f l H_0^2 . 10^{-7} \frac{\displaystyle\int_0^R \{r\,\mathrm{ber}'^2\,(mr) + r\,\mathrm{bei}'^2\,(mr)\}\,\partial r}{\mathrm{ber}^2\,(mR) + \mathrm{bei}^2\,(mR)}.$$

Again, integrating by parts,

$$\int \mathrm{ber}'\,(mr)\,.\,r\,\mathrm{ber}'\,(mr)\,\partial r = \frac{1}{m}\,\mathrm{ber}\,(mr)\,.\,r\,\mathrm{ber}'\,(mr)$$

$$-\frac{1}{m}\int \mathrm{ber}\,(mr)\,\frac{\partial}{\partial r}\,\{r\,\mathrm{ber}'\,(mr)\}\,\partial r,$$

and

$$\int \mathrm{bei}'\,(mr)\,.\,r\,\mathrm{bei}'\,(mr)\,\partial r = \frac{1}{m}\,\mathrm{bei}\,(mr)\,.\,r\,\mathrm{bei}'\,(mr)$$

$$-\frac{1}{m}\int \mathrm{bei}\,(mr)\,\frac{\partial}{\partial r}\,\{r\,\mathrm{bei}'\,(mr)\}\,\partial r.$$

By differentiating the series for $\mathrm{ber}\,(mr)$ and $\mathrm{bei}\,(mr)$, it is easy to show that

$$\frac{\partial}{\partial r}\,r\,\frac{\partial}{\partial r}\,\mathrm{ber}\,(mr) = -\,m^2 r\,\mathrm{bei}\,(mr),$$

and

$$\frac{\partial}{\partial r}\,r\,\frac{\partial}{\partial r}\,\mathrm{bei}\,(mr) = m^2 r\,\mathrm{ber}\,(mr).$$

The sum of the integrals on the right-hand sides of the two equations therefore vanishes when they are taken between the same limits, and thus, integrating between the limits, we have

$$W = (1/2m)\,\pi R\mu f l H_0^2\,\{Z\,(mR)\}\,10^{-7}/\{X\,(mR)\}.$$

In practice we want to know the average power expended per cubic centimetre of the conductor on account of eddy currents. Hence, dividing W by $\pi l R^2$ and noticing that m^2 equals $8\pi^2\mu f/\rho$, we find that the loss is

$$\frac{\sqrt{\mu f \rho}\,H_0^2 . 10^{-7}}{4\pi R\sqrt{2}} \cdot \frac{Z\,(mR)}{X\,(mR)}$$

watts per cubic centimetre of the core.

In this formula ρ, H_0 and R are in C.G.S. units. The numerical value of this expression, in any given case, can be found easily by means of the tables given on p. 509. We have to remember however that we made two assumptions when proving this formula, namely, that the magnetising force H follows the harmonic law and that μ, the permeability of the core, is a constant.

By the formula given above and (29), p. 211, we have, if W denote the eddy current loss per cubic centimetre,

$$W = \frac{\pi}{4} H_0^2 (\pi R^2) \frac{\mu^2 f^2}{\rho} 10^{-7} \left\{1 - \frac{11}{24}\left(\frac{mR}{2}\right)^4 + \dots\right\}.$$

In this case we also have

$$\tan \gamma_c = \frac{(mR)^2}{4}\left\{1 + \frac{(mR)^4}{72} - \dots\right\},$$

and

$$H_c = H_0\left\{1 - \frac{(mR)^4}{64} + \dots\right\}.$$

Hence, when mR is small, H_c is practically equal to H_0 and γ_c is very nearly zero. The magnetisation therefore will be practically uniform over the cross section of the core and we may write

$$B_{\text{max.}} = \mu H_0.$$

Thus in this case the formula may be written in the form

$$W'' = 0.785 \, (S/\rho) f^2 B^2_{\text{max.}} \cdot 10^{-7},$$

where W'' is the average loss due to eddy currents, in watts per cubic centimetre of the core, S is the area of the cross section of the core in square centimetres, ρ is the resistivity of the core in c.g.s. units, f is the frequency and $B_{\text{max.}}$ is the maximum value of the flux density in c.g.s. units.

This formula shows us that for a thin wire the higher the frequency and the greater the induction density the greater will be the core loss due to eddy currents. If we double both the induction density and the frequency, the eddy current loss will be increased sixteen times. But by using sixteen cores each having only one-quarter the diameter of the original one or by using a core of the same size but of sixteen-fold resistivity the loss in the second case would be the same as in the first.

We saw earlier in this chapter (p. 498) that the eddy current loss, per cubic centimetre, in thin sheets was given by the formula

$$W'' = 1.64 \, \{(2a)^2/\rho\} f^2 B^2_{\text{max.}} \cdot 10^{-7},$$

where $2a$ is the thickness of the sheets. If we used wire of diameter $2a$, the eddy current loss per cubic centimetre would be given by

$$W'' = 0.62 \, \{(2a)^2/\rho\} f^2 B^2_{\text{max.}} \cdot 10^{-7},$$

and thus the eddy current loss per cubic centimetre would only be about two-fifths as great as it was in the case of the sheet.

When mR is great $Z(mR)/X(mR)$ equals $1/\sqrt{2}$ nearly (p. 212) and thus the eddy current loss in watts per cubic centimetre of the core equals

$$\sqrt{\mu f \rho}\, H_0{}^2 . 10^{-7}/(8\pi R).$$

In making calculations in connection with choking coils and transformers, where the core consists of a bundle of insulated iron wires, it is sometimes assumed that to a first approximation the core loss is given by an equation of the form

$$W_0 = V\left\{\eta f B_{\text{max.}}^{1\cdot6} + \theta\,(S/\rho)f^2 B^2{}_{\text{max.}}\right\} 10^{-7},$$

where η is Steinmetz's coefficient, θ a constant and V the volume of the iron in the core in cubic centimetres. If the magnetic induction does not approximately follow the harmonic law, then the term in the above formula for the eddy current losses is certainly wrong. For example, if the flux density B be practically uniform at any instant in the metal, then we can see that the eddy current loss must vary as $(\partial B/\partial t)^2$ and the mean value of this expression for the complete cycle depends not only on $B_{\text{max.}}$ but also on the law of variation of B with regard to the time. Owing to hysteresis, even when the magnetic induction does follow the harmonic law, the flux density does not follow it and this must modify the second term considerably. Thus when μ is not constant, the formulae given for the eddy currents induced in masses of metal must be regarded only as rough approximations.

REFERENCES

OLIVER HEAVISIDE, 'The Induction of Currents in Cores.' *The Electrician*, Vol. 13, p. 583, May 1884.

Also, *Electrical Papers*, Vol. 1, p. 353.

LORD KELVIN, 'Ether, Electricity and Ponderable Matter.' *Journ. of the Inst. of El. Eng.* Vol. 18, p. 4, Jan. 1889.

—— 'On anti-effective Copper in Parallel Conductors or in Coiled Conductors for Alternate Currents.' *Brit. Ass. Report*, 1890, p. 736.

Also, *Mathematical and Physical Papers*, Vol. 3.

SIR J. J. THOMSON, 'On the Heat produced by Eddy Currents in an Iron Plate exposed to an Alternating Magnetic Field.' *The Electrician*, Vol. 28, p. 599, 1891.
Also, *Elements of the Mathematical Theory of Electricity and Magnetism.*
SIR J. A. EWING, 'On Magnetic Screening, Eddy Currents and Hysteresis in Transformer Cores.' *The Electrician*, Vol. 28, p. 631, 1891.
G. F. C. SEARLE and T. G. BEDFORD, 'On the Measurement of Magnetic Hysteresis.' *Philosophical Transactions*, Vol. 198, 1902.
A. RUSSELL, 'Methods of Computing the Ber and Bei and Allied Functions.' *Phil. Mag.* p. 524, April 1909.

Tables of $X(x)$, $V(x)$ and $Z(x)$, computed by Harold G. Savidge.

$$X(x) = \text{ber}^2 x + \text{bei}^2 x, \qquad V(x) = \text{ber}'^2 x + \text{bei}'^2 x,$$
$$\text{and} \quad Z(x) = \text{ber}\, x\, \text{ber}'\, x + \text{bei}\, x\, \text{bei}'\, x.$$

x	$X(x)$	$V(x)$	$Z(x)$	x	$X(x)$	$V(x)$	$Z(x)$
0	1·000	0·000	0·000	16	$6·752 \times 10^7$	$6·461 \times 10^7$	$4·561 \times 10^7$
1	1·031	$2·513 \times 10^{-1}$	$6·266 \times 10^{-2}$	17	$2·612 \times 10^8$	$2·506 \times 10^8$	$1·770 \times 10^8$
2	1·510	1·084	$5·209 \times 10^{-1}$	18	$1·014 \times 10^9$	$9·752 \times 10^8$	$6·887 \times 10^8$
3	3·803	3·240	2·054	19	$3·950 \times 10^9$	$3·806 \times 10^9$	$2·688 \times 10^9$
4	$1·183 \times 10$	$1·007 \times 10$	6·909	20	$1·543 \times 10^{10}$	$1·489 \times 10^{10}$	$1·052 \times 10^{10}$
5	$3·883 \times 10$	$3·375 \times 10$	$2·345 \times 10$	21	$6·041 \times 10^{10}$	$5·842 \times 10^{10}$	$4·127 \times 10^{10}$
6	$1·323 \times 10^2$	$1·177 \times 10^2$	$8·215 \times 10$	22	$2·371 \times 10^{11}$	$2·296 \times 10^{11}$	$1·622 \times 10^{11}$
7	$4·643 \times 10^2$	$4·203 \times 10^2$	$2·943 \times 10^2$	23	$9·326 \times 10^{11}$	$9·044 \times 10^{11}$	$6·390 \times 10^{11}$
8	$1·666 \times 10^3$	$1·526 \times 10^3$	$1·072 \times 10^3$	24	$3·675 \times 10^{12}$	$3·568 \times 10^{12}$	$2·521 \times 10^{12}$
9	$6·077 \times 10^3$	$5·621 \times 10^3$	$3·953 \times 10^3$	25	$1·451 \times 10^{13}$	$1·410 \times 10^{13}$	$9·966 \times 10^{12}$
10	$2·245 \times 10^4$	$2·093 \times 10^4$	$1·474 \times 10^4$	26	$5·736 \times 10^{13}$	$5·582 \times 10^{13}$	$3·945 \times 10^{13}$
11	$8·383 \times 10^4$	$7·863 \times 10^4$	$5·541 \times 10^4$	27	$2·271 \times 10^{14}$	$2·213 \times 10^{14}$	$1·564 \times 10^{14}$
12	$3·157 \times 10^5$	$2·977 \times 10^5$	$2·099 \times 10^5$	28	$9·007 \times 10^{14}$	$8·783 \times 10^{14}$	$6·207 \times 10^{14}$
13	$1·197 \times 10^6$	$1·134 \times 10^6$	$7·999 \times 10^5$	29	$3·576 \times 10^{15}$	$3·490 \times 10^{15}$	$2·467 \times 10^{15}$
14	$4·568 \times 10^6$	$4·344 \times 10^6$	$3·065 \times 10^6$	30	$1·422 \times 10^{16}$	$1·389 \times 10^{16}$	$9·799 \times 10^{16}$
15	$1·752 \times 10^7$	$1·672 \times 10^7$	$1·180 \times 10^7$	∞	∞	∞	∞

CHAPTER XXI

In geometry the method of reciprocal polars enables us, when
The method of duality. we are given a theorem concerning lines and points, to deduce a reciprocal theorem concerning points and lines. For example, this method enables us to deduce at once from the theorem that 'the three lines joining the angular points of a hexagon described about a circle meet in a point,' the reciprocal theorem, that 'the three points of intersection of the opposite sides of a hexagon inscribed in a circle lie on a straight line.' There are many reciprocal relations of this nature in geometry, and the method of reciprocating a theorem may be called the method of duality. In electrical theory there are also many reciprocal relations, and we shall show that the method of duality often leads to important results. When the solution of the reciprocal of a problem is known, the solution of the problem can always be written down at once. The only difference in the mathematical working in the two cases is that certain quantities in the one equation are the reciprocals of certain quantities in the other. The importance of the method, however, is not so much in the saving of mathematical labour effected, as in the suggestion of novel theorems which sometimes indicate more convenient methods of making measurements or even suggest novel instruments or machines of value in electro-technics. Instead of writing down the differential equations and giving a list of the reciprocal relations at once, it will be more instructive to gradually build

up these reciprocal relations and illustrate them by simple examples as we proceed. We shall first consider direct current theory.

Ohm's law may be expressed algebraically by

Ohm's law.

$$E = CR \dots\dots\dots\dots\dots(1),$$

or

$$C = E(1/R) \dots\dots\dots\dots(2).$$

If in (1) we write C for E, E for C and $1/R$ for R, we get (2). We may therefore regard (1) and (2) as reciprocal equations.

In general, if we are given any relation between E, C and R which holds for a particular theorem, we can write down at once a reciprocal relation, in which C, E and $1/R$ are connected in the same way, which holds for the reciprocal theorem.

The equation which gives us the sum of the potential differences across the terminals of n coils in series is

Series and parallel.

$$E = C(R_1 + R_2 + \dots + R_n).$$

Reciprocating this equation, we get

$$C = E(1/R_1 + 1/R_2 + \dots + 1/R_n),$$

which is the formula for the sum of the currents in n coils in parallel. Hence, in direct current problems, coils in series in the original theorem become coils in parallel in the reciprocal theorem.

The following elementary problems illustrate the method of reciprocating theorems. We shall denote the theorem by α and its reciprocal by β.

Kirchhoff's laws are reciprocal theorems.

α. For currents meeting at a point, we have

$$\Sigma C = 0.$$

β. For the potential differences round a circuit, we have

$$\Sigma E = 0.$$

α. The currents in conductors in parallel are given by

$$C_1 = \frac{1/r_1}{\Sigma 1/r} C.$$

β. The potential differences across conductors in series are given by

$$E_1 = \frac{r_1}{\Sigma r} E.$$

The reader will notice that power reciprocates into power.

α.   The formulae for power are
$$W = CE, \quad W = C^2 R \quad \text{or} \quad W = E^2/R.$$

β.   The formulae for power are
$$W = EC, \quad W = E^2/R \quad \text{or} \quad W = C^2 R.$$

α.   If a resistance x (Fig. 153 a) be placed in series with the terminals of a battery of constant electromotive force E, then the power expended in x is a maximum when x equals the internal resistance R of the battery, and the maximum power is $E^2/4R$. (Stokes's Theorem.)

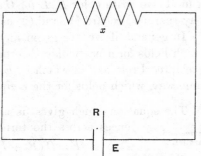

Fig. 153 a.   The power expended on x is a maximum when $x = R$ (E const.).

β.   If a resistance x (Fig. 153 b) be placed in parallel with a coil of resistance R and if the total current through the two coils in parallel have the constant value C, then the power expended in x is a maximum when x equals the resistance R of the coil, and the maximum power is $C^2 R/4$.

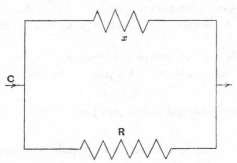

Fig. 153 b.   The power expended on x is a maximum when $x = R$ (C const.).

The preceding theorems may be written symbolically as follows :—

(α)   If $cE = ce + c^2 R$ (E, R constants),
then ce is a maximum when c equals $E/2R$.

(β)   If $eC = ec + e^2/R$ (C, $1/R$ constants),
then ec is a maximum when e equals $CR/2$.

α.   If the potential difference between A and B (Fig. 154 a) be constant, the current in x is a minimum when $2x$ equals R.

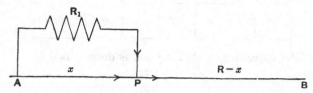

Fig. 154 a.   The current in x is a minimum when $2x$ equals R
(P.D. between A and B constant).

β.   If the current in the main be constant the voltage across x is a maximum when $2x$ equals R (Fig. 154 b).

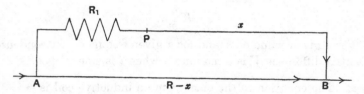

Fig. 154 b.   The voltage across x is a maximum when $2x$ equals R
(total current constant).

Capacity and inductance.
The method of duality is particularly suggestive when applied to alternating current theory.   We remind the reader that a choking coil is an ideal coil having inductance but no resistance.   The equations for a condenser and a choking coil are

$$i = K\frac{\partial e}{\partial t} \quad \text{and} \quad e = L\frac{\partial i}{\partial t}.$$

We may thus regard K as the reciprocal of L and that a condenser is the reciprocal of a choking coil.

By integrating the preceding equations we get

Flux and quantity.
$$q = Ke \quad \text{and} \quad \phi = Li.$$

Hence ϕ and q are reciprocal quantities.   We shall now illustrate the method by giving a few reciprocal theorems in which use is made of the following table of reciprocal quantities and connections.

(a)	e or V	r	K	φ	series
(β)	i or A	1/r	L	q	parallel

α. The current in a choking coil is determined by

$$e = L\frac{\partial i}{\partial t}.$$

For a given value of the effective voltage V, and for a given frequency, the effective value of the choking coil current A is a maximum when e is sine-shaped.

β. The voltage across a condenser is determined by

$$i = K\frac{\partial e}{\partial t}.$$

For a given value of A and for a given frequency the condenser potential difference V is a maximum when i is sine-shaped.

α. The equation to the current in an inductive coil is

Inductive coil and leaky condenser.
$$e = Ri + L\frac{\partial i}{\partial t}.$$

β. This reciprocates into

$$i = \frac{e}{R} + K\frac{\partial e}{\partial t}.$$

An inductive coil therefore reciprocates into a condenser shunted by a non-inductive resistance.

α. In sine curve theory, the impedance Z of an inductive coil is given by the formula

$$Z = \sqrt{R^2 + \omega^2 L^2}.$$

β. In sine curve theory, the impedance Z of a leaky condenser is given by the formula

$$1/Z = (1/R^2 + \omega^2 K^2)^{\frac{1}{2}}.$$

α. In sine curve theory, when the applied potential difference is constant, the mean power expended at a given frequency, in a variable resistance in series with a choking coil, is a maximum when R is $L\omega$. (J. Hopkinson.)

β. In sine curve theory, when the main current is constant the mean power expended at a given frequency, in a variable resistance shunting a condenser, is a maximum when $1/R$ is $K\omega$.

α. In sine curve theory, a condenser K may be replaced by a choking coil whose resistance is zero and inductance $-1/(K\omega^2)$. (Rayleigh.)

β. In sine curve theory, a choking coil L may be replaced by a condenser whose resistance is infinite and capacity $-1/(L\omega^2)$.

α. In sine curve theory, when the current in a resistance coil and choking coil in series lags behind the applied potential difference by an angle θ, then

$$\sin\theta = \frac{\omega L}{(R^2 + \omega^2 L^2)^{\frac{1}{2}}}; \quad \cos\theta = \frac{R}{(R^2 + \omega^2 L^2)^{\frac{1}{2}}}; \quad \tan\theta = \frac{\omega L}{R}.$$

β. In sine curve theory, when the potential difference between the terminals of a resistance coil and a condenser in parallel lags behind the main current by an angle θ, then

$$\sin\theta = \frac{\omega K}{(1/R^2 + \omega^2 K^2)^{\frac{1}{2}}}; \quad \cos\theta = \frac{1/R}{(1/R^2 + \omega^2 K^2)^{\frac{1}{2}}}; \quad \tan\theta = \omega K R.$$

α. In sine curve theory, if θ be the angle of phase difference between the current and the potential difference applied to a resistance coil, a choking coil and a condenser in series, then

$$\tan\theta = \frac{\omega\{L - 1/(K\omega^2)\}}{R}.$$

β. In sine curve theory, if θ be the angle of phase difference between the potential difference and the current supplying a resistance coil, a condenser and a choking coil in parallel, then

$$\tan\theta = \omega R\{K - 1/(L\omega^2)\}.$$

α. The power expended in n leaky condensers in series is

$$V_1^2/R_1 + V_2^2/R_2 + \ldots\ldots$$

β. The power expended in n coils in parallel is

$$A_1^2 R_1 + A_2^2 R_2 + \ldots\ldots,$$

whether the coils are inductive or not.

33—2

α. The power factor of an inductive coil is given by
$$\cos \phi = RA/V.$$

β. The power factor of a leaky condenser is given by
$$\cos \phi = V/RA.$$

α. The resonance of electromotive forces. When a condenser and a choking coil are in series and the effective value of the applied potential difference is constant, the effective potential difference across the terminals of either attains maximum values when
$$LK \{(2n + 1)\, \omega\}^2 = 1.$$

β. The resonance of currents. When a choking coil and a condenser are in parallel and the main current is constant, the currents in either of them attain maximum values when
$$KL \{(2n + 1)\, \omega\}^2 = 1.$$

α. In sine curve theory, when the effective value of the main current is constant, the effective current in an inductive coil (Fig. 155 *a*) shunted by a condenser is a maximum, when
$$K = \frac{L}{R^2 + \omega^2 L^2}.$$

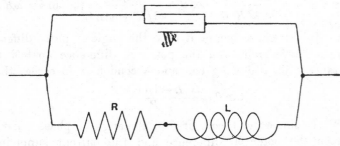

Fig. 155 *a*.   The current in the coil L is a maximum when $K = \dfrac{L}{R^2 + \omega^2 L^2}$.
The main current is constant.

β. In sine curve theory, when the applied potential difference is constant, the potential difference across a leaky condenser, which is put in series with a choking coil, is a maximum when the self inductance L is given by
$$L = \frac{K}{1/R^2 + \omega^2 K^2}.$$

The theorems given above, which are of importance in practical work, can easily be proved graphically. The main steps in the proof, by the method of Chap. XI, may be written as follows :—

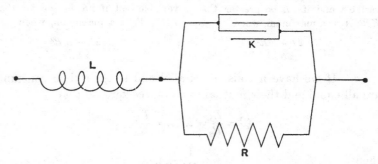

Fig. 155 b.   The P.D. across K is a maximum when it is put in series with a coil whose self inductance $L = \dfrac{K}{1/R^2 + \omega^2 K^2}$. The applied P.D. is constant.

α.

Inductive coil shunted by a condenser.

$$[i_1] = \frac{[\rho_2]}{[\rho_1 + \rho_2]} [i],$$

where

$$\rho_1 = R + L\omega \sqrt{-1}$$

and

$$\rho_2 = \frac{1}{K\omega \sqrt{-1}}.$$

Thus A_1

$$= \frac{1/K\omega}{\{R^2 + (L - 1/K\omega^2)^2 \, \omega^2\}^{\frac{1}{2}}} A$$

$$= \frac{A}{\{K^2\omega^2(R^2 + \omega^2 L^2) - 2LK\omega^2 + 1\}^{\frac{1}{2}}}.$$

Hence, if K be the only variable, A_1 is a maximum, when

$$K = \frac{L}{R^2 + \omega^2 L^2}.$$

β.

Leaky condenser in series with a choking coil.

$$[e_1] = \frac{[\sigma_2]}{[\sigma_1 + \sigma_2]} [e],$$

where

$$\sigma_1 = 1/R + K\omega \sqrt{-1}$$

and

$$\sigma_2 = \frac{1}{L\omega \sqrt{-1}}.$$

Thus V_1

$$= \frac{1/L\omega}{\{1/R^2 + (K - 1/L\omega^2)^2 \, \omega^2\}^{\frac{1}{2}}} V$$

$$= \frac{V}{\{L^2\omega^2(1/R^2 + \omega^2 K^2) - 2KL\omega^2 + 1\}^{\frac{1}{2}}}.$$

Hence, if L be the only variable, V_1 is a maximum, when

$$L = \frac{K}{1/R^2 + \omega^2 K^2}.$$

COR. I.

α.

If L be the only variable, A_1 is a maximum, when

$$L = 1/K\omega^2.$$

β.

If K be the only variable, V_1 is a maximum, when

$$K = 1/L\omega^2.$$

COR. II.

α.	β.
If the frequency be the only variable and if $2L$ be greater than KR^2, A_1 is a maximum, when	If the frequency be the only variable and if $2K$ be greater than L/R^2, V_1 is a maximum, when
$$\omega^2 = \frac{2L - KR^2}{2KL^2}.$$	$$\omega^2 = \frac{2K - L/R^2}{2LK^2}.$$

α. If we have n coils in parallel and if their time constants are all equal and their mutual inductances zero, so that

$$\frac{L_1}{R_1} = \frac{L_2}{R_2} = \ldots\ldots = \frac{L_n}{R_n},$$

then $$i_1 = \frac{1/R_1}{\Sigma 1/R}\, i,$$

where i_1 is the instantaneous value of the current in the coil (R_1, L_1) and i is the instantaneous value of the current in the main.

β. If we have n leaky condensers in series and if their time constants are all equal, so that

$$K_1 R_1 = K_2 R_2 = \ldots\ldots = K_n R_n,$$

then $$v_1 = \frac{R_1}{\Sigma R}\, v,$$

where v_1 is the instantaneous value of the potential difference across the condenser K_1, and v is the instantaneous value of the applied potential difference.

α. In sine curve theory, when a constant effective potential difference is applied at the terminals of a non-inductive resistance R in series with a shunted choking coil, the effective current in the main is a minimum when

$$x = \frac{\omega^2 L^2}{2R} + \frac{\omega L}{2R}\{\omega^2 L^2 + 4R^2\}^{\frac{1}{2}},$$

or $$\frac{1}{x} = \left\{ \frac{1}{4R^2} + \frac{1}{\omega^2 L^2} \right\}^{\frac{1}{2}} - \frac{1}{2R},$$

where x is the non-inductive resistance shunting the choking coil L. This is a particular case of a theorem given on page 296.

β. In sine curve theory, when the effective value of the current in the main is constant and we have two branch circuits, one being a resistance R and the other a condenser K in series with a variable resistance x, then the effective potential difference between the terminals of the fixed resistance is a maximum, when

$$\frac{1}{x} = \frac{\omega^2 K^2 R}{2} + \frac{\omega K R}{2} \{\omega^2 K^2 + 4/R^2\}^{\frac{1}{2}},$$

or

$$x = \left\{\frac{R^2}{4} + \frac{1}{\omega^2 K^2}\right\}^{\frac{1}{2}} - \frac{R}{2}.$$

The general theorem given on page 296 may be reciprocated in the same manner (see Figs. 156 a and 156 b).

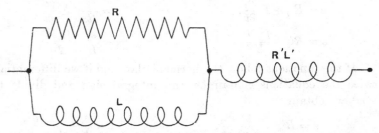

Fig. 156 a. The main current is a minimum when R has a certain value. The applied P.D. is constant.

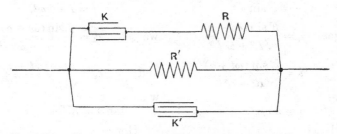

Fig. 156 b. The P.D. is a maximum when R has a certain value. The main current is constant.

α. In sine curve theory, if we have a condenser K in series with a resistance R, and if the combination be shunted by a choking coil L in series with a resistance R, then if $LK\omega^2 = 1$, the combination is equivalent to a non-inductive coil whose resistance R' is given by

$$R' = \frac{L}{2KR} + \frac{R}{2}.$$

β. In sine curve theory, if we have a choking coil L shunted by a resistance R, and if the combination be in series with a condenser K shunted by a resistance R, then if $KL\omega^2 = 1$, the combination is equivalent to a non-inductive coil whose resistance R' is given by

$$\frac{1}{R'} = \frac{KR}{2L} + \frac{1}{2R}.$$

The following is an outline of the analytical proof of the theorems above arranged so as to show that from the mathematical point of view the two problems are identical.

α. $\qquad\qquad\qquad\qquad\qquad\qquad\qquad$ β.

$$e = Ri_1 + L\frac{\partial i_1}{\partial t}, \qquad\qquad i = \frac{e_1}{R} + K\frac{\partial e_1}{\partial t},$$

$$e = Ri_2 + \frac{\int i_2\partial t}{K}. \qquad\qquad i = \frac{e_2}{R} + \frac{\int e_2\partial t}{L}.$$

If the functions follow the harmonic law and if we differentiate twice the equations containing the integral sign and divide by $-\omega^2$, we obtain

$$e = Ri_2 - \frac{1}{K\omega^2}\frac{\partial i_2}{\partial t}. \qquad\qquad i = \frac{e_2}{R} - \frac{1}{L\omega^2}\frac{\partial e_2}{\partial t}.$$

Assuming $\qquad\qquad\qquad\qquad\qquad\qquad$ Assuming

$$e = E\sin\omega t, \qquad\qquad\qquad\qquad i = I\sin\omega t,$$

we get $\quad i_1 = \dfrac{E\sin(\omega t - \alpha)}{\sqrt{R^2 + \omega^2 L^2}},$ $\qquad$ we get $\quad e_1 = \dfrac{I\sin(\omega t - \alpha)}{\sqrt{1/R^2 + \omega^2 K^2}},$

and $\quad i_2 = \dfrac{E\sin(\omega t + \alpha)}{\sqrt{R^2 + \omega^2 L^2}},$ $\qquad$ and $\quad e_2 = \dfrac{I\sin(\omega t + \alpha)}{\sqrt{1/R^2 + \omega^2 K^2}},$

where $\qquad\qquad\qquad\qquad\qquad\qquad\qquad$ where

$$\tan\alpha = L\omega/R. \qquad\qquad\qquad \tan\alpha = KR\omega.$$

Hence $\qquad\qquad\qquad\qquad\qquad\qquad\qquad$ Hence

$$\frac{E\sin\omega t}{i_1 + i_2} \qquad\qquad\qquad\qquad \frac{I\sin\omega t}{e_1 + e_2}$$

$$= \frac{\sqrt{R^2 + \omega^2 L^2}}{2\cos\alpha}, \qquad\qquad = \frac{\sqrt{1/R^2 + \omega^2 K^2}}{2\cos\alpha},$$

and $\quad R' = \dfrac{R^2 + \omega^2 L^2}{2R}$ $\qquad\qquad$ and $\quad 1/R' = \dfrac{1/R^2 + \omega^2 K^2}{2/R}$

$$= \frac{R}{2} + \frac{L}{2KR}. \qquad\qquad\qquad = \frac{1}{2R} + \frac{KR}{2L}.$$

a. When a condenser K (Fig. 157 *a*) shunted by a resistance R is placed in series with a choking coil L shunted by a resistance R, the combination will act like a non-inductive resistance R at all frequencies and whatever the shape of the wave of the applied potential difference, provided that L equals KR^2 (see page 143).

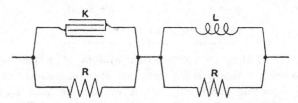

Fig. 157 *a*.   When L equals KR^2 the combination acts like
a non-inductive resistance R.

β. When a choking coil L (Fig. 157 *b*) in series with a resistance R is placed in parallel with a condenser K in series with a resistance R, the combination will act like a non-inductive resistance R at all frequencies and whatever the shape of the wave of the main current, provided that K equals L/R^2 (see page 142).

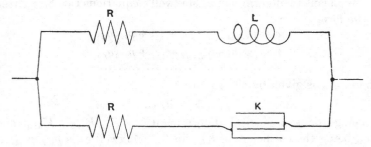

Fig. 157 *b*.   When K equals L/R^2 the combination acts like
a non-inductive resistance R.

Let us now consider whether the coefficients of self and mutual induction for electrostatic charges have the corresponding electromagnetic coefficients for their reciprocals. Maxwell's equations for the electrostatic charges in terms of the potentials of n conductors (page 147) are

Electrostatic and electro-magnetic coefficients.

$$q_1 = K_{1,1}v_1 + K_{1,2}v_2 + \ldots + K_{1,n}v_n,$$
$$q_2 = K_{2,1}v_1 + K_{2,2}v_2 + \ldots + K_{2,n}v_n,$$
$$\ldots\ldots\ldots\ldots\ldots\ldots\ldots\ldots\ldots\ldots\ldots\ldots\ldots\ldots\ldots$$

If ϕ_p be the flux through a circuit p, which has $L_{p.p}$ and $L_{p.q}$ for its self and mutual inductances respectively, and the current i_p entirely surrounds the flux ϕ_p, as it does when the circuit has infinite conductivity, then the electromagnetic equations to n circuits carrying currents i_1, i_2, ... respectively are

$$\phi_1 = L_{1.1}i_1 + L_{1.2}i_2 + \ldots + L_{1.n}i_n,$$
$$\phi_2 = L_{2.1}i_1 + L_{2.2}i_2 + \ldots + L_{2.n}i_n,$$
$$\cdots\cdots\cdots\cdots\cdots\cdots\cdots\cdots\cdots\cdots$$

We see, then, that an electrostatic system of n conductors at given potentials reciprocates into an electromagnetic system of n circuits carrying given currents, if we assume that $L_{p.p}$ is the reciprocal of $K_{p.p}$ and that $L_{p.q}$ is the reciprocal of $K_{p.q}$. In the equations above it is to be noticed that

$$K_{p.q} = K_{q.p} \quad \text{and} \quad L_{p.q} = L_{q.p}.$$

It is also to be noticed that when p and q are different $K_{p.q}$ is always negative but $L_{p.q}$ is not necessarily negative. We have supposed that the conductors and circuits or coils have perfect conductivity.

By means of determinants, Maxwell's equations can be written in the form

$$v_1 = p_{1.1}q_1 + p_{1.2}q_2 + \ldots + p_{1.n}q_n,$$
$$v_2 = p_{2.1}q_1 + p_{2.2}q_2 + \ldots + p_{2.n}q_n,$$
$$\cdots\cdots\cdots\cdots\cdots\cdots\cdots\cdots\cdots\cdots$$

where $p_{l.m}$ is given by the equation

$$p_{l.m}\,\Delta = M_{l.m},$$

Δ being the symmetrical determinant $\Sigma \pm K_{1.1}K_{2.2}\ldots K_{n.n}$, and $M_{l.m}$ being the coefficient of $K_{l.m}$ in Δ. Maxwell calls $p_{1.1}, p_{1.2}, \ldots$ coefficients of potential.

Similarly in the electromagnetic problem we have

$$i_1 = \lambda_{1.1}\phi_1 + \lambda_{1.2}\phi_2 + \ldots + \lambda_{1.n}\phi_n$$

and $(n-1)$ similar equations, where

$$\lambda_{l.m}\Delta' = M'_{l.m},$$

Δ' being the symmetrical determinant $\Sigma \pm L_{1.1}L_{2.2}\ldots L_{n.n}$ and $M'_{l.m}$ being the coefficient of $L_{l.m}$ in Δ'. We may call $\lambda_{1.1}, \lambda_{1.2}, \ldots$ coefficients of current. The above equations prove that $\lambda_{l.m}$ is the reciprocal of $p_{l.m}$.

α. The capacity K_v for equal potentials of n conductors is given by

$$K_v = \Sigma K_{1.1} + 2\Sigma K_{1.2},$$

for when they are all at the same potential v,

$$q_1 = (K_{1.1} + K_{1.2} + \ldots + K_{1.n})\, v,$$
$$q_2 = (K_{2.1} + K_{2.2} + \ldots + K_{2.n})\, v,$$
$$\ldots\ldots\ldots\ldots\ldots\ldots\ldots\ldots\ldots\ldots\ldots$$

and hence $\quad K_v = (\Sigma q)/v = \Sigma K_{1.1} + 2\Sigma K_{1.2},$

since $\quad\quad\quad K_{p.q} = K_{q.p}.$

The electrostatic energy of the system in this case is $\frac{1}{2}v\Sigma q$ or $\frac{1}{2}K_v v^2$. It follows from a theorem proved on p. 150 that, if v be maintained constant, the mutual electrical actions of the conductors will tend to move them so that $\frac{1}{2}K_v v^2$ increases. Thus as the conductors separate under the action of the electric forces, K_v continually increases.

β. The self inductance L_s of n coils in series is given by

$$L_s = \Sigma L_{1.1} + 2\Sigma L_{1.2},$$

for when they are all carrying the same current i,

$$\phi_1 = (L_{1.1} + L_{1.2} + \ldots + L_{1.n})\, i,$$
$$\phi_2 = (L_{2.1} + L_{2.2} + \ldots + L_{2.n})\, i,$$
$$\ldots\ldots\ldots\ldots\ldots\ldots\ldots\ldots\ldots\ldots\ldots$$

and hence $\quad L_s = (\Sigma\phi)/i = \Sigma L_{1.1} + 2\Sigma L_{1.2},$

since $\quad\quad\quad L_{p.q} = L_{q.p}.$

The electromagnetic energy in this case is $\frac{1}{2}i\Sigma\phi$ or $\frac{1}{2}L_s i^2$. If i be maintained constant, the mutual electrical actions of the various coils will tend to move them so that $\frac{1}{2}L_s i^2$ increases. Thus, as the coils separate under the action of the electromagnetic forces, L_s continually increases.

If we have two coils in parallel carrying currents i_1 and i_2 and if $\dfrac{\partial\phi}{\partial t}$ be the applied potential difference and the total current in each of the neighbouring circuits is zero, then

$$\phi = L_{1.1}i_1 + L_{1.2}i_2,$$

and $\quad\quad\quad\quad \phi = L_{2.1}i_1 + L_{2.2}i_2.$

Hence $\phi(L_{2.2} - L_{1.2}) = (L_{1.1}L_{2.2} - L^2_{1.2})\,i_1,$

and $\phi(L_{1.1} - L_{1.2}) = (L_{1.1}L_{2.2} - L^2_{1.2})\,i_2.$

Therefore $$\frac{\phi}{i_1 + i_2} = \frac{L_{1.1}L_{2.2} - L^2_{1.2}}{L_{1.1} + L_{2.2} - 2L_{1.2}}.$$

But $i_1 + i_2$ is the current in the main and therefore $\phi/(i_1 + i_2)$ is the self inductance of the coils in parallel. Now leaving the positions of the coils unchanged, reverse the connections of the coil 2, so that the applied potential difference is acting on the coil 1 in the same direction as in the last example but on the coil 2 in the opposite direction. The equations become

$$\phi = L_{1.1}i_1 + L_{1.2}i_2,$$
$$-\phi = L_{2.1}i_1 + L_{2.2}i_2,$$

and therefore $$\frac{\phi}{i_1 - i_2} = \frac{L_{1.1}L_{2.2} - L^2_{1.2}}{L_{1.1} + L_{2.2} + 2L_{1.2}}.$$

Now, in this case, $i_1 - i_2$ is the current in the main. Hence, when the two coils are connected in 'cross parallel,' the self inductance is

$$\frac{L_{1.1}L_{2.2} - L^2_{1.2}}{L_{1.1} + L_{2.2} + 2L_{1.2}}.$$

If we reciprocate this expression we get

$$\frac{K_{1.1}K_{2.2} - K^2_{1.2}}{K_{1.1} + K_{2.2} + 2K_{1.2}},$$

and this equals $q/(v_1 - v_2)$, that is, according to our definition on page 151, the capacity between the two conductors.

We see, therefore, that the self inductance of two coils in 'cross parallel,' when the total current in each of the neighbouring coils is zero, reciprocates into the capacity between two conductors when all neighbouring conductors are earthed.

If we reciprocate the formula for the inductance of two coils in parallel we get

$$\frac{q}{v_1 + v_2} = \frac{K_{1.1}K_{2.2} - K^2_{1.2}}{K_{1.1} + K_{2.2} - 2K_{1.2}}.$$

This formula is derived from the equations

$$q = K_{1.1}v_1 + K_{1.2}v_2,$$

and $$q = K_{2.1}v_1 + K_{2.2}v_2.$$

If two conductors, therefore, have equal charges and all the other conductors in the neighbourhood are at zero potential, the ratio of the charge on either conductor to the sum of the potentials of the two is a constant. In practice it is not convenient to give equal charges of the same sign to conductors and thus this constant ratio is, at present, mainly of theoretical importance.

α. The self inductance of n circuits in parallel is given by

$$1/L_p = (\Sigma i)/\phi = \Sigma\lambda_{1.1} + 2\Sigma\lambda_{1.2},$$

where $\lambda_{1.1}$, $\lambda_{1.2}$, ... are the current coefficients. The electromagnetic energy of the circuits is $\phi^2/2L_p$.

β. The capacity K_q for equal charges of n conductors is given by

$$1/K_q = (\Sigma v)/q = \Sigma p_{1.1} + 2\Sigma p_{1.2},$$

where $p_{1.1}$, $p_{1.2}$, ... are the potential coefficients. The electrostatic energy of the conductors is $q^2/2K_q$. It is easy to see that the mutual electric forces acting on the conductors tend to increase K_q.

α. When we have a system of n circuits in parallel, the ratio of ϕ to i_1 is constant and is called the effective inductance of the first circuit.

β. When we have n conductors each of which has a charge q, the ratio of q to v_1 is constant and is called the effective capacity of the first conductor.

It is worth noticing that, when all the charges on a system of conductors are equal and the charges on all neighbouring conductors are zero, the electrostatic energy is $q^2/2K_q$, and thus depends only on the charge, the geometrical configuration of the system and its position. Similarly, when all the potentials are maintained at a known value, the electrostatic energy can be written down at once when K_v is known. Now, in many cases, K_v can be measured easily experimentally, and in some cases K_p can be found. A knowledge of the value of these quantities will be helpful, therefore, in studying the electrical properties of a system of fixed conductors. Similarly a knowledge of L_s and L_p will be a help in studying a system of coils.

We have shown that the following quantities are reciprocals :—

$K_{p.p}$	$K_{p.q}$	The capacity between two conductors
$L_{p.p}$	$L_{p.q}$	The self-inductance of two coils in cross parallel

Denoting the coefficients of current and potential by $\lambda_{1.1}$, $\lambda_{1.2}$, ... and $p_{1.1}$, $p_{1.2}$, ... respectively, we also have :—

K_v	K_q	$p_{1.1}$	$p_{1.2}$
L_s	L_p	$\lambda_{1.1}$	$\lambda_{1.2}$

α.   Formula for the electromagnetic wattmeter :—

The electromagnetic and the electrostatic wattmeter.

$$P = ki_1 i_2.$$

One practically non-inductive coil shunts the load, and the inductive coil is in series with the load.

β.   Formula for the electrostatic wattmeter :—

$$P = ke_1 e_2.$$

One resistance which has practically no condenser effect is placed in series with the load, and a condenser (formed by the plate and quadrants) is placed shunting the load.

α.   The three voltmeter method of measuring power :—

$$P = (1/2R)(V^2 - V_1^2 - V_2^2).$$

β.   The three ammeter method of measuring power :—

$$P = R/2 (A^2 - A_1^2 - A_2^2).$$

These examples could easily be multiplied.

An inspection of Chapters XV and XVI will show that many of the theorems in two and three phase theory are reciprocals.   In a star connection Σi is zero and **Star and mesh.** in a mesh connection Σv is zero.   Hence theorems concerning currents in star systems reciprocate into theorems concerning potential differences in mesh systems.   The following are a few illustrations.

α. The mesh potential differences between the mains are to
Three phase examples. one another as the sines of the phase differences
between them. In symbols, we have

$$\frac{V_{2.3}}{\sin \alpha} = \frac{V_{3.1}}{\sin \beta} = \frac{V_{1.2}}{\sin \gamma}.$$

β. The currents in the arms of a star load are to one another
as the sines of the phase differences between them. In symbols,
we have

$$\frac{A_1}{\sin \alpha} = \frac{A_2}{\sin \beta} = \frac{A_3}{\sin \gamma}.$$

α. In a non-inductive star load, if p, q and r be the
resistances of the arms,

$$\frac{E_1}{p \sin \theta_{2.3}} = \frac{E_2}{q \sin \theta_{3.1}} = \frac{E_3}{r \sin \theta_{1.2}}.$$

β. In a non-inductive mesh load, if p, q and r be the
resistances of the arms,

$$\frac{I_1 p}{\sin \theta_{2.3}} = \frac{I_2 q}{\sin \theta_{3.1}} = \frac{I_3 r}{\sin \theta_{1.2}}.$$

α. If the load consist of three equal non-inductive resistances
connected mesh fashion, then

$$A_1{}^4 + A_2{}^4 + A_3{}^4 = 9 \left(I_1{}^4 + I_2{}^4 + I_3{}^4 \right),$$

where A_1, A_2 and A_3 are the currents in the three mains and
I_1, I_2 and I_3 are the mesh currents.

β. If the load consist of three equal non-inductive resistances
connected star fashion, then

$$V_{2.3}^4 + V_{3.1}^4 + V_{1.2}^4 = 9 \left(E_1{}^4 + E_2{}^4 + E_3{}^4 \right).$$

α. The formula for the potential differences across one of the
arms of a non-inductive star load is

$$E_1{}^2 \left(\frac{1}{r_1} + \frac{1}{r_2} + \frac{1}{r_3} \right)^2 = \left(\frac{V^2_{1.2}}{r_2} + \frac{V^2_{3.1}}{r_3} \right) \left(\frac{1}{r_2} + \frac{1}{r_3} \right) - \frac{V^2_{2.3}}{r_2 r_3}.$$

β. The formula for the current in one side of a non-inductive
mesh load is

$$I_1{}^2 (r_1 + r_2 + r_3)^2 = (A_3{}^2 r_2 + A_2{}^2 r_3)(r_2 + r_3) - A_1{}^2 r_2 r_3.$$

α. The mesh voltages being constant, the sum of the voltages between the centre of a star load and the three mains is a minimum when the currents in the arms are all equal.

β. The currents in the mains being constant the sum of the mesh currents is a minimum when the potential differences between the mains are equal.

An inspection of the formulae and constructions, given in Chapter XVI for two phase theory, will show how easy it is to reciprocate many of them.

Two phase examples.

α. Rule for finding the voltages across the arms of a non-inductive star load when the voltage tetrahedron and the resistances of the arms are given.

Find the centre of gravity G of masses $1/r_1$, $1/r_2$, $1/r_3$ and $1/r_4$ placed at the four angular points of the voltage tetrahedron; then, the lines joining G to the four angular points will give the magnitudes and the phase differences of the required voltages.

β. Rule for finding the currents in a non-inductive mesh load when the current tetrahedron and the resistances of the arms are given.

Find the centre of gravity G of masses r_1, r_2, r_3 and r_4 placed at the four angular points of the current tetrahedron; then, the lines joining G to the four angular points will give the magnitudes and the phase differences of the required currents.

(α) The power w expended on a star load (p. 394) is given by

$$w = v_1 a_1 + v_2 a_2 + v_3 a_3 + v_4 a_4.$$

(β) The power w expended on a mesh load is given by

$$w = i_{2.3} v_{2.3} + i_{3.4} v_{3.4} + i_{4.1} v_{4.1} + i_{1.2} v_{1.2}.$$

These examples could easily be multiplied, but sufficient have been given to prove that the method of duality is as useful in electrical theory as it is in geometry.

REFERENCE

H. Sire de Vilar, ' La Dualité en Électrotechnique.' *L'Éclairage Électrique*, Vol. 27, p. 252, 1901.

INDEX

Printed in the United States
By Bookmasters